Neuroscience in Education

Neuroscience in Education

The good, the bad and the ugly

Edited by

Sergio Della Sala

and

Mike Anderson

OXFORD
UNIVERSITY PRESS

OXFORD
UNIVERSITY PRESS

Great Clarendon Street, Oxford OX2 6DP
United Kingdom

Oxford University Press is a department of the University of Oxford.
It furthers the University's objective of excellence in research, scholarship,
and education by publishing worldwide. Oxford is a registered trade mark of
Oxford University Press in the UK and in certain other countries

First published 2012
Reprinted 2013

British Library Cataloguing in Publication Data
Data available

Library of Congress Cataloging in Publication Data
Data available

ISBN 978-0-19-960049-6

To our children and our teachers

Acknowledgements

This book could not have seen the light without the work and the insight of several people whom we would like to thank: the senior commissioning editor for Psychology, Psychiatry and Neuroscience Martin Baum; and the Assistant Commissioning Editor for Brain Sciences Charlotte Green at Oxford University Press; Moira MacRae, from the University of Edinburgh, who provided clerical support throughout and Corinne Reid for comments on the introduction. We are particularly grateful to Prof. Roberto Cubelli with whom we held several discussions on the topics covered in this volume, and who pointed us to the story of Gilles de Robien, which we refer to in our Introduction. We would also like to thank the School of Psychology at the University of Western Australia who provided funding for an Adjunct Chair for Sergio Della Sala that enabled our collaboration on this project. We would also like to thank the Australian Research Council for the award of a discovery grant (DP0665616) to Mike Anderson (and colleagues, Corinne Reid, Allison Fox and Dorothy Bishop) that provided much of the inspiration of how to think about brain development and education. Of course our gratitude goes to all the authors who kindly contributed to this volume.

Sergio Della Sala and Mike Anderson
May 2011

Contents

Editors' biographies

Sergio Della Sala is a trained clinical neurologist, Professor of Human Cognitive Neuroscience in the Psychology Department at the University of Edinburgh, UK and holds an adjunct chair at the Psychology Department of the University of Western Australia. He is a Fellow of the British Psychological Society, of the Association for Psychological Science and of the Royal Society of Edinburgh, and Editor of *Cortex*. His research focuses on memory and memory impairments and on the cognitive deficits associated with brain damage.

Mike Anderson is a Professor of Psychology and Director of the Neurocognitive Development Unit in the School of Psychology at the University of Western Australia. His research is based around his theory of intelligence and development and focuses most recently on the influence of the developing brain on intellectual functions in children.

Contributors

Mike Anderson
Neurocognitive Development Unit,
School of Psychology,
The University of Western Australia,
Perth, NSW, Australia

Daniel Ansari
Developmental Cognitive Neuroscience,
The University of Western Ontario,
London, ON, Canada

Jane Ashby
Department of Psychology,
Central Michigan University,
Mount Pleasant, MI, USA

Emma Ashwin
Department of Psychology,
University of Bath,
Bath, UK

Simon Baron-Cohen
Department of Experimental Psychology,
Trinity College,
Cambridge, UK

Catherine Beauchamp
School of Education,
Bishop's University,
Sherbrooke, QC, Canada

Miriam Beauchamp
Department of Psychology,
University of Montreal,
Montreal, QC, Canada

Timothy C. Bates
Department of Psychology,
University of Edinburgh,
Edinburgh, UK

Eric H. Chudler
Department of Bioengineering,
University of Washington,
Seattle, WA, USA

Donna Coch
Department of Education,
Dartmouth College,
Hanover, NH, USA

Max Coltheart
Macquarie Centre for Cognitive Science,
Macquarie University,
Sydney, NSW, Australia

Michael W. Connell
Institute for Knowledge Design,
Arlington, MA, USA

Michael C. Corballis
Department of Psychology,
University of Auckland,
Auckland, New Zealand

Nelson Cowan
Department of Psychological Sciences,
University of Missouri-Columbia,
Columbia, MO, USA

Sergio Della Sala
Human Cognitive Neuroscience,
Psychology Department,
University of Edinburgh,
Edinburgh, UK

Bridgid Finn
Psychology Department,
Washington University,
St Louis, MO, USA

Monika Finsterwald
Faculty of Psychology,
University of Vienna,
Vienna, Austria

Howard Gardner
Graduate School of Education,
Harvard University,
Cambridge, MA, USA

Ofer Golan
Department of Psychology,
Bar-Ilan University,
Ramat-Gan, Israel

Usha Goswami
Department of Experimental Psychology,
University of Cambridge,
Cambridge, UK

Elena L. Grigorenko
Child Study Center,
Department of Psychology,
Department of Epidemiology & Public Health,
Yale University,
New Haven, CT, USA

Paul Howard-Jones
Graduate School of Education,
University of Bristol,
Bristol, UK

Hideaki Koizumi
Hitachi, Ltd; and
Research Center for Advanced Science and
Technology,
The University of Tokyo,
Tokyo, Japan

Yulia Kovas
Department of Psychology,
Goldsmiths
University of London,
London, UK

Genevieve McArthur
Macquarie Centre for Cognitive Science,
Macquarie University,
Sydney, NSW, Australia

Robert D. McIntosh
Human Cognitive Neuroscience Psychology,
University of Edinburgh,
Edinburgh, UK

Mary Oliver
Graduate School of Education,
The University of Western Australia,
Crawley, WA, Australia

Domenico Parisi
Institute of Cognitive Sciences and
Technologies,
National Research Council,
Rome, Italy

Robert Plomin
Social, Genetic and Developmental Psychiatry
Centre,
Institute of Psychiatry,
King's College London,
London, UK

Keith Rayner
Department of Psychology,
University of California, San Diego,
La Jolla, CA, USA

Stuart J. Ritchie
Human Cognitive Neuroscience,
University of Edinburgh,
Edinburgh, UK

Henry L. Roediger III
Department of Psychology,
Washington University,
St Louis, MO, USA

Barbara Schober
Faculty of Psychology,
University of Vienna,
Vienna, Austria

Xavier Seron
Université catholique de Louvain,
Louvain-la-Neuve, Belgium

Christiane Spiel
Faculty of Psychology,
University of Vienna,
Vienna, Austria

Zachary Stein
Harvard Graduate School of Education,
Cambridge, MA, USA

Petra Wagner
School of Applied Health and Social Sciences,
Upper Austria University of Applied Sciences,
Linz, Austria

Yana Weinstein
Department of Psychology,
Washington University,
St Louis, MO, USA

If a neurologist or a psychologist (or a neuroscientist) had 30 people in their surgery or office every day, all of whom had different needs, and some of whom did not want to be there and were causing trouble, and they, without assistance, had to guide them through a curriculum of excellence consecutively for nine months, then they might have some conception of the classroom teacher's job.

(Paraphrased from a quote attributed to Donald D. Quinn)

Section 1

Introductions

Chapter 1

Neuroscience in education: an (opinionated) introduction

Mike Anderson and Sergio Della Sala

1.1 Neuroscience or cognition?

We begin with an apology. Our book title leads with 'Neuroscience in education' but after reading this book you might be forgiven for thinking this is deceptive. Our stance is that the scientific domain that has most to offer education is the study of cognition and that neuroscience itself has qualified value. Yet there is no intention to deceive in our use because it has entered the lexicon as the term that refers to the interaction of education and brain sciences (including cognitive psychology). For example, this is testified by the title of recent reports on the topic by the Royal Society (*Neuroscience: implications for education and lifelong learning*) and the ESRC (*Neuroscience and education: issues and opportunities*) and many current publications (*Education and neuroscience*—Howard-Jones, 2009; *The brain at school: educational neuroscience in the classroom*, Geake, 2009) and indeed these and many other contributions have spawned the new label of *neuroeducation*. So our book is targeted fairly and squarely at the centre of this new field. But the first and most important lesson we have learned in putting the book together is that while the use of the term 'neuroscience' is attractive for education it seems to us that it is cognitive psychology that does all the useful work or 'heavy lifting'. The reason for this is straightforward. We believe that for educators, research indicating that one form of learning is more efficient than another is more relevant than knowing where in the brain that learning happens. There is indeed a gap between neuroscience and education. But that gap is not filled by the 'interaction' of neuroscientists and teachers (nearly always constituted by the former patronizing the latter) or 'bridging' the two fields by training teachers in basic neuroscience and having neuroscientists as active participators in educating children. Rather what will ultimately fill the gap is the development of evidence-based education where that base is cognitive psychology. Of course this is not an uncontested view and some of our contributors disagree, but ultimately it is for you the reader to draw your own conclusion.

1.2 Use and misuse of neuroscience in education

A recent investigation in the US-based journal *Mind, Brain and Education* (Pickering and Howard-Jones, 2007) showed that almost 90% of teachers consider knowledge about brain functioning relevant for the planning of education programmes. Thanks to the upsurge of interest in neuroeducation over the past few years we have been asked to give a number of talks to educators with a view to sharing our professional knowledge. From the outset we were somewhat perplexed at the interest that educators seemed to have in hearing about the latest research in 'neuroscience'. Perhaps inevitably, poking around the field and lifting a few chosen stones revealed many instances of what we came to call 'the good, the bad and the ugly'. The 'good' was nearly always sound cognitive research that had clear implications for educational practice. The 'bad' was the exploitation of the enthusiasm of educators for neuroscience that opened the door to the adoption of programmes and teaching aids apparently based on neuroscience but which no reputable neuro-scientist would endorse. The 'ugly' was simplistic interpretation of cognitive theories leading to errors in their application.

1.3 The good

An excellent example of the *good* is recent research reported in *Science* showing that retrieval practice produces more learning than elaborative studying with concept mapping (Karpicke & Blunt, 2011). This is interesting because the process of 'retrieval' has had a bad press in education. It is often regarded as part of the old-fashioned rote-learning approach to teaching that is anti-thetical to modern notions of creativity and discovery-based learning. But as the authors point out, their results were counter-intuitive, showing that 'retrieval is not merely a read out of the knowledge stored in one's mind—the act of reconstructing knowledge itself [a necessary corollary of retrieval] enhances learning. This dynamic perspective on the human mind can pave the way for the design of new educational activities based on consideration of retrieval processes'. Even for the good, however, we would sound a note of caution. What works in the laboratory must still be tested in the classroom.

It is imperative to add at this point that in any case we do not believe the science to be prescriptive. Even solid scientifically-based results cannot be applied without consideration of the multiple factors that influence both student learning and the role that schooling plays within the culture. A particular culture may decide that the goal of schooling is the reduction of learning disadvan-tage in society whereas for another it may be the maximization of individual potential whatever the price in terms of social inequality. The point is that the science of learning is neutral with respect to any such *political* or *social* choice.

1.4 The bad

On the other hand it is nonsensical to follow any classroom innovations whose core ideas have not been pre-tested in the laboratory of empirical science. It seems to us that we live in a credulous world and education is no exception. Any half-baked idea is worth 'giving a go'. So listening to Mozart (not the Sex Pistols) surely makes us more intelligent, so let's give it a go. Neurolinguistic programming sounds pretty good and plausible (neuro is brains, linguistic is important for learn-ing and computers are programmed) so let's give it a go. We say no. If there are no sound empirical data supporting such claims (as there are not in these examples) they should not be given a go in the classroom. Sometimes ideas are intuitively attractive and easy to understand. But without evi-dence this should count for nothing. Moreover, if the claim that breathing through the left nostril to enhance creativity (right and left brains do different things, right?) not only has no empirical basis but flies in the face of well-established facts (in this case about anatomy), then it is especially not worth giving it a go. Education is too important to be hijacked by crazy ideas that gain currency only by hearsay. Lest you think we are exaggerating or tilting at windmills take the case of Lambeth Council's decision to spend £90,000 of their hard-earned education budget on providing foot mas-sage in school to curb unruly behaviour in children (reported in *The Guardian*, Lipsett, 2008). The company who provide this service, Bud-Umbrella, are a London-based charity and on their web-site (http://www.bud-umbrella.org.uk/) they list as one of their 'Aims and Objectives'—'To conduct research into the effectiveness of the service we offer and the positive impact on our clients'. There are a number of features that are worth emphasizing here: (1) 'doing research' is clearly a selling point; but (2) surely the research should be done *before* the product is sold; and (3) we should be careful who does said research and where it is reported because the clear implication is that the provider themselves has already determined that there is a positive impact on clients.

1.5 **The ugly**

A jewel among the *uglies* is where a well-researched and developed theory of reading (dual-route theory) is caricatured in education to justify what would otherwise be an ideological preference for the whole-word reading movement. Briefly, the dual-route theory says that single-word reading can be accomplished through a route of letter to sound conversion (think phonics) or through a route of direct visual recognition (think whole-word reading). It does *not* say that both are equally effective in teaching children to read. Further a great deal of research motivated by this cognitive theory has demonstrated beyond doubt that phonics is the much more effective method. That this can count for very little is evidenced by the case of Gilles de Robien, the French Minister for Education, who in 2006 announced the government's enthusiasm to adopt phonics-based teaching for reading in French schools. He made this announcement after being persuaded by the overwhelming scientific evidence in favour of phonics approaches compared with 'holistic' or whole-word strategies which Mr de Robien wanted French teachers to stop using. A number of educationalists opposed this move (largely on the grounds that it was old-fashioned) and a major dispute began between teachers and the government. In the UK there has been a long-running dispute between the government and education unions over this issue and currently this is focused on opposition to the introduction of a phonics test for 6-year-olds.

1.6 **Opinions versus data**

As illustrated by these cases, the discussion seems rarely informed by evidence but owes much to rhetoric and appeals to authority, sometimes from neuroscientists themselves. Note this quote from Professor Colwyn Trevarthen, a renowned neuroscientist/psychologist, about the proposal for a phonics test in schools:

> I am an expert on child development and learning. Children learn language to communicate and want to share tasks and knowledge. Instruction in elements of language, out of creative and meaningful communication is forced labour that can break a child's confidence. I am opposed to phonics teaching and testing. It might have use for linguists at university. The government must try to understand and value the natural talents of young children and curb the urge to instruct and measure performance. (http://www.gopetition.com/petition/42347/signatures.html)

So much for evidence-based interaction between teachers and neuroscientists. These campaigns do not resemble the usual debates that surround evidence-based practice as much as the process of litigation and the battle of expert testimonies. Consequently when teachers' sincere intuitions about the nature of the way children are learning is contradicted by science, the result is usually a fight not an interaction and discussion. And in the case of such fights the long-term winner is nearly always preconceived (albeit well-motivated) ideology not science.

1.7 **A fistful of Dollars**

Sadly, the *bad* has swamped both the good and the ugly (possibly because of the potential for monetary reward). The recently deceased DDAT (Dyslexia Dyspraxia Attention Treatment) program[1], which claimed to offer cures for both reading disorders and attention deficit hyperactivity

---

[1] Also known as the Dore Program.

disorder (ADHD), provides one of the richest case studies of this category of neuroscience in education (hear Prof. Dorothy Bishop's scepticism about the method here: http://www.dystalk. com/talks/60-evaluating-alternative-solutions-for-dyslexia). What made this programme such an illuminating case? First, there is a non-obvious and puzzling connection between reading disorders and ADHD. Second, there is a plausible-sounding explanation for the connection based in neuroscience (in this case, activity of the cerebellum). Third, there is an easily implemented (though costly) intervention based on 'stimulating' cerebellar activity through throwing bean-bags and the like. Fourth, there was an apparent scientific study that evaluated the effectiveness of the programme itself. A heady brew indeed. Needless to say, the programme was endorsed by the media and educators alike and academics who dared to voice their scepticism were threatened with lawsuits (see e.g. http://www.badscience.net/2008/05/dore-the-medias-miracle-cure-for-dyslexia/ for a detailed exposition). The result? A non-effective programme and (ultimately) a bankrupted company leaving parents out of pocket. Not to mention that the necessary corollary of any ineffectual programme is yet another experience of failure for a vulnerable child.

1.8 **The allure of everything 'neuro'**

All of this begs the question, why is it that neuroscience is so alluring for education? The most obvious answer is that it is alluring for our whole culture. We have neuro-economics, neuro-aesthetics, neuro-marketing, neuro-theology, neuro-law, neuro-anthropolgy and indeed neuro-politics (see *Neuromania* by Legrenzi, Umiltà & Anderson, 2011). We have recently had the 'decade of the brain' and there can be little doubt that neuroscience is one of the most exciting fields in 21st-century science and technology. There is optimism that scientific advances in neuroscience may solve what have been seen as chronic, intractable problems (irrational economic behaviour may be explained by motivational states of the brain, criminals can be diagnosed and treated by judicious poking around in the frontal lobes and the like). Indeed we have seen this optimism bear fruit already in medicine where such innovations drawn from progress in neuroscience, including stem cell therapy, deep brain stimulation or nanotechnology. So there is an element of a new dawn provoked by powerful new research and technology. However, this view is itself founded more on deep-seated attitudes to what is real and informative than it is to established fact (we doubt currently that neuroscientists have any fundamentally different explanations for 'irrationality' than philosophers have proposed for centuries). Even seasoned professionals find information about the brain more 'compelling'. Brain structures are somehow more real (more concrete and less abstract) than cognitive systems. Yet each has equal epistemological status. While at one level there is obviously a cerebellum in a way that there is not obviously a 'working memory' (you can touch a cerebellum), this is a trivial difference and what is important about a cerebellum or a working memory is what they do, i.e. their function. Their function is a theoretical matter: in the former case theories of cerebellar functioning in neuroscience and in the latter theories of working memory in cognitive psychology. Hence, in fact, cerebellar functioning and the functions of working memory are no more or less real than each other. Both are scientific constructs that owe their meaning to the theories in which they live and consequently have the same degree of concreteness, abstractness and, ultimately, reality.

Leaving this point from the philosophy of science to one side, research has shown that it is also the case that in common discourse information that includes vacuous reference to neuroscientific data or concepts is much more persuasive than the same information without it (for compelling experimental demonstrations see Weisberg, Keil, Goodstein, Rawson & Gray, 2008; McCabe & Castel, 2008).

1.9 **No ready-made recipes**

We emphasized earlier that science is not prescriptive. Scientific facts about cognition and neuroscience is a different domain to moral and political decisions about the nature and goals of education. Equally, because educational choices necessarily involve moral and political choices this does not mean that these choices change scientific facts. So once ethical or moral dimensions to choices about educational interventions are made, for example, whether children should be medicated to modulate their attentional capacities, we can still ask the scientifically more straightforward question—as a matter of fact did the intervention have an effect on outcomes or not? An answer in the affirmative does not absolve us from the moral choice of whether we 'should' use it. To take perhaps another, more nuanced, example—should children with learning disabilities be integrated in the mainstream population or should we create separate classes for them? Are we justified in conducting a study to see if such children fare better when segregated? Well the answer would depend on a moral calculus of educational benefit, social loss and the like. But nevertheless we can appropriately answer the empirical question—do they do better in some learning activity if they are integrated compared with when they are not? There may be questions for which we decide a priori that the benefits from the answers do not justify the moral compromise involved in any particular scientific endeavour (and we can all think of some of our own values that would rule out asking particular questions) but that does not justify the assertion that a scientific study cannot deliver those answers.

Perhaps the strongest claims for the benefits of a neuroscientific perspective are made in the case of learning pathologies. Many of our contributors would testify to the insights and educational potential provided by research in, for example, dyslexia, dyscalculia, attentional deficits or working memory deficits. (And one of the possible negative cultural consequences is the increasing tendency to seek a diagnosis for children who 'fail to learn' at school.) Yet the implicit assumption that the compensatory strategies proposed to alleviate deficits provide a recipe for education in general seems to us to be exactly that—an assumption. Whether or not this is the case is an empirical question that should be answered.

1.10 **Mind, the gap**

Now that the book has been assembled and we can survey the entire terrain it does appear to us that there are a few gaps. It is interesting that much of the discussion between neuroscientists and educationalists usually involves the presentation of some kind of basic data about the brain (often to do with plasticity or learning) and how this might be relevant, or not, for the classroom. In our reading at least very little is focused on what is probably the most relevant neuroscience—that of the *developing* brain. Of particular interest here is the developmental research linking individual differences in children's intellectual ability (usually measured by intelligence quotient (IQ) tests) and developmental changes in the cerebral cortex. Indeed it is the link with IQ that makes this research especially pertinent to the topic of our book. As a number of contributors have made clear, children do exhibit individual differences in their ability to learn from classroom instruction. The single most pervasive difference is in what is called IQ but should more properly be referred to as general intelligence. There is every reason to suppose that not only are these differences moderately heritable, they are likely to be based on some property of the brain most appropriately studied with the techniques available to modern cognitive neuroscience. However, we would also like to stress that this is not some kind of advocacy of some of the odious and simplistic views that have dogged the study of IQ differences in the history of psychology. The research we are interested in does not argue, say, for genetically-based race differences in IQ nor does it say that bigger brains cause higher intelligence (as has been said for centuries now; see e.g. *Über den psychologischen Schwachsinn des Weibes*, by Julius Moebius (1900) or von Bischoff's *Das Hirngewicht des Menschen*, 1880). Rather this much more nuanced research says that correlations of aspects of brain development with IQ scores may provide the clues of how the changing brain influences the child's capacity to learn.

1.11 Developmental neuroscience

Part of the reason for a relative dearth of developmental neuroscience studies is that it is difficult to study the child's brain *in vivo*. Many jurisdictions (Australia, for example) will not allow a number of techniques commonly used in studying adults (functional magnetic resonance imaging (fMRI), for example) to be used for research purposes in children. Consequently, many studies in the past have relied on postmortem anatomical data. However, when brain-imaging data can be collected they are invaluable for our understanding of development. One such exciting line of research is that of Philip Shaw and colleagues at the National Institute of Mental Health, Bethesda, Maryland who have amassed a large database on children that combines neuroanatomical imaging with a number of clinical assessments (often including IQ tests). The kinds of anatomical data collected include information on white and grey matter volumes and densities, all of which can be localized to particular anatomical structures in the brain (for example, the prelateral frontal cortex or the cerebellum). Crucially this research included longitudinal data so that changes in individual brains (rather than comparisons between averaged data from children of different ages) can be charted. In a seminal study in *Nature* in 2006, Shaw and colleagues showed that the thickness of the cortex is related to IQ differences amongst children. But the relationship is not a simple one. Rather it is changes in the trajectory of cortical thickness that is most related to IQ. Indeed in general there is a shift from a negative correlation between cortical thickness and IQ in early development (children with higher IQs have *thinner* cortices) to a positive correlation in later childhood (children with higher IQs have *thicker* cortices). This group has also reported exciting new research that shows that some developmental disorders may result from characteristic deviations in this typical developmental profile. For example, in the case of ADHD there is in effect a delay of the trajectory of attaining peak cortical thickness and this is most pronounced in the prefrontal cortex. Another research group (Eric Courchesne and colleagues at the University of California, San Diego) have shown that in autism there are abnormalities in brain development especially over the first 2 years of life. In this particular case it seems that, if anything, brain volumes in a number of cortical areas are larger than those found in typically developing children. And again in a recent review of the literature on the development of prefrontal cortex Sharon Thompson-Schill and colleagues at the Universities of Pennsylvania and Stanford argue that in some situations the delay in development of brain structures can convey benefits for children's learning. In their case they argue that prolonged maturation of prefrontal cortex allows easier learning of social and linguistic cultural conventions. These and other research efforts are new and not themselves devoid of interpretive problems (particularly cause and effect relationships— do the brain differences cause the characteristic symptoms of the disorders or are they the consequence?) but it seems to us to offer promising strategies for investigating developmental brain behaviour relationships in children and this must surely be the stuff of anything that we will come to call *neuroeducation*.

There are, of course, also gaps between our views and those of our contributors. We indicated earlier that several contributors may disagree with our views expressed in this introduction. By the same token we beg to differ with some of our contributors on a number of issues. Perhaps the most important is how the concept of plasticity is used as a seminal example of the influence of neuroscientific research for education. Yet it would be a mistake to assume that there is a consensus on what is meant by plasticity. Indeed plastic simply means that the brain changes, it does not mean that the brain increases its underlying capacity and yet this is the common interpretation. In our view what is interesting in education is functional plasticity, the ability of the brain to support behavioural changes not any notion of their anatomical counterpart.

1.12 **Brain plasticity**

As Marvin Minsky stated in his 1986 book *The Society of Mind*, 'the principal activities of brains are making changes in themselves'. Indeed the brain is a pliable organ, like plastic, hence the term brain plasticity or neuroplasticity. The brain can be moulded. Imagine it as a lump of porridge (same colour also), and imagine an external event, like a spoon falling into it, this will change the shape of the porridge. However, the metaphor, as all metaphors, is imprecise. In fact, only the internal structure of the brain will change following an event, only its microstructure will change, not its overall shape; that is, the neural networks within the brain reorganize in response to sensory stimulation. Not only will there be reorganization in response to sensory input, but these circuitries will reorganize also following internal events, including the effects of our own thoughts or visual imagery, of hormones and of course of our genes and following brain injuries.

It is relevant to note that plasticity does not refer solely to some increment, but, particularly in younger persons, it encompasses also the *loss* of brain cells and the pruning of their connections. Children will lose about 50% of the synapses that they had at birth. Why is this advantageous? Consider as an example the acquisition of the direction of writing and the issue of mirror writing, i.e. the production of individual letters, whole words or sentences in reverse direction often observed in children who are learning to write. A fetus would not know whether s/he will be born in the US, in Palestine or in China. Hence, it makes sense to believe that s/he gets ready to learn any writing direction functional to the society s/he will live in. Experience will then determine which is the relevant direction of writing s/he needs to learn and use. Before acquiring this specific knowledge, children would process words and letters exactly as they process other objects, orientation being irrelevant to the stimulus identity. It follows that children would show directional writing problems, producing letters either rightwards or reversed, until the knowledge of writing direction of their language stabilizes. Before the writing direction parameters are set by experience and exercise, the child's directional cognitive system is necessarily ambivalent at birth to permit compliance with leftwards or rightwards languages. Learning a writing direction would require stamping out the unwanted alternative rather than acquiring one anew. This would predict that children little exposed to writing would randomly compose asymmetrical letters and digits facing left or right with either hand, as observed in the current study. The corollary of this hypothesis implies that mirror writing is a transient phenomenon not predictive of any difficulties in learning to write or read in primary school.

The relevant issue is what exactly changes in the brain thanks to external or internal stimulation. It is certainly true that our brain rewires according to new experiences. This is testified by studies assessing the performance and the brain changes in people learning new skills. For instance, people learning to juggle show increased recruitment of brain cells in some specific motor, sensory and cognitive areas of their brain, so do people learning to type, or taxi drivers who learn to navigate through complex routes, or musicians who attempt a new instrument. Of course, if we are exposed to experience, our behaviour changes and the substrate of our behaviour, our brain, changes correspondingly. However, when discussing educational issues the implicit assumption is that since the brain is plastic, then by providing the appropriate stimulation new cells will grow. But this is not (entirely) correct. There is clear evidence of the possibility of cell growth, neurogenesis, solely in the olfactory regions and some evidence in non-human mammals that a similar growth can occur in the hippocampus (particularly in the dentate gyrus); however, very little growth occurs in the cortex of the brain, which is what matters to education as most cognition happens there. In short, the brain has the capacity of rewiring itself, via synaptogenesis, but not of growing cortical cells, or at least evidence for such cortical growing is still scant and debatable. This is relevant, as too often brain plasticity is used as a synonym for neurogenesis, which certainly it is not. The rhetoric that we are too often exposed to seems to imply that our brain has a vast reserve which is waiting to be exploited by apt education programmes; however, we do not use only 10% of our brain, this is a myth (see the chapters by Barry Beyerstein (1999a, 1999b; 2007a, 2007b) in *Mind Myths*, and in *Tall Tales about the Mind & Brain*).

In sum, what is interesting in education is functional plasticity, i.e. the ability to change behaviour following external or internal experiences, not where these changes occur in the brain. Experience will determine which connections will be strengthened and which will be pruned; hence plasticity refers to the growing, the dropping and the strengthening of these connections which through synapses and dendrites branch out to form an ever evolving neuronal network.

1.13 **Sensitive periods**

A concept related to brain plasticity is that of 'critical periods', i.e. particular time windows whereby children could learn a specific skill. However, the learning capacity of humans does not function in quantum leaps, rather it is a continuum. Think about our ability to acquire a second language; younger children will usually find it easier, but older children could learn it too. Thus, it would be better to refer to these time windows as 'sensitive periods'—periods of time during which it would be generally easier to apprehend a skill, which will need more effort in other periods. Again in neuroeducational debates this concept is often used rhetorically to support the notion that not only should the environment be stimulating and rich but also that one should not miss one's windows of opportunities. And this matters to education. The concept sprang from the observation that humans deprived of the opportunity to develop a given skill will never regain it; for example, if binocular vision is impeded in early childhood, it will never develop properly. However, these examples concern sensory deprivation or the development of perceptual capacities. We know little about time windows for the development of cognitive skills, indeed there is little evidence supporting this view. The usual examples given to support the notion of sensitive periods in higher cognition derive from the work on sound discrimination; these studies show, for instance, that infants can discriminate sounds from different languages but that this ability decreases in time. These sensitive periods though should be thought of as the exception rather than the rule (see, for a critique, Bailey et al., 2001). We can take for granted that normally rich environments are needed for higher-level skills to develop. However, the argument for enrichment per se comes from the logic that a deprived environment causes deprived development therefore an enriched environment improves development. But by enriched we really mean 'normal'. There is no evidence that supranormal stimulation has any additional benefit. It might even be that such hot-housing could be detrimental. Consider that food deprivation is bad for you. The solution is to be exposed to a normal healthy diet not to overeat and stuff yourself to death.

Moreover, the windows of time for learning these higher functions are not as constrained as those for more basic functions. Indeed it could be argued that the very concept of plasticity (the brain can change and adapt to any experience) contradicts the notion of critical (and maybe even sensitive) periods. However, as just argued, 'true' critical periods are based on functions that are fundamentally not themselves plastic. For the development of cognitive abilities such as literacy and numeracy what matters, or is 'critical', is not time per se but experiential contingencies.

So sometimes you just need to know something before you can learn something else and of course this generates characteristic sequences of change that it is important to understand for educational planning. The acceptance of this view would of course take to task the idea of learning as much as possible as early as possible. None of this denies that some contingencies might depend on the maturation of fundamental processing machinery in the developing brain, but that this is likely to be the exception rather than the rule for the contingencies that are important for educational development.

1.14 **Right to disagree**

In term of issues where there may be disagreements both amongst some of our contributors and with us as editors we have left perhaps the most important to last. What is the most fruitful relationship between neuroscientists and educators?

While we would all (or nearly all!) eschew the pontificating neuroscientist who would tell teachers how it should be in the classroom, we think there is a clear role for scientists who understand the data to be a source of information to teachers. Most obviously, scientists should take active roles as mythbusters. We should not presume to tell teachers how to teach but we should be prepared to point out situations where educational interventions are not based on any evidence.

We also think it is absurd to expect teachers to be educated in neuroscience (far less contributing themselves to neuroscientific research) especially if that came at the cost of being educated about psychology. We can think of no fact about the brain that a teacher might learn that might cause them to be better teachers (it might be of course that better teachers are interested but this reverses the causal relationship). Time spent learning arcane neuroscientific facts or concepts is only likely to be time taken away from learning something useful. We think this probably applies especially to knowledge of genetics. It is important that teachers understand the basic proposition (and come to accept) that children differ in their ability to learn and that some of this difference is caused by difference in genetic inheritance. This would take at most no more than two 2-hour lectures in teacher training college. But it seems to us ludicrous to expect teachers to be on top of the specialized methods and techniques that are the basis of modern behaviour genetics and even more so, molecular genetics. Again, this is not to say that research done in the coming decades on genetic influences on the ability to learn, or even more interestingly on genetic influences on the efficacy of different methods of teaching, will not be important for education—on the contrary, this may turn out to be one of the most exciting impacts of modern neuroscience. It is just to say that it is enough for educators to be informed about the results of that research.

Again nearly all our contributors recognize that when discussing the relationship between neuroscience and education we are talking about different levels of description and explanation (see Discussion Forum on Mind, Brain and Education in *Cortex*, 2009, volume 45, issue 4). In our view, a great deal of educational research can be conducted without any genuflection in the direction of neuroscience. The effect of class size, the consequences of streaming, the personality or the ability of teachers on outcomes, to give but three examples, does not need to have some kind of neuroscientific grounding to be interesting and informative in its own right. Indeed we think there may be a tendency to inflate the importance of *neuroeducation* for education in general. We believe of far greater importance is the concept of evidence-based education no matter what discipline may provide the inspiration. However, we hope that this book will contribute another small step towards understanding how developments in neuroscience might contribute to better education of our children and probably, most importantly, help teachers to recognize what might or might not be worth paying attention to. Hope springs eternal.

References

Bailey, D. B., Bruers, J. T., Symons, F. J., & Lichtman, J. W. (Eds.) (2001). *Critical thinking about critical periods*. Baltimore, MD: Paul H. Brookes Publishing.

Beyerstein, B. L. (1999a). Pseudoscience and the brain: tuners and tonics for aspiring superhumans. In S. Della Sala (Ed.) *Mind Myths: Exploring Everyday Mysteries of the Mind and Brain*, pp. 59–82. Chichester: John Wiley and Sons.

Beyerstein, B. L. (1999b). Whence cometh the myth that we only use ten percent of our brain? In S. Della Sala (Ed.) *Mind Myths: Exploring Everyday Mysteries of the Mind and Brain*, pp. 1–24. Chichester: John Wiley and Sons.

Beyerstein, B. L. (2007a). Graphology—a total write-off. In S. Della Sala (Ed.) *Tall Tales about the Mind & Brain: Separating Fact from Fiction*, pp. 233–70. Oxford University Press.

Beyerstein, B. L. (2007b). The neurology of the weird: brain states and anomalous experience. In S. Della Sala (Ed.) *Tall Tales about the Mind & Brain: Separating Fact from Fiction*, pp. 314–35. Oxford University Press.

Geake, J. G. (2009). *The Brain at School: Educational neuroscience in the classroom*. Maidenhead: McGraw Hill-Open University Press.

Howard-Jones, P. (Ed.) (2009). *Education and Neuroscience. London: Routledge.*

Karpicke, J. D. & Blunt, J. R. (2011). Retrieval practice produces more learning than elaborative studying with concept mapping. *Science*, published online 20 January 2011.

Legrenzi, P., & Umiltà, C. (Anderson, F. Trans.) (2011). *Neuromania: On the limits of brain science*. Oxford: Oxford University Press.

Lipsett, A. (2008). Reflexology for pupils who don't toe the line. *The Guardian*, 3 November.

Minsky, M. (1986). *The Society of Mind*. New York: Simon and Schuster.

Moebius, J. (1900). *Über den psychologischen Schwachsinn des Weibes* [On the Physiological Weakness of Women]. Halle: Carl Marhold.

Pickering, S. J. & Howard-Jones, P.A. (2007). Educators' views of the role of neuroscience, in education: A study of UK and International perspectives. *Mind, Brain and Education*, 1(3), 109–13.

Shaw, P., Greenstein, D., Lerch, J., Clasen, L., Lenroot, R., Gogtay, N., Evans, A., Rapoport, J., & J. Giedd, J. (2006). Intellectual ability and cortical development in children and adolescents. *Nature*, 440, 676–9.

von Bischoff, T. (1880). *Das Hirngewicht des Menschen* [The weight of the human brain]. Bonn.

Weisberg, D. S., Keil, F. C., Goodstein, J., Rawson, E., & Gray, J. R. (2008). The seductive allure of neuroscience explanations. *Journal of Cognitive Neuroscience*, 20(3), 470–7.

Chapter 2

Understanding the neuroscience and education connection: themes emerging from a review of the literature

Miriam Beauchamp and Catherine Beauchamp

Overview

The importance of creating a link between the findings of neuroscience research and the contexts in which teaching and learning take place has recently been established; however, issues have been raised regarding the gap between findings in neuroscience and their usefulness for educational contexts and suggestions have been made about how this gap could be bridged. We undertook a systematic review of the literature from 1970–2011 with the aim of identifying: (1) literature central to an understanding of the connection between neuroscience and education, and (2) prevalent themes in this literature related to the connection between neuroscience and education. Eighty-six works were selected for detailed analysis due to their focus on the problems, solutions and recommendations associated with connecting the disciplines of neuroscience and education (educational neuroscience). In this chapter, we describe our categorization of the content of these articles according to seven salient themes that emerge: *Misapplication, Multiple disciplines, Language, Knowledge development, Value, Collaboration, Research design*. We present a visual framework representing the relationships among these themes to consolidate an understanding of issues in the literature. We suggest how the conceptual themes provide direction for further defining what is evidently a valuable connection between these two fields of study.

2.1 Introduction

Modern neuroscience research has produced findings about the brain and its functioning that could potentially impact the ways teachers in school classrooms approach their learners. The need to establish a clear link between the findings of neuroscience research and the contexts in which teaching and learning take place has been documented in both education and neuroscience literature (e.g. Ansari, 2008; Goswami, 2006). Concerns expressed in the literature suggest that the impact of neuroscience research on education has not been as effective as one would hope and that further efforts should be made to strengthen the connections between the two fields (Willingham, 2009). Our individual backgrounds as neuroscientist and educator have prompted us to deepen our comprehension of the possibilities inherent in linking knowledge from neuroscience and education. In doing so, we are mindful that, as an emerging field, there is no clear consensus for a definition of what is termed *educational neuroscience*. One definition, proposed by Szucs and Goswami (2007), suggests it is 'the combination of cognitive neuroscience and behavioral methods to investigate the development of mental representations' (p. 114). Other terms have been used somewhat synonymously with educational neuroscience including 'mind, brain and education', defined as 'the integration of the diverse disciplines that investigate human learning and development—to bring together education, biology and cognitive science to form the new field of mind, brain and education' (Fischer, et al., 2007, p.1), as well as 'neuroeducation' (see Theodoridou & Triarhou, 2009, p. 119).

The definitions suggest a multifaceted linking of diverse yet related fields. In keeping with the range of domains involved in educational neuroscience, an initial cursory examination of obviously relevant literature made us aware of the challenges in developing a complete understanding of how neuroscience and education might be combined to produce powerful school learning contexts. For example, some literature refers specifically to the problems that plague educational neuroscience, recognizing the need for bridging the gap between the two fields (e.g. Ansari & Coch, 2006; Willis, 2008) or cautioning about the problem of 'neuromyths' that pervade educators' approaches to teaching and learning (e.g. Christodoulou & Gaab, 2009; Purdy, 2008). Other work focuses on explaining the origin of such problems, evoking, for instance, concepts related to the challenges of combining multiple disciplines (Samuels, 2009) or emphasizing the lack of effective communication between them (Goswami, 2006) and the paucity of opportunities for cross-disciplinary professional development (Coch, Michlovitz, Ansari & Baird, 2009). Still other literature underscores the value of the educational neuroscience endeavour (e.g. Caine & Caine, 1998) and makes specific suggestions about how the field can move forward, for example, through increased collaboration (e.g. Ansari & Coch, 2006) and better research design (e.g. Ablin, 2008).

Historical accounts indicate that attempts to combine education and neuroscience are not new. Theodoridou and Triarhou (2009) describe the early efforts of neurologist Henry Herbert Donaldson and educator Reubon Post Halleck to explore the application of neurobiological research to education. Mayer (1998) highlights the pioneering efforts of E. L. Thorndike in recognizing the importance of brain physiology for educational psychology. Peterson (1984) notes the importance of a 1975 meeting of a group of scientists to discuss 'the relationship of recent research in neurophysiology and brain biochemistry to learning' and the subsequent publication of their report entitled 'Neural Mechanisms of Learning and Memory' (p. 75).

Given that the educational neuroscience literature extends over a considerable period and that a number of works have contributed both critiques and recommendations for advancement of the field, an objective and complete overview of this body of work is important. We therefore undertook a systematic search of the literature over the last 40 years (1970–2011) to help us better understand the connection between neuroscience and education. In this chapter we briefly describe the systematic review process we followed as we focused on the following questions: (1) What literature is central to an understanding of the connection between neuroscience and education? (2) In this central literature, what are the salient themes related to the connection between neuroscience and education?

Drawing on established approaches for conducting systematic reviews (e.g. Petticrew & Roberts, 2006), the iterative search and analysis process involved three phases: (1) extraction of keywords and systematic search in education, psychology and science databases including PsycInfo, ERIC, Medline and Current Contents for full articles; conceptual, systematic or theoretical reviews; opinion pieces, editorials, and commentaries from peer-reviewed, English language journals; (2) article selection for analysis based on whether the focus of the work is on the connection between the fields of education and neuroscience; (3) analysis of emergent themes through iterative reading and identification of salient problems, issues and recommendations in the discussions about the link between neuroscience and education.

The search resulted in an initial total of 482 references, with a final selection of a core set of 86 articles providing relevant discussion of the link between the two fields. Seven broad themes emerged from the critical analysis of the contents of the 86 articles retained. These we named *Misapplication, Multiple disciplines, Language, Knowledge development, Collaboration, Research design* and *Value.*

2.2 Publication trends in educational neuroscience

In addition to extracting the main themes of the educational neuroscience literature, our systematic review permitted us to gain an overview of the trends in works over time and in terms of the nature of the journals that published them. Figure 2.1 indicates the significant years of publication of works exploring the link between the two fields selected for analysis.

Overall, these publications tend to be from 2008 and 2009. Some factors contributing to this focus may be the inauguration of the journal *Mind, Brain, and Education* in 2007 and the appearance of a special issue on the neuroscience and education link in *Cortex* in 2009. Earlier than this focus, it is noteworthy that no articles from the 1970s surfaced and very few from the 1980s. Although in general the 1990s also produced few works, 1998 appears to have been an important year. Nine of the 14 selected articles published between 1990 and 1999 were published in 1998. Significantly, seven out of nine of these appeared in *Educational Psychologist Review* (Volume 10, Issue 4) as commentaries in response to an article by Byrnes and Fox (1998) published in the same issue titled 'The educational relevance of research in cognitive neuroscience'.

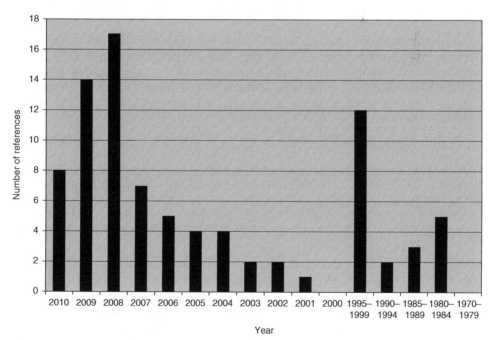

Fig. 2.1 Distribution of publications in the systematic review selected for analysis over time between 1970–2010.

A further point of interest is the nature of the publications in which the works appear. Of the 86 works which explore a link between neuroscience and education, 23 appear in clinical, science or neuroscience journals (e.g. *Science, Cortex, Trends in Cognitive Sciences*); these tend also to be high-impact science journals. In contrast, 63 works appear in journals related to education (e.g. *Phi Delta Kappan, Educational Researcher, Oxford Review of Education, Early Childhood Education Journal*), suggesting a strong interest on the part of educators in exploring ways that knowledge in neuroscience can be used in educational contexts. Publication trends also indicate that a significant set of works comes from a few journals in particular: 17 articles appear in the journal *Mind, Brain, and Education*, and another 10 appear in educational psychology journals (*Educational Psychology Review, Educational Psychologist, Journal of Educational Psychology*). For these latter journals, the overlapping of the fields of education and psychology inherent to their individual focal points positions them to comment on the links between neuroscience and education. In fact, Mason (2009) has suggested that the bridging of the gap between neuroscience and education could most effectively be made by educational psychologists.

2.3 A framework for understanding the educational neuroscience literature

To address our second research question concerning the salient themes in the literature, the content of the selected works was initially categorized by grouping similar arguments together. This resulted in the emergence of seven clear themes each representing a different underlying purpose or idea. For example, some themes highlight problems with the connection between neuroscience and education, while others represent solutions or proposals. The analysis also revealed that a number of themes are inter-related; we will show further in this chapter how some of the themes provide explanations for other themes. In order to represent the themes and their inter-relationships we created an integrative framework (see Figure 2.2).

The circular background of the framework represents the body of works that discuss the connection between neuroscience and education generated by the present search. The seven themes are depicted in graduated shades of gray from those addressing problems (*Misapplication*) and those that provide explanations for the problems (*Multiple disciplines, Language, Knowledge development*), to themes that suggest ways to overcome the acknowledged problems (*Collaboration, Research design*), leading ultimately to the broad theme of the importance of the connection between neuroscience and education (*Value*).

The most prevalent theme that emerges we named *Misapplication*. It groups together statements highlighting the problems associated with inappropriate interpretation and use of neuroscientific findings in education. Most of the criticism about the merging of neuroscience and education focuses on this topic. Three other themes can be seen as related to *Misapplication*, as they also highlight problems with the field, yet in addition they constitute explanations for the misapplication of neuroscientific findings. These three themes we have termed *Multiple disciplines* (problems associated with joining many disciplines each with their own fundamental theories, epistemologies, origins and methods), *Language* (problems associated with the fact that education and neuroscience have their own languages and that people within these disciplines may struggle to communicate), and *Knowledge development* (problems associated with gaps in knowledge among individuals attempting to breach the divide between the two disciplines).

The second most prevalent theme from the literature we named *Value*, as a theme that includes a variety of arguments related to the potential for neuroscience and education to contribute productively to each other, as well as the limitations of the applications the two fields can provide.

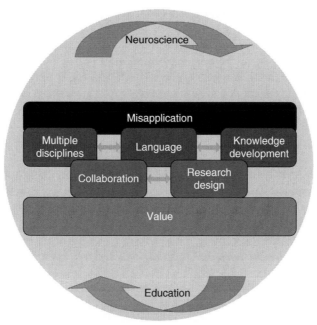

Fig. 2.2 Framework representing common themes derived from the systematic review of the educational neuroscience literature.

Despite scepticism on the part of some authors, in general arguments in this theme are positive, reflecting the potential of the field.

The final two themes, termed *Collaboration* and *Research design*, are directly related to *Value* as they include recommendations for increasing the value of the contributions made between neuroscience and education.

Together, these seven themes represent the common points of discussion raised in the works included in our systematic review of the educational neuroscience literature. An appreciation of these themes and their relationships is helpful in assessing the state of the field, as well as under-scoring documented problems and challenges associated with the field and recognizing agreed upon directions for further development of the field. The following sections describe the themes and the relationships among them.

2.4 Themes emerging from the educational neuroscience literature

Misapplication

This theme emerges from works referring to the misapplication of neuroscientific findings to the practice of education and the problems that ensue when findings from fundamental research on the neural bases of learning and cognition are oversimplified, misinterpreted, misunderstood or generalized (e.g. Coch, et al., 2009; Cubelli, 2009; Della Sala, 2009; Goswami, 2006; Theodoridou & Triarhou, 2009). The theme of misapplication is pervasive across the literature reviewed, evoked in half (49%) the articles analysed and appearing in both early discussions of the integration of neuroscience and education (Peterson, 1984; Sylwester, 1981, 1984) and accounts of the field as it evolves (e.g. Alferink & Farmer-Dougan, 2010; Carew & Magsamen, 2010). While there is a sense from the literature that neuroscience can provide important information on how children learn and guidance in terms of educational practice, there is a prevailing consensus that ineffective teaching strategies and policies have emerged from inappropriate interpretation of complex neuroscientific research (e.g. Ansari, 2008; Bruer, 2002; Coch, et al., 2009; Della Sala, 2009).

Central to this theme is the concept of *neuromyths* (e.g. Della Sala, 1999; Landau, 1988), a term used to signify popular and unrealistic beliefs about what neuroscience can do for education, and a phenomenon that has emerged as one of the important obstacles for educational neuroscience (e.g. Fischer, 2009; Goswami, 2004a, 2004b, 2008; Hirsh-Pasek & Bruer, 2007; Lindsey, 1998; Szucs & Goswami, 2007; Willis, 2008). Geake (2008), Goswami (2006) and Purdy (2008) provide detailed reviews of several of the most popular neuromyths and their origins, including the long-standing (see Peterson, 1984) laterality myth regarding left- and right-brain thinking, the idea that we only use 10% of our brain, and others (e.g. VAK learning styles, multiple intelligences, critical periods, whole brain learning/Brain Gym® and enriched environments/synaptogenesis).

While much of the criticism in works containing the *Misapplication* theme has been directed towards teachers who are implored to be more careful of their interpretations of neuroscience findings (e.g. Byrnes & Fox, 1998; Christoff, 2008; Purdy, 2008; Willis, 2007; see also Section 5, this volume), there is also caution directed towards neuroscientists who are often too quick to make links between the conclusions of their studies and educational practice and should themselves be careful not to misinterpret their results, speculate on the applied significance of their data or misconstrue the findings of their research (Bruer, 2002, 2006; Caine & Caine, 1998). A further caution regarding neuroethics, in particular ethical issues related to applying scientific research findings to education, is supplied by Fischer, Goswami and Geake (2010). Reviews of neuromyths such as those mentioned here and other critical commentaries (e.g. Ansari, 2008) can be seen as efforts to dispel incorrect beliefs about the brain and educate all parties, including the general public, educators and neuroscientists.

The perpetuation of neuromyths has resulted in the development of a number of brain-based teaching approaches, learning strategies and tools established on misrepresentation and oversim-plification of neuroscientific findings or unwarranted inferences (e.g. Byrnes & Fox, 1998). Defined by Stern (2005, p. 745) as 'pedagogical advice that is not grounded in scientific fact', and variously referred to as 'folk theory about the brain and learning' (Bruer, 2002, p. 1032), 'tall tales' (Della Sala, 2009, p. 443), 'professional folklore' (Sylwester, 1986, p. 163) and even 'educational malpractice' (Jorgenson, 2003, p. 368), brain-based pedagogy is generally portrayed in a negative light in the literature reviewed here. Many brain-based programmes and curricula have been criticized for their lack of empirical support, under evaluation, unscientific application and general reliance on claims that are unsubstantiated by science (Alferink & Farmer-Dougan, 2010; Howard-Jones, 2008a, 2008b; Samuels, 2009). The Brain Gym® programme in particular has been subject to disapproval for its unfounded claims that exercise will help improve interhemispheric information transfer (e.g. Della Sala, 2009; Goswami, 2006; Purdy, 2008; see also Chapter 13, this volume). Despite these criticisms, recommendations for more rigorous scientific and in-class-room testing of such programmes (Coch & Ansari, 2009) suggest the basing of teaching strategies on knowledge about the brain is not inherently wrong, but that incorporation of neuroscience findings in educational settings should not be naïve or simplistic (Mayer, 1998) and should avoid 'transgressions in sense making' (Howard-Jones, 2008b, p. 3). Similarly, Geake (2008) states that 'teachers should seek independent scientific validation before adopting brain-based products in their classroom' (p.123). Despite negative perceptions of brain-based programmes, Devonshire and Dommett (2010) caution against the removal of techniques that are accepted by educators, as this might discourage teachers from using techniques that work for them in their classrooms.

A similar concern is reflected in the work of Pickering and Howard-Jones (2007) who raise the following question suggesting the complications involved in making the connections between neuroscience and education: 'If something works in the classroom but lacks scientific validation, what consequences might this have for efforts to develop the interdisciplinary area of neuroscience and education?' (p. 111). However, others feel that direct application of findings to education is problematic and should not be encouraged (Ansari & Coch, 2006; Mason, 2009).

Finally, words of warning about neuromythologies are extended to other members of the educational neuroscience community. Willis (2007) evokes the problems that have surfaced when brain research is inappropriately used to support educational policies and the lack of formal guidelines that should guide researchers, curriculum publishers and educational consultants when promoting brain-based products and strategies. Varma, McCandliss and Schwartz (2008) describe the transfer of brain-based products into commercial interests and the desire of some to reap the financial benefits of brain-based education as 'neuromarketing' (p. 144), a trend referred to by Willis (2008) as the 'booming business in brain-booster products' (p. 424). Ansari (2008) and Samuels (2009) likewise recognize the growing availability of commercial products and the potential dangers of the commercial design and marketing of educational interventions.

Multiple disciplines

One potential explanation for the problem of misapplication of neuroscience findings is the fact that educational neuroscience involves two (or more) disciplines, each with their own histories, philosophies, epistemologies, theories and conceptualizations of knowledge (Samuels, 2009). This theme surfaces in 41% of the works reviewed. Educational neuroscience as a field necessitates a combination of knowledge from these disciplines and so members of the new, merged field may find themselves 'torn between the somewhat conflicting sets of responsibilities that have tradi- tionally been associated with the two domains' (Gardner, 2008, p. 167), a phenomenon referred to by Samuels (2009) as 'disciplinary polarity' (p.48). The ideas central to this theme stem from the notion that in a field such as educational neuroscience, knowledge about learning originates not only from a two-way integration of neuroscientific insight with educational expertise (Howard-Jones, 2008a; Mason, 2009), but also by incorporating and blending knowledge from psychology, cognitive science (Bruer, 1999; Carew & Magsamen, 2010) and other social sciences such as anthropology (Brandt, 1999). Goswami (2008) extends the notion of disciplinary integration to yet other domains such as biology, culture, cognition, emotion, perception and action. This idea resurfaces in Fischer's (2009, p. 3) description of the goal of the mind, brain and education (MBE) movement, which is to 'join biology, cognitive science, development and education to create a sound grounding of education in research,' and is also prevalent in Caine and Caine's (1998) plea for knowledge in education to be used in conjunction with the best knowledge available from other domains:

> Perhaps the most important basic rule to observe is that the results of neuroscientific research should rarely – perhaps never – be treated in isolation. Our own approach has always been to map findings in one domain over findings in others. A great deal is now known from the cognitive sciences and other sources about how people function, with input ranging from creativity theory to anxiety and stress literature. When coupled with the best of what is available about teaching and learning, we can find the basic patterns that are really valuable, and that enable us to be appropriately challenged by the brain research as well as find out how to capitalize on it (p.3).

Tommerdahl (2010) emphasizes the complex relationships that exist between education and neurosciences that have led to a disconnection between the two fields. She synthesizes some of the arguments via a theoretical model describing different types of knowledge, some of which are from different domains, which must be combined to contribute to new teaching methodologies that take into account a partnership between education and the brain sciences: neuroscience, cognitive neuroscience, psychology, educational theory and testing, and the classroom.

The idea that combining knowledge from a variety of fields may produce challenges for the field of educational neuroscience is more concretely stated in works by a number of authors who directly refer to notions of interdisciplinarity (Christodoulou, Daley & Katzir, 2009; Coch & Ansari, 2009; Fischer, et al., 2010; Fuller & Glendening, 1985; Geake, 2008; Gilkerson, 2001; Howard-Jones, 2008a; Hung, 2003; Immordino-Yang & Damasio, 2007; Katzir & Pare-Blagoev, 2006; McCandliss, 2010; Stern, 2005; Szucs & Goswami, 2007; Theodoridou & Triarhou, 2009; Willis, 2008), multidisciplinarity (Ansari & Coch, 2006; Coch, et al., 2009; Varma, et al., 2008), and transdisciplinarity (Ansari, 2008; Fischer, 2009; Koizumi, 2004; Samuels, 2009; Theodoridou & Triarhou, 2009). The works of Koizumi (2004) and Samuels (2009) may be particularly helpful in understanding the idea that educational neuroscience is not simply a combination of knowledge from many relevant disciplines, but rather the emergence of a novel conceptual field that goes beyond its original constituents (see also Chapter 20, this volume). Both these authors promote the term and framework of transdisciplinarity over those of inter- and multidisciplinarity because the former represents the synthesis of knowledge from multiple disciplines into a new discipline at a higher hierarchical level (see depiction by Koizumi, 2004, figure 1, p. 440). Transdisciplinarity recognizes that the knowledge taken from one discipline is not only the sum of individual knowledge from other groups (multidisciplinarity), nor is it the intersection of established disciplines (interdisciplinarity), but rather is the interaction and integration of diverse fields within a new group. As stated by Samuels (2009): 'What connects transdisciplinarity participants is not a common theoretical perspective or methodology or epistomology, but a common issue to which all apply their own particular expertise with the goal of reaching a holistic understanding of the issue' (p. 49).

Delving further into the factors that may be impeding the neuroscience and education connection, a number of authors refer to differences in levels of analysis (Berninger & Corina, 1998; Bruer, 2006; Katzir & Pare-Blagoev, 2006; Stanovich, 1998). This issue complicates interpretation of research on learning across domains such as biology, neuroscience, psychology and education, in that their experimental paradigms aim to understand learning at completely different levels of analysis. For example, the structure and function of the brain can be analysed at many different levels that are not necessarily directly related to each other (e.g. molecular basis of learning versus cognitive demonstration of learning versus child behaviour, Willingham & Lloyd, 2007, figure 1, p. 141). Practically speaking, differences in levels of analysis give rise to the question of how complex neuroscientific data should be integrated or mapped on to complex behavioural theory (Willingham & Lloyd, 2007). Fundamental differences in levels of research between disciplines (e.g. neuroscientists study neurons, cognitive psychologists study components of the mind, educators operate at the level of the entire mind of the child) are referred to as a 'vertical problem', whereas problems related to the question of how data from one level can be applied to another is considered a 'horizontal problem' (Willingham, 2009, p. 545). Two suggestions for addressing this issue emerge from the literature. First, experimental paradigms should be designed to capture the relationships across levels of analysis (Katzir & Pare-Blagoev, 2006) and second, neuroscientific data could contribute to behavioural data when a rich body of data and theory exist at the behavioural level (Willingham, 2009).

Language

A second theme incorporating arguments that contribute to explaining the problem of misapplication follows from the theme of *Multiple disciplines*: the idea that each of these disciplines has its own language and way of communicating. This theme is revealed in 34% of the works reviewed. Despite their common ground, differences in the expression of basic concepts and accompanying vocabulary between disciplines can create misconnections between researchers and educators (Ansari & Coch, 2006), can be a practical barrier to merging the two fields (Devonshire & Dommett, 2010) and can contribute to difficulties in translating terms and knowledge across fields (Blakemore & Frith, 2005; Byrnes & Fox, 1998; Ronstadt & Yellin, 2010; Tommerdahl, 2010). The issue of vocabulary is also reminiscent of the problems of *Multiple disciplines*, in that the meaning of certain words such as *learning* (e.g. Howard-Jones, 2008b) may differ significantly across disciplines and levels of analysis.

There is a consensus among authors addressing this issue that there is a need for an accessible, shared, common language and mode of communicating applicable to both researchers and educators, and that members of both disciplines should become 'bilingual' (Ablin, 2008; Byrnes & Fox, 1998; Carew & Magsamen, 2010; Coch & Ansari, 2009; Coch, et al., 2009; Geake, 2008; Hirsh-Pasek & Bruer, 2007; Pickering & Howard-Jones, 2007; Stanovich, 1998). It is thought that a new, shared language may help to 'transcend the borders' separating education and neuroscience (Koizumi, 2004, p. 440). Neuroscientists in particular have been criticized for not being the best people to communicate with teachers, in part because their language may be difficult to understand, but also because they tend to avoid broad, practical messages in order to preserve the scientific rigour of their experiments (Carew & Magsamen, 2010; Devonshire & Dommett, 2010; Goswami, 2006, 2008; Mason, 2009).

A common language could facilitate communication, transfer and translation of findings across disciplines by providing a frame of reference for contextualizing and understanding research results from scientific literature (Alferink & Farmer-Dougan, 2010; Christodoulou & Gaab, 2009; Hung, 2003; Purdy & Morrison, 2009; Theodoridou & Triarhou, 2009; Willis, 2008). Also noted is the lack of dissemination of neuroscience findings in education journals (Sylwester, 1981). At a more public level, it is suggested that publication of neuroscientific findings in more accessible forms, such as plain text or lay summaries that have the potential to reach a wide and more general audience rather than just scientists and academics would allow for a more effective dissemination of these findings (Purdy, 2008), facilitate responsible reporting (Varma, et al., 2008), and ensure that the public stays correctly informed of neuroscientific results (Stern, 2005, see also neuromyths).

Knowledge development

A third explanation for the misapplication of neuroscientific research, evoked in 28% of the works reviewed, is that there may be gaps in knowledge on both sides of the neuroscience–education divide. A portion of the work in this theme addresses the question of what kind of background and professional development is necessary for educators to benefit from neuroscientific research. In this regard, educators and practitioners are often targeted for their aversion to (Ansari, 2008) or lack of reliance on research for guidance (Brandt, 1999), also referred to as 'research avoidance' (Samuels, 2009, p. 47). In general, it is felt that educators need to 'upgrade their scientific literacy' (Sylwester, 1984, p. 73) and be more knowledgeable about the brain and neuroscience research so that they can be better informed and more critical in their evaluation of research findings (Ansari & Coch, 2006; Bruer, 1999; Byrnes & Fox, 1998; Carew & Magsamen, 2010; Devonshire & Dommett, 2010; Gilkerson, 2001; Jorgenson, 2003; Purdy, 2008; Sylwester, 1981, 1984, 1986, 2006; Willis, 2004, 2008). Ablin (2008) suggests that proper interpretation of scientific data should be modelled to educators so that they may learn to draw out the working principles and that the neuroscientific community should play a direct role in such professional development. Berninger and Corina (1998) feel that educational psychologists, in particular, should be offered courses about the brain.

A critical question is the issue of 'how, where, when and at what level' educators should be taught about neurosciences (Coch, et al., 2009, p. 28). Early in the span of literature analysed, Sylwester (1986) recognizes the need for increased preparation of teachers in natural sciences at both pre- and in-service levels. Others go a step further by suggesting that professional development and training should work bidirectionally; that is, educators need to become literate in basic neuroscience and researchers, neuroscientists and neurologists should become more familiar with education and pedagogical issues (Ansari, 2008; Ansari & Coch, 2006; Coch & Ansari, 2009; Fischer, et al., 2010; Pickering & Howard-Jones, 2007; Straham & Toepfer, 1983). Of note, these works are in part from advocates of MBE programmes, which incorporate neurosciences as an integral portion of the education curriculum and suggest that specially trained interdisciplinary 'neuroeducators' or 'hybrid professionals' (Pickering & Howard-Jones, 2007, p. 112) could be best placed to bridge the gap between disciplines. The MBE movement is therefore taking concrete steps to address the issue of knowledge development for educational neuroscientists (e.g. the undergraduate programme offered at Dartmouth College, Coch, et al., 2009).

Value

In contrast to the main theme of *Misapplication* and the themes (*Multiple disciplines*, *Language*, *Knowledge development*) that provide explanations for this problem, the theme of *Value* emerges in a range of works by authors either promoting the view that neuroscience and education can contribute significantly and productively to each other or discussing the ways in which they might connect. Like *Misapplication*, the theme of *Value* is prevalent in the literature reviewed, surfacing in 43% of the works reviewed.

There is ongoing discussion in a large portion of the literature on the value of making the connection between education and neuroscience and suggestions abound as to how this should be done (see discussion of the themes *Collaboration* and *Research design* that follow). Among those highlighting the value of a combined field of educational neuroscience, many make general comments about the potential of this emerging field and feel that neuroscience has a role to play in education (Bruer, 2002; Byrnes & Fox, 1998; Carew & Magsamen, 2010; Goswami, 2004b, 2006, 2008; Hinton, Miyamoto & Della-Chiesa, 2008; Hirsh-Pasek & Bruer, 2007; Howard-Jones, 2008a; Hung, 2003; Immordino-Yang & Damasio, 2007; Saunders & Vawdrey, 2002; Tommerdahl, 2010). Perspectives on the value of neuroscience for education come not only from authors of the works generated by the systematic search, but also directly from educators and practitioners. Pickering & Howard-Jones (2007) conducted a survey of educators' views on the role of the brain in education and found that the majority of the 189 teachers surveyed were open and enthusiastic to an increased understanding of the brain, especially in terms of 'how' they teach (as opposed to 'what' they teach). Specifically, the teachers surveyed were interested in teaching and learning approaches, cognitive and neuroscience knowledge, learning styles, educational kinesiology, ingestion and the brain, emotion and learning (Pickering & Howard-Jones, 2007).

Ways in which neuroscience can contribute to education include changing assumptions about what learning and teaching look like (Caine & Caine, 1998) and providing information on how the brain works, as well as validating theories of learning that can be translated into classroom strategies (Saunders & Vawdrey, 2002; Willis, 2007). Work by Carew and Magsamen (2010) suggests that the field of neuroscience is 'ripe for expanding its translational reach' (p. 685) and that its translational value has been underevaluated. Others remain more cautious, suggesting that the scope of the potential is not yet known or will perhaps be limited (Varma, et al., 2008; Willingham, 2008).

In assessing the value and potential of the field of educational neuroscience, many look more specifically to the practical value and applicability of neuroscientific findings for educational instruction. While Bruer (2002) and Howard-Jones (2008a) agree that there is some potential for neuroscience to contribute to education, they also critique the fact that neuroscience has little to contribute towards solving the problems that confront teachers in the classroom and cannot provide the type of explanations required for improving instruction and therefore lacks practical value (Bruer, 2002, 2006; Howard-Jones, 2008b). It is suggested that the richness of educational contexts and their social dimension are difficult to capture (Szucs & Goswami, 2007), although Howard-Jones (2008b) proposes a model for the integration of the social dimension into the connection between neuroscience and education.

Consistent with ideas about the value of making the neuroscience–education connection is the notion that conclusions and scientific theories drawn from neuroscience should be 'descriptive' for education rather than 'prescriptive' (Cubelli, 2009; Howard-Jones, 2008a; Mason, 2009; Willingham, 2009). This approach to educational neuroscience is illustrated in the following statement:

> Rather than seeking from neuroscience directives informing educational practice directly (i.e., how can neuroscience lead to better teaching or learning), neuroscience research is most valuable in describing aspects of educational concern (e.g., why do readers with dyslexia read differently) (Christodoulou & Gaab, 2009 p. 555).

Others relate the descriptive/prescriptive argument to the applicability of neuroscience findings to education, suggesting that direct or automatic extrapolation and applications of such findings to the classroom should be avoided (Ansari & Coch, 2006; Caine & Caine, 1998; Coch & Ansari, 2009; Mason, 2009; Tommerdahl, 2010) or are not always possible (O'Boyle & Gill, 1998). This view is in keeping with the belief that classroom applications should not be the focus or purpose of educational neuroscience studies (Szucs & Goswami, 2007). In contrast, there are some that feel that application is essential to the contribution of neuroscience to education and that educators should extract research findings from laboratory settings and put them to use in the classroom (e.g. Carew & Magsamen, 2010; Saunders & Vawdrey, 2002).

Directly related to the issues of practical value and applicability is the concept of 'usable knowledge,' a term from the field of MBE (as well as a regular special section of the journal by the same name) that relates to the 'efforts within and across fields of biology, cognitive science, human development and education that offer practical and applicable programmes of thought, research or action based on scientific research and inquiry' (Christodoulou, et al., 2009, p. 65). Consistent with the collaborative approach promoted by others in MBE (e.g. Ansari, 2008; Coch & Ansari, 2009), the goal for the production of usable knowledge is not only to translate scientific findings to applications in education, but also to recognize reciprocal relationships between research and practice (Christodoulou, et al., 2009). However, some feel that, to date, education has not systematically been grounded in practical research that has the potential to produce usable knowledge (e.g. Fischer, et al., 2010).

Though there is a prevalence of works and arguments for the value of educational neuroscience, there are a number of more reserved or pessimistic views that emerge from this systematic search and these balance the enthusiasm of some regarding the current state of the field. Several authors invoke temporal or 'geographical' factors as obstacles for educational neuroscience, including arguments that it is premature to merge the two fields (Alferink & Farmer-Dougan, 2010; Berninger & Corina, 1998; Hirsh-Pasek & Bruer, 2007; Peterson, 1984), that it will take many more years (Bruer, 2002; Varma, et al., 2008), that the relationship between the two disciplines remains distant (Mason, 2009), has not reached its full potential (Devonshire & Dommett, 2010) and that educational neuroscience is a 'bridge too far' (Bruer, 1997, p. 4) and even a 'dead-end street' (Mayer, 1998, p. 395). These views are tempered somewhat by those of Willis (2008) who, despite recognizing that brain research has not yet provided a direct connection between brain structure or function and classroom intervention, suggests this does not mean the work is irrelevant. Furthermore, Willis suggests that where education is concerned, it is not always possible to wait the required length of time until neuroscience provides solid proof for brain-based strategies and that some 'temporary leaps of faith' (p. 426) in trying interventions before research is complete may be necessary. A similar argument is put forward in the early literature by Sylwester (1981), who recognizes that educators may not be able to wait until all the problems of neuroscience are solved before trying classroom interventions.

Despite some negative views, it is clear from the literature reviewed that many contributors to the field of educational neuroscience see potential for the field in providing an added value to both research and practice, though readers are cautioned to take heed of the limitations of the field and types of knowledge that are likely to be of practical significance. In terms of the value of the field, two further themes surface in the literature as distinct ways that value can and should be created. These are *Collaboration* and *Research design*.

Collaboration

The need to develop partnerships between science and education is recognized early and persistently in the literature reviewed (36% of the works). In 1984, Peterson admonished that, at the time, little direct contact existed between brain researchers and educators. Several years later, in his influential 1997 paper, Bruer identified the need to develop 'an interactive, recursive relationship among research programs in education, cognitive psychology and systems neuroscience' (p. 15) and subsequently Berninger and Corina (1998) suggested that educational psychologists and cognitive neuroscientists in particular should establish bidirectional collaborations. The critical need for collaborations (rather than competition), and difficulties in establishing them, has been taken up by many others (Ansari & Coch, 2006; Christodoulou & Gaab, 2009; Coch & Ansari, 2009; Coch, et al., 2009; Fischer, 2009; Haier & Jung, 2008; Hinton, et al., 2008; Koizumi, 2004; Stern, 2005; Varma, et al., 2008). Drawing on associated themes of interdisciplinarity and value, Katzir and Pare-Blagoev (2006) advise that in order for such collaborations to be fruitful and truly interdisciplinary, they should meet two conditions. First, they must be 'guided by the goal of fostering interprofessional interactions that enhance the practice of each discipline' and second, 'interdisciplinary education should be based on mutual understanding and respect for the actual and potential contributions of the disciplines' (p.71). Varma et al. (2008) extend this notion beyond interdisciplinary collaboration to the need for narrowing the divide that separates the neuroscience and education communities, a concept that is taken up again by Christodoulou and Gaab ('collaborative community', 2009, p. 556) and by Carew and Magsamen (2010), who see in the merging of education and neuroscience the potential for a new community with common terminology, new traditions and learning outcomes. The latter report on the founding of the Neuro-Education Leadership Coalition in 2009, which 'seeks to blend the collective fields of neuroscience, psychology, cognitive science and education to create more effective teaching methods and curricula and, ultimately, to inform and transform educational policy' (p. 685).

One of the problems evoked in relation to the difficulties in establishing such collaborations is the lack of a common forum in which researchers and educators can interact (Ansari & Coch, 2006; Blakemore & Frith, 2005; Samuels, 2009) and the lack of opportunities for dialogue (Carew & Magsamen, 2010; Coch, et al., 2009). There is a general sense across the literature that more interdisciplinary exchange and ongoing dialogue are needed among neuroscientists, cognitive scientists, educators and educational researchers (Christoff, 2008; Fischer, et al., 2010; Geake, 2008; Gilkerson, 2001; Goswami, 2004b; Hirsh-Pasek & Bruer, 2007; Posner & Rothbart, 2005;

Purdy, 2008; Ronstadt & Yellin, 2010; Sylwester, 1981; Willis, 2008), with 'open and meaningful lines of communication' (O'Boyle & Gill, 1998, p. 407). Clearly, the intention of the MBE movement is to foster this type of ongoing dialogue.

A further recommendation for strengthening the relationship between education and neuroscience involves the adoption of shared theoretical/conceptual frameworks and integrated frameworks for practice that better reflect the contributions of the two fields (Coch & Ansari, 2009; Gilkerson, 2001; Ronstadt & Yellin, 2010; Szucs & Goswami, 2007). Samuel's (2009) proposal for a 'collaborative framework for knowledge creation' is in keeping with this recommendation, as he believes that the most likely theoretical framework for success in bridging the education–neuroscience gap is that of transdisciplinarity (p. 49). At a more practical level, Fischer (2009) suggests that common infrastructure, in the form of the creation of databases on learning and development, could foster connections between research and practice.

Research design

A second recurring theme (16% of the works) in relation to the value or potential of educational neuroscience involves research design. There is agreement in the literature that educators should become more active in the identification and formulation of research questions that may be relevant and appropriate for collaborations with neuroscience (Katzir & Pare-Blagoev, 2006) and that more research will need to be carried out in educational settings rather than solely in laboratories (Tommerdahl, 2010). As suggested by Ablin (2008), teachers should 'rethink the classroom as a place not of problem solving but rather problem design' (p. 52). However, there is also a sense that involvement in the generation of study questions and hypotheses should be more bidirectional and that to date, neuroscientists have mostly worked in isolation, usually with little consultation of the relevant educational literature to develop hypotheses and conduct studies (Coch & Ansari, 2009). Scientists need to be more attentive to the practical questions promoted by educators and educational policymakers (Hirsh-Pasek & Bruer, 2007) and 'spend more time in schools before beginning neuroscience projects with educational aims' (Pickering & Howard-Jones, 2007, p. 112). Such engagement of educators is central to Fischer et al.'s (2010) argument to move away from more traditional ways of collecting data in schools by involving educators and learners and paying attention to the 'ecology of schools' (p. 68). A barrier to this approach, however, is the lack of infrastructure in education for research on learning and teaching, which could be improved by the creation of research schools, shared databases on learning and the development of research designs based on neuroscience (Fischer, et al., 2010).

Some see the issue of research design and questions as an occasion for dialogue, collaboration and interdisciplinary work (see also *Multiple Disciplines* and *Collaboration* themes) whereby discussions between neuroscientists and educators could provide school staff with the opportunity to identify research questions of interest to be pursued in partnership (Christodoulou & Gaab, 2009; Hung, 2003; McCandliss, 2010). Such proposals for collaboration in establishing research questions are, however, linked with the previously described challenges surrounding the merging of multiple disciplines. In improving collaboration between the education and neuroscience communities and identifying a common research ground, Varma et al. (2008) make the following recommendation for putting the focus on problems rather than on individual disciplines:

> One strategy is to stop putting forward our disciplines as the basis for our identities and instead to put forward the problems we study. Problems can serve as neutral ground and can anchor intellectual exchange. (…) When researchers identify themselves by the problems they study, then it is valuable to travel to foreign disciplines in search of new insights and to bring back souvenirs – new methods, data and theories for answering the questions of one's native discipline (p.149).

This recommendation not only supports the notion that research methods from neuroscience may be useful to education and vice versa, it also relates to the theme of *Collaboration* between disciplines and the need to find a common ground for discussion.

Alternately, it may be that neither neuroscientists nor educators are ideally placed to transmit research questions across the disciplinary divide, but that an intermediary is needed in the form of a network of communicators capable of translating high-quality knowledge (e.g. ex-scientists with an interest in education, Goswami, 2006). This notion is consistent with the need for 'neuroscience-enriched educators' and 'pedagogically-enriched neuroscientists' who have the training and background to adapt educational questions to variables that can be measured meaningfully and realistically using neuroscientific design (Goswami, 2009, pp. 177, 182). More integration of design of experimental paradigms within educational neuroscience could address the question of what methods are optimal and appropriate for addressing traditional educational questions in new ways (Szucs & Goswami, 2007).

2.5 **Conclusion**

In summary, it is clear that many of the problems identified in the early literature on the connection between neuroscience and education (e.g. neuromyths, lack of common language, inappropriate communication style) continue to be evoked, suggesting that solutions have not yet been implemented or, if implemented, are not yet effective. However, some works from the literature that discuss these problems appear to offer specific attempts to rectify the situation. Endeavours already in place, stemming from work within the MBE movement and other organizations (e.g. Center for Educational Neuroscience, European Association for Research on Learning and Instruction (EARLI) Neuroscience and Education Special Interest Group, Learning & the Brain Society), such as the establishment of research schools (e.g. Centre for Neuroscience in Education, University of Cambridge; Mind, Brain and Education programme, Harvard Graduate School of Education; Dartmouth's Center for Cognitive and Educational Neuroscience); specific collaborations between educators and neuroscientists in the form of conferences (EARLI Zurich 2010; International Mind, Brain and Education Society, Philadelphia 2009, San Diego 2011); and the creation of specific journals (e.g. *Mind, Brain, and Education*) or special journal issues in educational neuroscience (*Cortex*, 2009, Forum on Mind, Brain, and Education, volume 45, issue 4; *The International Journal of Environmental and Science Education*, Mind, Brain and Education Special Issue, 2011, volume 6, issue 4; *Developmental Cognitive Neuroscience*, Education and Neuroscience Special Issue, forthcoming 2012, volume 2, issue 1) appear as viable opportunities for advancing the connection between the disciplines. It might also be granted that while the spread of neuromyths remains a concern in the literature, the large number of articles noting this problem indicates a recognition of the need to address it. The message, therefore, appears to be disseminated widely. Despite some reticence to endorse the connection fully, and some ongoing concern about providing too optimistic a view of what might be possible, there is an overall positive impression that there is value in merging the fields. This impression is accompanied with the sense that there are ways that this merging might take place, even if for some it might be in a more limited way than for others in terms of applicability of neuroscience findings to education.

The visual framework offered in this chapter is intended as a picture of the themes that predominate in the literature analysed as the focal points of discussion. The themes range across identification of problems and their associated explanations to recognition of the value of connecting the fields, with at least two possibilities for enhancing this value. As some articles contain elements of several of the seven themes, it is evident that some authors note not only problems, but also provide explanations for them, as well as recognize value in the connection between the fields, with specific ideas about the elements that will contribute further to this value. The framework may represent a starting point for those wishing to attain an understanding of where future efforts could be directed, in light of the existing concerns. The frequency of reference to transdisciplinarity as a lens through which to examine the connection is striking and may lead to a deepened appreciation of the problems and the possibilities inherent in the connection.

The results of the search establish a basis for further exploring this literature from a conceptual standpoint and suggest a direction for future attempts to clarify the link between education and neuroscience. Questions that could be posed include the following: How do the themes that emerge from this study drive ongoing understandings of educational neuroscience? What problems continue to hinder a smooth connection between education and neuroscience? In what ways can neuroscientists and educators establish fruitful collaborations to improve educational contexts?

Acknowledgement

We are grateful for the help of Dominic Desaulniers (EPC-Biology Library, Université de Montréal) in developing the initial search strategy.

References

(*References from the systematic review)

* Ablin, J. L. (2008). Learning as problem design versus problem solving: Making the connection between cognitive neuroscience research and educational practice. *Mind, Brain, and Education*, 2(2), 52–54.

* Alferink, L. A., & Farmer-Dougan, V. (2010). Brain-(not) based education: Dangers of misunderstanding and misapplication of neuroscience research. *Exceptionality*, 18(1 Special Issue), 42–52.

* Ansari, D. (2008). The brain goes to school: Strengthening the education-neuroscience connection. *Education Canada*, 48(4), 6–10.

* Ansari, D., & Coch, D. (2006). Bridges over troubled waters: Education and cognitive neuroscience. *Trends in Cognitive Sciences*, 10(4), 146–51.

* Berninger, V. W., & Corina, D. (1998). Making cognitive neuroscience educationally relevant—creating bidirectional collaborations between educational psychology and cognitive neuroscience. *Educational Psychology Review*, 10(3), 343–54.

* Blakemore, S.-J., & Frith, U. (2005). The learning brain: Lessons for education: A precis. *Developmental Science*, 8(6), 459–71.

* Brandt, R. (1999). Educators need to know about the human brain. *Phi Delta Kappan*, 81(3), 235–38.

* Bruer, J. T. (1997). Education and the brain: A bridge too far. *Educational Researcher*, 26(8), 4–16.

* Bruer, J. T. (1999). In search of . . . Brain-based education. *Phi Delta Kappan*, 80(9), 648–57.

* Bruer, J. T. (2002). Avoiding the pediatrician's error: How neuroscientists can help educators (and themselves). *Nature Neuroscience*, 5(Suppl S), 1031–33.

* Bruer, J. T. (2006). On the implications of neuroscience research for science teaching and learning: Are there any? A skeptical theme and variations-the primacy of psychology in the science of learning. *CBE Life Sciences Education*, 5(2), 104–110.

* Byrnes, J. P., & Fox, N. A. (1998). The educational relevance of research in cognitive neuroscience. *Educational Psychology Review*, 10(3), 297–342.

* Caine, R. N., & Caine, G. (1998). Building a bridge between the neurosciences and education: Cautions and possibilities. *National Association of Secondary School Principals Bulletin*, 82(598), 1–8.

* Carew, T. J., & Magsamen, S. H. (2010). Neuroscience and education: An ideal partnership for producing evidence-based solutions to guide 21(st) century learning. *Neuron*, 67(5), 685–88.

* Christodoulou, J. A., Daley, S. G., & Katzir, T. (2009). Researching the practice, practicing the research, and promoting responsible policy: Usable knowledge in mind, brain, and education. *Mind, Brain, and Education*, 3(2), 65–67.

* Christodoulou, J. A., & Gaab, N. (2009). Using and misusing neuroscience in education-related research. *Cortex*, 45(4), 555–57.

* Christoff, K. (2008). Applying neuroscientific findings to education: The good, the tough, and the hopeful. *Mind, Brain, and Education*, 2(2), 55–58.

* Coch, D., & Ansari, D. (2009). Thinking about mechanisms is crucial to connecting neuroscience and education discussion. *Cortex*, 45(4), 546–47.

* Coch, D., Michlovitz, S. A., Ansari, D., & Baird, A. (2009). Building mind, brain, and education connections: The view from the upper valley. *Mind, Brain, and Education*, 3(1), 27–33.

* Cubelli, R. (2009). Theories on mind, not on brain, are relevant for education. *Cortex*, 45(4), 562–64.

* Della Sala, S. (2009). The use and misuse of neuroscience in education. *Cortex*, 45(4), 443.

Della Sala, S. (Ed.). (1999). *Mind myths: Exploring popular assumptions about mind and body*. Chichester: John Wiley & Sons.

* Devonshire, I. M., & Dommett, E. J. (2010). Neuroscience: Viable applications in education? *Neuroscientist*, 16(4), 349–56.

* Fischer, K. W. (2009). Mind, brain, and education: Building a scientific groundwork for learning and teaching. *Mind, Brain, and Education*, 3(1), 3–16.

Fischer, K. W., Daniel, B. D., Immordino-Yang, M. H., Stern, E., Battro, A., & Koizumi, H. (2007). Why mind, brain, and education? Why now? *Mind, Brain, and Education*, 1(1), 1–2.

* Fischer, K. W., Goswami, U., & Geake, J. (2010). The future of educational neuroscience. *Mind, Brain, and Education*, 4(2), 68–80.

* Fuller, J. K., & Glendening, J. G. (1985). The neuroeducator: Professional of the future. *Theory into Practice*, 24(2), 135–37.

* Gardner, H. (2008). Quandaries for neuroeducators. *Mind, Brain, and Education*, 2(4), 165–69.

* Geake, J. (2008). Neuromythologies in education. *Educational Research*, 50(2), 123–33.

* Gilkerson, L. (2001). Integrating an understanding of brain development into early childhood education. *Infant Mental Health Journal*, 22(1-2), 174–87.

* Goswami, U. (2004a). Neuroscience and education. *British Journal of Educational Psychology*, 74(Pt 1), 1–14.

* Goswami, U. (2004b). Neuroscience, education and special education. *British Journal of Special Education*, 31(4), 175–83.

* Goswami, U. (2006). Neuroscience and education: From research to practice? *Nature Reviews Neuroscience*, 7(5), 406–411.

* Goswami, U. (2008). Principles of learning, implications for teaching: A cognitive neuroscience perspective. *Journal of Philosophy of Education*, 42(3-4), 381–99.

* Goswami, U. (2009). Mind, brain, and literacy: Biomarkers as usable knowledge for education. *Mind, Brain, and Education*, 3(3), 176–84.

* Haier, R. J., & Jung, R. E. (2008). Brain imaging studies of intelligence and creativity: What is the picture for education? *Roeper Review*, 30(3), 171–80.

* Hinton, C., Miyamoto, K., & Della-Chiesa, B. (2008). Brain research, learning and emotions: Implications for education research, policy and practice. *European Journal of Education*, 43(1), 87–103.

* Hirsh-Pasek, K., & Bruer, J. T. (2007). The brain/education barrier. *Science*, 317(5843), 1293.

* Howard-Jones, P. (2008a). Education and neuroscience. *Educational Research*, 50(2), 119–22.

* Howard-Jones, P. (2008b). Philosophical challenges for researchers at the interface between neuroscience and education. *Journal of Philosophy of Education*, 42(3-4), 361–80.

* Hung, D. (2003). Supporting current pedagogical approaches with neuroscience research. *Journal of Interactive Learning Research*, 14(2), 129–55.

* Immordino-Yang, M. H., & Damasio, A. (2007). We feel, therefore we learn: The relevance of affective and social neuroscience to education. *Mind, Brain, and Education*, 1(1), 3–10.

* Jorgenson, O. (2003). Brain scam? Why educators should be careful about embracing "Brain research." *Educational Forum*, 67(4), 364–69.

* Katzir, T., & Pare-Blagoev, J. (2006). Applying cognitive neuroscience research to education: The case of literacy. *Educational Psychologist*, 41(1), 53–74.

* Koizumi, H. (2004). The concept of 'developing the brain': A new natural science for learning and education. *Brain & Development*, 26(7), 434–41.

Landau, W. M. (1988). Clinical neuromythology i: The Marcus Gunn phenomenon: Loose canon of neuro-opthalmology. *Neurology*, 38, 1141–42.

* Lindsey, G. (1998). Brain research and implications for early childhood education. Research reviews. *Childhood Education*, 75(2), 97–100.

* Mason, L. (2009). Bridging neuroscience and education: A two-way path is possible. *Cortex*, 45(4), 548–49.

* Mayer, R. E. (1998). Does the brain have a place in educational psychology? *Educational Psychology Review*, 10(4), 389–96.

* McCandliss, B. D. (2010). Educational neuroscience: The early years. *Proceedings of the National Academy of Sciences USA*, 107(18), 8049–50.

* O'Boyle, M. W., & Gill, H. S. (1998). On the relevance of research findings in cognitive neuroscience to educational practice. *Educational Psychology Review*, 10(4), 397–409.

* Peterson, R. W. (1984). Great expectations: Collaboration between the brain sciences and education. *American Biology Teacher*, 46(2), 74–80.

Petticrew, M., & Roberts, H. (2006). *Systematic reviews in the social sciences*. Mulden, MA: Blackwell Publishing.

* Pickering, S. J., & Howard-Jones, P. (2007). Educators' views on the role of neuroscience in education: Findings from a study of UK and international perspectives. *Mind, Brain, and Education*, 1(3), 109–113.

* Posner, M. I., & Rothbart, M. K. (2005). Influencing brain networks: Implications for education. *Trends in Cognitive Sciences*, 9(3), 99–103.

* Purdy, N. (2008). Neuroscience and education: How best to filter out the neurononsense from our classrooms? *Irish Educational Studies*, 27(3), 197–208.

* Purdy, N., & Morrison, H. (2009). Cognitive neuroscience and education: Unravelling the confusion. *Oxford Review of Education*, 35(1), 99–109.

* Ronstadt, K., & Yellin, P. B. (2010). Linking mind, brain, and education to clinical practice: A proposal for transdisciplinary collaboration. *Mind, Brain, and Education*, 4(3), 95–101.

* Samuels, B. M. (2009). Can the differences between education and neuroscience be overcome by mind, brain, and education? *Mind, Brain, and Education*, 3(1), 45–55.

* Saunders, A. D., & Vawdrey, C. (2002). Merging brain research with educational learning principles. *Business Education Forum*, 57(1), 44–46.

* Stanovich, K. E. (1998). Cognitive neuroscience and educational psychology: What season is it? *Educational Psychology Review*, 10(4), 419–26.

* Stern, E. (2005). Pedagogy meets neuroscience. *Science*, 310(5749), 745.

* Straham, D. B., & Toepfer, C. F. (1983). The impact of brain research on the education profession: Agents of change. *Journal of Children in Contemporary Society*, 16(1), 219–33.

* Sylwester, R. (1981). Educational implications of recent brain research. *Educational Leadership*, 39(1), 6–19.

* Sylwester, R. (1984). The neurosciences and the education profession: Inserting new knowledge of a child's developing brain into an already well-developed school. *Journal of Children in Contemporary Society*, 16(1), 1–8.

Sylwester, R. (1986). Synthesis of research on brain plasticity: The classroom environment and curriculum enrichment. *Educational Leadership*, 44(1), 90–93.

* Sylwester, R. (2006). Cognitive neuroscience discoveries and educational practices. *School Administrator*, 63(11), 32.

* Szucs, D., & Goswami, U. (2007). Educational neuroscience: Defining a new discipline for the study of mental representations. *Mind, Brain, and Education*, 1(3), 114–27.

* Theodoridou, Z. D., & Triarhou, L. C. (2009). Fin-de-siecle advances in neuroeducation: Henry Herbert Donaldson and Reuben post Halleck. *Mind, Brain, and Education*, 3(2), 119–29.

* Thorndike, E.L. (1926). *The Measurement of Intelligence*, New York: Teachers College, Columbia University.

* Tommerdahl, J. (2010). A model for bridging the gap between neuroscience and education. *Oxford Review of Education*, 36(1), 97–109.

* Varma, S., McCandliss, B. D., & Schwartz, D. L. (2008). Scientific and pragmatic challenges for bridging education and neuroscience. *Educational Researcher*, 37(3), 140–52.

* Willingham, D. T. (2008). When and how neuroscience applies to education. *Phi Delta Kappan*, 89(6), 421–23.

* Willingham, D. T. (2009). Three problems in the marriage of neuroscience and education. *Cortex*, 45(4), 544–45.

* Willingham, D. T., & Lloyd, J. W. (2007). How educational theories can use neuroscientific data. *Mind, Brain, and Education*, 1(3), 140–49.

* Willis, J. (2004). Learning and the brain: How administrators can improve teacher effectiveness through instruction on how the brain learns. *AASA Journal of Scholarship and Practice*, 1(2), 13–15.

* Willis, J. (2007). Which brain research can educators trust? *Phi Delta Kappan*, 88(9), 697–99.

* Willis, J. (2008). Building a bridge from neuroscience to the classroom. *Phi Delta Kappan*, 89(6), 424–27.

Theoretical approaches for developing the good, removing the bad and giving the ugly a makeover in neuroscience and education

Chapter 3

Constructing connection: the evolving field of mind, brain and education

Donna Coch and Daniel Ansari

Overview

'Constructing connection: the evolving field of mind, brain and education' provides an overview of some of the challenges and benefits to building the evolving field of mind, brain and education (MBE). The authors critically discuss selected challenges associated with making meaningful connections across the fields of psychology, neuroscience and education (e.g. differing levels of explanation, different methodological research and design issues); address some of the mechanisms that need to be in place in order to make such connections (e.g. transdisciplinary training for educators and neuroscientists, blurring the traditional applied versus basic research distinction); and offer suggestions about the next steps to take in order to build and sustain an interdisciplinary science of MBE that provides a new, integrated, and multilevel perspective on child development, learning and teaching. As traditional boundaries among behavioural, psychological, neuroscientific, and educational research are increasingly crossed, the authors conclude that neuroscience should be considered one source of evidence that can contribute to evidence-based practice and policy in education, that the challenges to constructing meaningful connections between education and neuroscience are surmountable, and that the science and art of education can be connected and integrated in principled ways.

I say moreover that you make a great, a very great mistake, if you think that psychology, being the science of the mind's laws, is something from which you can deduce definite programmes and schemes and methods of instruction for immediate schoolroom use. Psychology is a science, and teaching is an art; and sciences never generate arts directly out of themselves. An intermediary inventive mind must make the application.

James (1899/2001, p. 3)

3.1 **Introduction**

If the brain is considered the 'organ for learning' (e.g. Hart, 1983, p. 10) and one of the goals of education is learning, it seems sensible that brain research might have something valuable to contribute to education and, in turn, that education could have something valuable to contribute to brain research. Indeed, in recent years, educators have become increasingly enthusiastic about what neuroscience can offer to education (e.g. Pickering & Howard-Jones, 2007). In turn, there has been a growing interest among scientists with respect to what they can offer to education, recognizing, over a century after James (1899/2001), that 'we must look at the "art" of education through the critical lens of science if we are to survive' (e.g. Alberts, 2009, p. 15). There has been much less discussion about what education can offer to neuroscience (e.g. Ansari & Coch, 2006; Varma, McCandliss & Schwartz, 2008).

What might such mutual contributions look like? What form could such contributions take? Educators might have intimate knowledge about the learning strengths and weaknesses of individual students, but how that knowledge can be translated into the realm of neuroscience is not always clear. Brain researchers might reveal how biological processes both constrain and facilitate learning, but the translation of such research into classroom practice and educational policy is not always obvious (e.g. Ansari & Coch, 2006; Blakemore & Frith, 2005; Goswami, 2004, 2006, 2009; Posner & Rothbart, 2005; Stern, 2005). This chapter is about the multifaceted issue of translation: how James' (1899/2001) inventive minds can apply the art of teaching to the science of neuroscience and the science of neuroscience to the art of teaching within a new, transdisciplinary science of learning (e.g. Koizumi, 2004; Meltzoff, Kuhl, Movallan & Sejnowski, 2009). We focus on how connections across mind, brain and education must be mindfully constructed and what mechanisms must be put into place to allow for such connections to develop. Constructing such connections across science, practice and policy is foundational for making meaningful 'cross-cultural' translations (Shonkoff, 2000).

Historically, the connection between empirical science and educational practice in general has been troubled (e.g. Condliffe Lagemann, 2000; Shavelson & Towne, 2002). Other disciplines, such as clinical psychology, have also struggled to base practice on scientific evidence (Baker, McFall & Shoham, 2009). Even in its short history, the field of MBE has encountered difficulties with making meaningful connections between educational practice and neuroscientific findings. Just over a decade ago, Bruer (1997) claimed that such connections were 'a bridge too far', yet so-called 'brain-based' commercial programmes and products have proliferated at an alarming rate nonetheless. Indeed, there seems to be a certain allure to explanations and products that are accompanied by 'neuro' or 'brain-based' language that makes them somehow more convincing than explanations without neuroscience—even, it turns out, when the neuroscience language adds nothing of value (e.g. Weisberg, Keil, Goodstein, Rawson & Gray, 2008; see also Section 5 this volume).

Many popularized, so-called 'brain-based' curricular materials rest on the notion that the results of laboratory neuroscience experiments can be applied directly to classroom teaching—from Petri dish to pedagogy, so to speak. In our experience, many teachers who have been exposed to such materials believe that neuroscience can tell them what to do in the classroom (see also, e.g. Goswami, 2006). We contend that such expectations for neuroscience-based, easy-to-follow recipes for practice are unrealistic. Much of the 'brain-based' learning material, upon closer inspection, is often based on very loose links and factually incorrect interpretations or gross generalizations of neuroscience data; in essence, pseudoscience rather than neuroscience. Such generalizations can lead to 'neuromyths' (e.g. Goswami, 2004; see also Section 5 this volume) such as that there are 'left-brained' and 'right-brained' learners who must be taught differently in order to reach their full potential. While there are reliable neuroscience data indicating that the left and right hemispheres of the human brain process information in different ways, many of these data come from carefully designed laboratory studies with adult patients who have undergone surgical disconnection of the two cerebral hemispheres (e.g. Gazzaniga, 1998). Such findings do not necessarily apply in a direct way to typically developing children with intact corpus callosum in an everyday classroom setting.

In contrast to the direct application view espoused by many in the 'brain-based' learning movement, we believe that connections between neuroscience and education are not so simple to make; that they are more nuanced, more sophisticated and more complex than direct application from lab to classroom. In part, the bridge metaphor so often used in the literature (e.g. Ansari & Coch, 2006; Bruer, 1997) may contribute to this mistaken view of a direct and linear connection. Perhaps a more fruitful visual metaphor would be a complex highway interchange or motorway junction in which various constraints emerge as different roadways at different levels are connected and integrated to lead seamlessly from one to another. As different domains of education and neuroscience are integrated in the common goal of improving student learning and development, a similarly complex network of connections will emerge. The construction of such connections will be facilitated by the training and development of a new generation of researcher-practitioners, the adaptation of existing and new research methodologies and research approaches that blur what is basic and what is applied, and the collaborations between educators and researchers that are necessary for a sustainable science of MBE.

In this view of links between neuroscience and education, strong and direct connections that resonate with what educators already know or believe in terms of craft knowledge or folk psychology may not be forthcoming. Instead, connections that are built incrementally and iteratively as part of the process of scientific inquiry and discourse may be more characteristic of links between neuroscience and education. In this view, decidedly less sexy, the blockbuster headlines are missing and the work of building a field and constructing connections based on collaboration, discussion and scientific design at multiple levels of analysis is at the forefront. Here, we critically discuss the emerging field of MBE, highlighting some challenges associated with both deconstructing obstacles and constructing principled, responsible connections between educational practice and policy and neuroscience research.

3.2 **Deconstruction: overcoming obstacles**

Levels of explanation

One of the challenges to making meaningful connections between education and neuroscience is that these two disciplines operate at different levels of explanation, a difficulty of scale or grain size (e.g. Ansari & Coch, 2006; Perkins, 2009; Willingham, 2009). Educators might be used to working at the level of individual students, the classroom or curriculum while neuroscientists might be used to working at the level of single neurons, neural assemblies or cognitive networks. Scaling across these grain sizes is not easy. Yet, even while this is a crucial challenge, spanning across grain size is also a remarkable strength of MBE; we believe that multilevel explanations of a given phenomenon can only lead to richer understanding. The multiple levels of analysis inherent in MBE allow for the possibility of translation across levels while each level of analysis itself contributes to an integrated understanding. For example, a student's poor performance on a reading assessment may be explicable by inattentive behaviours in the classroom involving differential activity in particular attentional neural networks, which may be better understood with additional knowledge about levels of specific neurotransmitters, such as dopamine, at critical synaptic junctions, which may be related to both the classroom environment and the student's genome. At the same time, psychological, educational, social and cultural factors are likely at play. For example, prior knowledge may be impoverished or relevant knowledge may be inaccessible to the student at the moment, motivation may be low, the curricular sequencing may not be appropriate for the student's developmental level, the student may be more concerned about her social position in class than about reading, and the form of thinking required by the task may be culturally unfamiliar. Being able to evaluate and consider learning across several grain sizes and using multiple lenses and perspectives allows for a more complex understanding than any one level, lens or perspective alone can provide. This ideal portrait of a deep and integrated understanding of the range of processes involved in learning can serve as a strong basis for individualized intervention, pedagogy and curriculum and research design.

Related to the notion of MBE spanning levels of explanation or analysis, we are often challenged to state what neuroscience studies add to the knowledge base on learning and development beyond corroborating behavioural evidence; that is, what does neuroscience contribute uniquely? In part, one response may simply be that neuroscience adds another level of analysis; but, in part, this response is not adequate or particularly fruitful, as what does it mean and why is it important that, for example, reading words increases blood flow in Wernicke's area? The strength of the MBE approach, in our view, is in attempting to integrate across levels of analysis, to constrain interpretation at each level with evidence from the other levels. As a basic example, consider that changes in brain activations can be observed in the absence of behavioural differences. Ansari et al. (2005) found no differences in the reaction times of children and adults in a simple number discrimination paradigm, but observed that different brain areas were recruited during the judgement of relative numerical magnitude in children and adults. The ability to compare numerical magnitudes is associated with individual differences in children's school-relevant mathematical abilities (e.g. Halberda, Mazzocco & Feigenson, 2008; Holloway & Ansari, 2009). More specifically, the authors measured the effect of numerical distance (the difference between two numbers) on the ability to judge which of the two numbers was numerically larger. Both adults and children were faster at the judgement when the numbers were separated by a large compared to a relatively small numerical distance. This 'distance effect' on reaction times did not differ between the children and adults. However, the effect of distance on brain activation was strongest in frontal regions in the children, but strongest in parietal regions in adults.

Such dissociations between behavioural and brain imaging data illustrate that neuroscience allows for observation of online processing (what is inside the 'black box') in addition to behavioural outcomes. That is, neuroscience allows researchers and educators to consider not only the observable product but also the previously unobservable process, not in terms of linear box-and-arrow models, but in terms of dynamic, cascaded networks and their interactions. However, most developmental cognitive neuroscience studies are very limited and do not span across all the levels relevant to learning and development, even those concerned with more 'macro' factors such as culture or socioeconomic status (e.g. Noble, McCandliss & Farah, 2007; Noble, Norman & Farah, 2005). One of the challenges of MBE is to design studies or sets of studies that address issues of grain size more adequately. It will be crucial for future studies to systematically compare and contrast brain level, behavioural level and sociocultural level changes in a wide range of domains relevant to education, with a focus on how interactions across levels can help to better understand the processes of learning and education and how to improve them. Likewise, it will be crucial to recognize that studies conducted at one level of analysis cannot be translated directly into another level of analysis (as often happens in 'brain-based' programmes); instead, construction needs to proceed at multiple levels in order to create useable interchanges.

Reduction

An important corollary to the levels of explanation issue is that a given level is not expected to be reduced to any other level. MBE as an emerging field does not seek to replace successful behavioural means of gleaning insight into the processes of learning and development with neuroimaging methods. Neither does it seek to simply provide another redundant or dualistic level of analysis for a given issue. Nor does an integration of education and neuroscience presuppose that brain or neuroimaging data are a priori a more valuable source of information than behavioural data. Indeed, it is almost always the case that neuroimaging findings can only be meaningfully understood and interpreted in conjunction with behavioural data. For example, the power of combining such data sources in educationally significant ways is nicely illustrated by the recent finding of an increase in predictability of elementary school reading ability with combined use of neuroimaging and behavioural measures (Hoeft et al., 2007).

Indeed, our choice of the terminology 'mind, brain and education' reveals our bias for conscious inclusion of the behavioural and psychological ('mind') levels within MBE. The alternative 'educational neuroscience', in our view, leaves out a critical component of the MBE endeavour. In short, MBE as an emerging field rests on the principle that the integration of methods, findings and questions from the developmental and learning sciences, neuroscience and education can provide greater explanatory power than each alone, with the whole being more than the sum of the parts. There is little doubt that educational, biological, cognitive, emotional, contextual and sociocultural aspects of development are interactive, interrelated and interdependent, such that constructing connections across methods and disciplines can only lead to a richer understanding of the complexities of child development and learning (Diamond, 2007; Stiles, 2009).

Research designs and methods

Another set of challenges to constructing connections between neuroscience and education involves research designs and methods. Recently, there has been much discussion in the United States about evidence-based educational practice and large, randomized controlled trials in educational research (e.g. Slavin, 2002). In contrast, developmental cognitive neuroscience research, for example, typically uses small, self-selected samples. The cost and time-intensiveness of neuroimaging research methods are often cited as key factors in this design issue. We believe that a large variety of research designs and methodologies are integral to MBE. That being said, research methods that are positioned to provide strong internal validity and generalizability of findings may, on their own, be ineffective for naturalistic studies of classroom learning and behaviour. A science of education requires research programmes that 'integrate basic research in "pure" laboratories with field work in "messy" classrooms' (Klahr & Li, 2005, p. 217). With the advent of more portable imaging systems (e.g. event-related potential and near-infrared spectroscopy systems) and the potential for lab schools adopting an MBE perspective (e.g. Hinton & Fischer, 2008), even greater integration across contexts may be possible in the near future.

Overall, one of the crucial methodological challenges to MBE is to integrate across research approaches. It is not difficult to envision combinations of these approaches that would, in principle, build on the strengths of each; for example, a large-scale intervention involving many schools and classrooms to document student performance, accompanied by an imaging component to address the neuroscience questions for some subset of students, and a classroom observation component to address the sociocultural questions. These sorts of multitiered studies would provide scientific evidence at multiple levels of analysis, producing findings that would mutually constrain interpretation across levels, and could potentially contribute to evidence-based teaching and curriculum design. Thus, with respect to methodological issues, while the goal of a better understanding of learning and development may be shared across fields, the approaches to meeting that goal are diverse. Converging evidence from multiple methodologies—approaches to research and teaching that diverge both conceptually and methodologically—can help to both develop and constrain theory and practice.

Another methodological challenge to linking educational practice and policy with neuroscience research is that only a small subset of neuroscience studies is conducted with school-age children and adolescents as participants, and many of these studies may make unwarranted developmental assumptions. For example, in the domain of cognitive neuroscience, while it seems relatively certain that the blood oxygenation level dependent (BOLD) contrast upon which most current functional magnetic resonance imaging (fMRI) studies is based is similar in children and adults (e.g. Kang, Burgund, Lugar, Petersen & Schlagger, 2003; Richter & Richter, 2003), it is an empirical question whether brain activation patterns (BOLD responses) observed in a given region in children reflect the same function as activation patterns observed in adults in the same region for the same task—yet this is an unspoken assumption of many developmental fMRI studies. It is also not clear how functional and anatomical neural development might relate and dynamically interact at different ages. Although theoretical frameworks have been posited (e.g. Fischer & Rose, 1994), there are few studies addressing changes in the interactions among brain systems over developmental time (e.g. see Brown & Chiu, 2006; Schlaggar & Church, 2009; Simos et al., 2001). Moreover, there are few developmental cognitive neuroscience studies that make age-group comparisons or have longitudinal designs, but many studies with large variability in the age of participants that does not affect the focus of the study but makes it difficult to draw developmental conclusions that are likely to be educationally significant. In short, developmental neuroscience research with humans may be in need of a more truly developmental perspective, both conceptually and methodologically, to be relevant to educational practice and policy in an integrated science of MBE.

3.3 Construction: building connections

Transdisciplinary training

If a complex highway interchange or motorway junction with various connected and integrated roadways that lead seamlessly from one to another is an adequate visual metaphor, researcher-practitioners in MBE must be able to effortlessly travel those multiple pathways. Yet students who are trained in education are often not trained in developmental or learning psychology or neuroscience and students who are trained in neuroscience are often not trained in learning sciences or education (e.g. see Ansari, 2005; Ansari & Coch, 2006; Eisenhart & DeHaan, 2005; McDevitt & Ormrod, 2008). Without common background and preparation, students who could potentially participate in and contribute to the field of MBE are left with few shared theoretical frameworks, methods, bodies of knowledge, or vocabulary, despite common interests and questions. For example, an educator might use the term 'number sense' to refer to a child's understanding of basic numerical operations (e.g. counting, place value, basic addition and subtraction; National Council of Teachers of Mathematics, 2003) while a psychologist might use the same term to mean the basic representation of numerical quantity and its psychophysical characteristics (e.g. Dehaene, 1997). Both practitioners are interested in mathematical development, but without interdisciplinary training these definitional differences can lead to potential misunderstandings in research, practice and policy.

Interdisciplinary training opportunities for undergraduate, graduate and post-doctoral students are central for developing researcher-practitioners in MBE. Previous reports have indicated that students in the sciences benefit from interactions with schoolchildren, and that schoolchildren benefit from interactions with scientists (e.g. Cameron & Chudler, 2003; Peplow, 2004). Thus, neuroscience students should benefit from experiencing children's learning and development in an educational context—experiences that can be brought back to the lab in the form of testable hypotheses (e.g. Klahr & Li, 2005). The MBE perspective affords neuroscientists and developmentalists an additional methodological tool and context—the classroom—that can be used to test hypotheses about interactions between biology and environment with respect to specific aspects of learning and development. Likewise, students of education should benefit from experiences in neuroscience that can set foundational knowledge about learning and development and be brought back to the classroom—not necessarily as a direct-to-classroom application or lesson plan, but perhaps as a more complex, multilevel, reflective understanding of the how and why behind a given phenomenon. That is, education students might benefit from the window into process, as opposed to product, that neuroscience can provide, as well as benefiting from developing the perspective of a 'teacher-as-researcher' (Postholm, 2009). In short, 'much is known about the brain and neurosciences that should be central to teacher preparation programs' (Society for Neuroscience, 2009, p. 4).

Integrating these sorts of experiences into teacher and neuroscientist training would, in many cases, involve re-conceptualizing programmes to include more interdisciplinary goals and varied experiences for students (e.g. see Berninger, Dunn, Lin & Shimada, 2004; Eisenhart & DeHaan, 2005). This might require new initiatives to make connections between departments of neuroscience, psychology and education that allow for interactions outside the traditional within-field divisions such as 'developmental' and 'educational' psychology. Even within fields, a wider view from the science of learning (e.g. development of mathematical reasoning skills and neural systems) must replace a narrow disciplinary focus (e.g. teaching mathematics; Hardiman & Denckla, 2009). Issues of educational literacy, such as an understanding of key issues including transfer, differentiated instruction, sociocultural context and the real-world constraints of the classroom must be addressed for neuroscientists contemplating making connections with education. Issues of scientific literacy, including knowledge of basic research designs and their limitations, an understanding of the strengths and limitations of neuroimaging techniques (e.g. Berninger & Richards, 2002), and an appreciation of the ethical issues raised by developmental neuroimaging research connecting with education (e.g. Coch, 2007; Illes & Raffin, 2005; Sheridan, Zinchenko & Gardner, 2006), will need to be better addressed not only in such training programmes but also by researchers, practitioners, policy-makers and the media (e.g. Racine, Bar-Ilan & Illes, 2005). The best translators from education to neuroscience and vice versa—the ideal inventive minds (James, 1899/2001)—would have a shared background in psychology, neuroscience and education. Perhaps, as was first suggested decades ago, a 'neuroeducator' (Cruickshank, 1981; Fuller & Glendening, 1985) who could help to guide questions from education into the design of useful neuroscience research and, iteratively, bring the results of such research back to the classroom.

Overall, training that can provide educators with a better understanding of empirical approaches to deciding what methods work best with which students and why, and training that can provide neuroscientists with insight into what questions and issues are crucial in the classroom so that they can design relevant studies in collaboration with educators is key. Such training will afford educators and neuroscientists alike with the skills and knowledge to be able to critically evaluate the extent to which commercial, 'brain-based' products and curricula are supported by rigorous, peer-reviewed efficacy studies that yield useable knowledge (e.g. Editorial, 2004, 2005, 2007). This will allow the field to move beyond the 'myth of brain-based pedagogy' (Hirsh-Pasek & Bruer, 2007, p. 1293). In order for such training to be successful and lead to tangible results, it needs to happen in education and teacher training programmes as well as in graduate and undergraduate programmes in developmental and learning psychology and neuroscience. Unfortunately, such integrative training that is foundational to advancing the field of MBE is still far from the norm. At present, to our knowledge, there are only a handful of programmes that offer the sort of integrative training that might support the development of 'neuroeducators'. Indeed, what is perhaps the first report of a graduate course in MBE was only recently published (Blake & Gardner, 2007).

Blurring boundaries: applying the basics

Encouraging interdisciplinary thinking in the developmental sciences results in a blurring of the traditional distinction between basic and applied research and leads to new advances and a more integrated science of learning (e.g. Bransford et al., 2006). While balancing and integrating explanation theories from neuroscience and action theories from education is challenging (Perkins, 2009), it should be at the core of MBE. Institutionally, such integration requires a revision of the traditional view of developmental psychology programmes in which a strong distinction between applied and basic research is frequently made, with the latter often promoted and favoured to a greater extent; and of education programmes, with the reverse. Funding for translational research that required relationships among basic and applied researchers and educators would also help to close the lab (basic) to classroom (applied) gap (Brabeck, 2008). Practitioners and researchers in multidisciplinary fields, especially those in an early stage such as MBE, are often faced with the challenges of integrating the basic and applied to an even greater extent than those in traditional fields. The availability of resources and training, spanning across research paradigms and working within institutional constraints are perhaps even more pressing issues for trans-disciplinarians.

In addition to these more practical concerns, an important step in blurring the boundary between applied and basic will be researchers recognizing that educators who guide children's learning every day have unique insight into learning and development that could complement and extend a scientific understanding of development (e.g. Coch, Michlovitz, Ansari & Baird, 2009). Research questions and study designs prompted by observations of classroom interactions along with familiarity with neuroimaging techniques are more likely to lead to results that would iteratively feed back into the classroom, obscuring both the distinction between basic and applied research and the translation gap from neuroscience laboratory to classroom and back.

Interestingly, many so-called 'brain-based' curricular materials have, in essence, blurred the boundary between basic and applied—but not in a principled way; they have simply taken basic neuroscience findings and claimed an educational application. Particularly in the case of basic neuroscience findings (e.g. results from studies with non-human animals), there is a non-negotiable necessity for rigorous, scientific testing in the classroom before any translation can occur in order to avoid vacuous claims about 'how the brain learns'. The findings of basic neuroscience studies rarely, if ever, speak to particular educational strategies or curricula; most of the translation involves unacceptable extrapolation that is heavily value-laden rather than truly evidence-based (e.g. see Sheridan et al., 2006). However, educational questions can guide neuroscience research in more principled ways so that MBE research is not exclusively driven by scaling up from more elemental research findings, but is also advanced by 'scaling down' from more macroscopic considerations (M. J. Nathan, personal communication, 5 January, 2009).

Collaboration: making meaningful connections

Educators, developmental and learning scientists and developmental neuroscientists often have many shared interests—for example, interests in the development of reading or mathematical skills—and these areas of overlap, once defined, can serve as the basis for mutually beneficial design experiments (McCandliss, Kalchman & Bryant, 2003). Shared interest in domains that are amenable to multidisciplinary, multilevel analyses (e.g. Katzir & Paré-Blagoev, 2006) would seem to be at the core of MBE; within these domains, a shared language and knowledge base can serve as the foundation for development of integrative research programmes, situated within the wider principles of the field of MBE. Opportunities for bidirectional collaboration in designing multi-method studies using different levels of analysis will increase the relevance of findings to both theory and practice (e.g. Berninger & Corina, 1998; Hardiman & Denckla, 2009). With respect to so-called 'brain-based' materials, the need for researchers to collaborate with educational practitioners is particularly pressing, so that researchers have an appreciation of the many facets of learning behaviours from the teacher's point of view and can consider specific behaviours in their natural environments; laboratory results alone are necessary but not sufficient for useable knowledge. Researchers need to carefully consider the unique constraints of the classroom in order to design contextually relevant studies and generate useable knowledge with educationally relevant implications, and educators need to carefully consider the unique constraints of laboratory research in order to pose neuroscience research questions that can lead to useable knowledge with educationally relevant implications. In general, scientific studies of practice can both examine and refine educators' craft knowledge; for example, in a recent instance of a science of learning approach to education, Pashler and colleagues found little evidence in support of the learning styles hypothesis although it is commonly thought that instructional style should match learning style (Mayer, 2009; Pashler, McDaniel, Rohrer & Bjork, 2009). Overall, practitioners and researchers, in concert, will need to explicitly think about how research studies can be designed and combined neuroscience and education findings can be used to make evidence-based curricular and instructional decisions that improve learning and teaching (e.g. Stanovich & Stanovich, 2003; Thomas & Pring, 2004); to that end, a rich, interdisciplinary dialogue will be necessary to reshape the study of learning (Society for Neuroscience, 2009, p. 3).

Such dialogue and collaboration, based on the shared interests of educators and scientists, should drive the future agenda of MBE. Rather than top-down application of neuroscience findings to the classroom (a model employed by many of the 'brain-based' curricula available today), questions that are central to education can provide some direction for research in neuroscience. For example, to our knowledge, there is no published neuroimaging research comparing children taught in classrooms using a whole language approach with children taught in classrooms using a phonics approach on tasks specifically designed to index the theoretical strengths and weaknesses of these pedagogical approaches. Although both of these approaches to teaching reading are being used in elementary school classrooms every day in the United States and each involves relatively well-defined instructional practices, how they affect the developing brain, comparatively, is virtually unknown. Such investigations are of course subject to the challenges of varying grain size: while precise and standardized measures of phonological awareness are available (e.g. Wagner, Torgesen & Rashotte, 1999), the building of mental models to support comprehension, for example, is more difficult to measure both behaviourally and in terms of fMRI; the former is a focus of phonics instruction, while the latter is a focus of whole language instruction. Knowing which instructional methods develop not only which skills but also which neural networks would help in making educational decisions. That is, neuroscience findings can play a role in evidence-based decisions about reading instruction (e.g. Eden & Moats, 2002), and may help to advance the use of research-based practices that are not typically currently taught in teacher preparation programmes (Walsh, Glaser & Wilcox, 2006).

Benefits: navigating the interchange

There are many benefits to this new perspective on the developmental and learning sciences; here, we briefly note three. First are the opportunities that will arise to pose and answer new questions about learning, development, the brain and education—questions that may emerge only from synergistic collaborations. Neuroscientists and psychologists will gain opportunities to generate new hypotheses based on classroom learning experiences (e.g. few, if any, studies exist investigating the neural correlates of typical, 'business as usual' classroom mathematics or reading instruction, for students or for teachers); to develop hypotheses, in concert with educators and learning scientists, to directly evaluate, compare and contrast potential educational implications of current and future developmental cognitive neuroscience studies; to design studies with greater ecological validity; to conduct collaborative work incorporating multilevel analyses that provide results with real-world significance; and to advance understanding of development in educational (not just laboratory) contexts. It is now possible to investigate how specific classroom experiences and teaching methodologies shape not only observable behaviours but also the human brain, making meaningful and sustainable connections among psychology, neuroscience and education. While there is a growing number of neuroscience studies investigating the effects of targeted educational interventions, revealing some mechanisms of how such interventions do or do not work (e.g. Shaywitz et al., 2004; Stevens, Fanning, Coch, Sanders & Neville, 2008; Temple et al., 2003), such studies of the neural correlates of typical pedagogical classroom experiences have yet to be conducted, for the most part.

A second benefit is the advancement of a multidisciplinary science of learning and development that allows for integrated, multilevel analysis within a field. Such approaches allow for the possibility that a child may demonstrate typical, age- or grade-level-appropriate behaviours but may be using different neural systems than the comparison group. For example, a child may appear to be at grade level given his scores on a standardized reading assessment, but the processes by which he has read the passages (what is happening inside the 'black box') may be quite different from the processes by which a classmate, also at grade level, has read the same passages. Further reading instruction intended to build on the shared foundation of grade-level reading skills may be compromised by the lack of such a shared foundation; thus, there are critical implications for later development of more complex skills, continued educational achievement and further targeted instruction. Having insight into both the product and the process can provide for a multifaceted understanding on which to base pedagogical decisions.

A third benefit is the mutually beneficial relationships between scientists and educators that will lead to the growth of researcher-practitioners poised to make real differences in theory, practice and policy in psychology, neuroscience and education. Much has been written about the lack of translation of research findings, in general, into educational practice, the so-called research-to-practice gap (e.g. Carnine, 1997). There is also a practice-to-research gap: much of what teachers do in classrooms has never been studied empirically, particularly with neuroscience methods. MBE allows for reciprocal connections between research and practice in which laboratory findings feed into classroom practice which in turn feeds into laboratory research in an iterative cycle. Such a cycle cannot be constructed without relationships among researchers, educators and scientist-practitioners, all of whom—in addition to students—benefit from this kind of research.

3.4 **Conclusion**

We believe that the challenges to constructing meaningful connections between education and neuroscience are surmountable; that viable, evidence-based interchanges can be established; and that inventive minds can build an infrastructure that links science and art in principled ways (e.g. see also Ansari & Coch, 2006; Varma et al., 2008). Many past attempts to formulate scientific approaches to education have failed or had very localized effects, in part because they did not succeed in immediately providing results that improved classroom practice (e.g. Condliffe Lagemann, 2000). With a broader view of successful connection beyond direct-to-classroom application as espoused by many 'brain-based' products and curricula, the new field of MBE is not just another educational fad. Neuroscience can be considered just one source of evidence that can contribute to responsible, evidence-based practice and policy in education (e.g. Huston, 2008; Slavin, 2002; Thomas & Pring, 2004) but should be used in concert with theory and behavioural data from the developmental and learning sciences (e.g. Stern, 2005). Indeed, traditional boundaries among behavioural, psychological, neuroscientific and educational research are increasingly being crossed in multidisciplinary studies of learning and human development. The scientific field of MBE provides a supportive, connective, interdisciplinary home for this new approach to understanding and improving learning and teaching.

Acknowledgements

We thank Mitchell Nathan and other colleagues for insightful discussions on many of the topics and issues discussed in this chapter.

References

Alberts, B. (2009). Making a science of education. *Science, 323,* 15.

Ansari, D. (2005). Time to use neuroscience findings in teacher training [Letter]. *Nature, 437,* 26.

Ansari, D., & Coch, D. (2006). Bridges over troubled waters: education and cognitive neuroscience. *Trends in Cognitive Sciences,* 10(4), 146–51.

Ansari, D., Garcia, N., Lucas, E., Hamon, K., & Dhital, B. (2005). Neural correlates of symbolic number processing in children and adults. *NeuroReport,* 16(16), 1769–73.

Baker, T. B., McFall, R. M., & Shoham, V. (2009). Current status and future prospects of clinical psychology: Toward a scientifically principled approach to mental and behavioral health care. *Psychological Science in the Public Interest,* 9(2), 67–103.

Berninger, V. W., & Corina, D. (1998). Making cognitive neuroscience educationally relevant: creating bidirectional collaborations between educational psychology and cognitive neuroscience. *Educational Psychology Review,* 10(3), 343–54.

Berninger, V. W., Dunn, A., Lin, S.-J. C., & Shimada, S. (2004). School evolution: scientist-practitioner educators creating optimal learning environments for all students. *Journal of Learning Disabilities,* 27(6), 500–508.

Berninger, V. W., & Richards, T. L. (2002). *Brain literacy for educators and psychologists.* Boston, MA: Academic Press.

Blake, P. R., & Gardner, H. (2007). A first course in mind, brain, and education. *Mind, Brain, and Education,* 1(2), 61–65.

Blakemore, S.-J., & Frith, U. (2005). *The learning brain: lessons for education.* Malden, MA: Blackwell Publishing.

Brabeck, M. (2008). Why we need 'translational' research: putting clinical findings to work in classrooms. *Education Week,* 27(38), 28, 36.

Bransford, J., Stevens, R., Schwartz, D., Meltzoff, A., Pea, R., Roschelle, J., et al. (2006). Learning theories and education: toward a decade of synergy. In P. Alexander & P. Winne (Eds.), *Handbook of educational psychology* (2nd ed., pp. 209–44). Mahwah, NJ: Erlbaum.

Brown, R. D., & Chiu, C.-Y. P. (2006). Neural correlates of memory development and learning: combining neuroimaging and behavioral measures to understand cognitive and developmental processes. *Developmental Neuropsychology,* 29(2), 279–91.

Bruer, J. T. (1997). Education and the brain: a bridge too far. *Educational Researcher,* 26(8), 4–16.

Cameron, W., & Chudler, E. (2003). A role for neuroscientists in engaging young minds. *Nature Reviews Neuroscience,* 4, 1–6.

Carnine, D. (1997). Bridging the research-to-practice gap. *Exceptional Children,* 63(4), 513–21.

Coch, D. (2007). Neuroimaging research with children: ethical issues and case scenarios. *Journal of Moral Education,* 36(1), 1–18.

Coch, D., Michlovitz, S. A., Ansari, D., & Baird, A. (2009). Building mind, brain, and education connections: the view from the Upper Valley. *Mind, Brain, and Education,* 3(1), 27–33.

Condliffe Lagemann, E. (2000). *An elusive science: the troubling history of education research.* Chicago, IL: University of Chicago Press.

Cruickshank, W. M. (1981). A new perspective in teacher education: the neuroeducator. *Journal of Learning Disabilities,* 24(2), 337–41, 367.

Dehaene, S. (1997). *The number sense: how the mind creates mathematics.* New York: Oxford University Press.

Diamond, A. (2007). Interrelated and interdependent. *Developmental Science,* 10(1), 152–58.

Eden, G. F., & Moats, L. (2002). The role of neuroscience in the remediation of students with dyslexia. *Nature Neuroscience,* 5, 1080–84.

Editorial. (2004). Better reading through brain research. *Nature Neuroscience,* 7(1), 1.

Editorial. (2005). Bringing neuroscience to the classroom. *Nature,* 435, 1138.

Editorial. (2007). A cure for dyslexia? *Nature Neuroscience,* 10(2), 135.

Eisenhart, M., & DeHaan, R. L. (2005). Doctoral preparation of scientifically based education researchers. *Educational Researcher,* 34(4), 3–13.

Fischer, K. W., & Rose, S. P. (1994). Dynamic development of coordination of components in brain and behavior: A framework for theory and research. In G. Dawson & K. W. Fischer (Eds.), *Human behavior and the developing brain* (pp. 3–66). New York: Guilford Press.

Fuller, J. K., & Glendening, J. G. (1985). The neuroeducator: professional of the future. *Theory into Practice,* 24(2), 135–37.

Gazzaniga, M. S. (1998). The split brain revisited. *Scientific American,* 279(1), 50–55.

Goswami, U. (2004). Neuroscience and education. *British Journal of Educational Psychology,* 74, 1–14.

Goswami, U. (2006). Neuroscience and education: from research to practice? *Nature Reviews Neuroscience,* 7, 406–413.

Goswami, U. (2009). Mind, brain, and literacy: biomarkers as usable knowledge for education. *Mind, Brain, and Education,* 3(3), 176–84.

Halberda, J., Mazzocco, M. M. M., & Feigenson, L. (2008). Individual differences in non-verbal number acuity correlate with maths achievement. *Nature,* 455, 665–68.

Hardiman, M., & Denckla, M. B. (2009). The science of education: informing teaching and learning through the brain sciences. *Cerebrum, November/December.*

Hart, L. A. (1983). *Human brain and human learning.* Oak Creek, AZ: Books for Educators.

Hinton, C., & Fischer, K. W. (2008). Research schools: grounding research in educational practice. *Mind, Brain, and Education,* 2(4), 157–64.

Hirsh-Pasek, K., & Bruer, J. T. (2007). The brain/education barrier. *Science,* 317, 1293.

Hoeft, F., Ueno, T., Reiss, A. L., Meyler, A., Whitfield-Gabrieli, S., Glover, G. H., et al. (2007). Prediction of children's reading skills using behavioral, functional, and structural neuroimaging measures. *Behavioral Neuroscience,* 121(3), 602–613.

Holloway, I. D., & Ansari, D. (2009). Mapping numerical magnitudes onto symbols: the numerical distance effect and individual differences in children's mathematics achievement. *Journal of Experimental Child Psychology,* 103(1), 17–29.

Huston, A. C. (2008). From research to policy and back. *Child Development,* 79(1), 1–12.

Illes, J., & Raffin, T. A. (2005). No child left without a brain scan? Toward a pediatric neuroethics. *Cerebrum,* 7(3), 33–46.

James, W. (1899/2001). Talks to teachers on psychology and to students on some of life's ideals. Mineola, NY: Dover Publications.

Kang, H. C., Burgund, E. D., Lugar, H. M., Petersen, S. E., & Schlagger, B. L. (2003). Comparison of functional activation foci in children and adults using common stereotactic space. *NeuroImage,* 19, 16–28.

Katzir, T., & Paré-Blagoev, J. (2006). Applying cognitive neuroscience research to education: the case of literacy. *Educational Psychologist,* 41(1), 53–74.

Klahr, D., & Li, J. (2005). Cognitive research and elementary science instruction: from the laboratory, to the classroom, and back. *Journal of Science Education and Technology,* 14(2), 217–38.

Koizumi, H. (2004). The concept of 'developing the brain': a new natural science for learning and education. *Brain & Development,* 26, 434–41.

Mayer, R. (2009). Advances in applying the science of learning and instruction to education. *Psychological Science in the Public Interest,* 9(3), i–ii.

McCandliss, B. D., Kalchman, M., & Bryant, P. (2003). Design experiments and laboratory approaches to learning: steps toward collaborative exchange. *Educational Researcher,* 32(1), 14–16.

McDevitt, T. M., & Ormrod, J. E. (2008). Fostering conceptual change about child development in prospective teachers and other college students. *Child Development Perspectives,* 2(2), 85–91.

Meltzoff, A. N., Kuhl, P. K., Movallan, J., & Sejnowski, T. J. (2009). Foundations for a new science of learning. *Science,* 325, 284–87.

National Council of Teachers of Mathematics. (2003). *Principles and standards for school mathematics.* Reston, VA: National Council of Teachers of Mathematics.

Noble, K. G., McCandliss, B. D., & Farah, M. J. (2007). Socioeconomic gradients predict individual differences in neurocognitive abilities. *Developmental Science,* 10(4), 464–80.

Noble, K. G., Norman, M. F., & Farah, M. J. (2005). Neurocognitive correlates of socioeconomic status in kindergarten children. *Developmental Science,* 8(1), 74–87.

Pashler, H., McDaniel, M., Rohrer, D., & Bjork, R. (2009). Learning styles: concepts and evidence. *Psychological Science in the Public Interest,* 9(3), 105–119.

Peplow, M. (2004). Doing it for the kids. *Nature,* 430, 286–87.

Perkins, D. (2009). On grandmother neurons and grandfather clocks. *Mind, Brain, and Education,* 3(3), 170–75.

Pickering, S. J., & Howard-Jones, P. (2007). Educators' views on the role of neuroscience in education: findings from a study of UK and international perspectives. *Mind, Brain, and Education,* 1(3), 109–113.

Posner, M. I., & Rothbart, M. K. (2005). Influencing brain networks: implications for education. *Trends in Cognitive Sciences,* 9(3), 99–103.

Postholm, M. B. (2009). Research and development work: developing teachers as researchers or just teachers? *Educational Action Research,* 17(4), 551–65.

Racine, E., Bar-Ilan, O., & Illes, J. (2005). fMRI in the public eye. *Nature Reviews Neuroscience,* 6, 159–64.

Richter, W., & Richter, M. (2003). The shape of the fMRI BOLD response in children and adults changes systematically with age. *NeuroImage,* 20, 1122–31.

Schlaggar, B. L., & Church, J. A. (2009). Functional neuroimaging insights into the development of skilled reading. *Current Directions in Psychological Science,* 18(1), 21–26.

Shavelson, R. J., & Towne, L. (2002). *Scientific research in education.* Washington, DC: National Academy Press.

Shaywitz, B. A., Shaywitz, S. E., Blachman, B. A., Pugh, K. R., Fulbright, R. K., Skudlarski, P., et al. (2004). Development of left occipitotemporal systems for skilled reading in children after a phonologically-based intervention. *Biological Psychiatry,* 55, 926–33.

Sheridan, K., Zinchenko, E., & Gardner, H. (2006). Neuroethics in education. In J. Illes (Ed.), *Neuroethics: defining the issues in theory, practice, and policy* (pp. 265–75). Oxford: Oxford University Press.

Shonkoff, J. P. (2000). Science, policy, and practice: three cultures in search of a shared mission. *Child Development,* 71(1), 181–87.

Simos, P. G., Breier, J. I., Fletcher, J. M., Foorman, B. R., Mouzaki, A., & Papanicolaou, A. C. (2001). Age-related changes in regional brain activation during phonological decoding and printed word recognition. *Developmental Neuropsychology, 19*(2), 191–210.

Slavin, R. E. (2002). Evidence-based education policies: transforming educational practice and research. *Educational Researcher, 31*(7), 15–21.

Society for Neuroscience. (2009). *The promise of interdisciplinary partnerships between brain sciences and education.* Irvine, CA: University of California at Irvine.

Stanovich, P. J., & Stanovich, K. E. (2003). *Using research and reason in education: how teachers can use scientifically based research to make curricular and instructional decisions.* Jessup, MD: National Institute for Literacy.

Stern, E. (2005). Pedagogy meets neuroscience [Editorial]. *Science, 310,* 745.

Stevens, C., Fanning, J., Coch, D., Sanders, L., & Neville, H. (2008). Neural mechanisms of selective attention are enhanced by computer training. *Brain Research, 1205,* 55–69.

Stiles, J. (2009). On genes, brains, and behavior: Why should developmental psychologists care about brain development? *Child Development Perspectives, 3*(3), 196–202.

Temple, E., Deutsch, G. K., Poldrack, R. A., Miller, S. L., Tallal, P., Merzenich, M. M., et al. (2003). Neural deficits in children with dyslexia ameliorated by behavioral remediation: Evidence from functional MRI. *Proceedings of the National Academy of Sciences, 100*(5), 2860–65.

Thomas, G., & Pring, R. (Eds.). (2004). *Evidence-based practice in education.* New York: Open University Press.

Varma, S., McCandliss, B. D., & Schwartz, D. L. (2008). Scientific and pragmatic challenges for bridging education and neuroscience. *Educational Researcher, 37*(3), 140–52.

Wagner, R. K., Torgesen, J. K., & Rashotte, C. A. (1999). *The comprehensive test of phonological processing.* Austin, TX: Pro-Ed.

Walsh, K., Glaser, D., & Wilcox, D. D. (2006). *What education schools aren't teaching about reading—and what elementary teachers aren't learning.* Washington, DC: National Council on Teacher Quality.

Weisberg, D. S., Keil, F. C., Goodstein, J., Rawson, E., & Gray, J. R. (2008). The seductive allure of neuroscience explanations. *Journal of Cognitive Neuroscience, 20,* 470–77.

Willingham, D. T. (2009). Three problems in the marriage of neuroscience and education. *Cortex, 45,* 544–45.

Chapter 4

Principles of learning, implications for teaching? Cognitive neuroscience and the classroom

Usha Goswami

Overview

Educational neuroscience is a long-term enterprise, and it may be some time before education can 'cash out' the promise that it offers. Nevertheless, advances in neuroscience are extremely important for education, because education is the most powerful means we have to enhance learning and to overcome limitations in biological and environmental conditions. Neuroscience seeks to understand the processes that underpin learning, and to delineate causal factors in individual differences in learning. These insights into the mechanisms of learning will be of value for education, as they will provide information about the drivers of development and the optimal targets for educational intervention. The evidence from neuroscience is not just interesting scientifically. It will eventually provide an evidence base for education in which mechanisms of learning will be precisely understood.

This evidence base will be at a new and complementary level of enquiry, the biological level. A biological approach will not *replace* social, emotional and cultural analyses of learning. Rather, it will provide research tools for these complementary analyses which will enrich the entire educational field. Currently many educators seem to see neuroscience as a challenge to the traditions of their discipline. In this chapter, I will argue that it should instead be seen as an opportunity. Education enables human beings to transcend the physical limits of biological evolution. The experiences that education provides work their changes via changing the brain. Neuroscience enables us to understand these brain-changing processes at a level of detail and specificity that will be extremely valuable for education. The effects of social, emotional and cultural practices on these brain-changing processes will then be amenable to rigorous empirical study.

4.1 Introduction

As shown by some of the contributors to this volume, there is a spirit of healthy scepticism about what neuroscience can offer to education. Despite the scepticism, and also despite more trenchant criticism that neuroscience is often asking misguided questions via misconceived experiments (Bennett & Hacker, 2003), I do not believe that education can afford simply to ignore neuroscience.

Of course there is 'bad' and 'ugly' in some attempts to apply neuroscience to education. However, a deeper understanding of brain physiology could also generate advances in learning and theory that could parallel the remarkable advances made in medicine as we have gained a deeper understanding of the physiology of the body. Accordingly, here I offer some *in principle* examples of how current cognitive neuroscience may eventually make a contribution to teaching and learning in the classroom.

First, some ground-clearing points. The negative critiques by educationists of neuroscience are often based upon a particular world view. This view suggests that approaches to education based on the empirical examination of brain physiology put at stake 'questions about the nature of knowledge and how we learn, of what constitutes human being and the good life' (Cigman & Davis, 2009, p. ix). In general, such commentators view the reductionism of neuroscience as running counter to understanding the essence of being human and the good life. True education has to involve development of mind, body *and* spirit, hence by focusing on the cognitive, neuroscience is unnecessarily narrow. Neuroscience is thus rejected as method of enquiry for helping educators to develop the potentialities of the human mind and spirit. Medical/neuroscience approaches are seen as simply one 'discourse' among many, based on professional identity and power relationships rather than science and knowledge. 'On precisely what basis might neuroscience have any authority to make claims about the nature of learning?' (Cigman & Davis, 2009, p. 76).

In my view, the claims of neuroscience are important simply because its discoveries about learning are open to empirical test and disconfirmation. Scientific discoveries are not about power relationships. They are part of an ongoing process of the discovery of knowledge. Although reductionism is viewed by some educational researchers as mutually exclusive to discovering the potentialities of the human mind and spirit, this is an unnecessarily divisive stance. A simple example is offered by the technology of the cochlear implant. Basic research into how the brain converts speech sounds into language may not give us insight into how meaning is derived from an acoustic signal. However, it does give us insight into how changes in sound pressure on the ear when receiving this acoustic signal are converted into electrical impulses that stimulate the brain tissue involved in audition. This research eventually discovered that a small implant utilizing relatively few frequency channels could provide electrical stimulation directly to the brain tissue of those born congenitally deaf, and that those experiencing this stimulation could hear and learn spoken language. In fact, deaf children who are implanted as infants can learn spoken language skills that are equivalent to those of hearing children. The effect on the potentiality of the minds and spirits of those concerned are likely to be enormous.

By offering an understanding of mechanism, neuroscience has the potential to transform our understanding of human learning and cognitive development, and, I would argue, ultimately it has the potential to transform education. In my view, educational neuroscience has to begin by careful basic research into mechanisms of learning. Indeed, the longer I spend in the field, the more basic I find the essential research questions to be. My own key aim is to study how sensory systems build cognitive systems over developmental time. Small initial differences in sensory function are likely to cause large differences in cognitive performance over the learning trajectory, and neuroscience will enable understanding of developmental mechanisms in fine-grained detail.

For example, mechanisms of neural information coding and transmission seem particularly important to understand (as in the cochlear implant example). These mechanisms are the likely building blocks of the cognitive systems critical for education, such as attention and language. The argument from some educational quarters that it is in principle impossible to uncover general laws of learning, as the learning and capacities of individuals are inextricably linked to the contexts and situations of learning, misses the point. Neuroscientists are trying to achieve an understanding of *how* learning occurs in the brain. As this will depend on physiological processes, at the neuroscience level of analysis there *are* likely to be general laws of learning. These discoveries can then be applied to studying learning in response to the various contexts and situations of learning that are of interest to educators. Neuroscience does not seek to *replace* understandings arising from social science. Rather, it seeks to make a contribution to education at a complementary level.

4.2 **In principle: cognitive neuroscience and classroom teaching**

In cognitive neuroscience, researchers both measure electrochemical activity directly and model neural activity using connectionist computational models. When electrochemical activity is measured directly, patterns of activity across large networks of neurons (called cell assemblies) are found. These activation patterns correspond to or are correlated with mental states such as remembering a telephone number. Brain imaging techniques can also reveal the time course of the electrochemical activity (e.g. which neural structures were activated in which order) and interactions and feedback processes that may occur within these large networks. Computational modelling of these interactions and feedback processes then enables in-principle understanding of how synchronized neuronal activity within cell assemblies results in learning and development. I will argue that these kinds of information are *in principle* of interest to the discipline of education, even though the field of cognitive neuroscience is still in its early stages. Education is concerned with how to enhance learning, and the discipline of neuroscience aims to understand the processes of learning.

Considerable brain development has already taken place when a baby is born. Most of the neurons (brain cells) that will make up the mature brain have already formed, and have migrated to the appropriate neural areas. Brain structures such as the temporal cortex (audition) and the occipital cortex (vision) are present, and will become progressively specialized as the infant and young child experiences environmental stimulation. Environmental stimulation determines specialization, as fibre connections grow between brain cells and within and between different neural structures in response to external input (this is 'synaptogenesis'). For our basic sensory systems, growth in fibre connections reflects 'experience expectant' processes. Here there is abundant early fibre growth in response to types of environmental stimulation (such as light) that the brain 'expects' (via evolution) to receive. Other fibre connections are 'experience dependent'. Here the brain is growing connections in order to encode unique information that is experienced by the individual. Every baby is born into a distinctive environment, even children growing up in the same family. Experience-dependent connections are the ones that make each brain subtly different. Experience-dependent synaptogenesis enables lifelong plasticity with respect to new learning and reflects the kind of learning mechanisms that may be of most interest to education.

The specialization of brain structures takes place within developmental trajectories, and these trajectories are constrained by both biology and environment. Neuroscientists study these developmental trajectories by asking questions about structure and function, as well as questions about information coding and transmission. Such questions include which brain structures are important for learning different educational inputs (e.g. reading versus arithmetic), which types of information coding or transmission are important for different educational inputs, what the temporal sequencing is between brain structures for a particular type of learning, and what can go wrong. Neuroscientists also try to distinguish cause from effect.

4.3 **Example 1: neural structures for learning**

A very busy area of research in cognitive neuroscience is the study of which brain structures are active as the brain learns different inputs or performs different tasks. The most frequent method of study is functional magnetic resonance imaging (fMRI). This technique measures changes in blood flow in the brain. As blood flows to neural structures that are active, this kind of neuroimaging can identify which parts of the brain are most involved in certain tasks. Such research is correlational, showing relationships between structure and function. The fact that groups of cells in a particular brain structure are active when language is heard does not mean that we have found the site of the mental lexicon. Nevertheless, if different groups of cells are active when meaning is conveyed in different ways, this tells us something useful. For example, it has been shown that *reading* action words like *lick* and *kick* causes activity in the parts of the brain that are also active when moving one's tongue versus moving one's legs respectively (Hauk, Johnsrude & Pulvermuller, 2004). This suggests that the cell assemblies involved in action also contribute to our conceptual understanding of what it means to 'lick' versus 'kick'. Hence the neural structures underpinning learning may not always be those that intuition may suggest. The structures that underpin the understanding of action words are in part the same structures that are active when actions are carried out by the body.

An educational example of structure–function correlations comes from the study of reading development. The brain structures that are active in novice readers as they perform tasks with print can be measured with fMRI. It turns out that the neural structures for *spoken* language are the most active (Turkeltaub et al., 2003). These data make it less likely that logographic theories of reading acquisition are correct (e.g. Frith, 1985; Seymour & Elder, 1986). Logographic theory argued that children first learn to read by going directly from the visual word form to meaning. The sounds of the words were not involved. Rather, holistic visual stimuli were associated with meanings in the same way as familiar symbols like £ and $ are associated with the meanings of 'pound' and 'dollar'. But if children can really go directly from print to meaning without recoding the print into sound first, only the neural structures active when viewing text and when understanding meaning should show activation.

Learning a mental dictionary of visual word forms is a slow and incremental process. Studies using fMRI show that as children are exposed to more and more printed words, a structure in the visual cortex of the brain becomes increasingly active (Cohen & Dehaene, 2004). This area has been labelled the 'visual word form area' (VWFA), and appears to store information about letter patterns for words and chunks of words and their connections with sound. It is experience dependent, storing the learning that results as children are exposed to more and more printed words and experience reading them aloud (recoding them into sound). Because this learning is experience dependent, the VWFA also responds to 'nonsense words' that have never been seen before. Sub-parts of these nonsense words are familiar from prior learning experiences (e.g. 'treen' is a nonsense word, but it is analogous to real words like 'seen' and 'tree'). So the brain is also able to recode these nonsense letter strings into sound, even though the string TREEN has never been encountered before.

Experiments using electrophysiology (electroencephalography, EEG) have also been performed to study children's reading. In EEG, the tiny electrical currents that move between neural structures during a cognitive task like reading are recorded by sensors that measure how this activity waxes and wanes over time. Experiments contrasting real words and nonsense words show that the child's brain responds differently to real versus nonsense words within one-fifth of a second (200 milliseconds). This difference implies that lexical access (contact between the visual word form and its meaning) occurs very rapidly during reading. The speed of this differentiation has been shown to be similar for both children and adults, across languages, suggesting that the time course of visual word recognition is very rapid (160–180 milliseconds, see Csepe & Szucs, 2003; Sauseng, Bergmann & Wimmer, 2004; this response is called the 'N170'). In such experiments, the visual/spatial demands linked to processing text are kept constant, and the only factor that varies is whether the target is a real word or not. The N170 has been replicated in many studies. Such information represents an objective fact about how the brain behaves during reading.

Some neuroscientists have suggested that the amount of activity in the VWFA is the best neural correlate that we have of reading expertise (Pugh, 2006). As might be expected, the VWFA shows reduced activation in developmental dyslexia (e.g. Shaywitz et al., 2005). More recently, neuroscientists have analysed how neural activity in the VWFA 'tunes itself' to print, namely how activity in these particular brain cells becomes specialized for the letter strings that are real words. Maurer, Brandeis and their colleagues (e.g. Maurer et al., 2005, 2007) have followed a sample of German-speaking children who either were at genetic risk for developmental dyslexia or who had no genetic risk for dyslexia, from the age of 5 years. They have used EEG to measure millisecond-level changes in the electrical activity associated with the recognition of word forms from before reading instruction began. The children were asked to detect the repetition of either real words or of meaningless symbol strings in a stream of consecutively-presented items. Before any reading instruction had commenced, none of the children showed an N170 to printed words, despite having considerable knowledge about individual letters. After approximately 1.5 years of reading instruction, the typically-developing children did show a reliable N170 to words. The children at risk for dyslexia showed a significantly reduced N170 to word forms, even though they had not yet been diagnosed as having reading difficulties. This is still correlational data. However, this finding raises the possibility of using a brain response like the N170, which is not under conscious control, to identify children who are at risk for reading difficulties before these difficulties are manifest in behaviour. Eventually, neuroscience will be able to offer education 'neural markers' of this type. These markers will be evidence of a processing difficulty rather than, say, evidence of 'being dyslexic'. Nevertheless, they will be useful to education, for example, in counteracting arguments that the child is being lazy, stupid or is not trying.

Clearly, the brain did not evolve for reading. Nevertheless, as this brief survey shows, neuroscience is revealing how fibre connections to support reading develop and encode print experience into the nervous system. This is partly achieved by recruiting neural structures that by evolution already performed similar functions, such as the neural tissue active during object recognition which is in and near to the VWFA. One prominent educational neuroscientist calls this feature of experience-dependent plasticity the 'neuronal recycling hypothesis' (Dehaene, 2008). Dehaene argues that our evolutionary history and genetic organization constrain new cultural acquisitions to some extent, as new learning must be encoded by a brain architecture that evolved to encode at least partially similar functions over primate evolution. Nevertheless, studies using fMRI have revealed that unexpected neural structures can be involved in educational performance.

A good example comes from studies of arithmetic and number processing. Early studies with adults found that an area in parietal cortex was particularly active whenever numerical magnitude had to be accessed (Dehaene, 1997). Nearby areas of parietal cortex were activated when judgements about size or weight were required. This led Dehaene and others to argue that the parietal cortex could be the location of an approximate, analogue magnitude representation in the human brain. In particular, they argued that activity in this brain structure in the horizontal intraparietal sulcus (IPS) enabled an intuitive understanding of quantities and their relations (Dehaene, 1997). It was surprising then to discover that brain activation in an exact addition task (e.g. 4 + 5 = 9) was not particularly high in this parietal structure. The IPS was only preferentially activated if the task involved approximate addition (e.g. 4 + 5 = 8) (Dehaene, Spelke, Pinel, Stanescu & Tsivkin, 1999). In the exact calculation task, the highest relative activation was in a left-lateralized area in the inferior frontal lobe, an area usually most active in language tasks. Dehaene et al. (1999) argued that this could be because exact arithmetic requires the retrieval of over-learned 'number facts', which are thought to be stored in the language areas.

Again, this correlational insight would seem to be of interest to educators. It implies that part of school mathematical learning is linguistic. The multiplication tables and 'number facts' that children are taught may be learned as verbal routines, like the months of the year. Admittedly, this supposition is based on the finding that the highest brain activity found when reciting the months of the year is in the same neural structure as the highest brain activity found when doing mental arithmetic. Again, these are correlations. But a priori, one might not have expected the brain to develop fibre connections in the cell assemblies most active during language processing in order to encode classroom activities assumed to develop neural structures for mathematics.

4.4 **Example 2: brain mechanisms of learning extract structure from input**

As we have seen, different neural structures are specialized to encode different kinds of information, and one important mechanism for this is the growth of fibre connections that record experiential input (such as the incremental learning of printed word forms that is encoded primarily by fibre growth in the visual cortex). Most environmental experiences are multisensory, and therefore fibre connections between modalities are ubiquitous. But this mechanism of ever-growing and branching fibre connections between structures carries with it an important corollary for our understanding of learning. Because learning is encoded cumulatively by large networks of neurons, a whole network of cells that have been connected because of prior experiences will still be activated even when a *particular* aspect of sensory information in a *particular* experience is *absent*. This is the documented ability of the brain to respond to *abstracted dependencies* of particular sensory constellations of stimuli. This fact about learning enables, for example, the brain to 'fill in' a missed word when someone coughs across another speaker. In sensory terms, the word was drowned out by the louder cough. Yet even though our brain received the sensory information about the cough rather than the sensory input comprising the phonemes in the missed word, prior learning of the statistical regularities between words in connected discourse has enabled the brain to 'fill in' the missing information and 'hear' the absent word (e.g. Pitt & McQueen, 1998, for a related example).

The fact that incremental learning yields abstracted dependencies is a powerful mechanism for learning and development (Goswami, 2008). The study of abstracted dependencies actually gives us an empirical handle on top-down learning—and is therefore relevant to the contexts and situations of learning that are of interest to educators. As the child's brain is exposed to particular sensory constellations of stimuli over multiple occasions, what is common across all these learning instances will naturally be encoded more strongly by the brain than what differs. In terms of mechanism, the fibre connections that encode what is common will become stronger than the fibre connections that encode details that may differ in each learning event. This neural mechanism effectively yields our conceptual knowledge (e.g. such as our 'basic level concepts', e.g. 'cat', 'dog', 'tree' and 'car'; Rosch, 1978). After the child has experienced 100 'cat' learning events, the strongest fibre connections will have encoded what has been consistent across all experienced instances, such as 'four legs', 'whiskers', 'tail' and so on. In this way, the brain will have developed a generic or 'prototypical' representation of a cat. This mechanism will apply to learning of all kinds, not just learning about perceptual objects in the real world. The same mechanism will underpin emotional reactions to frequently-experienced events, or physical reactions to these events, such as feeling nauseous.

The abstraction of dependencies is not the only learning mechanism used by the brain. In addition to the generalized learning on the basis of repeated experiences, there is also 'one trial' learning, when one experience is enough. This is shown, for example, by phobic reactions. Nevertheless, one effective intervention for overcoming maladaptive phobias is to 'flood' the brain with multiple instances of the phobic object in situations where negative consequences do not ensue. Eventually, this incremental learning becomes stronger than the learning underpinning the phobia, and the patient can function in the presence of the phobic object. Understanding more about these very different learning mechanisms in different environments seems likely to be of immense practical value to educators. One obvious example is in the education of children with emotional and behavioural disorders.

In principle, therefore, the infant's brain can construct detailed conceptual frameworks about objects and events from watching and listening to the world. Active experience then transforms learning once the infant becomes capable of self-initiated movement (see Goswami, 2008, for detail). As the child learns language and attaches labels to concepts, the neural networks underpinning learning become even more complex. These networks of cell assemblies will be the physiological basis of our conceptual knowledge, and eventually will be amenable to internal manipulation. As we learn new information via language, fibre connections will form in response that encode more abstract information and therefore more abstract concepts. Via thinking and 'inner speech' we can alter these connections ourselves, without external stimulation. Understanding how this occurs at the mechanistic level is not antithetical to understanding human learning in terms of mind, body *and* spirit. Indeed, mechanistic understanding can lead to valuable interventions, as when participants are taught to use biofeedback to self-calm in situations that engender crippling anxiety.

With respect to education, it is also important to point out that these learning mechanisms ensure that the brain will extract and represent structure that is present in the input *even when it is not taught directly*. This is evident from the implicit learning of many social phenomena, such as the documented effects of teacher expectations on pupil performance. An example that may be more amenable to physiological study comes from learning to read. Children will learn the higher-order consistencies in the spelling system of English without direct teaching. In research that I have previously described as showing 'rhyme analogies' in reading (Goswami, 1986), we have documented sensitivity to these higher-order consistencies without overt awareness (Goswami, Ziegler, Dalton & Schneider, 2003). Spelling-to-sound relations in English are often more reliable at the larger 'grain size' of the rhyme than at the smaller 'grain size' of the phoneme (Treiman et al., 1995; Ziegler & Goswami, 2005). For example, the pronunciation of a single letter like 'a' differs in words like 'walk' and 'car' from its pronunciation in words like 'cat' and 'cap'. The pronunciation in 'walk' or 'car' can be described as irregular, but it is quite consistent across other rhyming words (like 'talk' and 'star'). By giving children novel 'nonsense words' to read, we explored whether the brain is sensitive to these higher-level consistencies in letter patterning.

For example, we asked children to read aloud nonsense words matched for pronunciation, like *daik, dake, loffi* and *loffee*. Here the child could have learned the rhyme spelling patterns in items like *dake* and *loffee* from prior experiences with analogous real words, like 'cake' and 'toffee'. Chunks of print like '*ake*' and their connections with sound should be stored in the VWFA. As there are no real English words with letter chunks like '*aik*', even though there are many words with the graphemes '*ai*' and '*k*', these rhyme spelling patterns should be entirely novel. Hence if children are faster and more accurate at reading items like *dake* than items like *daik*, we would have evidence for implicit higher-order learning. We (Goswami et al., 2003) indeed found that English children showed a reliable advantage for reading aloud the analogous nonsense words like *dake* and *loffee*, despite the fact that the individual letter-sound correspondences were matched across word lists (i.e. the individual graphemes in *daik* were as familiar orthographically as the individual graphemes in *dake*). These data suggest that orthographic learning, presumably in the VWFA, recorded these higher-level consistencies even though 'rhyme analogy' reading strategies were not taught directly to participating children (see Ziegler & Goswami, 2005, for converging evidence from other paradigms).

Whether learning (i.e. reading performance) would have been even stronger if rhyme analogy strategies had been taught directly to participating children remains an open question. But this example suggests that using experiments to try and understand how the brain learns *is* relevant to education, even when the experiments are at a mechanistic level. The 'situated cognition' analyses popular in education show us that learning is 'embedded' in the experiences of the individual. Nevertheless, one goal of education is to help all individuals to extract the higher-order structure (or 'principles' or 'rules') that underpin a given body of knowledge. Although many in education currently question the existence of general pedagogical methods that would make education maximally and universally effective, the possible existence of such methods is open to empirical investigation. Neuroscience offers a method for posing such empirical questions, for example, by studying how different teaching regimes in initial reading actually affect the development of the VWFA.

Clearly, translating such research questions to education will be challenging. Deep understanding of a given educational domain is required in order to present cumulative information in the optimal sequence for the novice learner and to contrast different teaching regimes. Nevertheless, neural modelling will enable us to investigate whether certain classroom experiences result in previously distinct parts of a network becoming connected, or whether they enabled inefficient connections that were impeding understanding to be pruned away. In principle, this knowledge could be fed back into pedagogy, making it more effective. A deeper understanding of how the brain uses incremental experience to extract underlying structure does appear to be relevant to classroom teaching.

4.5 **The dangers of seeding neuromyths**

Nevertheless, it is critical to remain vigilant when evaluating neuroscience research (see Section 5, this volume). Correlations are still correlations, even when they involve physiological measures. Many correlational findings that reach the popular media are given causal interpretations and this is detrimental to popular understanding of neuroscience. One example is the data sets that have been interpreted to show that fatty acids such as fish oils play a potentially causal role in learning (see Chapter 15, this volume). Unsaturated fatty acids are known to be important in brain development and in neural signal transduction. This in itself does not mean that ingesting omega-3 and omega-6 highly unsaturated fatty acids is good for the brain. In a recent paper, Cyhlarova et al. (2007) went further, claiming that 'the omega-3/omega-6 balance is particularly relevant to dyslexia' (p. 116). Their study in fact measured the lipid fatty acid composition of red blood cell membranes in 52 participants, 32 dyslexic adults and 20 control adults. There were no significant differences between dyslexics and controls for any of the 21 different measures of membrane fatty acid levels taken by the researchers. However, a *correlation* was found between a total measure of omega-3 concentration and overall reading in the whole sample. This correlation could mean that omega-3 concentration is linked to reading efficiency, or it could reflect the influence of a third variable such as intelligence quotient. At any rate, what a correlation of this nature does not show is that fatty acids play a role in dyslexia.

Unfortunately, when physiological variables such as changes in the brain are involved, people tend to suspend their critical faculties. Weisberg and her colleagues gave adult students 'bad' explanations of psychological phenomena, either with or without accompanying neuroscience information (Weisberg et al, 2008). The neuroscientific details were completely irrelevant to the explanations given. Nevertheless, the adults rated the explanations as far more satisfying when such details were present. Weisberg et al. pointed out that our propensity to accept explanations that allude to neuroscience makes it absolutely critical for neuroscientists to think carefully about how neuroscience information is viewed and used outside the laboratory.

4.6 **Conclusion**

Educational neuroscience is a long-term enterprise, and there will be few immediate pay-offs (Goswami & Szücs, 2011). Nevertheless, advances in neuroscience are important for education because they deliver insights into the mechanisms of learning. Eventually, neuroscience will enable componential understanding of the complex cognitive skills taught by education. Although this understanding will be at a basic ('reductionist') level, it cannot be ignored by educators. This is because neuroscience offers an empirical foundation for investigating theories and ideas already present in pedagogy, and for disputing others. The evidence from neuroscience is not just interesting scientifically. It enables an evidence base for education in which mechanisms of learning can be precisely understood. This understanding will only be at one level of enquiry. Neuroscience will not replace social, emotional and cultural analyses of learning. But biological, sensory and neurological influences on learning are likely to be replicable and open to falsifiability. As such, they will offer tools for applying the same empirical rigour to social, emotional and cultural analyses of learning.

Again, medicine can offer us a positive analogy. Thirty years ago, it was acceptable to argue that autism in children was a product of 'refrigerator parenting' (Bettelheim, 1963). Autism was hypothesized to be caused by social factors, the family environments offered by professional parents who were focused on their jobs rather than their offspring. In response, the child was thought to withdraw emotionally from an environment that was cold and rejecting. No one would make that causal argument today, because the neural basis of autism is far better understood. Parenting does not cause autism. Instead, autism reflects atypical processing of cues important for social cognition, such as the information about mental states conveyed by the eyes. Further, this neural knowledge provides important clues about which features of family contexts might be most successful for children with autism, and offers measures for exploring relative benefits. For example, children with autism may prefer environments with lower demands for eye contact. They may also be better at learning about mental state information from non-biological kinds (for example, toy trains, see the Transporter videos for teaching children with autism about emotions, Golan et al., 2010). Information about neural underpinnings does not make research on the optimal educational and cultural/social environments for teaching children with autism obsolete. Rather, it provides factual knowledge which enriches social, emotional and cultural analyses. By analogy, the truly ambitious goal for education is to cross disciplinary boundaries and embrace the scientific method. The scientific method, and neuroscience in particular, may enable surprising discoveries about the optimal way to educate mind, body *and* spirit.

References

Bennett, M.R., & Hacker, P.M.S. (2003). *Philosophical foundations of neuroscience*. Oxford: Blackwell.

Bettelheim, B. (1963). *The Empty Fortress*. Chicago, IL: Free Press.

Chomsky, N. (1957). *Syntactic Structures*. The Hague/Paris: Mouton.

Cigman, R., & Davis, A. (2009). *New philosophies of learning*. Chichester: Wiley-Blackwell.

Cohen, L. & Dehaene, S. (2004). Specialization within the ventral stream: the case for the visual word form area. *NeuroImage, 22*, 466–76.

Csepe, V. & Szűcs, D. (2003). Number word reading as a challenging task in dyslexia? An ERP study. *International Journal of Psychophysiology, 51*, 69–83.

Cyhlarova, E., Bell, J. G., Dick, J. R., Mackinlay, E. E., Stein, J. F., & Richardson, A. J. (2007). Membrane fatty acids, reading and spelling in dyslexic and non-dyslexic adults. *European Neuropsychopharmacology, 17*, 116–21.

Dehaene, S. (1997). *The Number Sense: How the Mind Creates Mathematics*. New York: Oxford University Press.

Dehaene, S. (2008). Cerebral constraints in reading and arithmetic: Education as a 'neuronal recycling' hypothesis. In A.M. Battro, K.W. Fischer & P.J. Lena (Eds). *The Educated Brain: Essays in Neuroeducation* (pp. 232–247). New York: Pontifical Academy of Sciences and Cambridge University Press.

Dehaene, S., Spelke, E., Pinel, P., Stanescu, R., & Tsivkin, S. (1999). Sources of mathematical thinking: behavioral and brain-imaging evidence. *Science, 284*, 970–74.

Frith, U. (1985). Beneath the surface of developmental dyslexia. In K.E. Patterson, J.C. Marshall & M.Coltheart (Eds.), *Surface Dyslexia* (pp 301–30). London: Lawrence Erlbaum Associates.

Golan, O., Ashwin, E., Granader, Y., McClintock, S., Kay, K., Leggett, V., & Baron-Cohen, S. (2010). Enhancing emotion recognition in children with autism spectrum conditions: An intervention using animated vehicles with real emotional faces. *Journal of Autism and Developmental Disorders, 40*, 269–79.

Goswami, U. (1986). Children's use of analogy in learning to read: A developmental study. *Journal of Experimental Child Psychology, 42*, 73–83.

Goswami, U. (2008). *Cognitive Development: The Learning Brain*. Hove: Psychology Press, Taylor & Francis.

Goswami, U., & Szücs, D., (2011). Educational neuroscience: Developmental mechanisms; towards a conceptual framework. *Neuroimage, 57*(3), 651–58.

Goswami, U., Ziegler, J., Dalton, L. & Schneider, W. (2003). Nonword reading across orthographies: How flexible is the choice of reading units? *Applied Psycholinguistics, 24*, 235–47.

Hauk, O., Johnsrude, I. S., & Pulvermuller, F. (2004) Somatotopic representation of action words in human motor and premotor cortex. *Neuron, 41*, 301–07.

Maurer, U., Brem, S., Bucher, K. & Brandeis, D. (2005). Emerging neurophysiological specialization for letter strings. *Journal of Cognitive Neuroscience, 17*, 1532–52.

Maurer, U., Brem, S., Bucher, K., Kranz, F., Benz, R., Steinhausen, H-C., & Brandeis, D. (2007). Impaired tuning of a fast occipito-temporal response for print in dyslexic children learning to read. *Brain, 130*, 3200–10.

Pitt, M. A., & McQueen, J. M. (1998). Is compensation for coarticulation mediated by the lexicon? *Journal of Memory and Language, 39*, 347–70.

Pugh, K. (2006). A neurocognitive overview of reading acquisition and dyslexia across languages. *Developmental Science, 9*, 448–50.

Rosch, E. (1978). Principles of categorisation. In E. Rosch & B.B. Lloyd (Eds.), *Cognition and Categorisation* (pp. 27–48). Hillsdale, NJ: Erlbaum.

Sauseng, P., Bergmann, J., Wimmer, H. (2004). When does the brain register deviances from standard word spellings? An ERP study. *Cognitive Brain Research, 20*, 529–32.

Seymour, P.H.K., & Elder, L. (1986). Beginning reading without phonology. *Cognitive Neuropsychology*, 3, 1–36.

Shaywitz, S.E., & Shaywitz, B.A. (2005). Dyslexia (specific reading disability). *Biological Psychiatry*, 57, 1301–9.

Simos, P.G. et al. (2005). Early development of neurophysiological processes involved in normal reading and reading disability: A magnetic source imaging study. *Neuropsychology,* 19 (6), 787–98.

Szücs, D. & Goswami, U. (2007). Educational neuroscience: Defining a new discipline for the study of mental representations. *Mind, Brain and Education*, 1(3), 114–27.

Treiman, R., Mullennix, J., Bijeljac-Babic, R., & Richmond-Welty, E. D. (1995). The special role of rimes in the description, use, and acquisition of English orthography? *Journal of Experimental Psychology: General,* 124, 107–36.

Turkeltaub, P.E., Gareau, L., Flowers, D.L., Zeffiro, T.A. & Eden, G.F. (2003). Development of neural mechanisms for reading. *Nature Neuroscience*, 6(6), 767–73.

Vygotsky, L. (1978). *Mind in Society*. Cambridge, MA: Harvard University Press.

Weisberg, D.S., Keil, F.C., Goodstein, E.R. & Gray, J.R. (2008). The seductive allure of neuroscience explanations. *Journal of Cognitive Neuroscience*, 20(3), 470–77.

Ziegler, J. C., & Goswami, U. (2005). Reading acquisition, developmental dyslexia, and skilled reading across languages: a psycholinguistic grain size theory. *Psychological Bulletin,* 131(1), 3–29.

The contribution of cognitive neuroscience to understanding domains of learning

Chapter 5

Reading in alphabetic writing systems: evidence from cognitive neuroscience

Jane Ashby and Keith Rayner

Overview

This chapter discusses the behavioural, neurophysiological and computational evidence relevant for understanding reading. Because readers must recognize printed words in order to comprehend a passage, we focus on research that illuminates how word recognition happens. The first section of this chapter describes research on how skilled word recognition occurs. Adult skilled readers begin to identify a word parafoveally, on the fixation before their eyes land on a word and that automatic, parafoveal processing speeds reading by decreasing word recognition time. We discuss evidence that phonology has a strong influence on parafoveal word processing, and that phonological processes facilitate word recognition in the first tenth of a second of seeing a word. Evidence from behavioural and cognitive neuroscience experiments indicates that efficient, skilled phonological processing registers in measures of brain activity at least as quickly as orthographic processing under many conditions.

   In the second section, we discuss early reading development (in 5–8-year-olds), beginning with the technology of alphabetic writing systems, which is based on the mapping of speech sounds to letters. Before learning this alphabetic principle, children guess at printed words. As alphabetic awareness improves, children begin attending to the letters and mapping them onto speech sounds to independently read unfamiliar words. After several accurate readings of a new word, a child can usually recognize it quickly. These phases of learning to read track the development of reading circuits in the brain. Initially, neural activation during reading is widely distributed across brain regions. As reading develops, the orthographic-phonological brain circuit becomes active. As printed words become familiar, orthographic-semantic areas in the ventral circuit also become tuned to word recognition. Skilled adult readers and skilled early readers activate orthographic and phonological information to identify words during reading.

5.1 Introduction

Learning to read is a complex and difficult task. Children expend tremendous effort learning to read, yet for adults the process feels automatic and effortless. Beginning readers often read aloud in a plodding, staccato fashion, pointing as they pronounce each word. With practice, children gradually become faster at recognizing words, but many still misread and stumble over unfamiliar words for a few years. Eventually they can read aloud accurately and smoothly, with their intonation reflecting the patterns in speech. For skilled readers, silent reading feels no more effortful than any other act of perception (e.g. listening). Thus, we see the paradox of learning to read: a skill that is so easy for adults can be quite difficult for children to learn. For skilled reading to occur, many component processes must coordinate—processes that include recognizing familiar words, decoding new words, reading with intonation, making efficient eye movements, comprehending the text and learning the meaning of new words. The efficiency of these processes in skilled reading underlies the common intuition that reading is an easy task.

At its most fundamental level, reading involves identifying the printed words that deliver the meaning of a text. Because words deliver the meaningful message, reading comprehension depends on word recognition accuracy as well as on how quickly and easily readers identify words (i.e. reading fluency). A skilled reader can recognize words quickly and accurately, identifying which of the 300,000-odd English words is printed on a page in about a quarter of a second. The link between word reading accuracy and reading rate can be observed in the close relationship of both factors with reading comprehension. Much of what we know about skilled word recognition during silent reading comes from experimental studies that measure eye movements as readers silently read text for meaning and brain activity as readers recognize single words. This chapter brings those findings to bear on three central questions in reading education. How do skilled readers identify words in print? What is the alphabetic principle? What methods of reading instruction are consistent with findings about reading processes? We begin by discussing the word identification processes of skilled readers and how computational models contribute to our understanding of skilled reading.

5.2 Word recognition in skilled reading

Several research approaches contribute to our understanding of skilled word recognition processes. Behavioural studies provide the foundation for this understanding, neurophysiological studies are refining our grasp of the time course of word recognition processes and the brain networks involved, and computational research offers simulations of reading data that lend insight into how the brain acquires and executes skilled word recognition processes. We focus on the word recognition component of skilled reading because words are the fundamental units that deliver the meaning of a text. Writing systems operate by connecting print (orthography) with spoken word forms (phonology) and with word meanings (semantics). Therefore, word recognition involves processing several aspects of the word: the symbols that make up the letter string, the spoken form of the word and its meaning. Writing systems differ in terms of how they encode these units of language. Some rely on memorization of the visual form of each printed word (as Chinese does), whereas others encode written words using a small number of letters that represent speech sounds (as any alphabetic system does). In the process of understanding text, readers of all writing systems form connections between symbol, sound and meaning.

Writing systems differ in how they encode words, and one might expect that this affects the word identification processes of skilled readers. Alphabetic systems operate based on the alphabetic principle: the letters represent speech sounds. The consistency of letter sound relationships varies across languages from very consistent (as in Spanish, Italian and Korean) to fairly inconsistent (as in English). In contrast, the relationship between letters and meaning is arbitrary (Van Orden & Kloos, 2005). The result of this alphabetic principle, then, is that when reading a new word, the letters provide the best clues to the word form.

One of the main controversies about reading is a question about how skilled readers process the letters in a word (or its orthography) and the sound of a word (its phonology) to access the word's meaning. The behavioural, neuroimaging and computational research suggests that skilled readers identify words via rapid, automatic activation of high quality lexical representations that include orthographic and phonological streams of information (Pugh & McCardle, 2009). The next sections discuss how skilled readers process orthographic and phonological information to recognize printed words.

Orthographic processes

One of the early questions about orthographic processing was whether words are processed as wholes or as a specific string of letters. Cattell (1886) addressed this issue by asking subjects to report what they saw when words and letters were briefly exposed. They were better able to report words than letters. Based on this evidence, Cattell theorized that written words were recognized by retrieving whole-word representations. Reicher (1969) and Wheeler (1970) investigated this question again using an improved experimental design. They replicated Cattell's finding: letters in words were identified more accurately than letters presented in isolation or in non-words. This *word-superiority effect* indicates that words are not recognized by processing each letter in a serial fashion; if they were, then readers would have been more accurate with single letters than words (with several letters) given the limited processing time. However, the second finding of the Reicher and Wheeler experiments was somewhat unexpected: skilled readers had comparable recognition rates for letters in any position. If words were recognized as wholes, one would expect better recognition of the first and last letters (which are most salient). Finding comparable recognition rates across letter positions indicated that skilled readers processed all the letters in short words in parallel rather than serially. Thus, it appears that skilled readers build orthographic representations that include every letter, rather than holistic representations. How beginning readers form such specific representations is discussed in the final section of the chapter.

In general, the word-superiority effect has been taken as evidence that skilled readers process letters in parallel, rather than in a series. The effect ultimately led to the development of letter-level models of word recognition and powerful computer models designed to account for parallel processing during word recognition. These initial models, called the interactive activation model (McClelland & Rumelhart, 1981; Rumelhart & McClelland, 1982) and the verification model (Paap, Newsome, McDonald & Schvaneveldt, 1982), were the forerunners of even more elegant and powerful connectionist models (such as Harm & Siedenberg, 2004; Plaut, McClelland, Seidenberg & Patterson, 1996; Seidenberg & McClelland, 1989). Together, these models are referred to as the family of triangle models, discussed later in this chapter.

More recently, cognitive neuroscience experiments have investigated the time course of orthographic processing in skilled readers and where it occurs in the brain. Maurer, Brandeis and McCandliss (2005) measured brainwaves elicited as subjects silently read high frequency (or common) words, pronounceable non-words and non-letter strings. More electrical activity was measured when reading words than pronounceable non-words than symbol strings, which suggests that readers process orthographic familiarity around 170 milliseconds (ms) after the word appears. This activity appears strongest in the occipito-temporal (OT) area of the left hemisphere, suggesting that this area is involved in early orthographic processing. Brain-imaging studies provide converging evidence that orthographic processing occurs in the left OT area. Cohen and Dehaene (2009) review studies indicating that neural activity increases there as letter strings more closely resemble words. For example, this area is less active when skilled readers read *dcyqv*, more active for *doans*, and still more active for *brain*. Together, behavioural and cognitive neuroscience studies indicate that when skilled readers read single words, they rapidly process letters in parallel and are sensitive to the familiarity of letter strings within one-fifth of a second.

Phonological processes

Reading research conducted over the past few decades indicates that skilled readers also use phonological (or speech-based) codes to identify words when reading. Intuitively, one might think of phonological coding as the slow sounding out process that a skilled reader may use to pronounce a word not seen before. However, behavioural and cognitive neuroscience research has identified fast, automatic phonological processes that readers engage when reading connected text silently for meaning and when reading single words preceded by quickly presented primes (see Halderman, Ashby & Perfetti (in press) for a review).

Behavioural studies that measure single-word reading time have demonstrated that readers can process phonological information when materials are presented for only a fraction of a second (Ferrand & Grainger, 1994; Lukatela & Turvey, 1994; Perfetti, Bell & Delaney, 1988; Tan & Perfetti, 1999; Van Orden, Johnston & Hale, 1988). These phonological effects are sensitive to task variables, and do not always appear in single-word reading tasks (Berent & Perfetti, 1995; Drieghe & Brysbaert, 2002; Rastle & Brysbaert, 2006; Van Orden & Kloos, 2005). However, consistent evidence for phonological processing in skilled word recognition is provided by studies that measure eye movements during silent reading.

Eye movement studies provide information about how skilled readers recognize words as they are reading connected text (as opposed to single words). Because eyetrackers measure gaze position within half a letter every millisecond (Rayner, 1998, 2009; Rayner & Pollatsek 1989), this technology provides a sensitive measure of the timing of word recognition processes. Classic studies have shown that skilled readers process the word they are looking at (or fixating), and some information about the word to its right (in languages that are read left-to-right) (Dodge, 1907; Rayner, 1975). While the eyes are completing their fixation of the previous word, attention shifts ahead to begin processing form information from the next word (Reichle, Pollatsek, Fisher & Rayner, 1998). This advance word form processing occurs parafoveally (or before the word is fixated) and without the reader's conscious awareness.

The high spatial and temporal resolution of eye trackers also allows experimenters to unobtrusively manipulate what text readers see and when they see it (McConkie & Rayner, 1975; Rayner, 1975). The parafoveal preview paradigm (Rayner, 1975) manipulates the relationship of the parafoveal information to the upcoming word (see Figure 5.1). A preview is presented parafoveally, and it changes to another word before the reader looks directly at it. Because the change occurs during a saccade (when vision is suppressed), most readers are unaware that a change has occurred.

Display 1 He claims he won't | bail if his team loses.

Display 2 He claims he won't | bawl if his team loses.

Fig. 5.1 Display 1: a fixation cross on the far left-hand side of the screen dissolves into a sentence display that initially includes the preview string. A reader begins reading the sentence and the hypothetical fixation points are indicated (*). They process the preview (either *ball* or *bail*) parafoveally when they fixate the word just left of the target region (*won't*). During the saccade into the target region, the eyes cross an invisible boundary (|) and Display 1 changes to Display 2. Display 2: the target word (*bawl*) displays before the eyes begin the fixation, and the text appears completely normal in most cases.

This technique allows researchers to manipulate the similarity of the preview and the word it replaces, and measure the effect of this on reading time. If readers process phonological information parafoveally, then congruent previews should facilitate recognition once the word is fixated relative to the incongruent condition. Thus, parafoveal preview studies measure effects of the initial information that skilled readers process during word recognition and provide singular insights into early lexical access processes.

Pollatsek, Lesch, Morris and Rayner (1992) conducted the initial parafoveal preview study of phonological processing using such materials. Shorter first fixation durations on words such as *beech* in a homophone preview condition (*beach*) compared to an orthographic control condition (*bench*) indicated that skilled readers use parafoveal phonological information to facilitate word recognition. More recently, parafoveal phonological effects have appeared in experiments using nonword and partial word previews (Ashby & Martin, 2008; Ashby & Rayner, 2004; Ashby, Trieman, Kessler & Rayner, 2006) as well as in experiments conducted in French and Chinese (Miellet & Sparrow, 2004; Pollatsek, Tan & Rayner, 2000). As congruency effects appear in word, non-word and partial-word previews, these phonological preview effects are generally considered to be pre-lexical.

Parafoveal preview studies make two notable contributions to the question of how phonological processes operate in skilled word recognition. First, the findings suggest that skilled readers routinely assemble phonological information during word identification in the course of silent reading. Because the critical word is not visibly different from the others in the sentence, it is likely that the preview data reflect automatic (as opposed to intentional) word recognition processes. Another sense of routine is that parafoveal phonological processing occurs for words in general during silent reading, not only for a subset of unfamiliar words (Ashby, 2006; Ashby et al., 2006; Miellet & Sparrow, 2004; Rayner, Sereno, Lesch & Pollatsek, 1995).

A second notable finding relates to the function of parafoveal phonological processes in reading. Readers typically recognize words faster when a parafoveal preview is available, as it is during normal reading, than when preview is denied (Rayner, 2009; Rayner, Liversedge & White, 2006). The shorter fixation durations observed in the phonologically-similar preview conditions indicate that parafoveal phonological processing contributes to the rate of skilled reading, or reading fluency (Ashby et al., 2006; Ashby & Rayner, 2004; Pollatsek et al., 1992). Eye movement data that link parafoveal phonological processing and reading rate are consistent with observations of individual differences in reading skill, where research indicates that skilled readers benefit from parafoveal phonological information, but less skilled readers do not (Chace, Rayner & Well, 2005). The probable connection between efficient phonological processes and fluency also helps explain the function of phonological processing in skilled reading: the automatic, parafoveal activation of phonological forms serves to speed word recognition.

In addition to insights from single-word reading studies and eye movement studies of text reading, neurophysiological measures such as event-related potentials (ERPs) and magnetoencephalography (MEG) provide unique information about how word recognition processes unfold over time, by measuring the electrical fields inherent in brain activity as it occurs. ERP studies have demonstrated phonological effects in word recognition using a masked priming paradigm that taps automatic word recognition processes (e.g. Grainger, Kiyonaga & Holcomb, 2006). Using a similar technique, several ERP experiments by Ashby and colleagues found phonological priming effects within the first tenth of a second of word reading (Ashby, Sanders, Kinston, 2009; Ashby, 2010). Phonologically similar targets elicited a reduced early negativity (or N1) compared to targets in incongruent conditions, indicating that the brain processes the phonological information in printed words as early as 100 ms after target onset. These congruency effects arose in response to initial syllable primes and phonemic feature primes, which demonstrates that skilled readers process several layers of phonological information automatically when silently reading single words. The phonological effects observed were coincident with the timing of lexical effects found in other experiments (see Pulvermüller, 2001).

Wheat, Cornelissen, Frost and Hansen (2010) used a masked priming paradigm and MEG technology to test the time course of phonological processing in relation to the brain's reading network. They reported phonological congruency effects around 100 ms in left hemisphere frontal areas in the proximity of Broca's region (i.e. the inferior frontal gyrus). Therefore, the Wheat et al. (2010) data provide converging evidence for the early time course of phonological activation during skilled word recognition and help to localize that early activity to the inferior frontal gyrus (IFG).

In summary, the evidence from behavioural and neurophysiological experiments that tap automatic word recognition processes indicates that skilled readers activate phonological information in the initial phase of word recognition. The eye movement data indicate skilled readers routinely use phonological processes to recognize words during silent reading. ERP studies indicate that readers activate several types of phonological information, including sub-phonemic information, within one tenth of a second. These phonological effects register too quickly to be considered intentional or strategic. Rather, phonological processing seems to occur automatically when advance word form information is available to skilled readers (i.e. either in the parafovea during silent reading or as a foveal masked prime). Parafoveal phonological processing facilitates reading fluency, since shorter fixation times support faster reading overall. Therefore, these studies indicate that phonological processes operate routinely in the initial phases of word recognition and contribute to the speed of skilled reading.

Orthography and phonology in skilled word recognition

Now that we have discussed classic and recent research that examines the orthographic and phonological processes in skilled word recognition, we turn to examine a schematic version of one of the primary models of word recognition and the data that support it. Understanding this simplified schematic and how it applies to the brain's reading network provides a foundation for later discussions about how reading develops in the brain and approaches to early reading instruction. Before focusing on one type of model, let's review one of the central debates in reading research: the controversy over how orthographic and phonological processes contribute to skilled word recognition.

In principle, skilled readers could identify words in two ways: by using the orthographic form to access meaning directly (*direct access*) or by using the orthographic form to compute a phonological code and use that information to access meaning (*phonologically-mediated access*). Most contemporary theories of word reading assume that both direct and phonologically-mediated mechanisms are available to skilled readers, but they make different assumptions about the division of labour between them, the timing of phonological processes and the level of interaction between the routes (Coltheart, 1978; Frost, 1998; Seidenberg, 1995).

The concept of two routes to word recognition has been discussed as a way of proposing distinct processing routes for identifying two types of English words: phonically regular words (e.g. *mint, hint, stint*) whose pronunciation can be determined by letter-sound correspondence rules and irregular words (e.g. *pint*) that are often exceptions to those rules (e.g. Baron & Strawson, 1976; Carr & Pollatsek, 1985; Coltheart, 1978). Some two-route theories assume the primacy of the direct pathway in skilled silent reading (e.g. Coltheart, 1978), whereas others assert the centrality of phonological processing in word recognition (Frost, 1998; Van Orden, Pennington & Stone, 1990). The timing of phonological processes in word recognition has proved to be one key issue that differentiates two-route models from connectionist, triangle models of reading. Two-route models of reading claim that the indirect route to word recognition is slower than the direct route, due mainly to slow phonological processing. The triangle models make no such claim about the relative speed of the two routes.

Triangle-type models of skilled word recognition

In addition to laboratory research, one of the main approaches to understanding basic reading processes is the development of computational models that can be run on computers to simulate the reading data collected from skilled readers. One category of computational model is the connectionist or 'triangle' model of word recognition (e.g. Harm & Seidenberg, 2004; Plaut et al., 1996; Seidenberg & McClellend, 1989). Connectionist models can be used to predict reading data from behavioural and neuroimaging experiments and, thereby, contribute to our understanding of how words are identified during reading, reading development and dyslexia. For our purposes, we will refer to a single triangle model which captures the basics of the connectionist approach (see Figure 5.2). The triangle model encodes the idea that skilled readers recognize words by perceiving the visual information available on a printed page and processing the letters, sounds and meaning(s) to access a specific word in their lexicon. As in two-route models of word recognition, there is a pathway in the triangle model for both direct and phonologically mediated lexical access; the former involves the orthography to semantics pathway and the latter involves the orthography to phonology to semantics pathway. Another similarity is that the lexical properties of a word, such as how common it is or how regularly it is spelled (frequency and regularity), determine which route will have the most accurate recognition rates.

The triangle model differs from two-route models of word recognition (e.g. Coltheart, Rastle, Perry, Langdon & Ziegler, 2001) in two fundamental ways. One difference is that triangle models do not assume that phonological processes operate slowly. Therefore, the triangle model of reading is consistent with data that demonstrate fast phonological processing (see Halderman et al., in press for review). Most two-route models conceptualize the phonological-mediated route as slower than the direct route, with phonological information contributing to word recognition primarily when the direct route fails to find a lexical entry. Secondly, the triangle model claims that identification of a unique lexical item results from inputs received from all the processors; word recognition is the product of cooperative activation. In contrast, the direct and phonologically-mediated routes in two-route models function essentially as distinct paths, with each route computing its output (e.g. Coltheart et al., 2001).

Rueckl and Seidenberg (2009) review successful simulations of behavioural reading data that indicate the benefits of parallel, cooperative activation: the model-generated data fit the human reading data better when the model assumes cooperative orthographic and phonological processing, as compared to separate routes (Harm & Seidenberg, 2004). Also, measurements of the inputs to the model's semantic processor indicated that both the orthographic and phonological processors contributed to semantic activation for nearly every word tested (Harm & Seidenberg, 2004).

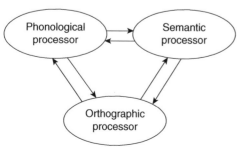

Fig. 5.2 A triangle model of word recognition processes.

This confirmed that semantic activation in the model is driven by the cooperative interaction of the orthography–semantics pathway and the orthography–phonology–semantics pathway. Further, Harm and Seidenberg demonstrated that cooperation among the routes activates word meaning more efficiently than does the single contribution of one route or the other. The superior performance of the triangle model that resulted from the interaction of all three processors is consistent with theories of skilled reading which claim that phonological processing plays a central role (e.g. Van Orden et al., 1990).

In summary, computer simulations demonstrate the triangle model's capacity to account for the human word recognition data that were collected using several different paradigms (Harm & Seidenberg, 2004). These simulations best account for the human reading data through a mechanism of cooperative activation among the orthographic, phonological and semantic processors. Now we turn the discussion toward the neuroimaging data that identify which areas of the brain are active during word recognition and the role(s) each plays.

Neuroimaging evidence and the triangle model

Neuroimaging research has shown that skilled readers utilize several brain areas in the left hemisphere to form a network for word recognition (see Figure 5.3): an anterior circuit centred around the IFG (i.e. Broca's area) and a posterior network that includes a dorsal circuit that comprises temporo-parietal cortex and the angular gyrus as well as a ventral circuit in OT and temporal cortex (Frost et al., 2009; Pugh et al., 2001). Experiments indicate that within these circuits, particular brain areas are tuned to the different streams of linguistic information that are processed during isolated word recognition (Cornelissen, Tarkianien, Helenius & Samelin, 2003; Fiebach, Friederici, Muller, von Cramon, 2002; Price, Moore, Humphreys & Wise, 1997; Rossell, Price & Nobre, 2003). Although the specifics of this brain network for reading may be difficult for a novice to grasp at first, two characteristics of the network are fairly obvious. The first is the distributed nature of the brain's word recognition network, which involves several areas in the left hemisphere.

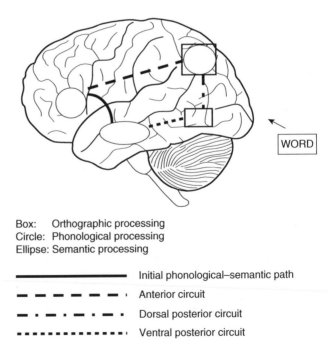

Box: Orthographic processing
Circle: Phonological processing
Ellipse: Semantic processing

——————————— Initial phonological–semantic path

— — — — — — Anterior circuit

— · — · — · — · Dorsal posterior circuit

▪▪▪▪▪▪▪▪▪▪▪▪▪▪ Ventral posterior circuit

Fig. 5.3 Brain networks involved in reading WORD.

The second characteristic is the redundancy in the circuits, with both the anterior and posterior circuits responding to semantic and phonological information. The triangle model reflects these neural characteristics in its distributed representations (i.e. separate processors for different layers of information about a word) and the parallel, interactive activation that allows for cooperation between pathways.

The triangle model also provides a way of organizing knowledge about the neural networks involved in skilled reading (Rueckl & Seidenberg, 2009). The orthographic–semantic path in the triangle model conceptualizes activation in the posterior ventral circuit, in areas such as OT and the middle and inferior temporal gyri. The orthographic–phonological pathway conceptualizes activation along the dorsal circuit in areas such the angular gyrus, supramarginal gyrus and Wernicke's area. In addition, the phonology–semantics path may involve activation in the lateral inferior prefrontal cortex area of the frontal circuit, which appears to have regions activated by phonological and semantic information (Poldrack & Wagner, 2004). The posterior-ventral network may constitute the direct path to visual word recognition whereas the dorsal-anterior network could instantiate the indirect path that computes phonology as well as meaning.

Given the organization of brain networks into direct and indirect pathways, it may be tempting to interpret the neuroimaging data in terms of a two-route model of word recognition (e.g. Fiebach et al., 2002). However, two-route theories of reading make several assumptions that seem inconsistent with much of the neuroimaging data. One frequent assumption is that phonological processing is slow, and we have discussed behavioural and neurophysiological data that refute that claim. Another assumption of two-route models is that the direct and indirect paths operate largely independently. Several recent neuroimaging studies report concurrent activation in the anterior and posterior circuit that suggests cooperation between the two pathways during skilled word recognition (Cornelissen et al., 2009; Pammer et al., 2004).

The Strain effect provides further evidence of cooperation between the direct and indirect routes (Rueckl & Seidenberg, 2009). Strain, Patterson and Seidenberg (1995) observed that consistent words were named faster and more accurately than inconsistent words when they were low in both frequency and imageability, accompanied by an imageability effect that appeared when words were low in frequency and consistency. Together, these two behaviour patterns constitute the Strain effect, which suggests that word identification involves both the indirect orthography-phonology-semantics path and the direct orthography–semantics path. The Strain effect evokes the following pattern of brain activity: more activation in the middle temporal gyrus for high than low imageability words combined with reduced activation in the IFG area that registers spelling-sound consistency (Frost et al., 2005).

In summary, behavioural and neuroscience research has identified particular brain regions with activation patterns that correspond to the three processors in the triangle model. Roughly speaking, the anterior route centres around the inferior frontal gyrus and is responsible for phonological processing. The posterior route involves orthographic processing in the OT region. Print-to-sound mapping occurs in the supramarginal and angular gyrus and semantic processing occurs primarily along the ventral route in the middle temporal gyrus. MEG data provide high-resolution measurements of the timing of these processes as they unfold during skilled word recognition. Visual information is picked up by the retina and projected to the visual cortex in occipital lobe, with activation progressing from the back of the brain forward through the other lobes. MEG studies indicate concurrent activation of orthographic and phonological processing areas when word recognition processes have been underway for 100–150 ms, and this activation seems to persist for about 200 ms in most studies. In addition to concurrent activation, fMRI research on the Strain effect indicates that skilled word recognition involves the interaction of orthographic, phonological and semantic processors. Therefore, the bulk of the cognitive neuroscience data indicate that skilled readers recognize single words through interactive, cooperative processing of several streams of information that yields the identification of a unique lexical item. Note the difference between this view of word recognition and the traditional view of two dissociable routes to word recognition. Case studies of brain-injured patients suggest that the two routes may be dissociated by trauma to the brain, but the experimental data indicate cooperative orthographic and phonological activation in healthy, skilled readers. Having completed our discussion of skilled word recognition, we now explore reading development.

5.3 Learning to read in alphabetic systems

In this section, we focus on the neuroimaging evidence of how reading networks develop in the brain and the implications of that development for reading instruction. First, some basic information about how alphabetic writing systems operate is discussed.

Alphabets

Alphabetic writing systems (English, Italian, Korean) use a limited number of graphemes (i.e. written symbols) to encode the basic phonemes of that language. In contrast, logographic writing systems (Chinese, Japanese Kanji) use a large numbers of graphemes to record the semantic and phonetic properties of words. The different alphabets of the world are based on the same principle: print encodes spoken word forms through the learned associations between written symbols and speech sounds. Letters represent phonemes across all alphabetic systems, in which different combinations of letters combine to allow one to write a large number of words from a small set of reusable symbols (*top, pot, spot, stop*). This alphabetic principle is the key to a productive writing system (Perfetti, 1985). The productivity of alphabets is economical for skilled readers, and relatively easy to learn in comparison with the process of memorizing thousands of Chinese characters. The consistency of letter to sound associations is the major advantage of a productive alphabet, as it allows readers to pronounce words never before seen.

The consistency of grapheme-phoneme associations in a particular alphabet is referred to as the script's *transparency*. Highly consistent written languages are orthographically transparent, which is to say that the mapping of letters to sounds is one-to-one. For example, in Spanish most sounds are written with distinct letters. In contrast, writing systems that are less transparent have multiple sounds mapping onto a single letter (e.g. *c* in English) as well as multiple letters mapping onto one sound (e.g. *ph*). English is typically considered an opaque writing system because of the number of multiple mappings and irregular patterns it contains.

When Seymour, Aro and Erskine (2003) compared reading development in several European languages, they found that the transparency of the writing system affects the difficulty of learning to read in that system, consistent with the orthographic depth hypothesis (Katz & Frost, 1992). First and second grade readers (i.e. 6–8-year-olds) in transparent writing systems (e.g. Spanish, Italian, Finnish) had accuracy rates above 85% when reading unfamiliar letter strings, whereas accuracy rates were much lower in inconsistent, opaque writing systems such as Danish (54%) and English (29%). Other cross-linguistic studies report similar findings (Aro & Wimmer, 2003), and indicate that it is common for children to be able to read most of the words in a transparent system accurately by the end of first grade (7 years of age) (see Share, 2008 for further discussion).

Ziegler and Goswami (2005) explain the role of orthographic depth in reading development with their *psycholinguistic grain-size theory*. The theory states that learning to read in more transparent systems is easier because it involves reliable sub-lexical grapheme-phoneme correspondences that operate at the level of the phoneme. Therefore, children learning to read in transparent orthographies learn letter-sound mappings at one grain size (i.e. the phoneme level), and they master this skill fairly quickly, as indicated by high accuracy rates at the end of first grade (Seymour et al., 2003). In contrast, English children must learn to process multiple phonological grain sizes such as the syllable, rhyme and phoneme in order to read and spell words accurately. For example, notice the consistency in the/a/sound in the words *hall*, *wall* and *small*. Here, words sharing the same rhyme unit (the vowel + the following consonants) are spelled consistently (Kessler & Treiman, 2001; Treiman, Mullenix, Bijeljac-Babic & Richmond-Welty, 1995). The need to learn mappings at several grain sizes may slow initial progress in learning to read English, and it takes children several years to be able to read fluently (Goswami, Ziegler, Dalton & Schneider, 2003; Ziegler & Goswami, 2005).

Understanding the alphabetic principle and learning to apply it efficiently takes time. One obstacle that young children face is that they have an imperfect idea of what phonemes are. Identifying individual phonemes in a word can be difficult for some children because phonemes are not distinct acoustic segments in the speech stream (Liberman, Cooper, Shankweiler & Studdert-Kennedy, 1967). The abstract nature of phonemes helps to account for beginning readers' difficulty with segmenting, storing and manipulating the phonemes in words they hear. This skill, termed *phonemic awareness,* predicts success in learning to read, presumably because it interferes with consistent access to letter–sound relations. Whereas the abstract quality of the phoneme can make learning to read a challenge in any alphabetic system, the complexity of the mappings between letters and phonemes in English makes learning to read in this alphabet especially challenging. Even though the mappings in English can be complex, letters provide the most reliable cues available for word recognition when learning to read.

Learning to read

In this section, we address questions such as (a) what is it that children learn when they learn to read? and (b) how does such learning come about?

The nature of learning to read

Learning to read shares similarities with learning other complex skills, such as how to play an instrument. A musical instrument is the technology that a child learns to operate in order to produce a tune. Learning to play an instrument is initially awkward and time-consuming, but getting the basic skills automatized through practice is necessary to become a musician. In learning to read, the print is the technology that the child must learn in order to receive the linguistic message a text encodes. Understanding how an alphabet operates is the foundation of basic decoding skills that will enable children to read unfamiliar words independently and become fluent readers. The alphabetic principle refers to the understanding that letters encode speech sounds, and that by assigning the right sounds to a printed string of letters, one can identify any printed word. Numerous studies indicate that mastery of the alphabetic principle is the key to successfully learning to read in the early grades (Backman, 1983; Bradley & Bryant, 1978; Bruce, 1964; Calfee, Chapman & Venezky, 1972; Calfee, Lindamood & Lindamood, 1973; Fox & Routh, 1975; 1976; 1984; Helfgott, 1976; Juel, Griffith & Gough, 1986; Liberman, 1973; Lundberg, Olofsson & Wall, 1980; Tornéus, 1984; Wagner & Torgesen, 1987; Zifcak, 1981).

Several investigators have described the phases children typically go through when beginning to read (Ehri, 1999, 2002; Gough & Hillinger, 1980; Marsh, Friedman, Welch & Desberg, 1981; Mason, 1980). Levels of phonemic awareness seem to affect the rate at which children progress through these phases. During the initial pre-alphabetic phase, children become aware of print in their environment. In this phase they can recognize common labels and logos, but they use default visual cues (such as colour or shape) to identify words rather than the actual letters (Byrne, 1992). When children begin identifying words based on some of the letters, they have entered the *partial alphabetic phase*, in which default visual cues are complemented by letter cues to word identification (De Abreu & Cardoso-Martins, 1998; Ehri & Wilce, 1985; Scott & Ehri, 1989; Treiman & Rodriguez, 1999). In this phase, children use the first letter of the word (and later the last letter as well) to constrain possible candidates. For example, the child might read *Bob liked to road* [ride]. Thus, nonsensical errors reflect the child beginning to attend to alphabetic information. In these first two phases, basic phonological awareness of syllables and rhymes will support some growth in reading. However, progress into *the full alphabetic phase* benefits from the ability to isolate and manipulate single phonemes. In the full alphabetic phase, word recognition is grounded in letter–sound relations, which enables children to decode new words and recognize them later when they appear in print, irrespective of any differences in letter case or font. Attending to letter–sound relations enhances memory for letter sequences within words beyond what can be achieved by visual memorization alone (Cunningham, Perry, Stanovich & Share, 2002). Paradoxically, it is in the full alphabetic phase, when children are beginning to fully apply the alphabetic principle that they may appear to struggle the most. They may read fairly accurately, but be slow and tire quickly. However, decoding words becomes faster with practice and continued reading exposes them to words repeatedly, so their speed gradually improves. Thus, frequent reading of easy texts eventually ushers in the last of Ehri's phases: the *consolidated alphabetic phase*. Children in this phase recognize words quickly, becoming sensitive to frequent letter sequences such as *-nt* or *-ing*.

As mentioned earlier, English has many words with simple letter-to-sound mappings (e.g. *map*, *admit*, *magnet*), but it also contains many words with complex mappings such as *foot* or *wrong*. The complex mappings in English, and in other less transparent alphabets, can cause confusion for beginning readers and make learning to read a difficult process. Different cultures attempt to solve this problem in different ways. For example, the Hebrew writing found in adult texts is very opaque (as most vowel letters are omitted), however, the text used to teach reading in Hebrew includes the vowel letters for children in the first few grades. In this way, children get initial practice with letter–sound mappings at the phoneme level. In fourth grade (9–10-year-olds), the vowels are no longer encoded in the text and children transition to using morphological information from consonant roots to identify words. Therefore, using a modified writing system for early reading instruction is one way that opaque orthographies may simplify the task of learning to read.

Unfortunately, how to best simplify the task of learning to read English is still disputed. Therefore, how early reading instruction should proceed is the foundation of a long-standing debate. On one side of this debate are those who subscribe to a whole language approach to reading instruction, and on the other side are those who advocate the explicit teaching of phonics (i.e. the letter–sound mappings that constitute the alphabetic principle) (Adams, 1990; Goodman, 1986; Grundin, 1994; McKenna, Stahl & Reinking, 1994; Rayner, Forman, Pesetsky, Perfetti & Seidenberg, 2001, 2002; Weaver, 1990). In both approaches, children experience storybooks read aloud by the teacher in order to build attention, awareness of plot and spoken vocabulary. The two approaches primarily differ in how they teach children to recognize words in print.

Whole language is presently the most common approach used to teach reading in the US today (Goodman, 1970, 1986; Smith, 1971, 1973; Smith & Goodman, 1971). Essentially, advocates of whole language instruction believe that children will learn to read naturally through immersion in a literacy-rich environment and that systematic phonics lessons are detrimental to reading development (Routman, 1991; Weaver, 1994). Based on these assumptions, whole language instruction emphasizes the meaning of written language over the technology of how writing encodes spoken word forms. For example, Pinnell and Fountas (2009) recommend that teachers help children in grades K–3 (UK schools' Years 1–4, ages 5–9 years) to identify unfamiliar words in text by first asking them to guess which word would make sense in the sentence (looking at the first letter of the word is the third recommended prompt). Rather than teaching focused lessons that sequentially develop specific decoding skills, whole language teachers are more likely to encourage the prediction of upcoming words based on context. New words are introduced by being pronounced by the teacher and then posted on a word wall in the classroom that is usually organized by semantic categories, rather than by spelling pattern. For example, *bread* would appear in the 'foods' category and *head* would appear in the 'my body' category.

On the other side of the reading debate from the whole language philosophy are those who view learning to read as a process of acquiring a skill that generally requires instruction (Gough & Hillinger, 1980). As young readers begin with stronger skills in spoken language comprehension than in reading comprehension, the initial roadblock to understanding print is the struggle to translate words on the page into their spoken forms (Curtis, 1980). This perspective favours reading instruction that teaches the alphabetic principle explicitly to children, rather than leaving its discovery to chance. In these code-emphasis programmes, a substantial portion of the reading lesson is spent teaching the letter–sound mappings that underlie the alphabetic principle, in order to help children enter the full alphabetic phase and develop decoding skills they can use to read new words independently. Phonics begins with a limited set of letters which can be combined into many different kinds of words. For example, children who learn two vowels and three consonants can read and spell *top, pot, pat, tap, sat, stop, pots, spat*, etc. Gradually more letters are added and then the children learn more complex patterns, such as consonant digraphs (e.g. *th, sh*) and vowel teams (e.g. *oi, ai*). With continued phonics instruction, children learn more letter–sound patterns that help them read and spell many more words. In this way, teaching phonics directly with plenty of practice draws children's attention to the productivity of our alphabetic writing system. Phonics advocates contend that the mapping of letters onto sounds marks the beginning of reading development, and mastery of that alphabetic principle is a critical bridge to getting meaning from print. Just as the dribble is necessary for scoring in basketball, automatic decoding is considered a fundamental skill for reading fluently and with comprehension (Beck, 1998).

Research points to several benefits of systematic phonics instruction in early reading instruction: increasing familiarity with the alphabetic principle, building reading independence and preventing early reading difficulties. Initial decoding practice draws children's attention to letter sequences in the middle of words, and helps usher in the full alphabetic phase of reading, which is necessary to develop the accurate and quick word recognition skills that enable reading fluency and support comprehension (Ehri, 1980, 1992; 1998, 1999; Perfetti, 1992; Rayner et al., 2001, 2002). Mastery of letter sound correspondences provides a powerful tool for translating print into speech that children use when reading independently (Jorm & Share 1983). In contrast, children who cannot decode unfamiliar words either skip over them or guess them. Inattention to word specific orthographic and phonological information can hinder the building of high-quality lexical representations that are fundamental to accurate reading (Perfetti, 2007). Systematic phonics instruction also has been shown to reduce the number of children who struggle with beginning to read (Bos, Mather, Dickinson, Podhajski & Chard, 2001; Ehri et al. 2001; Foorman, Francis, Fletcher, Schatschneider & Mehta, 1998).

Models of skilled word recognition are invoked to support both sides of the reading debate. For example, triangle models of word recognition propose that word recognition results from the cooperative activation of the direct and indirect pathways. Therefore, the triangle model is offered as support for code-emphasis approaches to reading instruction that encourage children to practice letter–sound mappings and apply decoding skills. In contrast, two-route theories propose a direct access account of word recognition, in which skilled readers rely primarily on the orthographic–semantic route to identify most words. Thus, two-route models are sometimes invoked to provide justification for reading instruction methods that emphasize sight-word memorization over systematic decoding. For example, Smith (1973) forcefully argued that direct access is necessarily more efficient because it does not require the extra phonological-recoding step, and this perspective is fundamental to the whole language approach, which opposes the systematic teaching of phonics. The greater efficiency of the direct (orthographic–semantic) route seems intuitively obvious, as one path involves three processors and the other involves two. However, that intuition does not fit well with the behavioural and neuroscience data described in section 5.2 of this chapter, which indicate that skilled readers activate orthographic and phonological processes within the first 100 ms of word recognition (Ashby et al., 2009; Pammer et al., 2004; Wheat et al., 2010). Furthermore, computational research with triangle-type models has demonstrated its ability to simulate human reading data fairly well based on the assumption of parallel processing (Harm & Seidenberg, 2004), which circumvents intuitions that gauge route efficiency based on the number of processors engaged. Therefore, the neuroscience and computational data available at this time do not indicate that more efficient word recognition processes are accomplished by the orthographic–semantic path.

Still, an important pedagogical question remains. If skilled readers use direct access, then why should teachers emphasize laborious decoding processes? The answer to this question depends on how one construes the meaning of 'use'. If the orthographic–semantic route was sufficient for skilled word recognition, then it might be appropriate to focus on whole word memorization peppered with incidental phonics as needed. Although some researchers and educators claim that skilled readers rely primarily on the direct route, the data do not seem to support that conclusion. Rather, the eye movement and ERP studies that found fast effects of phonological processing indicate that skilled readers do typically engage the orthographic–phonological pathway en route to word recognition during silent reading (Ashby, 2010, Ashby & Rayner, 2004; Ashby et al., 2009; Lee, Rayner & Pollatsek, 1999; Lee, Binder, Kim, Pollatsek & Rayner, 1999; Sereno & Rayner, 2000). The initial MEG data demonstrate concurrent activation of both the anterior orthographic–phonological pathway and the ventral orthographic–semantic pathway (Pammer et al., 2004; Wheat et al., 2010). In addition, computer simulations demonstrated superior word recognition performance in the case of cooperative input from both pathways (see Rueckl & Seidenberg, 2009 for a review).

Yet, it is also the case that skilled readers tend to show phonological effects for low- but not high-frequency words (Ashby, 2006; Jared & Seidenberg, 1991; Sereno & Rayner, 2000). Given that frequency effects are a result of word recognition, readers cannot use frequency information a priori to determine when to engage phonology to assist with word recognition. Therefore, it seems likely that both pathways are activated during the recognition of most words, with phonological effects only being measurable in the case of low-frequency words. Further complications emerge from imaging studies showing reduced activation for skilled readers relative to dyslexic readers along the anterior phonological route (Pugh et al., 2001). Initially, these data were interpreted as revealing skilled readers' shift from orthographic–phonological–semantic processing to orthographic–semantic processing. However, it is not clear how to interpret these reductions in the BOLD (blood oxygen level–dependent) response, as several perceptual/motor experiments have observed a reduced BOLD signal when skills are practiced to the point of automaticity (Poldrack & Gabrieli, 2001; Ungerleider, Doyon & Karni, 2002; Wang, Sereno, Jongman & Hirsch, 2003).

Likewise, Sandak et al. (2004) found a similarly reduced BOLD response in a neuroimaging study that examined how reading circuits become tuned for word recognition. Semantic training effects appeared as increased activation in the temporal lobe (which stores word meanings), whereas phonological training effects appeared to *reduce* activation along the dorsal (SMG) and anterior (IFG) phonological pathways. Therefore, the Sandak et al. study indicates that decreased activation along the dorsal and anterior routes marks the increased efficiency of phonological processing, rather than a decreased reliance on the orthographic–phonological path. Establishing that decreased activation in the anterior-phonological pathway indicates greater efficiency (rather than less use) informs the interpretation of studies such as Katz et al. (2005), in which word repetition resulted in reduced activation of the anterior circuit. In summary, the Sandak et al. training study suggests that skilled readers integrate information from both pathways to efficiently recognize most words, and that practice with phonological processing leads to increased processing efficiency that manifests as reduced activation of anterior-dorsal phonological areas as letter strings become more familiar.

Of course, demonstrating skilled readers' use of the orthographic–phonological route does not address concerns about the developmental appropriateness of directly teaching letter–sound relationships to young children (5–8-year-olds; kindergarten through second grade, Years 1–3 in the UK). Pedagogical concerns about phonics instruction take several forms. There is concern that children find phonics too difficult to learn and there is concern that even when learned, phonics will not be helpful. The concern that young children do not benefit from phonics instruction dates back to research conducted several decades ago. An article by Dolch and Bloomster (1937) is often cited as evidence that children under the age of seven will not benefit from phonics instruction (Dolch, 1948). However, those children were taught phonics incidentally rather than systematically, which suggests that the Dolch and Bloomster finding merely indicates the ineffectiveness of *embedded* phonics instruction for children in first grade (Brown, 1958; Chall, 1967). In contrast, several meta-analyses of reading studies have confirmed the effectiveness of teaching kindergarten and first graders to read with *systematic* phonics instruction, which methodically teaches the letter-sound relations in a simple-to-complex order (Adams, 1990; Ehri, Nunes, Stahl & Willows, 2001).

Neuroimaging studies that track the development of reading circuitry in the brain give some hints about why early phonics instruction is more effective than later instruction. These studies indicate that dorsal and anterior systems involved in orthographic–phonological processing are most active in beginning readers (Frost et al., 2009). For example, Shaywitz et al. (2002, 2007) conducted an fMRI study to measure brain activity as children read real words and pronounceable non-words. They found that younger readers used a diffuse network of brain areas in both the right and left hemispheres. Activation became lateralized to the left hemisphere as reading skill developed, and the dorsal and anterior systems were most active when reading words until age 10. After 10 years of age, the OT area in the posterior, ventral circuit becomes tuned specifically to process print. This route corresponds roughly to the orthographic–semantic pathway proposed by the triangle model, and activity in this circuit was positively correlated with reading skill in older children.

These neuroimaging data make two important contributions to our discussion of reading development. Finding activation primarily in the anterior and dorsal circuits involved in orthographic–phonological processing indicates that when children begin to read, their brains develop the circuitry to process the letter-sound mappings that are central to phonics instruction. Secondly, these data suggest that children who can read accurately and fluently develop the neural circuitry to access whole word forms through the ventral pathway. As reading fluency typically develops by age eight, it seems to precede the tuning of the OT to print that enables activation of the reputed visual word form area (VWFA) observed in skilled readers (Dehaene, 2009). Therefore, it appears that early phonics instruction is compatible with the developmental course of reading circuits in the brain.

Although the evidence indicates that phonics instruction is appropriate for beginning readers, there remains a concern that systematic phonics instruction will not benefit children's reading given the perceived irregularity of English spelling. However, alphabetic systems operate on the principle of letters encoding speech sounds. Therefore, by definition, the letter sounds will be the most reliable cue available for identifying written words. Phonics knowledge benefits children's reading in several ways. Once children understand the alphabetic principle, they use letter–sound mappings as a self-teaching mechanism to bootstrap their recognition of unfamiliar words and to consolidate their lexical representations of familiar words (Ehri, 2002; Jorm & Share, 1983; Share 1995). Children who have the opportunity to practice decoding text soon become proficient at mapping letters to sounds accurately and quickly, which means they can easily read unfamiliar words and maintain focus on the meaning of the text they are reading. After learning to read, children learn complex letter–sound patterns through exposure to print (see Kessler, 2009 for a review). By noticing that a spelling pattern is more common in some contexts than others, most children learn such patterns implicitly. However, exposure to print depends on developing reading fluency to a level that supports wide reading. Thus, initial systematic phonics instruction contributes to reading and spelling development in several ways.

Given these benefits of phonics instruction, it is not surprising to find that reading instruction that includes systematic phonics improves some components of reading. But do these skills translate into gains in reading comprehension? Several large-scale studies indicate the effectiveness of systematic phonics instruction (Foorman et al., 1998; Torgesen, Wagner, Rashotte, Rose, Lindamood, Conway & Garvan, 1999). Congress convened a National Reading Panel (NRP) to conduct a meta-analysis of the school-based research conducted thus far, and to make recommendations for effective practices for teaching children to read. The results of the NRP meta-analysis were fairly clear: Systematic phonics was more effective at helping children learn to read than any other approach, including whole language and basal reading programmes. Phonics instruction benefited both isolated word reading and text comprehension. In addition, phonics helped young children understand the alphabetic principle, as evidenced by progress in phonetic spelling and pseudoword decoding. In many cases, phonics instruction continued to support growth in decoding, isolated word reading and comprehension even after the intervention finished. Thus, there seems to be no way of getting around the conclusion that systematic phonics is a necessary component of reading instruction in grades K–2 (UK schools' Years 1–3, 5–8-year-olds) (Rayner et al., 2001). The NRP data indicate that this approach is currently the most effective approach for beginning reading instruction. While many children may discover some letter–sound correspondences without phonics instruction, teaching methods that make the alphabetic principle explicit provide a key to understanding our writing system that produces better readers overall.

5.4 **Conclusion**

In this chapter, we reviewed the behavioural, neurophysiological and computational evidence relevant to learning to read. Because readers must recognize printed words in order to comprehend text, we focused on research that illuminates how that recognition happens. At the end point of reading development, skilled readers process letters in parallel and engage phonological processes parafoveally to obtain advance word form information that speeds word recognition during reading. The phonological processes that skilled readers initially engage operate quite quickly, as indicated by research finding sub-lexical phonological activation in the brains of skilled readers as early as 80 ms after a word is presented. Fast phonological processing allows for the concurrent activation of brain circuits responsible for orthographic and phonological processing during skilled word recognition.

We discussed the triangle, connectionist model of skilled word recognition to examine the relationship between the orthographic–semantic path and the orthographic–phonological–semantic path. Triangle models have successfully simulated the Strain effect (Strain et al., 1995) in human reading data, which indicates that the two pathways work cooperatively as well as concurrently in skilled word recognition. Neuroimaging evidence identifies the neural signature of skilled word recognition: reduced activation in the frontal reading circuit indicates efficient phonological processing, whereas heightened activation in the ventral network indicates skilled semantic processing.

We also discussed the process of learning to read, beginning with the nature of alphabets and how they encode speech sounds. Before learning this alphabetic principle, children guess at printed words based on pictures and context cues. As alphabetic awareness improves, children begin to attend to all the letters in a word as they practice mapping letters onto speech sounds to independently read unfamiliar words. After several accurate readings of a new word, children can form a specific, high-quality lexical representation that supports fluent reading. These phases of learning to read track the development of reading circuits in the brain. Initially, neural activation during 'reading' (which is really just print perception at this phase) is bilateral and widely distributed. As reading develops, this activation becomes left-lateralized and focused mainly in the orthographic–phonological areas in the dorsal circuit. After children achieve reading fluency, the OT area in the ventral circuit becomes tuned to word recognition and correlated with reading skill.

The evidence reviewed indicates that the orthographic–phonological circuit is very important both in learning to read and skilled reading. We believe that the bulk of the available evidence indicates that methods of reading instruction that make the alphabetic principle explicit are most effective for helping children develop early reading behaviours that support their future reading development.

Acknowledgments

Preparation of this chapter was supported by Grant HD26765 from the National Institute of Child Health and Human Development and by an Early Career Investigator Grant from Central Michigan University.

References

Adams, M.J. (1990). *Beginning to read: Thinking and learning about print*. Cambridge, MA: MIT Press.

Aro, M., & Wimmer, H. (2003). Learning to read: English in comparison to six more regular orthographies. *Applied Psycholinguistics, 24*(4), 621–35.

Ashby J. (2006). Prosody in skilled silent reading: Evidence from eye movements. *Journal of Research in Reading, 29*, 318–33.

Ashby, J. (2010). Phonology is fundamental in skilled reading: Evidence from ERPs. *Psychonomic Bulletin & Review, 17*, 95–100.

Ashby, J., & Martin, A. E. (2008). Prosodic phonological representations early in visual word recognition. *Journal of Experimental Psychology: Human Perception & Performance, 34*, 224–36.

Ashby, J., & Rayner, K. (2004). Representing syllable information during silent reading: Evidence from eye movements. *Language and Cognitive Processes, 19*, 391–426.

Ashby J., Sanders L. D., & Kingston J. (2009). Skilled readers begin processing phonological features by 80 msec: Evidence from ERPs. *Biological Psychology, 80*, 84–94.

Ashby, J., Treiman, R., Kessler, B., & Rayner, K. (2006). Vowel processing during silent reading: Evidence from eye movements. *Journal of Experimental Psychology: Learning, Memory, and Cognition, 32*, 416–424.

Backman, J. (1983). The role of psycholinguistic skills in reading acquisition: A look at early readers. *Reading Research Quarterly, 18*(4), 466–79.

Baron, J., & Strawson, C. (1976). Use of orthographic and word-specific knowledge in reading words aloud. *Journal of Experimental Psychology: Human Perception and Performance, 2*(3), 386–93.

Beck, I. L. (1998). Understanding beginning reading: A journey through teaching and research. In J. Osborn & F. Lehr (Eds.), *Literacy for All: Issues in Teaching and Learning* (pp. 11–31). New York: Guilford Publications.

Berent, I., & Perfetti, C.A. (1995). A rose is a REEZ: The two-cycles model of phonology assembly in English. *Psychological Review*, 102, 146–84.

Bos, C., Mather, N., Dickson, S., Podhajski, B., & Chard, D. (2001). Perceptions and knowledge of preservice and inservice educators about early reading instruction. *Annals of Dyslexia*, 51, 97–120.

Bradley, L., & Bryant, P. E. (1978). Difficulties in auditory organisation as a possible cause of reading backwardness. *Nature,* 271(5647), 746–47.

Brown, R.W. (1958). *Words and things*. New York: Free Press.

Bruce, D. J. (1964). The analysis of word sounds by young children. *British Journal of Educational Psychology,* 34(2), 158–70.

Byrne, B. (1992). Studies in the acquisition procedure for reading: Rationale, hypotheses, and data. In P. B. Gough, L. C. Ehri, & R. Treiman (Eds.), *Reading acquisition* (pp. 1–34). Hillsdale, NJ: Lawrence Erlbaum.

Carr, T. H. & Pollatsek, A. (1985). Recognizing printed words: A look at current models. In D. Besner, T. G. Waller, and G. E. MacKinnon (Eds.), *Reading research: Advances in theory and practice 5* (pp. 1–82). San Diego, CA: Academic Press.

Calfee, R., Chapman, R., & Venezky, R. (1972). How a child needs to think to learn to read. In L. W. Gregg (Ed.), *Cognition in learning and memory* (pp. 139–82). Oxford: John Wiley & Sons.

Calfee, R. C., Lindamood, P., & Lindamood, C. (1973). Acoustic-phonetic skills and reading: Kindergarten through twelfth grade. *Journal of Educational Psychology,* 64(3), 293–98.

Cattell, J.M. (1886). The time it takes to see and name objects. *Mind*, 11, 63–65.

Chace, K. H., Rayner, K., & Well, A. D. (2005). Eye movements and phonological preview benefit: Effects of reading skill. *Canadian Journal of Experimental Psychology*, 59, 209–217.

Chall, J.S. (1996/1967). *Learning to read: The great debate* (3rd ed.). Fort Worth, TX: Harcourt Brace.

Cohen, L., & Dehaene, S. (2009). Ventral and dorsal contributions to word reading. In M.S. Gazzaniga (Ed.), *Cognitive neuroscience* (4th ed.). Cambridge, MA: MIT Press.

Coltheart, M. (1978). Lexical access in simple reading tasks. In G. Underwood (Ed.), *Strategies of Information Processing* (pp. 151–216). London: Academic Press.

Coltheart, M., Rastle, K., Perry, C., Langdon, R. & Ziegler, J. (2001). DRC: A dual route cascaded model of visual word recognition and reading aloud. *Psychological Review*, 108, 204–56.

Cornelissen, P. L., Kringelbach, M. L., Ellis, A. W., Whitney, C., Holliday, I. E., & Hansen, P. C. (2009). Activation of the left inferior frontal gyrus in the first 200 ms of reading: Evidence from magnetoencephalography (MEG). *PLoS ONE*, 4, e5359.

Cornelissen, P., Tarkiainen, A., Helenius, P., & Salmelin, R. (2003). Cortical effects of shifting letter position in letter strings of varying length. *Journal of Cognitive Neuroscience,* 15(5), 731–46.

Cunningham, A. E., Perry, K. E., Stanovich, K. E., & Share, D. L. (2002). Orthographic learning during reading: Examining the role of self-teaching. *Journal of Experimental Child Psychology*, 82(3), 185–99.

Curtis, M. E. (1980). Development of components of reading skill. *Journal of Educational Psychology,* 72(5), 656–69.

De Abreu, M. D., & Cardoso-Martins, C. (1998). Alphabetic access route in beginning reading acquisition in Portuguese: The role of letter-name knowledge. *Reading and Writing*, 10(2), 85–104.

Dehaene, S. (2009). *Reading in the brain*. Toronto, ON: Viking Press.

Dodge, R. (1907). Studies from the psychological laboratory of Wesleyan University: An experimental study of visual fixation. *Psychological Monographs,* 84(4), 1–95.

Dolch, E. W. (1948). *Helping handicapped children in school*. Oxford: Garrard Press.

Dolch, E. W., & Bloomster, M. (1937). Phonic readiness. *The Elementary School Journal,* 38, 201–205.

Drieghe, D., & Brysbaert, M. (2002). Strategic effects in associative priming with words, homophones, and pseudohomophones. *Journal of Experimental Psychology: Learning, Memory, and Cognition*, 28, 951–61.

Ehri, L.C. (1980). The role of orthography in printed word learning. In J.G. Kavanagh & R.L. Venezky (Eds.), *Orthography, reading, and dyslexia* (pp. 155–70). Baltimore, MA: University Park Press.

Ehri, L.C. (1992). Reconceptualizing the development of sight word reading and its relationship to recoding. In P.B. Gough, L.C. Ehri, & R. Treiman (Eds.), *Reading acquisition* (pp. 107–43). Hillsdale, NJ, England: Lawrence Erlbaum.

Ehri, L. C. (1998). Grapheme-phoneme knowledge is essential to learning to read words in English. In J. L. Metsala, & L. C. Ehri (Eds.), *Word recognition in beginning literacy* (pp. 3–40). Mahwah, NJ: Lawrence Erlbaum.

Ehri, L. C. (1999). Phases of development in learning to read words. In J. Oakhill, & R. Beard (Eds.), *Reading development and the teaching of reading: A psychological perspective.* (pp. 79–108). Oxford: Blackwell Science.

Ehri, L.C. (2002). Phases of acquisition in learning to read words and implications for teaching. In R. Stainthorp and P. Tomlinson (Eds.), *Learning and teaching reading.* London: British Journal of Educational Psychology Monograph Series II.

Ehri, L. C., Nunes, S. R., Stahl, S. A., & Willows, D. M. (2001). Systematic phonics instruction helps students learn to read: Evidence from the national reading panel's meta-analysis. *Review of Educational Research,* 71(3), 393–447.

Ehri, L. C., & Wilce, L. S. (1985). Movement into reading: Is the first stage of printed word learning visual or phonetic? *Reading Research Quarterly,* 20(2), 163–79.

Ferrand, L. & Grainger, J. (1994). Effects of orthography are independent of phonology in masked form priming. *Quarterly Journal of Experimental Psychology,* 47A, 365–82.

Fiebach, C., Friederici, A. D., Müller, K., & von Cramon, D. Y. (2002). fMRI evidence for dual routes to the mental lexicon in visual word recognition. *Journal of Cognitive Neuroscience,* 14(1), 11–23.

Foorman, B. R., Francis, D. J., Fletcher, J. M., Schatschneider, C., & Mehta, P. (1998). The role of instruction in learning to read: Preventing reading failure in at-risk children. *Journal of Educational Psychology,* 90, 37–55.

Fox, B., & Routh, D. K. (1975). Analysing spoken language into words, syllables, and phonemes: A developmental study. *Journal of Psycholinguistic Research,* 4(4), 331–42.

Fox, B., & Routh, D. K. (1976). Phonemic analysis and synthesis as word-attack skills. *Journal of Educational Psychology,* 68(1), 70–74.

Fox, B., & Routh, D. K. (1984). Phonemic analysis and synthesis as word attack skills: Revisited. *Journal of Educational Psychology,* 76(6), 1059–64.

Frost, R. (1998). Toward a strong phonological theory of visual word recognition: True issues and false trails. *Psychological Bulletin,* 123, 71–99.

Frost, S. J., Mencl, W. E., Sandak, R., Moore, D. L., Rueckl, J. G., Katz, L., Fulbright, R. K., & Pugh, K. R. (2005). A functional magnetic resonance imaging study of the tradeoff between semantics and phonology in reading aloud. *NeuroReport,* 16(6), 621–24.

Frost, S. J., Sandak, R., Mencl, W. E., Landi, N., Rueckl, J. G., Katz, L., & Pugh, K. R. (2009). Mapping the word reading circuitry in skilled and disabled readers. In K. Pugh, & P. McCardle (Eds.), *How children learn to read: Current issues and new directions in the integration of cognition, neurobiology and genetics of reading and dyslexia research and practice* (pp. 3–19). New York: Psychology Press.

Goodman, K. S. (1970). Reading: A psycholinguistic guessing game. In H. Singer & R. B. Ruddell (Eds.), *Theoretical models and processes of reading* (pp. 259–72). Newark, DE: International Reading Association.

Goodman, K. S. (1986). *What's whole in whole language?* Portsmouth, NH: Heinemann.

Goswami, U., Ziegler, J. C., Dalton, L., & Schneider, W. (2003). Nonword reading across orthographies: How flexible is the choice of reading units? *Applied Psycholinguistics,* 24(2), 235–47.

Gough, P. B., & Hillinger, M. L. (1980). Learning to read: An unnatural act. *Bulletin of the Orton Society*, 30, 179–96.

Grainger, J., Kiyonaga K., Holcomb P. J. (2006). The time course of orthographic and phonological code activation. *Psychological Science*, 17, 1021–26.

Grundin, H. U. (1994). If it ain't whole, it ain't language—or back to the basics of freedom and dignity. In F. Lehr, & J. Osborn (Eds.), *Reading, language, and literacy: Instruction for the twenty-first century* (pp. 77–88). Hillsdale, NJ: Lawrence Erlbaum.

Halderman, L.K., Ashby, J., Perfetti, C. (in press). Phonology: An early and integral role in identifying words. In J.S. Adelman (Ed.), *Visual Word Recognition*. New York: Psychology Press.

Harm, M. W., & Seidenberg, M. S. (2004). Computing the meanings of words in reading: Cooperative division of labor between visual and phonological processes. *Psychological Review*, 111, 662–720.

Helfgott, J., (1976). Phonemic segmentation and blending skills of kindergarten children: Implications for beginning reading acquisition. *Contemporary Educational Psychology*, 1, 157–69.

Jared, D., & Seidenberg, M. S. (1991). Does word identification proceed from spelling to sound to meaning? *Journal of Experimental Psychology: General*, 120(4), 358–94.

Jorm, A.F., & Share, D.L. (1983). Phonological recoding and reading acquisition. *Applied Psycholinguistics*, 4, 311–42.

Juel, C., Griffith, P. L., & Gough, P. B. (1986). Acquisition of literacy: A longitudinal study of children in first and second grade. *Journal of Educational Psychology*, 78(4), 243–55.

Katz, L., & Frost, R. (1992). Reading in different orthographies: The orthographic depth hypothesis. In R. Frost & L. Katz (Eds.), *Orthography, phonology, morphology, and meaning* (pp. 67–84). Amsterdam: Elsevier North Holland Press.

Katz, L., Lee, C. H., Tabor, W., Frost, S. J., Mencl, W. E., Sandak, R., Rueckl, J., & Pugh, K. R. (2005). Behavioral and neurobiological effects of printed word repetition in lexical decision and naming. *Neuropsychologia*, 43(14), 2068–83.

Kessler, B. (2009). Statistical learning of conditional orthographic correspondences. *Writing Systems Research*, 1(1), 19–34.

Kessler, B., & Treiman, R. (2001). Relationship between sounds and letters in English monosyllables. *Journal of Memory and Language*, 44(4), 592–617.

Lee, Y., Binder, K.S., Kim, J., Pollatsek, A., & Rayner, K. (1999). Activation of phonological codes during eye fixations in reading. *Journal of Experimental Psychology: Human Perception and Performance*, 25(4), 948–64.

Lee, H., Rayner, K., & Pollatsek, A. (1999). The time course of phonological, semantic, and orthographic coding in reading: Evidence from the fast-priming technique. *Psychonomic Bulletin & Review*, 6(4), 624–34.

Liberman, I. Y. (1973). Segmentation of the spoken word and reading acquisition. *Bulletin of the Orton Society*, 23, 65–77.

Liberman, A. M., Cooper, F. S., Shankweiler, D. P., & Studdert-Kennedy, M. (1967). Perception of the speech code. *Psychological Review*, 74(6), 431–61.

Lukatela, G. and Turvey, M. T. (1994). Visual lexical access is initially phonological: 2. Evidence from phonological priming by homophones and pseudohomophones. *Journal of Experimental Psychology: General*, 123(4), 331–53.

Lundberg, I., & Tornéus, M. (1978). Nonreaders' awareness of the basic relationship between spoken and written words. *Journal of Experimental Child Psychology*, 25(3), 404–412.

Lundberg, I., Olofsson, Å., & Wall, S. (1980). Reading and spelling skills in the first school years predicted from phonemic awareness skills in kindergarten. *Scandinavian Journal of Psychology*, 21(3), 159–73.

McKenna, M.C., Stahl, S.A., & Reinking, D. (1994). Critical commentary on research, politics, and whole language. *Journal of Reading Behavior*, 26, 211–233.

Marsh, G., Friedman, M., Welch, V., & Desberg, P. (1981). The development of strategies in spelling. In U. Frith (Ed.), *Cognitive strategies in spelling* (pp. 339–53). New York: Academic Press.

Mason, J. (1980). When do children begin to read: An exploration of four year old children's letter and word reading competencies. *Reading Research Quarterly,* 15, 203–27.

Maurer, U., Brandeis, D., & McCandliss, B.D. (2005). Fast, visual specialization in English revealed by the topography of the N170 response. *Behavioral & Brain Functions,* 1, 1–12.

McClelland, J.L., & Rumelhart, D.E. (1981). An interactive activation model of context effects in letter perception: Part I. An account of basic findings. *Psychological Review,* 88, 375–407.

McConkie, G., & Rayner, K. (1975). The span of the effective stimulus during a fixation in during reading. *Perception & Psychophysics,* 17, 578–86.

Miellet, S., & Sparrow, L. (2004). Phonological codes are assembled before word fixation: Evidence from boundary paradigm in sentence reading. *Brain & Language,* 90, 299–310.

Paap, K.R., Newsome, S.L., McDonald, J.E., & Schvaneveldt, R.W. (1982). An activation-verification model for letter and word recognition: The word superiority effect. *Psychological Review,* 89, 573–94.

Pammer, K., Hansen, P. C., Kringelback, M. L., Holliday, I., Barnes, G., Hillebrand, A., Singh, K. D., Cornelissen, P. L. (2004). Visual word recognition: The first half second. *NeuroImage,* 22, 1819–25.

Perfetti, C. A. (1985). *Reading ability.* New York: Oxford University Press.

Perfetti, C.A. (1992). The representation problem in reading acquisition. In P.B. Gough, L.C. Ehri, & R. Treiman (Eds.), *Reading acquisition* (pp. 145–74). Hillsdale, NJ: Erlbaum.

Perfetti, C. (2007). Reading ability: Lexical quality to comprehension. *Scientific Studies of Reading,* 11(4), 357–83.

Perfetti, C.A., Bell, L., Delaney, S. (1988). Automatic phonetic activation in silent word reading: Evidence from backward masking. *Journal of Memory and Language,* 27, 59–70.

Plaut, D.C., McClelland, J.L., Seidenberg, M.S., & Patterson, K.E. (1996). Understanding normal and impaired word reading: Computational principles in quasi-regular domains. *Psychological Review,* 103, 56–115.

Poldrack, R. A., & Gabrieli, J. D. E. (2001). Characterizing the neural mechanaisms of skill learning and repetition priming. evidence from mirror-reading. *Brain: A Journal of Neurology,* 124, 67–82.

Poldrack, R. A., & Wagner, A. D. (2004). What can neuroimaging tell us about the mind? Insights from prefrontal cortex. *Current Directions in Psychological Science,* 13, 177–81.

Pollatsek, A., Lesch, M., Morris, R. K., & Rayner, K. (1992). Phonological codes are used in integrating information across saccades in word identification and reading. *Journal of Experimental Psychology: Human Perception and Performance,* 18, 148–62.

Pollatsek, A., Tan, L. H., & Rayner, K. (2000). The role of phonological codes in integrating information across saccadic eye movements in Chinese character identification. *Journal of Experimental Psychology: Human Perception and Performance,* 26, 607–33.

Price, C. J., Moore, C. J., Humphreys, G. W., & Wise, R. J. S. (1997). Segregating semantic from phonological processes during reading. *Journal of Cognitive Neuroscience,* 9(6), 727–33.

Pugh, K., & McCardle, P. (Eds.) (2009). *How children learn to read: Current issues and new directions in the integration of cognition, neurobiology and genetics of reading and dyslexia research and practice.* New York: Psychology Press.

Pugh, K. R., Mencl, W. E., Jenner, A. R., Lee, J. R., Katz, L., Frost, S. J., Shaywitz, S. E., & Shaywitz, B. A. (2001). Neuroimaging studies of reading development and reading disability. *Learning Disabilities Research & Practice. Special Issue: Emergent and Early Literacy: Current Status and Research Directions,* 16(4), 240–49.

Pulvermüller, F. (2001). Brain reflections of words and their meaning. *Trends in Cognitive Sciences,* 5, 517–24.

Rastle, K., Brysbaert, M. (2006). Masked phonological priming effects in English: Are they real? Do they matter? *Cognitive Psychology,* 53, 97–145.

Rawson, M.B. (1995). *Dyslexia over the lifespan: a fifty-five year longitudinal study.* Cambridge, MA: Educator's Publishing Service.

Rayner, K. (1975). The perceptual span and peripheral cues in reading. *Cognitive Psychology,* 7, 65–81.

Rayner, K. (1998). Eye movements in reading and information processing: 20 years of research. *Psychological Bulletin*, 124, 372–422.

Rayner, K. (2009). The 35th Sir Frederick Bartlett Lecture: Eye movements and attention during reading scene perception and visual search. *Quarterly Journal of Experimental Psychology*, 62, 1457–1506.

Rayner, K., Foorman, B.R., Perfetti, C.A., Pesetsky, D., & Seidenberg, M.S. (2001). How psychological science informs the teaching of reading. *Psychological Science in the Public Interest*, 2(2), 31–74.

Rayner, K., Foorman, B.R., Perfetti, C.A., Pesetsky, D., & Seidenberg, M.S. (2002). How should reading be taught. *Scientific American*, 286, 85–91.

Rayner, K., Liversedge, S. P., & White, S. J. (2006). Eye movements when reading disappearing text: The importance of the word to the wright of fixation. *Vision Research*, 46, 310–23.

Rayner, K., Liversedge, S. P., White, S. J., & Vergilino-Perez, D. (2003). Reading disappearing text: Cognitive control of eye movements. *Psychological Science*, 14, 385–89.

Rayner, K., & Pollatsek, A. (1989). *The psychology of reading*. Englewood Cliffs, NJ: Prentice-Hall.

Rayner, K., Sereno, S.C., Lesch, M.F., & Pollatsek, A. (1995). Phonological codes are automatically activated during reading: Evidence from an eye movement priming paradigm. *Psychological Science*, 6, 26–32.

Reicher, G.M. (1969). Perceptual recognition as a function of meaningfulness of stimulus material. *Journal of Experimental Psychology*, 81, 275–80.

Reichle, E.D., Pollatsek, A., Fisher, D.L., & Rayner, K. (1998). Toward a model of eye movement control in reading. *Psychological Review*, 105, 125–57.

Rossell, S. L., Price, C. J., & Nobre, A. C. (2003). The anatomy and time course of semantic priming investigated by fMRI and ERPs. *Neuropsychologia*, 41(5), 550–64.

Routman, R. (1991). *Invitations: Changing as teachers and learners K-12*. Portsmouth, NH: Heinemann.

Rueckl, J. G., & Seidenberg, M.S. (2009). Computational modeling and the neural bases of reading and reading disorders. In K. Pugh & P. McCardle (Eds.), *How children learn to read: Current issues and new directions in the integration of cognition, neurobiology and genetics of reading and dyslexia research and practice* (pp.101–34). New York: Psychology Press.

Rumelhart, D.E., & McClelland, J.L. (1982). An interactive activation model of context effects in letter perception: Part 2. The contextual enhancement effect and some tests and extensions of the model. *Psychological Review*, 89, 60–94.

Sandak, R., Mencl, W. E., Frost, S. J., Rueckl, J. G., Katz, L., Moore, D. L., Mason, S. A., Fulbright, R. K., Constable, R. T., & Pugh, K. R. (2004). The neurobiology of adaptive learning in reading: A contrast of different training conditions. *Cognitive, Affective & Behavioral Neuroscience*, 4(1), 67–88.

Scott, J.A. & Ehri, L.C. (1989). Sight word reading in prereaders: Use of logographic vs. alphabetic access routes. *Journal of Reading Behavior*, 22, 149–66.

Seidenberg, M.S. (1995). Visual word recognition. In J.L. Miller & P.D. Eimas (Eds.), *Handbook of perception & cognition: Vol. 11. Speech, language & communication* (pp. 137–79). San Diego, CA: Academic Press.

Seidenberg, M.S., & McClelland, J.L. (1989). A distributed, developmental model of word recognition and naming. *Psychological Review*, 96, 523–68.

Sereno, S. C., & Rayner, K. (2000). Spelling–sound regularity effects on eye fixations in reading. *Perception & Psychophysics*, 62(2), 402–409.

Seymour, P. H., Aro, M., Erskine, J. M. (2003). Foundation literacy acquisition in European orthographies. *British Journal of Psychology*, 94(2), 143–74.

Share, D. L. (1995). Phonological recoding and self-teaching: Sine qua non of reading acquisition. *Cognition*, 55(2), 151–218.

Share, D. L. (2008). On the anglocentricities of current reading research and practice: The perils of overreliance on an outlier orthography. *Psychological Bulletin*, 134, 584–615.

Shaywitz, B. A., Shaywitz, S. E., Pugh, K. R., Mencl, W. E., Fulbright, R. K., Skudlarksi, P., Constable, R. T., Marchione, K. E., Fletcher, J. M., Lyon, G. R., & Gore, J. C. (2002). Disruption of posterior brain systems for reading in children with developmental dyslexia. *Biological Psychiatry*, 52(2), 101–110.

Shaywitz, B. A., Skudlarksi, P., Holahan, J. M., Marchione, K. E., Constable, R. T., Fulbright, R. K., Zelterman, D., Lacadie, C., Shaywitz, S. E. (2007). Age-related changes in readings systems of dyslexic children. *Annals of Neurology,* 61(4), 363–70.

Smith, F. (1971). *Understanding reading: A psycholinguistic analysis of reading and learning to read.* New York: Holt, Rinehart and Winston.

Smith, F. (1973). *Psycholinguistics and reading.* New York: Holt, Rinehart and Winston.

Smith, F., & Goodman, K.S. (1971). On the psycholinguistic method of teaching reading. *Elementary School Journal,* 71, 177–81.

Strain, E., Patterson, K., & Seidenberg, M. S. (1995). Semantic effects in single-word naming. *Journal of Experimental Psychology: Learning, Memory, and Cognition,* 21(5), 1140–54.

Tan, L.H., & Perfetti, C.A. (1999). Phonological and associative inhibition in the early stages of English word identification: Evidence from backward masking. *Journal of Experimental Psychology: Human Perception and Performance,* 25, 382–93.

Torgesen, J. K., Wagner, R. K., Rashotte, C. A., Rose, E., Lindamood, P., Conway, T., & Garvan, C. (1999). Preventing reading failure in young children with phonological processing disabilities: Group and individual responses to instruction. *Journal of Educational Psychology,* 91(4), 579–93.

Tornéus, M. (1984). Phonological awareness and reading: A chicken and egg problem? *Journal of Educational Psychology,* 76(6), 1346–58.

Treiman, R., Mullennix, J., Bijeljac-Babic, R., & Richmond-Welty, E. D. (1995). The special role of rimes in the description, use, and acquisition of English orthography. *Journal of Experimental Psychology: General,* 124(2), 107–36.

Treiman, R., & Rodriguez, K. (1999). Young children use letter names in learning to read words. *Psychological Science,* 10(4), 334–38.

Ungerleider, L.G., Doyon, J., & Karni, A. (2002). Imaging brain plasticity during motor skill learning. *Neurobiology of Learning and Memory,* 78(3), 553–64.

Van Order, G. C., Johnston, J. C., & Hale, B. L. (1988). Word identification in reading proceeds from spelling to sound to meaning. *Journal of Experimental Psychology: Learning, Memory and Cognition,* 14, 371–86.

Van Orden, G. C., & Kloos, H. (2005). The question of phonology and reading. In M. J. Snowling & C. Hulmes (Eds.), *The science of reading* (pp. 61–78). Malden, MA: Blackwell.

Van Orden, G.C., Pennington, B.F., & Stone, G.O. (1990). Word identification in reading and the promise of a subsymbolic psycholinguistics. *Psychological Review,* 97, 488–522.

Wagner, R. K., & Torgesen, J. K. (1987). The nature of phonological processing and its causal role in the acquisition of reading skills. *Psychological Bulletin,* 101(2), 192–212.

Wang, Y., Sereno, J. A., Jongman, A., & Hirsch, J. (2003). fMRI evidence for cortical modification during learning of mandarin lexical tone. *Journal of Cognitive Neuroscience,* 15(7), 1019–27.

Weaver, C. (1990). *Understanding whole language.* Portsmouth, NH: Heinemann.

Weaver, C. (1994). Reconceptualizing reading and dyslexia. *Journal of Childhood Communication Disorders,* 16(1), 23–35.

Wheat, K. L., Cornelissen, P. L., Frost, S. J., & Hansen, P. C. (2010). During visual word recognition, phonology is accessed within 100 ms and may be mediated by a speech production code: Evidence from magnetoencephalography. *Journal of Neuroscience,* 30(15), 5229–33.

Wheeler, D.D. (1970). Processes in word recognition. *Cognitive Psychology,* 1, 59–85.

Ziegler, J. C., & Goswami, U. (2005). Reading acquisition, developmental dyslexia, and skilled reading across languages: A psycholinguistic grain size theory. *Psychological Bulletin,* 131(1), 3–29.

Zifcak, M. (1981). Phonological awareness and reading acquisition. *Contemporary Educational Psychology,* 6(2), 117–26.

Chapter 6

Can teachers count on mathematical neurosciences?

Xavier Seron

Overview

Recent advances in cognitive and neuroscience research suggests that humans are equipped very soon after the birth with neural networks that allow us to discriminate and compare numerosities, as well as to anticipate the result of simple transformation of numbers (addition and subtraction). It has been suggested that these non-symbolic numerical and arithmetical abilities are sustained by two separate cognitive systems: an approximate number system able to discriminate and compare large quantities in a rough fashion and an accurate system that operates precisely but only on small numerosities limited to three elements. While some authors consider that the approximate system is the foundation of our symbolic arithmetic abilities, others give a functional priority to the accurate system in the transition from non-symbolic to symbolic arithmetic. Functional neuroanatomy research has identified a set of brain structures involved in arithmetic and highlighted the role of the superior intraparietal sulcus (IPS) as the headquarters of the semantic representation of numbers. The parietal lobe may play a key role in processing of different magnitudes (size, space, time . . .) and it is here that number knowledge relates with spatial processing, representation of finger and hand movements and processing of order.

Knowing that we are equipped with specialized neural systems that deal with numerosities and with other related basic skills can lead educators to focus more attention on those basic numerical abilities and to their connection with space, use of fingers, processing of order and of other magnitudes. However, it should be underlined that the different hypotheses regarding the relationships between numerical processing and these other basic skills are still in their infancy and in the current state of our knowledge are not yet well understood. Thus, before developing educational programmes inspired by the observation of these relationships, we suggest caution and that any modifications of teaching programmes are empirically validated within an evidence-based approach.

6.1 The big ambiguity

Five years ago I was invited to give a talk at a workshop on *neuropédagogie* organized by the Belgian Minister for Education. There were four speakers: an eminent cognitive psychologist who presented the different lexical models of reading and stressed the dangers of promoting any single strategy for teaching reading; the second speaker gave an excellent talk on arithmetical problem-solving by emphasizing some specific working-memory difficulties related to the phrasing of the problem; the third took more than an hour to explain why some Piagetian concepts are still important in mathematical education; then I had to give the last talk, on calculation and number processing. I began by pointing out that my three colleagues had all given excellent talks, but they were cognitive psychologists, and none of them had presented an image of the brain, or any ideas actually based on research in the neurosciences.

At around the same time, a Parisian psychologist introduced a double page spread on la neuropédagogie in a popular French review *Le Monde de l'intelligence* (The World of Intelligence). Neuropédagogie (or neuropedagogy in English) was seen as a new discipline emerging at the confluence of neurosciences and educational sciences, which could be defined as 'a new approach that integrates developmental psychology—from the baby up to the adult—cognitive neurosciences and pedagogy' (Houdé, 2006). It was suggested that neuropedagogy would inform educators about the capacities of and constraints on the learning brain, and that it might help to explain 'why some learning situations are so efficacious whereas others are not'. Neuropedagogy was also presented as a discipline for the future since its programme—not yet realized—would consist of studying 'from an anatomical (aMRI [magnetic resonance imaging]) and a functional (fMRI) point of view the construction of intelligence in the brain and the effects of scholarly teaching in reading, calculation and so on'.

At this period, I asked myself whether we were actually witnessing the birth of a new discipline or if it was a 'naked emperor'. Are there good reasons to think that by looking inside children's brains we will learn more about ways of teaching reading or mathematics? Although I am still not convinced that a new discipline of neuropedagogy or educational neuroscience[1] actually exists, I was quickly convinced that it provided a new and juicy market for training programmes. The French word neuropédagogie produces more than 3000 items on a Google search. On some commercial sites you can load (free-of-charge for a trial period), some representative items of apparently revolutionary exercises called *stimulation corticale gauche* (stimulation of the left cortex), *stimulation corticale droite* (stimulation of the right cortex) and even—thanks to the discovery of emotion by neuroscientists—*stimulation limbique gauche* and *stimulation limbique droite* (stimulation of the left and right limbic system). The Internet is already full of pseudoscientific information about the virtue of neuroscientific approaches in education, echoing old and new neuromyths about the left and right sides of the brain (see also Chapter 13, this volume), the role of sleep in learning and the different types of memory (see also Chapter 7, this volume).

Although I am not convinced that we are actually witnessing the birth of a new discipline, I am willing to concede that the neurosciences are important, and will in the near future become essential for educationalists who wish to develop more efficient teaching methods. A crucial question is what bits of neuroscientific knowledge will be of interest to educators, and precisely what neuroscientific information can influence teaching that is different from the information already collected by cognitive psychologists. If we hope to create a new discipline of educational

---

[1] There is however some indication for the emergence of a new interdisciplinary field connecting neuroscientists and educators. Such recent developments are illustrated by the foundation of Research Centre for Neuroscience in Education at the Universities of Cambridge and London, by the formation of a new International Mind, Brain and Education Society as well as by the development of specific educational courses such as the Graduate Program in Mind, Brain and Education at the Graduate School of Education at Harvard University and an undergraduate programme in Education focused on Education and Neuroscience at Dartmouth College. For a more optimistic view, see De Smedt and colleagues (2010).

neurosciences, we have at least to demonstrate that some knowledge based on the observation of brain activities and structures in relation to cognitive processing adds something to what is known on the basis of cognitive psychological models alone. In this chapter, I will examine the question of the influence of neuroscience on education in the particularly active domain of cognitive arithmetic.

In cognitive neuroscience, numbers and arithmetic have been the objects of intense and fruitful scientific research during the last 20 years. This has resulted in rapid and significant progresses in the description of the processes sustaining our mathematic abilities and their cerebral correlates (Ansari, 2008; Butterworth, 1999; Dehaene, 1997, 2009). On the educational side, numeracy constitutes a crucial domain of knowledge. It can be argued that a level of mathematical competence is a condition for the exercise of effective citizenship in a numerate society (Butterworth, 2005); it constitutes an important vector for the development of technologies and science, which in turn strongly influence the economic level, and the welfare of the population. In the US for example, some alarmist observations in the 1960s, indicated that the achievement scores of children in mathematics at school had begun to decline (Bishop, 1989).[2] The question of the level of mathematical development of US children, compared to Chinese or Japanese children, has been the object of numerous studies (Geary, Salthouse, Chen & Fan, 1996; Miller, Kelly & Zhou, 2005; Peak, 1996). In Europe too, the need to improve children's level of achievement in math is an educational priority. In such a competitive educational context, any scientific progress that could positively influence teaching methods and programme applications is not only welcome but is actively being sought. This may explain why those responsible for education are so interested by any advances in neuroscience.

However before considering the potential impact of neuroscience on educational practices it is first necessary to briefly summarize the advances that have been made in recent years in cognitive neurosciences. To avoid misunderstandings and illusions it is also essential to stress that many questions are still unresolved. I will therefore first present some essential details of cognitive neuroscience research on arithmetic, highlighting its achievements and those (more numerous) areas which remain under discussion.

6.2 **The cognitive neuroscience of arithmetic**

The existence of multiple number-code systems

It is largely accepted by the scientific community that human competence in number processing and in arithmetic rests on different representational systems (Fayol & Seron, 2005). Competencies in precise domains of arithmetic such as calculation or the precise quantification of large numbers are dependent on symbolic representations and are specific to humans. This symbolic knowledge has only developed in numerate cultures and is learned through explicit teaching. As well as this cultural invention, recent research suggests that humans are equipped with more basic numerical abilities that are sustained by two different systems: an approximate number system shared by non-human animals, infants and adults that permits them to discriminate and approximately represent visual and auditory numerosities without verbal counting; and another representational system that facilitates the exact representation of small numerosities (Butterworth, 1999; Dehaene,

---

[2] It was estimated that the poor level of maths and reading achievement costs the US economy $170 billion every year. As a result the development of educational programmes to improve the maths level of children is considered by many governments as an educational priority. As early as 1991, the amelioration of maths achievement and reading was adopted as one of six priorities by the US Department of Education (America 2000: An Education Strategy, US Department of Education, 1991, p. 3).

1992; Feigenson, Dehaene & Spelke, 2004; Nieder & Miller, 2003, 2004; Wynn, 1995). In the approximate system, the representation of numerosities becomes increasingly imprecise as a log function of the target size, with large numerosities represented less precisely than small ones. This fuzzy representation is captured by two behavioural effects: the distance effect (when normal adults have to indicate which of two numerosities (Arabic numerals or sets of dots) are larger, their reaction times and their error rates increase as the distance between the two numbers decreases) and the size effect (when the distance between two numbers is constant, but their absolute size increases, both reaction times and error rates increase) (Moyer & Landauer, 1967; Restle, 1970).

In addition to this approximate representation, there is some evidence that very small numerosities (from 1–3) are processed in different way, called subitizing (Chi & Klahr, 1975; Kaufman, Lord, Reese & Volkman, 1949). When the numerosities 1, 2 and 3 are presented visually they can be quickly and accurately identified and named without the intervention of a serial (and attention-demanding) counting process. The same effect has been found for auditory (Camos & Tillmann, 2008) and tactile (Riggs et al., 2006) presentation. The precise representation of small numerosities involves a distinct representational subsystem called the object-file system (Kahneman, Treisman & Gibbs, 1992; Simon, 1997; Trick & Pylyshyn, 1994). In contrast, to the analogue magnitude, the object file is a limited-capacity system only able to represent sets of three or fewer objects. In the object-file system the numerosity is implicitly represented through a one-to-one correspondence between the items in the world and the object files.

Human adults are thus equipped with two non-symbolic systems, shared with some other animal species: one able to represent large numerosities approximately and the other devoted to the rapid and exact identification of very small numerosities. In numerate populations, two main symbolic numeric systems are also acquired: number words (mainly used for communication) and Arabic numerals (sustaining calculation and arithmetic). The debate over the developmental relationship between the two non-symbolic systems and symbolic arithmetic is still continuing. The dominant view is that our symbolic number system (verbal numerals, Arabic numerals and exact calculation) are rooted in the approximate number system (Dehaene, 1997); a less widespread theory is that the exact non-symbolic representation activated by small numbers is the precursor of our symbolic representations (Carey, 2001). It is, of course, possible that the two systems interact in a way not yet understood.

Neural correlates of the representation of numerosities

On the neuronal side, a large amount of data provides strong evidence that the human brain (and the brains of some animal species) has dedicated neural circuits for numerosity processing and calculation (Ansari, 2008; Dehaene, 2009; Zammarian, Ischebeck & Delazer, 2009). This neural capacity depends of a network of parietal, prefrontal and cingulated neuronal structures. One of the most critical areas, systematically and bilaterally activated when subjects engage in calculation, is the parietal lobe (Chochon, Cohen, van de Moortele & Dehaene, 1999; Dehaene, Spelke, Pinel, Stanescu & Tsivkin, 1999; Fias, Lammertyn, Reynvoet, Dupont & Orban, 2003; Gruber, Indefrey, Steinmetz & Kleinschmidt, 2001; Lee, 2000; Menon, Rivera, White, Glover & Reiss, 2000; Pesenti, Thioux, Seron & De Volder, 2000; Pinel, Dehaene, Riviere & Le Bihan, 2001; Rickard et al., 2000; Simon, Mangin, Cohen, Le Bihan & Dehaene, 2002; Stanescu-Cosson et al., 2000; Zago et al., 2001).

Inside the parietal lobe, the IPS, and in particular its horizontal segment, seems to play a specific role in number representation. For instance, in number-comparison experiments in which participants had to decide if a given quantity was larger or smaller than a standard, the IPS was often the only activated region. IPS activations have also been observed when subjects have to detect Arabic numerals, but not letters or colours, (Eger, Sterzer, Russ, Giraud & Kleinschmidt, 2003) and when numerals were presented subliminally (Naccache & Dehaene, 2001). This suggests that quantity information can be accessed automatically from number symbols. The role of the IPS has been confirmed by fMRI adaptation studies showing a neural adaptation for the repetition of the same numerosity, and a rebound effect for deviant numerosities proportional to the numeric distance between the numbers (Nieder & Miller, 2003). This effect has been observed for concrete numerosities such as dot patterns (Cantlon, Brannon, Carter & Pelphrey, 2006; Piazza, Izard, Pinel, Le Bihan & Dehaene, 2004) as well as for symbolic numbers (Piazza, Pinel, Le Bihan & Dehaene, 2007), suggesting that the IPS coding system for numbers is abstract (but see Cohen Kadosh & Walsh, 2009 and Cohen Kadosh, Muggleton, Silvanto & Walsh, 2010).

Neuroscience had also made important advances in exploring the processes involved in the extraction of numerosities. Any mental representation of a numerosity requires the physical properties of the elements in a set to be ignored (three elephants and three mice must have the same numerosity, whatever their size). Some mechanism is thus required to extract the numerosity from a set of objects or events. This extraction process has been modelled for a visual set of objects (Dehaene & Changeux, 1993; Verguts & Fias, 2004). In the model, the numerosities go through two successive representations. In the first step, visual numerosities activate an object-location map where each neuron signals the presence of an object at a given location, independently of the physical appearance of that object. Since every object is represented by a firing unit, the more objects are presented the more intense is the map firing; at this map level numerosities are thus represented by a number-intensive code. In the second step, activation from the number-intensive maps is converted into a number-selective code in a structure in which some units are 'number selective', i.e. tuned to a specific numerosity.

This model and simulation are supported by brain activation studies. The neurobiological evidence for a numerosity summation coding system has been demonstrated by single-cell recording in the lateral intraparietal area (LIP) of the macaque monkey (Roitman, Brannon & Pratt, 2007). The existence of a number-selective neuron system had previously been observed by two groups of researchers within and near the intraparietal cortex of macaque monkey (Nieder, Freedman & Miller, 2002; Nieder & Miller, 2004; Sawamura, Shima & Tanji, 2002). Recent research in humans has also described two different cerebral pathways of activation for linguistic and non-linguistic numerosities. Concrete numerosities follow a two-stage code analysis (a number-summation code system followed by a number-selective system), while linguistic numerosities activate principally and more directly the number-selective system (Santens, Roggeman, Fias & Verguts, 2009).

The proposal that there is a specific representation for symbolic numbers and exact arithmetic is supported by many other brain-activation studies. There is converging evidence that the left angular (AG) and the left superior (LS) temporal gyri are highly involved in symbolic arithmetic. The AG, whose lesion result in the Gerstmann syndrome (Gerstmann, 1930, 1957) is indeed activated in calculation (Burbaud et al., 1995; Rueckert et al., 1996), is more involved in exact than in approximate calculations[3] (Dehaene, Spelke, Pinel, Stanescu & Tsivkin, 1999; Venkatraman, Siong & Chee, 2006), and is also involved in arithmetical fact retrieval (Delazer et al., 2003; Grabner et al., 2009). While the AG is assumed to help in the retrieval of arithmetical facts, it does not mediate quantity-based operations such as number-comparison estimation, and approximation.

[3] On the contrary, the IPS is more activated in approximate than in exact calculations (Stanescu-Cosson et al., 2000) and more strongly activated during subtraction than multiplication (Chochon et al., 1999; Lee, 2000)

There are also several observations suggesting that some anterior brain structures are activated in arithmetical tasks (Pesenti et al., 2000). In addition, a negative correlation has been observed between the numerical distance and the level of activation in the prefrontal and precentral regions, suggesting that these structures play a role in the treatment of numerical magnitude (Ansari, Garcia, Lucas, Hamon & Dhital, 2005; Pinel et al., 2001).

In summary, a lot of structures in the human brain have been identified as participating in number processing and arithmetic. But, as will be seen later in this chapter, their precise roles and their functional arithmetic specificity, are still under discussion.

The fundamental question of neuroscience: why the parietal lobes?

Activations of regions of the parietal cortex during number processing and arithmetical tasks have generated speculation among neuroscientists about the selection of this structure.[4] The intraparietal cortex is also activated by activities such as grasping, pointing, eye movement, the orientation of attention, and by more general attention- or response-selection mechanisms (Shuman & Kanwisher, 2004). The specificity of IPS activations for numerosities has also been called into question by its involvement in tasks requiring the treatment of other physical dimensions such as size, location, angle and luminance (Cohen Kadosh & Henik, 2006; Cohen Kadosh et al., 2005; Fias et al., 2003; Kaufmann et al., 2005; Pinel, Piazza, Le Bihan & Dehaene, 2004; Zago et al., 2008). Although some IPS regions are activated more strongly by numerical than non-numerical magnitudes, there is a considerable overlap of activations between number and size (Kaufmann et al., 2005; Pinel et al., 2004), and between number and location (Zago et al., 2008). Such overlaps have been interpreted in two different ways: either as indicating the existence, within the IPS, of a sub-system dedicated to the processing of numerosities that is only partially independent of, and is intermingled with other specific magnitude systems (Dehaene, 2009); or by suggesting that the IPS is the seat of a general system dedicated to the evaluation of the magnitude of time, space and numerosities (Walsh, 2003) and that such a magnitude system sustains the key role of the parietal lobes in the control of actions that require the calculation of the size, precise location and number of objects (Andres, Olivier & Badets, 2008; Milner & Goodale, 1995).

The precise role of the IPS in the processing of numbers has also been the subject of controversies. Although most studies have emphasized its role in the processing of the magnitude of numbers, some researchers have stressed its role in the treatment of order. A recent fMRI study showed that both numerical and non-numerical order (such as letters and months) activate areas in the anterior part of the IPS (Fias, Lammertyn, Caessens & Orban, 2007). These data suggest that the IPS could be a critical structure for the representation of order, and thus question the degree to which IPS activations principally reflect magnitude processing. These observations are consistent with the suggestion that the IPS plays a more general role in working memory for order information: IPS activations systematically decrease with increasing inter-item order differences (Marshuetz, Reuter-Lorenz, Smith, Jonides & Noll, 2006; Marshuetz & Smith, 2006).

Although both order and magnitude may be processed in the IPS some data point to subtle difference between the two dimensions. Studies by Turconi and Seron have shown that the ordinality

---

[4] This inquiry was first initiated by Brian Butterworth who asked whether there was any reason for the representation of numbers to be located in a part of the brain that take also controls the hands and fingers and spatial relationships (Butterworth, 1999). The same question was addressed by Stanislas Dehaene, from a phylogenetic perspective. In his original 'neuronal recycling theory' he suggested that a cultural invention (symbolic arithmetic) had colonized this specific structure in the human brain because it was already involved in non-symbolic numerical activities and in spatial transformations (Dehaene & Cohen, 2007).

and cardinality meanings of numbers, although often associated, can be dissociated at a behavioural level. Using event-related potential (ERP) recording, they observed different processing of time-courses in the left and right IPS and in the prefrontal regions (Turconi, Campbell & Seron, 2006; Turconi, Jemel, Rossion & Seron, 2004;). Double dissociation between ordinality and cardinality has also been observed in single case studies of patients with Gerstmann syndrome (Delazer & Butterworth, 1997; Turconi & Seron, 2002).

Number and space

The parietal localization has also been considered in the context of the strong connections that seem to exist between representations of number and space. It has been observed that, when asked to make judgements about the parity of numbers, people respond faster on the left side for small numbers and on the right for large numbers. Dehaene named this the SNARC (spatial number association of response codes) effect (Dehaene, Bossini & Giraux, 1993; Dehaene, Dupoux & Mehler, 1990). Such a spatial bias has been observed in many experimental situations. For example, it has been shown that when numbers are presented in the centre of the visual field, small numbers provoke an orientation of attention to the left, whereas large numbers result in orientation to the right (Fischer, Castel, Dodd & Pratt, 2003). These biases have also been observed with hemineglect patients in number bisection tasks. If asked, for example, for the number midway between 12 and 19, such patients tend to reply 17, as if they neglected the left side of their internal spatial representation of numbers (Zorzi, Priftis & Umilta, 2002). Some normal subjects also indicate that they see numbers in a particular visuospatial representation and this internal structure is generally (but not always) oriented from left to right (see Seron, Pesenti, Noël, Deloche & Cornet, 1992).

Such spatial effects in number processing could be interpreted in the context of the close and often overlapping areas involved in number processing and in the coding of different spatial dimensions such as size, location and gaze direction (Hubbard, Piazza, Pinel & Dehaene, 2005). In an initial interpretation of this spatial bias it has been proposed that numbers are represented as local activations of a mental spatial frame conceptualized as a compressed 'mental number line' with a left-to-right orientation derived from Western reading habits. In such an interpretation, the lateralized bias results from a spatial-compatibility effect between the mental number frame and the response side (Hubbard et al., 2005). Some evidence also suggests that the specific left-to-right orientation of this spatial bias is linked to the cultural environment, and more precisely to the direction of reading. It is reduced and even reversed in right-to-left readers (Dehaene et al., 1993).

The existence of a stored mental spatial structure to represent number is, however, difficult to reconcile with the existence of various contextual effects. For example, if subjects are asked to imagine digits on a clock face, an inverse relation between magnitude and left–right responses is observed (Bächtold, Baumükller & Brugger, 1998). Orienting the response keys along a vertical axis may lead some subjects to demonstrate a vertical bias, with small numbers being responded to faster with the bottom key, and large numbers with the top one (Gevers, Lammertyn, Notebaert, Verguts & Fias, 2006). Furthermore it has been shown in bilingual subjects that they may present opposite spatial biases related to the language used. For instance, Shaki and Fischer (2008) showed that reading either Hebrew or Cyrillic script for just 10 minutes was sufficient to modulate the spatial mapping of numbers. The very same bilingual Russian–Hebrew adults showed a normal SNARC when they performed a parity task after reading a few paragraphs in Cyrillic script (i.e. from left to right) but a significantly smaller (yet still significant) SNARC after reading some paragraphs in Hebrew script (i.e. from right to left). In the same direction these authors were also able to show that the reading of a single word printed in Cyrillic or Hebrew script may suffice to influence the direction of the SNARC in Russian–Hebrew readers asked to judge the parity of numbers. The observation of a modified SNARC for digit classification by the language of the immediately preceding number word constitutes strong evidence for an extremely rapid remapping of numbers in space during reading and indicates that the SNARC is not solely depending on a long-term stored association derived from reading habits (Fisher, Shaki & Cruise, 2009). Hung, Hung, Tzeng and Wu (2008) also demonstrated the influence of context on the SNARC effect, when they showed that the spatial association of numbers is notation-specific, with a horizontal mapping for Arabic number symbols (which most often appear in horizontally written text) and a vertical mapping for Chinese number symbols (which most often appear in vertically written text).

Such variations and contextual influences have led some researchers to argue that the SNARC effect does not tap a fixed long-term stored representation, but reflects a dynamic working-memory representation shaped by the task in hand, and linked to linguistic cueing and response selection strategies (Fias & Fischer, 2005; Fisher, 2006; Seron & Pesenti, 2001). It has also been suggested that the SNARC effect is linked to number order rather than number magnitude, since a weak SNARC effect has also been observed with non-numerical order such as letters or months (Gevers, Reynvoet & Fias, 2003, 2004).

Numbers, fingers and hand movements

The identification of the parietal lobe as a critical area for the representation of numbers has produced a renewal of interest in the well-known association between hands and fingers, and numbers and arithmetic. The existence of a specific association between finger use and numbers has been known for a long time in developmental psychology. Children use their fingers for counting and pointing activities (Fuson, 1988; Gelman & Gallistel, 1978), and in some cultures they also count on other body parts (Ifrah, 1994). Finger use is identifiable in the etymology of some number names in Indo-European languages (Butterworth, 1999), has been used for many centuries has a toolkit for addition and complex multiplication, and could be at the origin of our base-10 number system. The association with the parietal lobe is evidenced by patients suffering from a left brain parietal lesion, who often experience severe acalculia and digit agnosia (Gertsmann syndrome) (Cipolotti, Butterworth & Denes, 1991; Gertsmann, 1930, 1957; Mayer et al., 1999).

The central location of the IPS in the parietal lobe and its dense connectivity with the surrounding areas may create opportunities for the creation of associations between numbers and other spatial and movement-related domains in the course of arithmetical development. Researchers from the Catholic University of Louvain (UCL) in Belgium have observed several interactions between the hand and finger motor systems and numerosities. For example, Andres and colleagues (Andres, Davare, Pesenti, Olivier and Seron (2004) showed that processing the magnitude of Arabic numbers interferes with finger movements, leading to compatibility effects when responses required a grip closing or opening. Grip closing is initiated faster in response to small numbers, whereas grip opening is initiated faster in response to large numbers. Similar results have been reported for object-directed hand grips (Lindemann, Abolafia, Girardi & Bekkering, 2007), where the magnitude of the numerals, although not relevant for the grasping task, interferes with the size of the grip. The effect of number magnitude is higher during the first stages of object-reaching movements and decreases as the hand gets closer to the object (Andres, Ostry, Nicol & Paus, 2008). The influence of number magnitude on actions has also been observed when subjects are not required to execute the action, but are asked to mentally estimate its feasibility (Badets, Andres, Di Luca & Pesenti, 2007). In a more recent study, Badets and Pesenti (2010) observed an inverse relation between actions and numbers: observing a grip closing slows down the processing of large-magnitude numbers. Interestingly this interference is only observed for a biological hand movement (not for an artificial one). These results suggest that the motor system responsible for grasping is endowed with processes that automatically take into account the magnitude of numbers or of the graspable objects.

In addition to these data on grasping, there is other evidence of the relationship between fingers and numbers. At the developmental level, group studies have demonstrated bi-directional links between the presence of digital agnosia and an arithmetical deficit (Kinsbourne & Warrington, 1962, 1963; Rourke, 1993; Strauss & Werner, 1938) as well as in a single case study (Pebenito, 1987). In a longitudinal study, Fayol, Barrouillet and Marinthe (1998) showed that the perceptivo-tactile performances of children at age five are better predictors than their general development scores of subsequent arithmetical performance at 6 and 8 years of age. This result has been confirmed in a study showing that the ability of 6-year-old children to identify the fingers touched by the experimenter is a better predictor of their mathematical skills than standard developmental tests (Noël, 2005). These results suggest a link between perceptivo-tactile skills and the ability to represent and manipulate quantities; they also underline its long-term predictive character.

Recent experiments have suggested that some associations are still present in adults. In time-reaction tasks, it has been shown that when participants were asked to press a key on a computer keyboard in response to Arabic numerals from 1 to 10 using a different finger for each number, they responded faster when the finger assigned to each number matched their usual finger-counting strategy (Di Luca, Grana, Semenza, Seron & Pesenti, 2006). This finding was corroborated by an electrophysiological study showing that the amplitude of the motor twitch induced by transcranial magnetic stimulation (TMS) in a right-hand muscle during verbal judgements on the odd/even status of numbers ranging from 1 to 4 was significant; however no such increase was found during the processing of numbers between 6 and 9 (Sato, Cattaneo, Rizzolatti & Gallese, 2007). All the participants reported starting to count on their right hand, and so these excitability changes have been interpreted as reflecting the existence of overlapping representations for small numbers and the fingers of the right hand.

Fingers are also used to communicate a numerosity non-verbally, and cultural habits about the fixed order in which fingers are raised determine which configurations become canonical and which non-canonical. Various aspects of this activity have been studied by Di Luca and Pesenti (Di Luca, Lefevre & Pesenti, 2010; Di Luca & Pesenti, 2008). They have shown that number naming is faster when participants are presented with canonical finger configurations (i.e. configurations conforming to their personal finger-counting habits) rather than non-canonical ones. This has also been observed with 6-year-olds (Noël, 2005). Di Luca and Pesenti (2008) observed more extended priming in an Arabic-digit comparison task when the digit display was preceded by sub-threshold canonical finger configuration than when the preceding configuration was non-canonical. Finally, using a sophisticated priming design, Di Luca and his colleagues, (2010) showed that canonical and non-canonical configurations activate different semantic representations. The canonical configuration activates place-coding representations such as symbolic numbers, whereas non-canonical configurations behave more like concrete numerosities since they activate a summation code. This result could indicate that canonical finger representations (like number words and Arabic numerals) have a semantic representation in long-term memory.

There are evidently a variety of links between numbers and hand movement and finger postures. Some associations are explicit and related to counting, enumeration or communication activities about numerosities; others, such as those related to grasping movements, seem more implicit or are used to transmit information about continuous quantities.

Brain activations and development

Although cognitive neuroscientists have made substantial progress in the identification of the cerebral networks sustaining number processing and calculation, most such research has been carried out on adults. Information on adults' brains is not directly useful for educators, who are mainly concerned with the development of mathematical cognition and arithmetical learning. However studies addressing the question of the evolution of brain activations in children during mathematical development are now beginning to appear.

Developmental neuroimaging studies of number processing have generally shown that, in comparison to adults, children rely more on the prefrontal regions in tasks involving numerical symbols such as digits (e.g. Ansari, 2008; Kaufmann et al., 2006; Kucian, von Aster, Loenneker, Dietrich & Martin, 2008). As they get older, distinct parietal circuits began to emerge in the bilateral IPS and the left temporo-parietal cortex. For example, in a cross sectional study in which subjects aged from 8 to 19, had to check simple additions and subtractions, an age-related increase in brain activation was found in the left supramarginal gyrus and anterior IPS, as well as in the left lateral temporo-occipital cortex together with a decrease in activation in the frontal brain areas and in some subcortical structures. A very similar shift of activation was observed in an fMRI study comparing the brain activation of children (aged 9–14) and adults (aged 40–49) when they are solving simple one-digit addition, subtraction and multiplication problems (Kawashima et al., 2004). Evidence for a shift from frontal to parietal activations has also been shown in symbolic and non symbolic comparison tasks (Ansari & Dhital, 2006), and in an fMRI adaptation experiment it was observed that deviant numerosities activated the IPS bilaterally in adults but only the right IPS in 4-year-old children (Cantlon et al., 2006). Moreover, an ERP study in 3-month-old babies sensitivity to numerical deviants in a right-lateralized fronto-parietal network of regions. Taken together, these findings indicate that learning and development result in the recruitment of distinct parietal circuits for numerical-magnitude processing and calculation, with specialization of the left temporo-parietal cortex for calculation and the bilateral IPS for numerical-magnitude processing (Izard, Dehaene-Lambertz & Dehaene, 2008).

Brain activations and mathematics learning

Although not directly comparable to the developmental studies, research by Margareth Delazer and her colleagues at Innsbruck with young adults learning new arithmetical topics, has shown the same general tendencies: when subjects first encounter new arithmetical knowledge, important brain activations are observed in the IPS, the inferior parietal lobule and the inferior frontal gyrus; as they progress, stronger activations begin to appear in the left angular gyrus. This shift in brain activations seems to reflect the different learning phases (from initial quantity-based procedures at the early stages, to direct memory retrieval of the correct solution as learning becomes more complete). This shift from frontal to parietal areas illustrates a more general development of learning processes, which change from controlled and executive processing toward task-specific processes. Interestingly the Austrian team also compared the brain activations related to different learning processes to determine whether they are sustained by different brain networks. In a seminal study Delazer and her colleagues (2005) trained adults in two new arithmetical operations: one was learned through the application of an algorithm (strategy condition) while the other was learned by rote (drill condition). After the learning phase, the drill condition resulted in strong activation in the angular gyrus (AG) while the strategy condition activated a larger network. Some indication of a differential representation of operations has also been observed in a study comparing the acquisition of complex subtraction and multiplication (Ischebeck et al., 2006).

However the increase in activations in the AG observed after intensive training is probably not specific to arithmetic since it has been observed in the acquisition of untrained items in both complex multiplication and complex figural-spatial problems (with a larger peak of activation in the left AG for multiplication problems and in the right AG for figural-spatial problems) (Grabner et al., 2009). This indicates that the AG may play a substantial role in connecting problems with their solutions (Ansari, 2008).

6.3 **What is of interest to teachers?**

The cognitive and neuroscience studies summarized in section 6.2 show that neuroscientists have made considerable progress in the domain of numerical cognition. They have identified some of the critical cerebral structures involved in processing concrete numerosities, symbolic numbers and arithmetic, produced a body of research exploring the functional significance of these brain areas and collected data about the changes that occur in them during the learning of symbolic numbers and arithmetic in children and in young adults. These studies are, of course, fascinating in themselves and constitute an important step in our understanding of how the neural circuitry sustains arithmetical cognition and how neural networks are modified during the learning of symbolic arithmetic. I will now turn to the consideration of the impact of this impressive body of research, which is still ongoing, for education. To examine this question it is useful to distinguish three different aspects: the multiple parietal network functions, development and learning studies and developmental dyscalculia.

A parietal localization: so what?

Some parieto-temporal structures, in connection with other brain regions, play a critical role in mathematical cognition. At first sight it may not appear very useful for educators to know which part of the brain is activated by particular cognitive functions. However neuroscientists have not restricted their research to the localization question, they have also examined the many other functions of the parieto-temporal structures. The fact that the IPS and other structures (the pre-central gyrus and the left AG) are activated by numbers, but not only by numbers, has led them to question the functional signification of such overlapping activations. Thus the information produced by neuroscience does not simply concern the location of the neural activations responsible for the processing of numbers; it also indicates that the same brain areas are activated by several tasks involving other cognitive process or movements, which raises questions about the potential significance of these brain activations at a functional level. In the present state of knowledge, there is no consensus on the relations between the processes involved in mathematical cognition and those activated within the same brain structures by other tasks. Several hypotheses—some complementary and some contradictory—have been advanced to explain this phenomenon. Briefly, the main hypotheses are that:

- The parietal lobe (IPS) is a critical structure for the processing of different types of magnitude, including numerosities (Walsh, 2003).
- The IPS is a critical area for the processing of approximate numerosities in babies (Dehaene, 2009), and that the left IPS later becomes the seat of the magnitude representation for symbolic numbers.
- The IPS is a key structure in a network (which includes some anterior areas of the brain) controlling the representation of fingers and movements used in counting and in pointing to objects (Butterworth, 1999; Pesenti et al., 2000).

- The IPS plays a crucial role in the representation of order and is thus also involved in ordinality (Marshuetz & Smith, 2006).

- The role of the IPS in space is important in the development of a spatially-oriented mental number line (Dehaene, 2009).

Note that these proposals are not necessarily contradictory. It could, for example, be that the parietal lobe houses an approximate representation of numerosities and also its connections with some hand-movement and finger-pointing activities. However, the crucial question for the educator is to determine what the relevance of these suggestions for maths education is.

To be of interest to educationalists these neuronal activations must have a functional interpretation. The present state of knowledge in cognitive arithmetical neurosciences falls well short of being able to provide such an interpretation. The observation of overlapping or proximal neural activations does not necessarily imply the existence of a commonality of processing. For instance, overlapping activations in the IPS during the processing of numbers and other continuous magnitudes (such as location and size) have been interpreted in different ways. Some researchers have argued in favour of a general magnitude system (Walsh, 2003); others for the existence of different specialized neuronal systems within the same structure, one dedicated to numerosities and the others to other quantitative parameters (Dehaene, 2009).

We are probably confronted here with the limitations of the large-scale signal and of the present investigation procedures. As rightly pointed out by Nieder (2004):

> ...fMRI detects a continuous mass signal that needs to be smoothed and averaged over a relatively large tissue volume and time frame (at least compared to single-cell recordings) and even across subjects. Several cubic millimeters of brain tissue most likely harbor more than one neural network, and such networks may be specialized but partly overlapping or intermingled (Kleinschmidt, 2004; Pinel et al., 2004) so that they cannot be dissociated spatially or temporally by the relative macroscopic resolution of fMRI (p. 408).

Nieder also stressed that in single cell recordings, only 10–20% of all IPS neurons were concerned with visual numerosity (Nieder & Miller, 2004). This implies that the remaining 80–90% of neurons in this region are involved in other functions. There are still many unknowns about the functional interpretation of these brain activations.

Neuroscience observations do not stand alone; they are regularly extended and supplemented by behavioural data showing interference between the processing of numbers and the processing of the other associated dimensions (space, duration, order, fingers, hand movements, etc.). However, very similar difficulties are emerging at the behavioural level, since the existence of functional commonality of processing cannot be automatically inferred from interference between a numerical and a non-numerical task. For example, the fact that, in some elegant experimental designs, it has been shown that grasping movements and even the representation of movements are influenced by the presentation of a number whose magnitude is not pertinent to the realization of the action, does not necessarily means that the two processes are part of the same functional architecture. Other possibilities have to be considered: for example, it could be that two parallel but different processing interact at the level of a response selection stage; alternatively there could be an optional functional relationship between hand grasping and number magnitude estimation; or the co-activation recorded at the brain level and the interference observed at the behavioural level might constitute a residual trace of a functional relationship that is important at one stage of development, but then loses its functional pertinence.

In summary, recent neuroscience research on numerical cognition, as well as our knowledge of the function of some temporo-parietal structures, lead us to the conclusion that these structures are active in many different tasks: these include numerical activities, as well as various other actions and cognitive domains. These co-activations and behavioural interferences are appealing, but their functional significance needs to be more clearly established. This will require the development of functional hypothesis which are sufficiently precise to permit the development of cognitive models.

Of course, these reservations do not imply that the work done by neuroscientists should be neglected. However, given the many functional indeterminacies, educators have to consider the data collected by neuroscientists simply as starting points for new lines of thinking. If they decide to be more ambitious, and to suggest the introduction of new content or strategies in maths programmes on the basis of some neuroscientific information, educators must remain empirical and compare this new method with those which already exist, so as to investigate their effectiveness empirically.

Mathematical development

The data on mathematical development gathered by neuroscientists are, a priori, more important for education. Neuroscience research in mathematical development has led to the proposal of a dominant scenario that can be schematized as follows: very early in life, when infants are in contact with concrete numerosities, bilateral brain activations are observed in the IPS; later on, when children began to encounter symbolic numbers, the IPS is also the main structure activated, but there is a progressive refinement of its activation, with a lateralization of symbolic numerosities to a left temporo-parietal network. This indicates that the IPS—which is considered to be the seat of approximate number representation—is also a critical brain structure for the acquisition of the symbolic number code system.

This development and refinement of the IPS activations suggests that the innate approximate system for numerosities forms the basis for the construction of our symbolic arithmetical competence. This hypothesis has recently been reinforced by a lot of behavioural data. It has, for example, been shown that the acuity of the non-verbal approximate number system in 14-year-old children correlates with their past scores on standard maths-achievement tests as far back as kindergarten, even when the influence of other cognitive variables (such as lexical access or general intelligence) have been controlled (Halberda, Mazzocco & Feigenson, 2008). Given the retrospective aspect of this study it is, however, not possible to determine the nature of the causal relations: either the efficiency of the approximate system is responsible for the development of the symbolic system; alternatively the mastery of the symbolic system increases the acuity of the approximate number system. This difficulty has been surmounted by recent prospective studies which indicate positive relations between approximate arithmetic and later mathematical achievements. For instance, using a longitudinal design, De Smedt, Verschaffel and Ghesquière (2009) provided evidence that the size of a child's distance effect (as measured by their reaction time) at the start of formal schooling was predictive of their general mathematics achievement score recorded a year later; regression analyses showed that this association was independent of age, intellectual ability and the speed of number identification. A positive relation was also observed between the performances of 5- and 6-year-old children on large number non-symbolic additions at the beginning of their first year at school and on their school's mathematic curriculum at the end of that year. In this study too, the relationship was still present after controlling for literacy achievement and intelligence (Gilmore, McCarthy & Spelke, 2010). Other significant relations between symbolic and non-symbolic arithmetic have been observed in cross-sectional studies (see Holloway & Ansari, 2009). However, these relations seem to be more important at some ages than at others. A recent study by Schneider, Grabner and Paetsch (2009) found no significant association between the distance effect and children's understanding of fractions at 10–11 years of age.

Considered together, these brain imaging data and their behavioural correlations or predictions suggest the existence of a causal relation between the primitive approximate magnitude system and the symbolic arithmetical competence. However, it has to be stressed that the pattern of relationships that have been observed are not easy to understand from a functional perspective. For example, in Gilmore et al.'s study a relation was observed between the approximate addition of *large* numbers and the acquisition of *small* symbolic numerals (number words and Arabic numerals) (Gilmore et al., 2010). De Smedt and his collaborators found a relationship between *the speed of answer* (but not the *acuity* of the distance effect) in approximate number comparisons and the math test achievement (De Smedt et al., 2009). Furthermore, in these studies, the measure of math achievement is an ensemble of composite measures tapping very different numerical abilities such as number knowledge, simple arithmetic, word problems and measurement. Taken together, these data do indeed indicate the existence of some relations between non-symbolic approximations and symbolic arithmetic, but their nature remains mysterious and no functional description of the kind of processes that favour the transition from one competence to the other has ever been proposed.[5] Educators should also be aware that some researchers think it unlikely that the approximate innate performance on large numbers actually forms the basis of the development of exact symbolic arithmetic, which usually begins with the mastery of small numbers: the one-digit Arabic numerals and the first number words.

Whatever the relationships between non-symbolic and symbolic arithmetic, the emphasis of neuroscience research on innate human abilities has produced an increasing interest in the basic mathematical abilities which are already present before children enter elementary school. It is expected—as has already been shown for reading—that the screening of early mathematical abilities may have a predictive value. This could lead to the development of effective support programmes in kindergarten. Diverse screening batteries that have clearly been influenced by recent research in neuroscience have already been created, such as the 'number sense battery' (Jordan, Kaplan, Locuniak & Ramineni, 2007), the Dyscalculia screener (Butterworth, 2003), and the Tedi-Math battery (Van Nieuwhenhoven, Grégoire & Noël, 2001). These look at some very basic mathematical abilities such as verbal counting, the counting of objects, numerical magnitude judgements, simple calculations with concrete tokens, and number words.

Another outcome of the work on calculation undertaken in neuroscience is the renewal of interest in topics already studied in mathematical development that had became progressively less active. The study by Jordan and her colleagues (Jordan, Kaplan, Ramineni & Locuniak, 2008) is particularly illustrative of such a revival. Taking into account the new interest in the use of fingers in calculation, these authors studied the frequency of finger use in children's calculations, and related this to their accuracy scores. In this study the children's performance was tracked longitudinally over 11 time points, from the beginning of kindergarten (mean age = 5.7 years) to the end of second grade (n = 217). The numerous repeated observations over the critical time period allowed the growth (and decline) of the target behaviour to be analysed in relation to changes in levels of accuracy. This work was embedded in a model which allowed the trajectories of different subgroups (gender, high- and low-income groups) to be identified. The resulting growth curves provide a readily accessible and detailed representation of the developmental process. As rightly pointed out by Donlan (2008), such an approach 'reveals significant differences between gender and income groups, observed not simply as outcomes, but as processes of learning, of importance

---

[5] The transition from non-symbolic to symbolic numerosity was simulated in a neural network by Verguts and Fias (2004). Interestingly they observed that training a learning network to numerosity paired with symbols resulted in a transformation of the pre-existing representations which progressively became more precise, and changed from a logarithmic towards a more linear representation.

to educators and theorists alike'. Such studies represent, in my opinion, one of the best ways to use some of the hypotheses emerging from neuroscience as a starting point for educational research.

Dyscalculia

Disorders in the acquisition of numbers and arithmetic are not unusual in children. Approximately 3–6% of schoolchildren suffer from dyscalculia. In the domain of maths cognition, dyscalculia probably presents the best opportunity for research in neuroscience to have a substantial impact on education. It is clearly important at the level of diagnosis and aetiology. Neuroscience research using modern brain imagery methods (fMRI diffusion tensor MRI, anatomical analysis of the microstructural parameters of white-matter organization) in cognitive analysis and genetic, twin and hereditary family studies will surely help to refine the future diagnosis and identification of children at risk of developing specific mathematical learning problems. These developments are considerable and I do not intend to review them here. Let me just say that twin (Alarcon et al., 1997) and family (Shalev et al., 2001) studies already suggest that at least some dyscalculia is highly heritable, and genetic anomalies, such as Turner's syndrome, indicate an important role for genes on the X chromosome (Mazzocco & McCloskey, 2005).

At a functional level, recent research in neuroscience has had considerable influence on the interpretation of the causes of developmental dyscalculia (DD). Until recently, developmental research on DD had been interpreted as indicating that non-numerical factors lay the origin of the disorder: it was, for example, postulated that dyscalculia resulted from working memory limitations (Geary, Brown & Samaranayake, 1991; Hitch & McAuley, 1991; Passolunghi & Siegel, 2001, 2004), visuospatial and phonological deficits (Hecht, Torgensen, Wagner & Rashotte, 2001; Rourke, 1993), or inhibition weaknesses (Barouillet, Fayol & Lathulière, 1997; Passolunghi & Siegel, 2001). The emphasis of recent neuroscience research on innate abilities has produced a re-orientation towards the identification of impaired processing which is specific to the number domain. It has, for example, been suggested that DD may be due to a deficit in an inherited core system for representing numerosities, associated with functional or structural abnormalities in the neural systems that support these abilities (Butterworth, 1999, 2005; Wilson & Dehaene, 2007).

From this viewpoint, the core deficit affects the learner's ability to enumerate sets and to order sets by magnitude, which in turn make it difficult to understand arithmetic, and very hard to produce a meaningful structure for arithmetical facts. There is some evidence for a deficit in DD children in the processing of non-symbolic numerosities in the comparison of small (Mussolin, Mejias & Noël, 2010; Price, Holloway, Räsäsnene, Vesterinen & Ansari, 2007) and large (Piazza et al., 2010) numerosities. This approach is supported by an fMRI study showing that children aged 9–11 with DD do not show the same modulation of the bilateral IPS as a function of the numerical distance between pairs of numbers, as their paired controls (Mussolin et al., 2010). It has also been found that children with DD showed greater interindividual variability and weaker activation for approximate calculation, in almost the entire neuronal network (including the intraparietal sulcus and the middle and inferior frontal gyri of both hemispheres) (Kucian et al., 2006). These data have been interpreted has indicating that dyscalculia is associated with impairment in the brain areas involved in number magnitude processing.

An alternative hypothesis is that DD children do not have difficulties in processing number magnitude, but rather in accessing the numerical magnitude associated with number symbols such as Arabic numerals or number words (Rousselle & Noël, 2007). This proposal is also supported by empirical data showing that 7–8-year-old DD children have a deficit in Arabic number comparison and calculation, but not in tasks requiring them to compare, add or subtract concrete items such as collections of dots (Iuculano, Tang, Hall & Butterworth, 2008; Rousselle & Noël, 2007). There are thus some discrepancies in the empirical data currently available. It could be that the differences are due to the age of the children being studied (Noël & Rousselle, in press) but it is also possible that—as with dyslexia—there are different types of dyscalculia, which have different causes.

The idea that DD is the result of a deficit in an innate number sense can be interpreted in different ways depending on what precisely is considered the 'core' of arithmetic. As rightly pointed out by Dolan (2008, p. 700), '"Number sense" figures large. This phrase can be frustrating. Its power as a metaphor can obscure operational definition'. Some researchers think that it corresponds to the approximate magnitude system for large numbers, while others point to the object-file representation of small numbers. The basic functions to be checked in DD will be rather different, depending on which characteristics of the cognitive system are considered as critical for the development of symbolic mathematical abilities.

There are two good reasons not to disregard the hypothesis that the ability to process small numerosities exactly (the object–file representation) plays an important role in the development of symbolic mathematical abilities: firstly the size of this representation falls into the range of numerosities usually involved in the very first earliest counting behaviour; secondly, given its discrete and sequential structure, it is a good candidate to represent the 'plus one' relation associated with the 'next-word' in counting behaviour. From this perspective, other basic numerical abilities are linked to the object-file representation; abilities such as the quantification of small numerosities and the emergence of counting behaviour have to be considered more basic than the estimation of large numerosities. The role of finger counting also deserves more attention, since it has been reported that children who exhibit impaired performance in the discrimination of their fingers also achieve less well in mathematics (Fayol et al., 1998; Gracia-Bafalluy & Noël, 2008).

Irrespective of the many uncertainties which remain in the interpretation of DD disorders and the possible existence of different types of DD, it is indisputable that recent advances in neuroscience have been an important source of renewal for the methods used to identify DD children and to predict their mathematical achievements in the future. It may appear premature to construct remediation programmes based on neuroscience, but some such programmes have already appeared.

Wilson, Revkin, Cohen and Dehaene (2006) developed 'The Number Race', which is completely computer-based, and aims to enhance the number sense of DD children by training them on numerical comparison tasks. It contains exercises to reinforce the associations between the different number codes, and to emphasize the links between numbers and space. At its highest difficulty level, it includes addition and subtraction exercises. One of the virtues of this approach is that the level of difficulty is adapted to the child's performance. The programme has been trialled with some success on nine DD children, but so far its effectiveness has not been compared to that of other programmes. Furthermore, given the composite nature of the exercises, it is difficult to identify the specific role of the various exercises.

Other programmes constructed around the concept of the number line have also been developed recently. Ramani and Siegler's (2008) approach is to get children to move a token on a horizontal line. In the number version, Arabic numbers (from 1 to 10) are written on the line, and the child has to move the token according to the number that appears on a spinner. In the colour version, used as a control, he or she has to move the token according to a selected colour. Interestingly, when the two versions of the game were used with low-income children, only the children who had played with the number version improved their performance in a series of numerical tasks. Villette and her collaborators have also developed an elegant game also based on the concept of the number line. This programme (Villette, Mawart & Rusinek, 2010, reported by Noël & Rousselle, in press) is intended to improve the links between approximate and precise representations in addition and subtraction. Children are presented with addition and subtraction problems in Arabic numerals, and have to answer by indicating the estimated position of the answer on a horizontal number line. If the estimation is correct (within an accepted range of deviation) the result of the calculation appears on the screen; if the answer is incorrect, the child is asked to suggest a new location and this is repeated until a satisfactory point is chosen. This programme has been shown to be more effective than an exact calculation computer game with DD children aged 10 to 11.

These programmes constitute interesting avenues for the future, but only if considered as experiments, and not as ready-made solutions. Given the present state of our knowledge about the development of mathematical cognition and its impairment in cases of DD, a better understanding of typical and atypical mathematical development is a crucial underpinning for the design of both the mainstream mathematics curriculum and for helping those who fail to keep up. It is particularly important to develop longitudinal studies, to try to capture in a more fine-grained way, the development of some important aspects of basic mathematical cognition. In this respect, the work of Holloway and Ansari (2008), who, in a cross-sectional study, analysed the distance effect in children aged 6–8 years as well as in adult college students, is interesting. They found a main effect of age on the reaction time of the distance effect, and they observed the same general development of accuracy in non-numerical tasks (comparison of brightness and height of bars). This raises the question of the specificity of the distance effect to numerosity estimation: it may be that children who have difficulty estimating numerosities also have difficulty processing other magnitudes.

The work of Siegler and his colleagues is also very interesting. They used tasks where the children had to locate a verbal number on a line segment marked 1 on the left and 100 on the right, and found that the children progressively shifted from a logarithmic to a linear mapping as their experience with numbers increased (Booth & Siegler, 2006; Siegler & Booth, 2004; Siegler & Opfer, 2003). This suggests that maths education can produce a change from an approximate logarithmic feel for numbers to a more precise and linear representation. Jordan et al.'s longitudinal study of the frequency of finger-use in simple arithmetic and its relation to accuracy is also worth mentioning (Jordan et al., 2008). These authors explored the use of fingers according to age, gender and socioeconomic status. The results indicate that the relationship between finger-use and success in calculation changes with age: it is relatively strong and positive in kindergarten (5–6-year-olds); in first (age 6–7) and second grades (age 7–8) the correlations are smaller, but still positive; but by the end of second grade there is a small but significant negative correlation. These data are critical because they suggest that fingers probably play a role in calculation at one stage in development, but later on their use may indicate the presence of difficulties. This suggestion is supported by the frequent observation that dyscalculic adults continue to calculate on their fingers. This kind of longitudinal study may help educators to understand at which ages, and up to which developmental stages, the use of fingers should be encouraged or discouraged. The interest of these developmental approaches resides in the fact that they try to describe the stage at which changes occur.

In summary, neuroscience research on mathematical cognition has re-activated different avenues of research in mathematical development. It has also reinforced the idea that humans are equipped with an innate 'number sense', which is present in animals and infants as well as adults, and is responsible for basic knowledge about numbers and their relationships. This system is active in both children and adults during basic numerical tasks, and corresponds to a cerebral network with the parietal lobe in each hemisphere making a critical contribution. However the term 'number sense' or 'number module' is not easy to operationalize: it covers a range of diverse processes and representations, which are not necessarily identified in the same way by different researchers. The role of the parietal lobe in non-numerical functions had also generated a lot of theories about the associations between such functions and number processes and arithmetic.

It remains difficult to identify the influence of neuroscience on current developments in education. What is actually new, and what is a revisiting of ideas which had already been present in the fields of developmental psychology and education but had been gradually abandoned? For instance, the proposition that humans possess an approximate representation of number magnitude is not new; it can be traced back to the seminal paper by Moyer and Landauer published in *Nature* in 1967. Similarly, the idea that we are equipped with an innate system for processing numerosities was advocated by Gelman and Gallistel in their important book on *Children's Understanding of Number*, published in 1978.

But perhaps it is not really important to identify what is actually new and what is a revival or revisiting of old ideas. The main question is whether we are now in a better position to address the problems of maths education, because of the contribution of neuroscience. It is probably true that cognitive neuroscience is opening up avenues for reflection, but that this does not immediately lead to practical applications in the field of learning. To properly address the problems of learning we need to develop cognitive models that shed light on the mechanisms involved in the critical stages of the mastery of arithmetic (such as the transition from non-symbolic to symbolic calculation, the learning of arithmetical facts and the understanding of fractions).

6.4 **Conclusion**

Finally, I realize at the end of this chapter that I have not answered the question I raised at the beginning: what contribution does the study of brain functions make to educational practice, and what is the contribution of functional and cognitive analysis? It is of course not easy to precisely delineate the origin of an influence. However, in neuropsychology, the cognitive approach to maths cognition began with Deloche and Seron's transcoding model (Deloche & Seron, 1982; Seron & Deloche, 1984) and with McCloskey's classic modular architecture (McCloskey, 1992). These models introduced architectures previously developed in the cognitive domain of reading and lexical processing in the number domain. Concerning number processing, these architectures were effective in describing some important aspects of symbolic arithmetic (transcoding, arithmetic facts retrieval, algorithms of written calculation . . .) and they have identified several patients with specific deficits affecting selectively some subcomponents. The explanatory power of these architectures, however, was limited because they do not cover the treatment of numerosity and the broader non-symbolic arithmetic. We are indebted to Butterworth (1999) and Dehaene (1997), who have broadened the field by linking non-symbolic and symbolic arithmetic. More generally they also tried to integrate data from the observation of babies, acquired and developmental pathology, genetics, functional imagery, development, animals and non-numerate cultures to understand the biological basis of our arithmetical abilities. In this inquiry, it became increasingly clear that the brain structures involved in maths and number processing are also involved in several other functional systems such as attentional control, magnitude estimation, representation of the fingers and representation of space. While the classic functional modular models and the description of single cases of brain-lesioned subjects lead to observed dissociations; the finding in brain functional imagery studies that many brain structures are activated in different tasks and in different cognitive domains points, on the contrary, to the observation of associations between different functional systems. However, from a functional point of view, the interpretation of these many activation overlappings remains to be done. Are we dealing with systems that retain some degree of modularity and cohabiting while remaining functionally separated, or are we dealing with a real resource sharing of the same neural structures involved in different functions depending on the network in which they are inserted? There are a lot of important debates on these issues in neurociences. To some extent at least educators can participate in this discussion by testing the effectiveness of some learning programmes inspired by cognitive neuroscience to enhance the capacities of normally developing children or to help DD children. They can take advantage of the neural overlappings that have been functionally reinforced by behavioural experiments showing the existence of interferences between different cognitive domains—to examine if by training some no numerical functions they have (or not) an effect on arithmetic or number processing. However, given the present state of the art in cognitive neuroscience, it should be done on an exploratory basis, with great caution and with the utmost rigour. This means that they should follow the principles of evidence-based medicine, which are classically used in the health and educational sciences to establish the effectiveness of any intervention: they should create randomized groups, make pre- and post-intervention assessments, measure the long-term effects of the programmes and so on. If they also hope to understand why a programme is effective, they should build programmes with specific exercises rather than using a multidimensional approach.

6.5 **A peculiar future?**

Where will these developments take us in the future? While writing this chapter I had a dream— or perhaps it was a nightmare:

> Mrs Smith enters the Neuropsychomedical Center with her son John aged five. They have received an official summons, and they have to go to Room 247 on the eighth floor. There, a clinical neuroscientist greets them, and tells John to sit down opposite a large screen on a very special chair equipped with various sophisticated transducers; a technician connects these to his skull. After making sure that he is not afraid, different pictures, sounds, smells and tactile stimulations are delivered to John. For some tasks, John has just to be attentive; for others he has to answer orally or by pressing a button. He does his best. After 2 hours of testing (with a few pauses when he gets tired) John and his mother are given an appointment with an advisor in neuro-education. Mrs Smith receives a neuro-cognitive-psychogram of her son, and discovers that he will never be a good painter or writer, but that he can do well in music, mathematics and cooking. He is at risk of developing a strong phobia for snakes, and his inhibition mechanisms in conflict situations are weak. He is therefore recommended to follow, in addition to his normal schooling, a short training course on conflict resolution.

This story is of course a pure fiction. Some readers may fear such a situation, while others will find in it exactly what they hope from the advance of neuroscience. The 'dream' is that with the advent of increasingly sophisticated technology, which could be applied very early in childhood, we will be able to detect children's cognitive and emotional abilities and so help them select appropriate career paths, and, more generally, to develop in harmony with their individual abilities. Some neuroscientists hope that the widespread application of neuroscience predictive technology will allow society to identify the gifted individuals who will be needed to cope with the dramatic situations we will be experience in the near future: pollution, rapid population growth, depletion of natural resources, global warming, etc. Such dreams are not new; Skinner's (1976) proposal that behaviour should be regulated by controlling the interactions between responses and stimuli is another version of the same story. However a child's future is not determined by either his or her brain or environment (Figure 6.1); there is a continual interaction between the brain and the environment. As soon as a child is born, his or her brain comes into contact with the family, the social environment and the culture. Neuroscientists will only be able to make partial and conditional predictions, since development does not follow simple biological paths! But of course, once this interactivity has been recognized, knowledge from neuroscience can be used for the common good!

Fig. 6.1 'And now for the second part of the course, put your Mickey to the right please!' © Emilie Seron, 2011.

References

Alarcon, M., DeFries, J. C., Light, J. G., & Pennington, B. F. (1997). A twin study of mathematics disability. *Journal of Learning Disabilities*, 30, 617–23.

Andres, M., Davare, M., Pesenti, M., Olivier, E., & Seron, X. (2004). Number magnitude and grip aperture interaction. *Neuroreport*, 15, 2773–77.

Andres, M., Olivier, E., & Badets, A. (2008). Actions, words and numbers. *A motor contribution to semantic processing. Current Directions in Psychological Sciences*, 17, 313–317.

Andres, M., Ostry, D.J., Nicol, F., & Paus, T. (2008). Time course of number magnitude interference during grasping. *Cortex*, 44, 414–419.

Ansari, D. (2008). Effects of development and enculturation on number representation in the brain. *Nature Reviews Neurosciences*, 9, 278–91.

Ansari, D., & Dhital, B. (2006). Age-related changes in the activation of the intraparietal sulcus during non symbolic magnitude processing: an event-related functional magnetic resonance imaging study. *Journal of Cognognitive Neuroscience*, 18, 1820–28.

Ansari D., Garcia, N., Lucas, E., Hamon, K., & Dhital, B. (2005). Neural correlates of symbolic number processing in children and adults. *Neuroreport*, 16, 1769–73.

Bächtold, D., Baumüller, M., & Brugger, P. (1998). Stimulus–response compatibility in representational space. *Neuropsychologia*, 36, 731–35.

Badets, A., Andres, M., Di Luca, S., & Pesenti, M. (2007). Number magnitude potentiates action judgements. *Experimental Brain Research*, 180, 525–34.

Badets, A., & Pesenti, M. (2010). Creating number semantics through finger movement perception. *Cognition*, 115(1), 46–53.

Barrouillet, P., Fayol, M., & Lathuliere, E. (1997). Selecting between competitors in multiplications tasks: An explanation of the errors produced by adolescents with Learning difficulties. *International Journal of Behavioral Development*, 21, 253–75.

Bishop, A. J. (1989). *Mathematical Enculturation: a cultural perspective on Mathematics Education.* Dordrecht: Kluver Academic Press.

Booth, J. L., & Siegler, R. S. (2006). Developmental and individual differences in pure numerical estimation. *Developmental Psychology*, 42(1), 189–201.

Burbaud, P., Degreze, P., Lafon, P., Franconi, J. M., Bouligand, B., Bioulac, B., & Allard, M. (1995). Lateralization of prefrontal activation during internal mental calculation: a functional magnetic resonance Imaging study. *Journal of Neurophysiology*, 74, 2194–2200.

Butterworth, B. (1999). *The mathematical brain.* London: Macmillan.

Butterworth, B. (2003). *Dyscalculia screener.* London: NFER-Nelson (software & manual).

Butterworth, B. (2005). The development of arithmetical abilities. *Journal of Child Psychology and Psychiatry*, 46, 3–18.

Camos, V., & Tillmann, B. (2008). Discontinuity in the enumeration of sequentially presented auditory and visual stimuli. *Cognition*, 107, 1135–43.

Cantlon, J. F., Brannon, E. M., Carter, E. J., & Pelphrey, K. A. (2006). Functional imaging of numerical processing in adults and 4-year-old children. *PLoS Biology*, 4(5) e125, 1–11.

Carey, S. (2001). Cognitive foundations of arithmetic: evolution and ontogenesis. *Mind & Language*, 16, 37–55.

Chi, M.T.H., & Klahr, D. (1975). Span and rat of apprehension in children and adults. *Journal of Experimental Child Psychology*, 19, 434–39.

Chochon, F., Cohen, L., van de Moortele, P. F., & Dehaene, S. (1999). Differential contributions of the left and right inferior parietal lobules to number processing. *Journal of Cognitive Neuroscience*, 11(6), 617–30.

Cipolotti, L., Butterworth, B., & Denes, G. (1991). A specific deficit for numbers in a case of dense acalculia. *Brain*, 114(Pt 6), 2619–37.

Cohen Kadosh, R. & Henik, A. (2006). A common representation for semantic and physical properties: a cognitive-anatomical approach. *Experimental Psychology*, 53(2), 87–94.

Cohen Kadosh, R., Henik, A., Rubinstein, O., Mohr, H., Dori, H., Van de Ven, V., et al. (2005). Are numbers special? The comparison system of the human brain investigated by fMRI. *Neuropsychologia*, 43(9), 1238–48.

Cohen Kadosh, R., Muggleton, N., Silvanto, J., & Walsh, V. (2010). Double dissociation of format-dependent and number-specific neurons in human parietal cortex. *Cerebral Cortex*, 20(9), 2166–71.

Cohen Kadosh, R., & Walsh, V. (2009). Numerical representation in the parietal lobes: abstract or not abstract? *Behavioural Brain Sciences*, 32(3–4), 313–28.

De Smedt, B., Ansari, D., Grabner, R. H., Hannula, M. M., Schneider, M., & Verschaffel, L. (2010). Cognitive neuroscience meets mathematics education. *Educational Research Review*, 5, 97–105.

De Smedt, B., Verschaffel, L., & Ghesquière, P. (2009). The predictive value of numerical magnitude comparison for individual differences in mathematics achievement. *Journal of Experimental Child Psychology*, 103, 469–79.

Dehaene, S. (1992). Varieties of numerical abilities. *Cognition*, 44, 1–42.

Dehaene, S. (1997). *The number sense.* New York: Oxford University Press.

Dehaene, S. (2009). Origins of mathematical intuitions. The case of arithmetic. The year in cognitive neurosciences 2009. *Annals of the New York Academy of Science*, 1156, 232–59.

Dehaene, S., Bossini, S., & Giraux, P. (1993). The mental representation of parity and number magnitude. *Journal of Experimental Psychology: General*, 122, 371–96.

Dehaene, S., & Changeux, J. P. (1993). Development of elementary numerical abilities. *Journal of Cognitive Neuroscience*, 5, 390–407.

Dehaene, S., & Cohen, L. (2007). Cultural recycling of cortical maps. *Neuron*, 56, 384–98.

Dehaene, S., Dupoux, E., & Mehler, J. (1990). Is numerical comparison digital? Analogical and symbolic effects in two-digit number comparison. *Journal of Experimental Psychology: Human Perception and Performance,* 16(3), 626–41.

Dehaene, S., Spelke, E., Pinel, P., Stanescu, R., & Tsivkin, S. (1999). Sources of mathematical thinking: behavioral and brain-imaging evidence. *Science,* 284(5416), 970–74.

Delazer, M. & Butterworth, B. (1997). A dissociation of number meanings. *Cognitive Neuropsychology,* 14, 613–36.

Delazer, M., Domahs, F., Bartha, L., Brenneis, C., Lochy, A., Trieb, T., & Benke, T. (2003). Learning complex arithmetic—an fMRI study. *Brain Research Cognitive Brain Research,* 18(1), 76–88.

Delazer, M., Ischebeck, A., Domahs, F., Zamarian, L., Koppelstaetter, F., Siedentopf, C.M., et al. (2005). Learning by strategies and learning by drill—evidence from an fMRI study. *NeuroImage,* 25, 838–49.

Deloche, G., & Seron, X. (1982). From one to 1: An analysis of the transcoding process by means of neuropsychological data. *Cognition,* 12, 119–49.

Di Luca, S., Grana, A., Semenza, C., Seron, X., & Pesenti, M. (2006). Finger-digit compatibility in Arabic numeral processing. *Quarterly Journal of Experimental Psychology*, 59, 1648–63.

Di Luca, S., Lefevre, N., & Pesenti, M. (2010). Place and summation coding for canonical and non canonical finger numeral representations. *Cognition*, 117, 95–100.

Di Luca, S., & Pesenti, M. (2008). Masked priming effect with canonical finger numeral configurations. *Experimental Brain Research*, 185, 27–39.

Dolan, C. (2008). Special section: the development of mathematical cognition: Commentary: Uncovering processes in mathematical cognition. *Developmental Science,* 11, 700–702.

Eger, E., Sterzer, P., Russ, M. O., Giraud, A. L., & Kleinschmidt, A. (2003). A supramodal number representation in human intraparietal cortex. *Neuron,* 37(4), 719–25.

Fayol, M., Barrouillet, P., & Marinthe, C. (1998). Predicting arithmetical achievement from neuropsychological performance: A longitudinal study. *Cognition,* 68, 63–70.

Fayol, M., & Seron, X. (2005). About numerical representations: Insights from neuropsychological, experimental and developmental studies. In J. I. D. Campbell (Ed.), *Handbook of mathematical cognition* (pp. 3–22). New York: Psychology Press.

Feigenson, L., Dehaene, S., & Spelke, E. (2004). Core systems of number. *Trends in Cognitive Sciences*, 8, 307–314.

Fias, W., & Fischer, M. H. (2005). Spatial representation of numbers. In J. I. D. Campbell (Ed.), *Handbook of mathematical cognition* (pp. 43–54). New York: Psychology Press.

Fias, W., Lammertyn, J., Caessens, B., & Orban, G.A. (2007). Processing of abstract ordinal knowledge in the horizontal segment of the intraparietal lobe. *Journal of Neurosciences*, 27, 8952–56.

Fias, W., Lammertyn, J., Reynvoet, B., Dupont, P., & Orban, G. A. (2003). Parietal representation of symbolic and nonsymbolic magnitude. *Journal of Cognitive Neuroscience,* 15(1), 47–56.

Fischer, M. H. (2006). The future for SNARC could be stark. *Cortex*, 42(8), 1066–68.

Fischer, M. H., Castel, A.D., Dodd, M.D., & Pratt, J. (2003). Perceiving numbers causes spatial shifts of attention. *Nature Neurosciences*, 6, 555–56.

Fisher, M. H., Shaki, S., & Cruise, A. (2009). It takes only one word to quash the SNARC. *Experimental Psychology*, 56(5), 361–66.

Fuson, K. S. (1988). *Children's counting and concepts of numbers.* New York: Springer Verlag.

Geary, D.C., Brown, S.C., & Samaranayake, V.A. (1991). Cognitive addition: A short longitudinal study of strategy choice and speed-of-processing differences in normal and mathematically disabled children. *Developmental Psychology*, 27, 787–97.

Geary, D.C., Salthouse, T.I.A., Chen, G.P. & Fan, L. (1996). Are east Asian versus American differences in arithmetical ability a recent phenomenon? *Developmental Psychology*, 32, 254–62.

Gelman, R., & Gallistel, C. R. (1978). *The child's understanding of number*. Cambridge, MA: Harvard University Press.

Gerstmann, J. (1930). Zur symptomatologie der hirnläsionen im übergangsgebiet der unteren parietal-und mittleren occipitalwindung. *Nervenarzt*, 3, 691–95.

Gerstmann, J. (1957). Some notes on the Gerstmann syndrome. *Neurology*, 7(12), 866–69.

Gevers, W., Lammertyn, J., Notebaert, W., Verguts, T., & Fias, W. (2006). Automatic response activation of implicit spatial information: Evidence for the SNARC effect. *Acta Psychologica*, 122, 221–33.

Gevers, W., Reynvoet, B., & Fias, W. (2003). The mental representation of ordinal sequence is spatially organized. *Cognition*, 87, 87–95.

Gevers, W., Reynvoet, B., & Fias, W. (2004). The mental representation of ordinal sequence is spatially organized: evidence from days of the week. *Cortex*, 40, 171–72.

Gilmore, C. K., McCarthy, S. E., & Spelke, E. S. (2010). Non-symbolic arithmetic abilities and mathematics achievement in the first year of formal schooling. *Cognition*, 115, 394–406.

Grabner, R. H., Ischebeck, A., Reishofer, G., Koschutnig, K., Delazer, M., Ebner, F., & Neuper, C. (2009). Fact learning in complex arithmetic and figural-spatial tasks: the role of the angular gyrus and its relation to mathematical competence. *Human Brain Mapping*, 30(9), 2936–52.

Gracia-Baffaluy, M., & Noël, M. P. (2008). Does finger training increase young children's numerical performance? *Cortex*, 44, 368–75.

Gruber, O., Indefrey, P., Steinmetz, H., & Kleinschmidt, A. (2001). Dissociating neural correlates of cognitive components in mental calculation. *Cerebral Cortex*, 11(4), 350–59.

Halberda, J., Mazzocco, M.M., & Feigenson, L. (2008). Individual differences in non verbal number acuity correlate with maths achievement. *Nature*, 455, 665–68.

Hecht, S. A., Torgesen, J. K., Wagner, R. K., & Rashotte, C. A. (2001). The relations between phonological processing abilities and emerging individual differences in mathematical computation skills: A longitudinal study from second to fifth grades. *Journal of Experimental Child Psychology*, 79, 192–227.

Hitch, G. J., & McAuley, E. (1991). Working memory in children with specific arithmetical learning difficulties. *British Journal of Psychology*, 82, 375–86.

Holloway, I. D., & Ansari, D. (2008). Domain-specific and domain-general changes in children's development of number comparison. *Developmental Science*, 1, 644–49.

Holloway, I. D., & Ansari, D. (2009). Mapping numerical magnitudes onto symbols: The numerical distance effect and individual differences in children's mathematics achievement. *Journal of Experimental Child Psychology*, 103, 17–29.

Houdé, O. (2006). Neuropédagogie? Entretien pour *Le Monde de l'Intelligence*, 3, 24–25.

Hubbard, E. M., Piazza, M., Pinel, P., & Dehaene, S. (2005). Interactions between number and space in parietal cortex. *Nature Review, Neuroscience*, 6(6), 435–48.

Hung, Y., Hung, D. L., Tzeng, O. J. L., & Wu, D. H. (2008). Flexible spatial mapping of different notations of numbers in Chinese readers. *Cognition*, 106(3), 1441–50.

Ifrah, G. (1994). *Histoire universelle des chiffres*, Vol I et II. Paris: Robert Laffont.

Ischebeck, A., Zamarian, L., Siedentopf, C., Koppelstätter, F., Benke, T., Felber, S., & Delazer, M. (2006). How specifically do we learn? Imaging the learning of multiplication and subtraction. *NeuroImage*, 30, 1365–75.

Iuculano, T., Tang, J., Hall, C. W. B. & Butterworth, B. (2008). Core information processing deficits in developmental dyscalculia and low numeracy. *Developmental Science*, 11, 669–80.

Izard, V., Dehaene-Lambertz, G., & Dehaene, S. (2008). Distinct cerebral pathways for object identity and number in human infants. *PLoS Biology*, 6, e11.

Jordan, N. C., Kaplan, D., Locuniak, M. N., & Ramineni, C. (2007). Predicting first-grade math achievement from developmental number sense trajectories. *Learning Disabilities Research & Practice*, 22(1), 36–46.

Jordan, N. C., Kaplan, D., Ramineni, C., & Locuniak, M. N. (2008). Development of combination skill in the early school years: when do fingers help? *Developmental Science*, 11, 662–68.

Kahneman, D., Treisman, A., & Gibbs, B. J. (1992). The reviewing of object-files: object specific integration of information. *Cognitive Psychology*, 24, 174–219.

Kaufmann, L., Koppelstaetter, F., Delazer, M., Siedentopf, C., Rhomberg, P., Golaszewski, S., et al. (2005). Neural correlates of distance and congruity effects in a numerical Stroop task: an event-related fMRI study. *Neuroimage*, 25(3), 888–98.

Kaufmann, L., Koppelstaetter, F., Siedentopf, C., Haala, I., Haberlandt, E., Zimmerhackl, L.B., Felber, S., & Ischebeck, A. (2006). Neural correlates of the number-size interference task in children. *Neuroreport*, 17, 587–91.

Kaufman, E. L., Lord, M. W., Reese, T. W., & Volkman, J. (1949). The discrimination of visual number. *American Journal of Psychophysics*, 18, 373–78.

Kawashima, R., Taira, M., Okita, K., Inoue, K., Tajima, N., Yoshida, H., et al. (2004). A functional MRI study of simple arithmetic—a comparison between children and adults. *Brain Research. Cognitive. Brain Research*, 18, 227–33.

Kinsbourne, M., & Warrington, E. K. (1962). A study of finger agnosia. *Brain*, 85, 47–66.

Kinsbourne, M., & Warrington, E. K. (1963). The developmental Gerstmann syndrome. *Archives of Neurology*, 8, 490–501.

Kleinschmidt, A. (2004). Thinking big: many modules or much cortex. *Neuron*, 41(6), 842–44.

Kucian, K., Loenneker, T., Dietrich, T., Dosch, M., Martin, E., & von Aster, M. (2006). Impaired neural networks for approximate calculation in dyscalculic children: a functional MRI study. *Behavioral and Brain Functions*, 2, 31.

Kucian, K., von Aster, M., Loenneker, T., Dietrich, T., & Martin, E. (2008). Development of neural networks for exact and approximate calculation: a FMRI study. *Developmental Neuropsychology*, 33, 447–73.

Lee, K. M. (2000). Cortical areas differentially involved in multiplication and subtraction: a functional magnetic resonance imaging study and correlation with a case of selective acalculia. *Annals of Neurology*, 48(4), 657–61.

Lindemann, O., Abolafia, J. M., Girardi, G., & Bekkering, H. (2007). Getting a grip on numbers: Numerical magnitude priming in object grasping. *Journal of Experimental Psychology: Human Perception and Performance*, 33, 1400–1409.

Marshuetz, C., Reuter-Lorenz, P. A., Smith, E. E., Jonides, J., & Noll, D. C. (2006). Working memory for order and the parietal cortex: an event related functional magnetic resonance imaging study. *Neuroscience*, 139, 311–316.

Marshuetz, C., & Smith, E. E. (2006). Working memory for order information: multiple cognitive and neural mechanisms. *Neuroscience*, 139, 195–200.

Mayer, E., Martory, M. D., Pegna, A. J., Landis, T., Delavelle, J., & Annoni, J. M. (1999). A pure case of Gerstmann syndrome with a subangular lesion. *Brain*, 122(6), 1107–20.

Mazzocco, M. M. M., & McCloskey, M. (2005). Math performance in girls with Turner or fragile X syndrome. In J. I. D. Campbell (Ed.), *Handbook of Mathematical Cognition* (pp. 269–97). New York: Psychology Press.

McCloskey, M. (1992). Cognitive mechanisms in numerical processing: Evidence from acquired dyscalculia. *Cognition*, 44, 107–57.

Menon, V., Rivera, S. M., White, C. D., Glover, G. H., & Reiss, A. L. (2000). Dissociating prefrontal and parietal cortex activation during arithmetic processing. *Neuroimage*, 12(4), 357–65.

Miller, K.F., Kelly, M. & Zhou, X. (2005). Learning mathematics in China and the United States. cross-cultural insights into the nature and course of preschool mathematical development. In J.I.D. Campbell (Ed.). *Handbook of Mathematical Cognition* (pp. 163–78). New York: Psychology Press.

Milner, A. D., & Goodale, M. A. (1995). *The visual brain in action*. Oxford: Oxford University Press.

Moyer, R.S., & Landauer, K. (1967). Time required for judgements of numerical inequality. *Nature*, 215, 1519–20.

Mussolin, C., De Volder, A., Grandin, C., Schlögel, X., Nassogne, M-C., & Noël, M-P. (2010). Neural correlates of symbolic number comparison in developmental dyscalculia. *Journal of Cognitive Neuroscience*, 22(5), 860–74.

Mussolin, C., Mejias, S., & Noël, M.P. (2010). Symbolic and non-symbolic number comparison in children with and whithout dyscalculia. *Cognition*, 115, 10–25.

Naccache, L., & Dehaene, S. (2001). The priming method: imaging unconscious repetition priming reveals an abstract representation of number in the parietal lobes. *Cerebral Cortex*, 11(10), 966–74.

Nieder, A. (2004). The number domain—Can we count on parietal cortex? *Neuron*, 44(3), 407–408.

Nieder, A., Freedman, D. J., & Miller, E. K. (2002). Representation of the quantity of visual items in the primate prefrontal cortex. *Science*, 297(5587), 1708–1711.

Nieder, A., & Miller, E. K. (2003). Coding of cognitive magnitude: compressed scaling of numerical information in the primate prefrontal cortex. *Neuron*, 37(1), 149–57.

Nieder, A., & Miller, E.K. (2004). A parieto-frontal network for visual numerical information in the monkey. *Proceedings of the National Academy of Sciences*, 101, 7457–62.

Noël, M.P. (2005). Finger gnosia: a predictor of numerical abilities in children? *Child Neuropsychology*, 11, 413–30.

Noël, M.P., & Rousselle, L. (in press). How understanding math learning disabilities could guide math teaching, *British Journal of Educational Psychology*, Conference Monographs Series.

Passolunghi, M. C., & Siegel, L. S. (2001). Short term memory, working memory, and inhibitory control in children with specific arithmetic learning disabilities. *Journal of Experimental Child Psychology*, 80, 44–57.

Passolunghi, M. C., & Siegel, L. S. (2004). Working memory and access to numerical information in children with disability in mathematics. *Journal of Experimental Child Psychology*, 88, 348–67.

Peak, L. (1996). *Pursuing excellence: A study of US eighth-grade mathematics and science achievement in international context*. Washington, DC: US Government Printing Office.

Pebenito, R. (1987). Developmental Gerstmann syndrome: Case report and review of the literature. *Developmental and Behavioral Pediatrics*, 8, 229–32.

Piazza, M., Facoetti, A., Trussardi, A.M., Berteletti, I., Conte, S., Lucangeli, D., . . .Zorzi, M. (2010). Developmental trajectory of number acuity reveals a severe impairment in developmental dyscalculia. *Cognition*, 116(1), 33–41.

Piazza, M., Izard, V., Pinel, P., Le Bihan, D., & Dehaene, S. (2004). Tuning curves for approximate numerosity in the human intraparietal sulcus. *Neuron*, 44(3), 547–55.

Piazza, M., Pinel, P., Le Bihan, D., & Dehaene, S. (2007). A magnitude code common to numerosities and number symbols in human intraparietal cortex. *Neuron*, 53, 293–305.

Pesenti, M., Thioux, M., Seron, X., & De Volder, A. (2000). Neuroanatomical substrates of Arabic number processing, numerical comparison, and simple addition: a PET study. *Journal of Cognitive Neuroscience*, 12(3), 461–79.

Pinel, P., Dehaene, S., Riviere, D., & LeBihan, D. (2001). Modulation of parietal activation by semantic distance in a number comparison task. *Neuroimage*, 14(5), 1013–26.

Pinel, P., Piazza, M., Le Bihan, D., & Dehaene, S. (2004). Distributed and overlapping cerebral representations of number, size, and luminance during comparative judgments. *Neuron*, 41(6), 983–93.

Price, G., Holloway, I., Räsäsnene, P., Vesterinen, M., & Ansari, D. (2007). Impaired parietal magnitude processing in developmental dyscalculia. *Current Biology*, 17, R1042–1043.

Ramani, G. B., & Siegler, R. S. (2008). Promoting broad and stable improvements in low-income children's numerical knowledge through playing number board games. *Child Development*, 79(2), 375–94.

Restle, F. (1970). Speed of adding and comparing numbers. *Journal of Experimental Psychology*, 83, 274–78.

Rickard, T. C., Romero, S. G., Basso, G., Wharton, C., Flitman, S., & Grafman, J. (2000). The calculating brain: an fMRI study. *Neuropsychologia*, 38(3), 325–35.

Riggs, K. J., Ferrand, L., Lancelin, D., Fryziel, L., Dumur, G., & Simpson, A. (2006). Subitizing in tactile perception. *Psychological Science, 17,* 271–72.

Roitman, J. D., Brannon, E. M., & Platt, M. L. (2007). Monotonic coding of numerosity in macaque lateral intraparietal area. *PLoS Biology,* 5(8), e208.

Rourke, B. P. (1993). Arithmetic disabilities, specific and otherwise: A neuropsychological perspective. *Journal of Learning Disabilities,* 26, 214–26.

Rousselle, L. & Noël, M-P. (2007). Basic numerical skills in children with mathematics learning disabilities: A comparison of symbolic versus nonsymbolic number magnitude processing. *Cognition,* 102, 361–95.

Rueckert, L., Lange, N., Partiot, A., Appolonio, I., Litvan, I., Le Bihan, D., & Grafman, J. (1996). Visualizing cortical activation during mental calculation with functional MRI. *NeuroImage,* 3, 97–103.

Santens, S., Roggeman, C., Fias, W., & Verguts, T. (2009). Number processing pathways in human parietal cortex. *Cerebral Cortex,* 20, 77–88.

Sato, M., Cattaneo, L., Rizzolatti, G., & Gallese, V. (2007). Numbers within our hands: Modulation of corticospinal excitability of hand muscles during numerical judgment. *Journal of Cognitive Neuroscience,* 19, 684–93.

Sawamura, H., Shima, K., & Tanji, J. (2002). Numerical representation for action in the parietal cortex of the monkey. *Nature,* 415(6874), 918–22.

Schneider, M., Grabner, R. H., & Paetsch, J. (2009). Mental number line, number line estimation, and mathematical achievement: Their interrelations in Grades 5 and 6. *Journal of Educational Psychology,* 101, 359–72.

Seron, X., & Deloche, G. (1984). From 2 to two: An analysis of a transcoding process by means of neuropsychological evidence. *Journal of Psycholinguistic Research,* 13(3), 215–35.

Seron, X., & Pesenti, M. (2001). The number sense theory needs more empirical evidence. *Mind and Language,* 16(1), 76–88.

Seron, X., Pesenti, M., Noel, M. P., Deloche, G., & Cornet, J. A. (1992). Images of numbers, or "When 98 is upper left and 6 sky blue". *Cognition,* 44(1–2), 159–96.

Shaki, S., & Fischer, M.H. (2008). Reading space into numbers—a cross-linguistic comparison of the SNARC effect. *Cognition,* 108, 590–99.

Shalev, R. S., Manor, O., Kerem, B., Ayali, M., Badichi, N., Friedlander, Y., & Gross-Tsur, V. (2005). Developmental dyscalculia: a prospective six-year follow-up. *Developmental Medicine and Child Neurology,* 47, 121–25.

Siegler, R. S., & Booth, J. L., (2004). Development of numerical estimation in young children. *Child Development,* 75(2), 428–44.

Siegler, R. S., & Opfer, J. E., (2003). The development of numerical estimation: evidence for multiple representations of numerical quantity. *Psychological Science,* 14(3), 237–43.

Shuman, M., & Kanwisher, N. (2004). Numerical magnitude in the human parietal lobe; tests of representational generality and domain specificity. *Neuron,* 44, 557–69.

Simon, T. J. (1997). Reconceptualizing the origins of number knowledge: A 'non-numerical' account. *Cognitive Development,* 12, 349–72.

Simon, O., Mangin, J. F., Cohen, L., Le Bihan, D., & Dehaene, S. (2002). Topographical layout of hand, eye, calculation, and language-related areas in the human parietal lobe. *Neuron,* 33(3), 475–87.

Skinner, B.F. (1976). *Walden Two.* Revised edition. Indianapolis: Hackett Publishing Company.

Stanescu-Cosson, R., Pinel, P., van De Moortele, P. F., Le Bihan, D., Cohen, L., & Dehaene, S. (2000). Understanding dissociations in dyscalculia: a brain imaging study of the impact of number size on the cerebral networks for exact and approximate calculation. *Brain,* 123(11), 2240–55.

Strauss, A., & Werner, H. (1938). Deficiency in the finger schema in relation to arithmetic disability. *The American Journal of Orthopsychiatry,* 8, 719–25.

Trick, L.M., & Pylyshyn, Z.W. (1994). Why are small and large numbers enumerated differently? A limited-capacity preattentive stage in vision. *Psychological Review,* 101(1), 80–102.

Turconi, E., Campbell, J. I., & Seron, X. (2006). Numerical order and quantity processing in number comparison. *Cognition*, 98, 273–85.

Turconi, E., Jemel, B., Rossion, B., & Seron, X. (2004). Electrophysiological evidence for differential processing of numerical quantity and order in humans. *Brain Research Cognitive Brain Research*, 21(1), 22–38.

Turconi, E., & Seron, X. (2002). Dissociation between quantity and order meanings in a patient with Gerstmann syndrome. *Cortex*, 38, 911–914.

Van Nieuwhenhoven, C., Grégoire, J., & Noël, M-P. (2001). *Tedi-math: test diagnostique des compétences de base en mathématiques.* Paris: Editions du centre de psychologie appliqué.

Venkatraman, V., Siong, S. C., & Chee, M. W. V. (2006). Effect of language switching on arithmetic a bilingual FMRI study. *Journal of Cognitive Neuroscience*, 18, 64–74.

Verguts, T., & Fias, W. (2004). Representation of number in animals and humans: a neural model. *Journal of Cognitive Neuroscience*, 16(9), 1493–1504.

Walsh, V. (2003). A theory of magnitude: common cortical metrics of time, space and quantity. *Trends in Cognitive Sciences*, 7, 483–88.

Wilson, A. J., & Dehaene, S. (2007). Number sense and developmental dyscalculia. In D. Coch, G., Dawson, & K. Fisher (Eds.), *Human behavior, learning, and the developing brain: Atypical development* (2nd ed., pp. 212–37). New York: Guilford Press.

Wilson, A. J., Revkin, S. K., Cohen, L., & Dehaene, S. (2006). An open trial assessment of 'the number race', an adaptive computer game for remediation of dyscalculia. *Behavioral and Brain Functions*, 2, 1–16.

Wynn K. (1995). Infants possess a system of numerical knowledge. *Current Directions in Psychological Science*, 4, 172–76.

Zago, L., Pesenti, M., Mellet, E., Crivello, F., Mazoyer, B., & Tzourio-Mazoyer, N. (2001). Neural correlates of simple and complex mental calculation. *Neuroimage*, 13(2), 314–27.

Zago, l., Petit, L., Tuberlin, M. R., Andersson, F., Vigneau, M., & Tzourio-Mazoyer, N. (2008). How verbal and spatial manipulation networks contribute to calculation: An fRMI study. *Neuropsychologia*, 46(9), 2403–2414.

Zammarian, L., Ischebeck, A., & Delazer, M. (2009). Neuroscience of larning arithmetic—Evidence from brain imaging studies. *Neurosciences and Biobehavioral Reviews*, 33, 909–25.

Zorzi, M., Priftis, K., & Umilta, C. (2002). Brain damage: neglect disrupts the mental number line. *Nature*, 417, 138–39.

Chapter 7

Working memory: the seat of learning and comprehension

Nelson Cowan

Overview

Working memory is the small amount of information kept in mind at any time. It is needed for various sorts of learning, comprehension, problem-solving and goal-directed thinking. Humans have a small working memory limit but it can be overcome using strategies such as grouping items together and rehearsing them. Children's working memory capabilities grow with maturity, and educational practices should be based on an understanding of both the limitations and the educational possibilities. These limitations of working memory, and means to overcome the limitations, will be examined with emphasis on the developmental changes from the early elementary school years through adulthood. Educational principles will be proposed to make the most of working memory for optimal learning throughout development.

The first two sections of the chapter provide a definition of working memory and a description of how it is used in information processing. This is followed by a discussion of several types of working memory limitations (item limits, goal maintenance limits and time limits) and the strategies that are used to overcome the limits. Within this framework, the childhood development of working memory is examined. To gain a complete picture of development one must separately consider the growth of storage capacity, goal maintenance processes, memory persistence and strategy use. Finally, lessons for maximizing education are presented, focusing on the use of materials and instructions that are challenging and stimulating but not beyond the working memory capabilities of the child.

7.1 Introduction

Working memory is the temporary retention of a small amount of information, which is used in practically every kind of cognitive task. It is critical for learning and comprehension, and it has become a major area of brain research (e.g. Baddeley, 2003; Chein & Fiez, 2010; Cowan, 1995, 1999, 2009; D'Esposito, Postle, Jonides & Smith, 1999; Jonides et al., 2008; Majerus et al., 2010; McNabb & Klingberg, 2008; Todd & Marois, 2004; Xu & Chun, 2006). The present review, however, summarizes the behavioural research on working memory. It includes a more detailed discussion of what working memory is and how it is used; limitations in working memory, and how one can use strategies to overcome them; the childhood development of working memory abilities, in which the basic abilities increase along with increasing uses of strategies to overcome the limits; and suggestions for how this information about working memory can be used to improve educational practices (see also Chapter 8, this volume).

7.2 **Working memory and how it is used**

Back in the days of the first few laboratories of psychological experimentation, Wilhelm Wundt was extensively studying conscious perception and thought in Leipzig, Germany. His work is still germane to modern psychology but much of it has still not been translated into English (only into Hungarian and Russian, the other two leading academic languages of the time and locale). William James, working from Harvard University, summarized the gist of much of the early research by Wundt and others in his 1890 text, *Principles of Psychology*. He noted two basic kinds of memory, primary and secondary. Primary memory referred to a lingering residue of our consciousness of recent events; that is, conscious awareness of what is and what recently was. In contrast, secondary memory referred to the nearly limitless amount of information stored away in the brain from a lifetime of experience. Primary memory is the core of what we now call working memory.

American psychology went into a long era of concentrating on the lawful relationship between stimuli and responses, with little use for speculation about what happened in the brain or mind between those points. This attitude was furthered by what is known as Watson's (1913) behaviourist manifesto, and it continued until the cognitive revolution that began in the middle 1950s (Gardner, 1985). As part of that cognitive revolution, George Miller (1956) wrote a seminal review paper on the basis of his own research in several areas, which did suggest that we can use experiments on behaviour to make important inferences about the properties of the mind and brain. He noted that there are severe limits in humans' ability to process information.

In some ways, the limits described by Miller (1956) resemble channel capacities of electronic communication devices that were studied in depth in the process of fighting the First and Second World Wars. For example, one cannot transmit a message as efficiently over a telegraph as one can over a radio because any one telegraph signal carries only a limited amount of information; a single dash or a dot only narrows down the choices for the letter that is about to be conveyed. Somewhat analogously, humans can identify a simple property such as the length of a line or the loudness of a tone only in a limited fashion: they can identify it out of about seven distinct choices, but not much more. The transmission of information is limited by the properties of the electronic device or human mind.

In a critical way, though, Miller (1956) also found that the human mind has a means to leap beyond the basic limits. If one is asked to repeat a list of, say, random letters, one can do so if there are at most about seven letters. If, however, one can capitalize on past knowledge, the letters can be organized into larger chunks in order to reduce the load on primary memory. Suppose a list of nine letters forms three well-known initialisms, as in the series *IBM, CIA, FBI*. Then it becomes an easy matter to retain the letters. So, the human mind is a device that organizes information and thereby allows much more efficient transmission or storage of information, compared to basic electronic devices.

The limits of working memory and the ability to overcome these limits both have profound implications for how learning is accomplished. A new concept requires that memory be put to use to form new amalgams. Take the concept of a fraction. The ratio 1:2 is the same as the ratio 3:6 and both pairs of numbers form an equivalent fraction, with a value of 0.5. When a child first endeavours to understand the concept of a fraction, the data that must be kept in mind (in working memory) include the first number, the second number and the comparative magnitude of the two (cf. Hecht, Vagi & Torgesen, 2007). If the child loses sight of that principle, he or she might be swayed by the magnitude of one of the numbers rather than the comparative magnitude.

The formation of new ideas requires that old ideas be combined, metaphorically speaking, in the working memory cauldron.

Further thoughts into how the mind uses information were articulated in a short book by Miller, Galanter and Pribram (1960), who appear to have coined the term working memory. They noted that human activities must be planned and executed on many levels. In order to get to school in the morning, a child must think of all the activities that need to be finished first. These may include collecting homework and books, getting dressed, eating breakfast, brushing teeth and so on. Any one of these activities requires further thought, such as remembering where the homework was placed the night before, where the backpack was placed, perhaps a check as to whether the name is on the homework and so on. Miller and colleagues proposed that information about the overall plan (getting to school) and about subplans as they are enacted (loading the backpack or eating breakfast) are kept in a working memory as needed. One would test to see if the task was completed, at which point the details could be released from working memory, freeing it up for other parts of the overall plan. In this discussion, working memory was not considered a certain area in the brain or very specific mechanism, but rather the ensemble of whatever mental devices one can use to hold the information as needed for success in planning and carrying out daily activities.

In subsequent years, the concept of working memory was developed further. Atkinson and Shiffrin (1968) did not use that exact term, but they provided a seminal discussion of a short-term memory resembling James' (1890) primary memory. One important contribution they made was a careful consideration of how strategies might be used to enhance the limited amount of information in short-term memory. They suggested that the individual largely controls his or her brain's own flow of information from sensory memory to short-term memory to long-term memory, and back again (e.g. when long-term information is used to enrich the encoding of information in short-term memory). They modelled strategic processes such as *covert maintenance rehearsal*, in which an item is mentally re-entered into short-term or primary memory repeatedly to keep its representation from decaying. Also, *covert elaborative rehearsal*, the mental imposition of order and meaning to the materials in primary memory, helps with the retention of information in primary or working memory and also is a major way in which high-quality learning occurs. People using different strategies can often perform the same task in very different ways, leading to different patterns of results (Logie, Della Sala, Laiacona, Chalmers & Wynn, 1996).

Baddeley and Hitch (1974) wrote a book chapter entitled 'Working memory' that changed the field dramatically. They showed that it is an untenable oversimplification to think of working memory as a single entity as in James's (1890) primary memory or Atkinson and Shiffrin's (1968) short-term memory. Instead, they argued, different kinds of materials are stored separately. Verbal materials are stored in a faculty that is sensitive to phonological properties; for example, it is difficult to remember the order of words in a random list if the words all rhyme, as in *cat, bat, hat, mat* . . . etc. Visual, non-verbal materials, similarly, are stored in a faculty that is sensitive to visual or spatial properties. Therefore the interference between two stimuli in working memory depends to a large extent on whether they are encoded using similar features (both verbal or both visual-spatial) or different features (one verbal and one visual-spatial). These working memory stores, now called the *phonological loop* and the *visuo-spatial sketchpad*, were thought to preserve information automatically. It was thought that these stores hold information automatically, but only for a short time, about 2 seconds, unless it is rehearsed. (Lists of long words were not retained as well as lists of shorter words, suggesting that slowing down the rehearsal process allowed the information to decay from working memory before it could be rehearsed.)

Baddeley and Hitch suggested that there also would be a central working memory store that could hold ideas with the help of attention, but Baddeley (1986) later omitted that kind of storage until he later was convinced of a definite need for it. Then he essentially re-introduced it as the *episodic buffer*, a kind of storage in which disparate components could be held, bound together, to form ideas in working memory (Baddeley, 2000). This is the kind of working memory most closely aligned with information retrieved from long-term memory.

Cowan (1988, 1999, 2001) raised the possibility that information is held in a manner that is less modular than in Baddeley's conception. There may be many types of working memory store (memory for the spatial locations of tones, memory for the body locations of touches and so on). For the time being, not knowing the true taxonomy, Cowan considered them to be myriad instances of a temporarily activated portion of long-term memory. Similar items interfere with one another more than different items do, as a general principle within this type of memory. In this conceptualization, instead of the episodic buffer, it is said that the focus of attention holds only a small number of items in an integrated form, whereas many features in long-term memory outside of the focus of attention can be temporarily activated at the same time, but in a less-integrated form. According to this view, working memory is a composite that is centred on the few items or chunks of information currently in the focus of attention, plus a great deal more information that is readily accessible but is not at present within the person's focus of attention and awareness (recently spoken ideas, events within the last few seconds, etc.).

Given that learning depends heavily on working memory, it is important to know what aspect of working memory changes with childhood development and to incorporate this information to formulate the best educational practices. Recent work suggests that more items can be held in something like the focus of attention, and other recent work suggests that the process of rehearsing or refreshing information to keep its representations active increases in speed and efficiency with age. Before we get into all that, let us discuss in greater depth the limitations of working memory and how they can be overcome.

7.3 Limitations in working memory and how they are overcome

Item limits

The scientific understanding of working memory depends a great deal on finding the laws by which it's capabilities are limited. How do we know that there exists a working memory in the brain that functions separately from long-term or secondary memory? We can establish this, it seems, only by showing that there is a temporary quality of working memory that is subject to some limits that can be specified. Miller (1956) pointed to an item limit, in particular about seven items. However, the work of Baddeley and Hitch (1974) showed that this limit was not so absolute, but varied with properties of the items to be remembered, such as how similar they were to one another and how long it took to pronounce each one. The inability to state fixed limits of working memory even has provided a space for some investigators to suggest that there is no separate working memory mechanism per se, just memory with a general set of principles and the attention processes that operate upon memory (e.g. Nairne, 2002).

In a way, too, Miller's (1956) paper consumes itself, in that the chunking principle casts doubt on the meaning of the capacity limit. On one hand, it is suggested that people can remember about seven items. On the other hand, it is suggested that people can combine items to come up with larger chunks, which allows more items to be recalled. When people recall about seven items, how do we know that they are not rapidly forming chunks to assist in their performance? This would, for example, explain why it is helpful that telephone numbers are presented with a dash between the first three and the remaining four digits. Perhaps this is accomplished mentally for lists that are presented without grouping. The experimenter may present to the research participant the digits *3452168*, and the participant may think something like, *34–52–168*. Indeed, Broadbent (1975) suggested that when one looks for lists that elicit perfect performance rather than performance that is correct only some of the time, one finds that lists of three items usually fit the bill. Cowan (2001) greatly extended this reasoning, suggesting that when grouping and rehearsal strategies are not possible, adults usually recall only three to five items, not the five to nine that Miller suggested.

Chen and Cowan (2009) recently have had considerable success in showing that this line of reasoning from Broadbent (1975) is correct. Chen and Cowan taught individuals novel written-word parings (like *brick–dog*) to a 100% correct criterion, so that these word pairings could be considered two-item chunks. Other words were introduced as one-item chunks for a fair comparison. Then the words were presented in lists, which varied in several ways. They could comprise four, six, eight, or 12 singletons. They also could comprise four or six learned pairs. This study showed that it is necessary to remove rehearsal in order to see the chunk capacity limit. When nothing was done to remove rehearsal, the number recalled was somewhat variable from one condition to the next. Participants in a different group, though, were required to recite a simple word repeatedly, twice per second (*the, the, the. . .*) while reading the stimuli. What was observed in this group was that almost exactly three chunks were recalled, disregarding the correctness of the order of words in recall. This was the case no matter whether these three chunks were singletons or learned pairs. This finding is exactly what Broadbent (1975) would have predicted. The review of Cowan (2001) may have overestimated the limit just slightly, by including procedures in which strategies were not completely eliminated.

Cowan and colleagues (2005) showed that measures of the capacity limit are fairly well correlated with indices of successful learning, such as high school class rank and achievement scores, as well as with more basic tests of intelligence.

It is probably not be only the number of items in working memory that is limited, but also the number of relationships between items. Suppose I tell you that in Spain, cookies have more flour than sugar whereas in Italy, cookies have more sugar than flour (it is a completely fabricated example). Understanding this two-way interaction of country with the ratio of ingredients is difficult, as it requires the mental coordination of two of these variables: the relation between sugar and flour, and the dependence of that relation on the country. Halford, Baker, McCredden and Bain (2005) demonstrated that there is a strict limit in the human ability to understand higher-level interactions, even when all of the elements appear together on a printed page so that only the relationships have to be held in memory. Perhaps the basic limit in the understanding of complex ideas depends on the same memory faculty as the limit in chunks (Halford, Cowan & Andrews, 2007). Regardless, the important point is that the complexity of thought is highly dependent in some way on working memory capacity.

Goal maintenance limits

Adults' ability to hold in working memory only about three chunks of information seems rather restrictive. Making matters worse, it is not a foregone conclusion that working memory will always be filled with the most relevant information. A child in the classroom is often supposed to be using working memory to form new long-term knowledge. For example, in a history class, the various details are supposed to be used to paint a mental picture of an era, a series of events, or a movement over time. But here the child's motivation and ability to concentrate is critical. To the extent that irrelevant thoughts such as goings-on outside of the classroom, a classmate's whispering, or daydreams are allowed to enter the focus of attention and dominate working memory, the effort to acquire a new concept or idea will be sabotaged as some of the needed information is edged out of the focus of attention.

Two studies of individual differences illustrate well this principle of the potentially damaging effects of irrelevant information. One study (Kane & Engle, 2003) made use of the Stroop effect, based on a task in which participants are to name the colour in which colour words are written. If, say, the word *red* is written in blue on the computer screen, the assigned task is to say 'blue' but the tendency in adults is to want to say 'red' because word reading has become more automatic and faster than colour naming. The task is especially difficult if the word and colour match for most of the stimuli, which can lull the participant into reading words instead of maintaining the goal of naming the actual colours in which those words are printed. The task was administered to individuals with high and low working memory spans, as tested on a standard *operation span* task that requires concurrent storage and processing of information (words and mathematics problems, respectively). Low-spans did poorly on the Stroop task compared to high-span individuals because low-spans failed to maintain the goal as consistently. They more often were swayed by the word in place of its printed colour.

In another study on inhibition with a very different experimental technique (Vogel, McCollough and Machizawa, 2005), arrays of bars were presented and the task was to keep in mind the orientation of each bar. Actually, some bars were presented on the left and others on the right, and a cue indicated which side was to be retained in memory (without actually moving the eyes away from a centre fixation). This arrangement allowed the measurement of an electrical brain response, or event-related potential, to the task of holding the information in working memory. The more items held in working memory, the larger the magnitude of an event-related response component called *contralateral delay activity* (CDA). On the side of the display to be remembered, moreover, there were often some bars that could be neglected (perhaps the red ones) and other bars that had to be remembered (perhaps the green ones). The CDA magnitudes indicated that low-span individuals kept in mind both the relevant and the irrelevant items, whereas the high-span individuals excluded the irrelevant items from memory and therefore kept more of their working memory capacity free. In other situations, this freed-up working memory in the high-spans presumably could be put to use for the sake of learning more.

Kane et al. (2007) showed that this ability to exclude irrelevant information plays an important role in actual life. Participants carried around small devices on which they were cued to respond from time to time regarding whether they were paying attention to what they were doing or their minds were wandering. Low-span individuals reported their minds wandering, against their will, more often than did high-spans. There was no difference between the two when they deliberately were allowing their minds to wander; in low-spans, it was involuntary mind-wandering that was in excess.

Time limits

The press towards the belief that there are time limits in addition to, or instead of, item limits started early within the field of cognitive psychology. Peterson and Peterson (1959) found that even something as simple as a consonant trigram was forgotten precipitously across 18 seconds filled with a demanding distracting task to prevent rehearsal, counting backwards. Some recent studies, though, would attribute this loss to the presence of interference from the digits presented in the backward-counting task. Lewandowsky, Duncan and Brown (2004) have presented lists to be recalled, in conditions in which the participant was to repeat one word once (*super*) or several times in a row (*super, super, super*) before recalling each word. This was done to vary the time before recall without introducing new interference in the more delayed recall condition. No effect of the time before recall was obtained. Even when a third, attention-demanding task was added (Oberauer & Lewandowsky, 2008), little difference was seen in the loss of memory from one word to the next; the recall results did not show a fan-shaped pattern, which would have indicated more forgetting for slower recall.

In contrast to this research finding, Barrouillet and colleagues (e.g. Barrouillet, Bernardin, Portrat, Vergauwe & Camos, 2007; Portrat, Barrouillet & Camos, 2008) have findings that seem to require that there is loss of memory over time; but that is indirect reasoning. They have manipulated the time between items during their presentation and have shown that they can measure the proportion of time between items that is taken up with a distracting task. In one task, for example, letters are to be recalled but numbers between them are to be read. The time from the beginning of the presentation of each number until it is pronounced by the participant is assumed to be engaged in that task, and therefore unavailable for refreshing or rehearsing the letters to be recalled. It turns out that it is not the amount of time between items to be recalled that is critical, but rather the *cognitive load* defined as the proportion of time taken up by the distracting task. The higher the cognitive load, the lower the span, in a close linear relation. This was taken to suggest that information is lost over time unless it is refreshed, and that it cannot be refreshed during the distracting task.

Tests are under way to adjudicate between these views on the presence versus absence of decay. Meanwhile, though, it has become clear that there is some sort of memory loss over time in certain other situations. Ricker and Cowan (2010) presented a concurrent array of three unconventional characters followed by a pattern mask to eliminate sensory memory and then, in one condition, a blank interval of 1.5, 3, or 6 seconds. This was followed by a probe to test recognition of one randomly-selected item from the three-item array. The basic finding was that memory declined across retention intervals despite the absence of any interference. In contrast, though, arrays of six English letters were not forgotten across a 6-second interval. Zhang and Luck (2009) found that array memory loss over a number of seconds is not a loss of the precision of each item but rather a loss of items, and that it occurs suddenly for items that are lost. It appears from these studies that the loss over time may occur for items that are not easily categorized, and that categorical information such as letter arrays may persist longer over time.

A basic educational implication of these findings is that it is utterly important for the learner to have good categories for the information that is being presented, or it may well tend to fade from memory within a matter of seconds.

Overcoming the limits

What makes human cognition capable is its flexibility. We have already discussed the fact that learning to chunk items together increases the amount of information that can be saved by increasing the number of item per chunk, so that more information can be packed into the same limited number of chunks (Chen & Cowan, 2009; Miller, 1956). In processing ordinary language, a structure is built up as one listens to or reads the sentence, which allows quite a bit of material to fit into the limited number of chunks in working memory (see Tulving & Patkau, 1962). A dramatic demonstration of this principle is the work of Ericsson, Chase and Faloon (1980). They taught a participant to increase his digit memory span over the course of a year from six or seven items, a normal ability, to more than 80 items. Subjective reports suggested that he was able to do this because he started with knowledge of many athletic records. He was able to recode the digits into chunks based on such records (e.g. 3:57.2, a former world record for running a mile). In this way, he raised his span to about 20 items. After that, he learned how to make higher-level chunks out of the primary chunks and, in that way, raised his span to about 80 items. Yet, this increase was very specific and, when he was switched to letter materials, he again could recall only about six of them. Further studies have indicated that, despite the fantastic performance, memory experts still base their performance on a basic memory capacity that is around three or four chunks; it is just that each chunk is very densely packed with information (Ericsson, Delaney, Weaver & Mahadevan, 2004; Wilding, 2001).

Another strategy for remembering information is to rehearse it covertly. If the information is in a verbal form and the serial order of the information has to be retained, this retention is aided considerably by rehearsal, provided that the items to be rehearsed are phonologically distinct (e.g. not composed of rhyming words) and can be recited in about 2 seconds (Baddeley, Thomson & Buchanan, 1975). This was attributed at the time to a phonological memory trace that decays in about 2 seconds unless it is refreshed through rehearsal, though others would now suggest that the basis of the effect is phonological interference (e.g. Neath & Nairne, 1995; Lewandowsky & Oberauer, 2008). If the order of items does not have to be retained, what is found is that the first few items receive special rehearsal that is carried on until the list ends; this has been observed in studies in which the participants must rehearse aloud (Rundus, 1971; Tan & Ward, 2000).

Perhaps grouping and rehearsal work together, or perhaps they are completely separate strategies. In any case, these strategies account for why memory span tests typically show people remembering about seven items, not three items as Chen and Cowan (2009) found with rehearsal prevented and grouping controlled.

It is possible to refresh items in memory without using a verbal form of rehearsal; not all materials lend themselves to verbal rehearsal. Another mnemonic device is to use attention to refresh the information in working memory, presumably by re-entering the information into the focus of attention (Cowan, 1999; Raye, Johnson, Mitchell, Greene & Johnson, 2007).

As mentioned earlier, various kinds of thinking depend heavily on working memory. It is noteworthy that new or abstract problems make a higher demand on working memory than problems in which one's past knowledge can be applied, as the knowledge allows the problem to be represented with fewer separate chunks of information. Take the famous example of the four-card task (Wason & Shapiro, 1971). An abstract example of this task might be as follows. You have four cards, each with a number on one side and a letter on the other side. The cards are displayed with 4, 7, E and X showing. You wish to determine whether the following rule describes these four cards: *If a card has an even number on one side, it has a vowel on the other side.* Which cards need to be turned over to test the rule? This would be a good time to come up with an answer on your own before reading on.

This is actually a difficult problem for most people. Turning over the 4 card is important because the rule requires that it have a vowel on the other side. People generally get that. Turning over the 7 card is unimportant because the rule says nothing about cards with an odd number on one side. Turning over the E card is unimportant because the rule does not state what is on the other side of cards with an odd number; therefore, neither an even nor an odd number on the other side would disconfirm the rule. Finally, turning over the X is important because an even number on the other side would disconfirm the rule. People often mistakenly think that it is the E, and not the X, that needs to be turned over.

It can be argued that working memory is really challenged by this problem. It strains working memory to retain the rule while figuring out what each possible outcome would indicate in relation to the rule. This difficult problem becomes a lot easier, though, if common knowledge is used. Suppose, for example, that the problem is as follows. A certain car repair shop services Hondas and Buicks (as has been the case in Columbia, Missouri). A recall announcement indicates that the accelerator pedals on the Hondas can stick, potentially leading to accidents and therefore warranting repair. Four customer cards are sitting on the desk. Each one indicates the customer's name and the brand of car on one side and, on the other side, whether the accelerator pedal has been checked. The four cards as placed on the desk show *Honda, Buick, checked, and not checked*. Which cards need to be turned over?

It is obvious upon a moment's thought that the *Honda* and the *not-checked* cards need to be turned over. This would be obvious even if this problem were given before the abstract one. However, it is logically the same problem as the more abstract one. As Wason and Shapiro (1971) showed, a problem based in knowledge is much easier than the same problem expressed in an abstract manner. Making the best use of students' knowledge, as well as encouraging retention strategies, is critical for fostering learning and comprehension.

Often, abstract terms cannot be avoided because the domain of a problem is new. In fact, it could be argued that the point of a test of fluid intelligence is to examine how well an individual is able to cope with a problem that he or she has not encountered before (a working–memory–intensive enterprise). One reason that intelligence test scores appear to have increased over time worldwide, for example from the 1950s to the 1990s, may be that people these days have varied experiences that make the tests invalid inasmuch as the types of problems on the intelligence tests are not as unfamiliar as they once were in a previous generation (cf. Flynn, 2007).

When a problem is new or unfamiliar and working memory is insufficient to allow an individual to solve it easily, the individual may fall back upon several very common thinking strategies (albeit often unknowingly). These strategies often work but they also are fraught with danger. One strategy is to use a heuristic, a rule of thumb that yields the correct answer often, though certainly not always (Kahneman, 2003). For example, it can be effortful and working-memory-intensive to evaluate a political argument; less effortful to typecast the speaker based on political affiliation, emotional tone or the similarity of the speaker to oneself. Another strategy that is often used is to rely on social consensus. It may not give you the correct answer, but at least it gives you the answer that is easiest to obtain and that results in little friction with those around you. Even one's basic perceptual judgements (such as judging the length of a line) apparently can be influenced by what others say (Asch, 1956). One of the most important tasks of the educational system may be to make students aware of these cognitive pitfalls and train them to exert more effort so that their working memory can be put to best use in obtaining the most valid answers, rather than the answers that are simply easiest to obtain or are the most comfortable.

7.4 Development of working memory

It has been clear for a very long time that there is profound developmental growth in what is now called memory span (Bolton, 1892). It has, however, been much more difficult from an analytic point of view to figure out exactly which basic working memory mechanisms change with development (Cowan & Alloway, 2009). Nevertheless, some important progress has been made recently.

Growth of capacity

Neo-Piagetian psychologists have long believed that the fundamental basis of cognitive development is an increase in working memory ability, expressed either as increasing energy that allows more items to be held in the attention-based part of working memory as children develop (Pascual-Leone, 2005; Pascual-Leone & Smith, 1969) or in an increasing ability to keep in mind the relations between elements and therefore comprehend information of greater complexity (Andrews, Halford, Bunch, Bowden & Jones, 2003).

One difficulty in supporting the premise that capacity, expressed as the number of chunks in memory, increases with age is that there are other explanations of basic performance increases over age. It could be that what increases with age is only the ability to build larger chunks, in order to use the same basic chunk capacity more efficiently. However, Gilchrist, Cowan and Naveh-Benjamin (2009) were able to show that, in one circumstance at least, this is not the case; there is true development of chunk capacity. They presented for verbatim, spoken recall lists of short, simple spoken sentences that were unrelated to one another (e.g. take your paper and pencil; our neighbour sells vegetables; thieves took the painting; drink all of your milk). The length of the materials made it impractical to use verbal rehearsal to remember the items (Baddeley, 1986), which presumably forced participants to use a chunk-limited capacity. Presumably, each sentence forms a separate chunk of information, with strong links between words in a chunk and very weak links, if any, between chunks.

The beauty of this stimulus set-up is that it was possible to estimate not only the number of chunks, but also the integrity of chunks in memory. It turned out that the children from the first grade (aged 6–7 years), children from the sixth grade (aged 11–12 years) and adults all formed sentence chunks that were equally good. In particular, for participants of any age, if any substantive part of a short sentence was recalled, about 80% of the sentence was recalled. Yet, the number of such sentences recalled increased from an average of about 2.5 in first-grade children to about 3.0 in adults. This reflects a small but probably rather important growth in basic working memory capacity. Of course, capacity may be smaller in children too young to test in this procedure. Increase in the basic capacity might account for developmental increase in the complexity of propositions and materials that can be retained and understood during development (e.g. Andrews et al., 2003; Halford, Wilson & Phillips, 1998; Pascual-Leone & Smith, 1969).

Growth of goal maintenance

One of the most important aspects of the development of working memory is the increasing ability to ignore irrelevant information and focus on more relevant information. For example, Elliott (2002) visually presented lists of digits for recall, either in silence or in the presence of various speech backgrounds. When the background consisted of a variety of spoken words (red, blue, green, yellow, white, tall, big, short, long) this background had an effect that was quite devastating in second-grade (age 7–8) children, reducing their proportion correct on the digit recall task from about 70% to about 30% (Elliott, 2002, figure 1). Each participant was tested with lists of a length based on his or her own pre-determined span. Despite that adjustment, the effect of irrelevant speech diminished with development so that, in adults, the effect of irrelevant speech was less than a 10% reduction.

There are two slightly different interpretations of this finding by Elliott (2002), and other findings like it. For the young children was the distracting speech just intrusive? Or did the distracting speech actually make the children lose sight of the goal, which was to remember the printed digits and not the spoken words? That is not clear but, at least in some circumstances at least, young children actually do lose the goal. Zelazo (2003) has summarized a series of studies in which children must sort a deck of cards according to some rule (e.g. rabbits here, whether red or blue; cars over there, whether red or blue). After a while, the task switches to sorting on a different basis (e.g. red things here, whether cars or rabbits; blue things over there, whether cars or rabbits). The striking finding is that 3-year-old children can explain the new sorting rule, but nevertheless continue to sort according to the old rule. The old rule has become a too-compelling habit to overcome, given the apparently relatively weak support from the new rule, which may temporarily drop out of working memory.

Given these results, and many others on children's difficulty in inhibiting irrelevant information to focus on the most relevant information and task goals (e.g. Bjorklund & Harnishfeger, 1990; Williams, Ponesse, Schachar, Logan & Tannock, 1999), one might wonder if we really know that chunk capacity also increases with age. Perhaps young children have just as much capacity, but fill it with irrelevant materials. A study by Cowan, Morey, AuBuchon, Zwilling and Gilchrist (2010) argues against that as a general interpretation; a growth of the inhibition or filtering out of irrelevant information cannot be the entire explanation of developmental change. Cowan et al. presented arrays with both more-relevant and less-relevant items to be remembered. Participants were sometimes, for example, to remember the colours of the circles and ignore the colours of the triangles. Occasionally, however, they were tested on the less-relevant shape. With two more-relevant and two less-relevant items in the array, children in first and second grades (6–8-year-olds) allocated their attention as efficiently as older children and adults. All groups favoured the more-relevant items by very similar amounts. However, the younger children simply remembered fewer items than did older children or adults, which held true for both the better-remembered, more-relevant items and the poorly-remembered, less-relevant items. Just as Gilchrist et al. (2009) showed that chunking ability could not explain away an apparent growth in basic capacity, Cowan et al. showed that the allocation of attention to filter out irrelevant items cannot explain away this growth in basic capacity.

The smaller capacity and poorer goal maintenance of younger children together have profound implications for education. Gathercole (2008) describes applied studies in the classroom in which she and her colleagues have found that what appeared on the surface to be cases of children's disobedience, rebellion or incomprehension often must be attributed instead to working-memory-based failures, such as forgetting the instructions, not keeping in mind the task goal or not recalling information critical to the successful completion of an assigned task or academic activity.

Despite the poignant observations and anecdotes contributing to this connection between working memory and school performance, one might wonder whether the contribution of working memory is specific. After all, various kinds of working memory tests are highly correlated with the intelligence quotient (IQ) and other aptitude and achievement tests in children (e.g. Cowan et al., 2005). The contributions of working memory and general intelligence can be disentangled, however, using regression techniques. Alloway and Alloway (2010) examined children at 5 years of age and again at 11 years. They found that literacy and numeracy at 11 years were predicted by working memory even after taking into account IQ. It was also the case that literacy and numeracy were predicted by IQ even after taking into account working memory, but the contribution of IQ was smaller than that of working memory. Alloway (2009) tested children between 7 and 11 years of age who were identified as having learning difficulties, and found even more striking results. The children were tested at intervals 2 years apart. In this sample, Alloway found that working memory at the first test accounted for subsequent learning even after taking into account IQ, but the reverse was not true: after taking into account working memory, IQ made no further contribution. It seems safe to say that educators must attend to issues involving working memory in order to understand academic performance in children.

Growth of memory persistence over time

There are a few studies indicating that memory for acoustic sensation, exactly how a noise or word sounds, may persist longer in older children than in young children. Keller and Cowan (1994) found that the duration of tone pitch memory increased from first grade (6–7-years-olds) to adulthood. Cowan, Nugent, Elliott and Saults (2000) presented lists of spoken digits to be ignored while a silent rhyming game was carried out and occasionally presented a cue to switch from the game to recall of the last spoken list, which had ended 1, 5, or 10 seconds before the cue. Lists were adjusted for the child's memory span for attended digits. It was found that forgetting of ignored digits across the retention intervals was equivalent for all ages, for most of the list; but that forgetting of the very last digit in the list across the retention intervals was much more severe for younger children than for older participants. This suggests that the sensory memory, which is uninterrupted only for the final list item, fades more quickly in younger children. More research is still needed to provide an understanding of the implications of such findings for education but it does provide one more reason why younger children have more trouble following spoken directions. These instructions may remain vivid in sensory memory for a shorter time in younger children.

Growth of strategies

Finally, there has been a great deal of research showing that the ability to use mnemonic strategies to overcome the limitations increases markedly across age groups in childhood. It is not fully clear exactly why this growth occurs but there are leads. Many studies have shown that young children are unable to engage in covert verbal rehearsal in a sophisticated manner (Flavell, Beach & Chinsky, 1966; Ornstein & Naus, 1978). One possibility is that they are just unaware that rehearsal would help performance. Against that interpretation, attempts to train children to rehearse, or to recite items more quickly, have not been very successful (e.g. Cowan, Saults, Winterowd & Sherk, 1991; Hulme & Muir, 1985).

Instead, it could be that a certain working memory capacity is necessary in order to make these strategies work effectively. Two different findings in extremely different age groups support this hypothesis. First, in memory span tasks, young children recite items more slowly than young adults, and it was possible that this slower recitation could be allowing time for decay. Against that possibility, Cowan, Elliott et al. (2006) were able to get children to speed up so that they recalled digits in a list at an adult-like rate, but still their memory span did not improve. Second, at the other end of the age spectrum, Kynette, Kemper, Norman and Cheung (1990) found that older adults cannot rehearse as quickly and do not remember as much as young adults, even though there is no reason to assume that aging causes a loss of knowledge about the efficacy of particular mnemonic strategies. It seems reasonable to conclude that the growth of working memory capacity in children and its decline in old age both may affect the ability to use certain mnemonic strategies effectively, in opposite directions; and that the effects of age cannot be explained by the appreciation of the potential use of such a mnemonic strategy.

Barrouillet et al. (2007) found that memory in young adults depended on the cognitive load imposed by items in a secondary task between the items to be recalled. They suggested that this occurs because free time is used to attentionally refresh items in memory (Raye et al., 2007). This leaves open several potential bases of developmental growth. The basis may depend on the age range in childhood. Barrouillet, Gavens, Vergauwe, Gaillard and Camos (2009) found that 5-year-old children were sensitive to the presence or the absence of any distraction between items to be recalled, but were not sensitive to the cognitive load. This would be expected if these young children do not use the attentional-refreshing process (or any kind of rehearsal) that would be interrupted by the cognitive load. In contrast, 7-year-olds were sensitive to cognitive load. Moreover, the quality of attentional refreshing activities apparently continued to improve with age inasmuch as14-year-olds' spans increased with increasing free time to refresh items more quickly than did younger children's spans. Moreover, some factor other than cognitive load appears to have contributed to age differences in recall, as well. That other factor may be the basic working memory capacity.

Finally, Cowan, Saults and Morey (2006) found that the simple strategy of verbal rehearsal sometimes can be put to good use in sophisticated ways to enhance performance in a complex task. Illustrating their procedure, suppose you need to remember temporarily the names and locations of five cities in the eastern United States. That is a lot of name-place associations to hold in working memory. Sometimes, however, adults use a simpler strategy. It is not difficult for them to form a mental route from one location to the next; and it is not difficult for them to rehearse five city names. Then it is possible for them to use these two relatively effortless forms of working memory together; the first name goes with the first map location, and so on. That memorization strategy, breaking up a complex situation into simpler, verbal and spatial sub-problems, of course is unavailable to children too young to use rehearsal well.

7.5 Working memory and maximized education

The lessons from this review of working memory and its development are fairly clear. It is necessary to pitch any kind of education to a level that the student can comprehend, and the ability to assess this comprehension is improved by an appreciation of how closely it may depend on working memory abilities. Failure of working memory must be kept in mind as a possible explanation of any cognitive shortcoming in the classroom or as a basis, at least in part, for learning disabilities. A student's existing knowledge can be used to overcome working memory limits in many situations. One also can try to teach better learning skills and mnemonic strategies, all the while keeping in mind that the successful implementation of a strategy itself may depend on a certain working memory capacity.

In recent studies, there has been a suggestion that challenging working memory tasks, carried out repeatedly and with the challenges increasing as the participant improves, can be used to train working memory and attention. It is also said to have positive benefits for other aspects of performance, raising intelligence tests and helping to overcome attention deficit disorders (e.g. Klingberg, 2010; Tang et al., 2007). This has been in contrast to most prior studies in cognitive psychology, in which the benefits of training and practice have been found to be somewhat narrow (e.g. Boron, Willis & Schaie, 2007; Ericsson et al., 1980). Perhaps what is learned in the newer training studies is that, by applying effort and attention repeatedly and persistently, one can improve one's performance on just about any task. It is thought, for example, that a difference in attitude helps to explain why Asian children learn mathematics with much more proficiency on average than children in the United States. They learn that if one does not do well in mathematics, one must work harder; US children, generally, tend to believe that some people are just not good at math (Chen & Stevenson, 1995).

If working memory training or some other challenging training in a game-like form can help to overcome biases such as the belief that one is just not good at certain tasks, such training will be well worth while. Perhaps it might even help children to become citizens who are willing to use their effort and working memory to formulate and carry out the best possible decisions for our society, without an over-reliance on simplistic heuristics, biases or prejudices, and social convention or popular approval.

References

Alloway, T. P. (2009). Working memory, but not IQ, predicts subsequent learning in children with learning difficulties. *European Journal of Psychological Assessment,* 25, 92–98.

Alloway, T. P. & Alloway, R. G. (2010). Investigating the predictive roles of working memory and IQ in academic attainment. *Journal of Experimental Child Psychology,* 106, 20–29.

Andrews, G., Halford, G.S., Bunch, K. M., Bowden, D., & Jones, T. (2003). Theory of mind and relational complexity. *Child Development,* 74, 1476–99.

Asch, S. E. (1956). Studies of independence and conformity: A minority of one against a unanimous majority. *Psychological Monographs,* 70, (Whole no. 416).

Atkinson, R. C., & Shiffrin, R. M. (1968). Human memory: A proposed system and its control processes. In K. W. Spence & J. T. Spence (Eds.), *The psychology of learning and motivation: Advances in research and theory* (Vol. 2, pp. 89–195). New York: Academic Press.

Baddeley, A. D. (1986). *Working memory.* Oxford: Clarendon Press.

Baddeley, A. D. (2000). The episodic buffer: a new component of working memory? *Trends in cognitive sciences,* 4, 417–23.

Baddeley, A. D. (2003). Working memory: looking back and looking forward. *Nature Reviews: Neuroscience,* 4, 829–39.

Baddeley, A. D., & Hitch, G. (1974). Working memory. In G. H. Bower (Ed.), *The psychology of learning and motivation* (Vol. 8, pp. 47–89). New York: Academic Press.

Barrouillet, P., Bernardin, S., Portrat, S., Vergauwe, E., & Camos, V. (2007). Time and cognitive load in working memory. *Journal of Experimental Psychology: Learning, Memory, and Cognition,* 33, 570–85.

Barrouillet, P., Gavens, N., Vergauwe, E., Gaillard, V., & Camos, V. (2009). Working memory span development: A time-based resource-sharing model account. *Developmental Psychology,* 45, 477–90.

Bjorklund, D. F., & Harnishfeger, K. K. (1990). The resources construct in cognitive development: Diverse sources of evidence and a theory of inefficient inhibition. *Developmental Review,* 10, 48–71.

Baddeley, A. D., Thomson, N., & Buchanan, M. (1975). Word length and the structure of short-term memory. *Journal of Verbal Learning and Verbal Behavior,* 14, 575–89.

Bolton, T. L. (1892). The growth of memory in school children. *American Journal of Psychology,* 4, 362–80.

Boron, J. B., Willis, S. L., & Schaie, K. W. (2007). Cognitive training gain as a predictor of mental status. *The Journals of Gerontology: Series B: Psychological Sciences and Social Sciences,* 62B, 45–52.

Broadbent, D. E. (1975). The magic number seven after fifteen years. In A. Kennedy & A. Wilkes (Eds.), *Studies in long-term memory* (pp. 3–18). Oxford: John Wiley & Sons.

Chein, J. M., & Fiez, J. A. (2010). Evaluating models of working memory through the effects of concurrent irrelevant information. *Journal of Experimental Psychology: General,* 139, 117–37.

Chen, C., & Stevenson, H. W. (1995). Motivation and mathematics achievement: A comparative study of Asian-American, Caucasian-American, and East Asian high school students. *Child Development,* 66, 1215–34.

Chen, Z., & Cowan, N. (2009). Core verbal working memory capacity: The limit in words retained without covert articulation. *Quarterly Journal of Experimental Psychology,* 62, 1420–29.

Cowan, N. (1988). Evolving conceptions of memory storage, selective attention, and their mutual constraints within the human information processing system. *Psychological Bulletin,* 104, 163–91.

Cowan, N. (1995). *Attention and memory: An integrated framework* (Oxford Psychology Series, No. 26). New York: Oxford University Press.

Cowan, N. (1999). An embedded-processes model of working memory. In A. Miyake & P. Shah (Eds.), *Models of Working Memory: Mechanisms of active maintenance and executive control* (pp. 62–101). Cambridge: Cambridge University Press.

Cowan, N. (2001). The magical number 4 in short-term memory: A reconsideration of mental storage capacity. *Behavioral and Brain Sciences,* 24, 87–185.

Cowan, N. (2009). Working memory from the trailing edge of consciousness to neurons. *Neuron,* 62, 13–16.

Cowan, N., & Alloway, T. (2009). Development of working memory in childhood. In M. L. Courage & N. Cowan (Eds.), *The development of memory in infancy and childhood* (pp. 303–42). Hove: Psychology Press.

Cowan, N., Elliott, E. M., Saults, J. S., Morey, C. C., Mattox, S., Hismjatullina, A., & Conway, A. R. A. (2005). On the capacity of attention: Its estimation and its role in working memory and cognitive aptitudes. *Cognitive Psychology,* 51, 42–100.

Cowan, N., Elliott, E. M., Saults, J. S., Nugent, L. D., Bomb, P., & Hismjatullina, A. (2006). Rethinking speed theories of cognitive development: Increasing the rate of recall without affecting accuracy. *Psychological Science,* 17, 67–73.

Cowan, N., Morey, C. C., AuBuchon, A. M., Zwilling, C. E., & Gilchrist, A. L. (2010). Seven-year-olds allocate attention like adults unless working memory is overloaded. *Developmental Science,* 13, 120–33.

Cowan, N., Nugent, L. D., Elliott, E. M., & Saults, J. S. (2000). Persistence of memory for ignored lists of digits: Areas of developmental constancy and change. *Journal of Experimental Child Psychology, 76,* 151–72.

Cowan, N., Saults, J. S., & Morey, C. C. (2006). Development of working memory for verbal-spatial associations. *Journal of Memory and Language, 55,* 274–89.

Cowan, N., Saults, J. S., Winterowd, C., & Sherk, M. (1991). Enhancement of 4-year-old children's memory span for phonologically similar and dissimilar word lists. *Journal of Experimental Child Psychology, 51,* 30–52.

D'Esposito, M., Postle, B. R., Jonides, J., & Smith, E. E. (1999). The neural substrate and temporal dynamics of interference effects in working memory as revealed by event-related functional MRI. *Proceedings of the National Academy of Sciences USA, 96,* 7514–7519.

Elliott, E. M. (2002). The irrelevant-speech effect and children: Theoretical implications of developmental change. *Memory & Cognition, 30,* 478–87.

Ericsson, K. A., Chase, W. G., & Faloon, S. (1980). Acquisition of a memory skill. *Science, 208,* 1181–82.

Ericsson, K. A., Delaney, P. F., Weaver, G., & Mahadevan, R. (2004). Uncovering the structure of a memorist's superior 'basic' memory capacity. *Cognitive Psychology, 49,* 191–237.

Flavell, J. H., Beach, D. H., & Chinsky, J. M. (1966). Spontaneous verbal rehearsal in a memory task as a function of age. *Child Development, 37,* 283–99.

Flynn, J. R. (2007). *What is Intelligence?: Beyond the Flynn Effect.* Cambridge: Cambridge University Press.

Gardner, H. (1985). *The mind's new science: A history of the cognitive revolution.* New York: Basic Books.

Gathercole, S. E. (2008). Working memory in the classroom. *The Psychologist, 21,* 382–85.

Gilchrist, A. L., Cowan, N., & Naveh-Benjamin, M. (2009). Investigating the childhood development of working memory using sentences: New evidence for the growth of chunk capacity. *Journal of Experimental Child Psychology, 104,* 252–65.

Halford, G. S., Baker, R., McCredden, J. E., & Bain, J. D. (2005). How many variables can humans process? *Psychological Science, 16,* 70–76.

Halford, G. S., Cowan, N., & Andrews, G. (2007). Separating cognitive capacity from knowledge: A new hypothesis. *Trends in Cognitive Sciences, 11,* 236–42.

Halford, G. S., Wilson, W. H., & Phillips, S. (1998). Processing capacity defined by relational complexity: Implications for comparative, developmental, and cognitive psychology. *Behavioral and Brain Sciences, 21,* 803–65.

Hecht, S. A., Vagi, K. J., Torgesen, J. K. (2007). Fractional skills and proportional reasoning. In D. B. Berch & M. M. Mazzocco (Eds.), *Why is math so hard for some children? The nature and origins of mathematical learning difficulties and disabilities* (pp. 121–32). Baltimore, MD: Paul H Brookes Publishing.

Hulme, C., & Muir, C. (1985). Developmental changes in speech rate and memory span: A causal relationship? *British Journal of Developmental Psychology, 3,* 175–81.

James, W. (1890). *The principles of psychology.* New York: Henry Holt.

Jonides, J., Lewis, R. L., Nee, D. E., Lustig, C. A., Berman, M. G., & Moore, K. S. (2008). The mind and brain of short-term memory. *Annual Review of Psychology, 59,* 193–224.

Kahneman, D. (2003). A perspective on judgment and choice: Mapping bounded rationality. *American Psychologist, 58*(9), 697–720.

Kane, M. J., Brown, L. H., McVay, J. C., Silvia, P. J., Myin-Germeys, I., & Kwapil, T. R. (2007). For whom the mind wanders, and when: An experience-sampling study of working memory and executive control in daily life. *Psychological Science, 18,* 614–21.

Kane, M. J., & Engle, R. W. (2003). Working-memory capacity and the control of attention: The contributions of goal neglect, response competition, and task set to Stroop interference. *Journal of Experimental Psychology: General, 132,* 47–70.

Keller, T. A., & Cowan, N. (1994). Developmental increase in the duration of memory for tone pitch. *Developmental Psychology, 30,* 855–63.

Klingberg, T. (2010). Training and plasticity of working memory. *Trends in Cognitive Sciences,* 14(7), 317–24.

Kynette, D., Kemper, S., Norman, S., & Cheung, H. (1990). Adults' word recall and word repetition. *Experimental Aging Research,* 16, 117–21.

Lewandowsky, S., Duncan, M., & Brown, G.D.A. (2004). Time does not cause forgetting in short-term serial recall. *Psychonomic Bulletin & Review,* 11, 771–90.

Lewandowsky, S., & Oberauer, K. (2008). The word-length effect provides no evidence for decay in short-term memory. *Psychonomic Bulletin & Review,* 15, 875–88.

Logie, R. H., Della Sala, S., Laiacona, M., Chalmers, P., & Wynn, V. (1996). Group aggregates and individual variability: The case of verbal short-term memory. *Memory & Cognition,* 24, 305–21.

Majerus, S., Argembeau, A. D., Perez, T. M., Belayachi, S., Van der Linden, M., Collette, F., Salmon, E., Seurinck, R., Fias, W., and Maquet, P. (2010). The commonality of neural networks for verbal and visual short-term memory. *Journal of Cognitive Neuroscience,* 22(11), 2570–93.

McNab, F., & Klingberg, T. (2008). Prefrontal cortex and basal ganglia control access to working memory. *Nature Neuroscience,* 11, 103–107.

Miller, G. A. (1956). The magical number seven, plus or minus two: Some limits on our capacity for processing information. *Psychological Review,* 63, 81–97.

Miller, G. A., Galanter, E., and Pribram, K. H. (1960). *Plans and the structure of behavior.* New York: Holt, Rinehart and Winston, Inc.

Nairne, J. S. (2002). Remembering over the short-term: The case against the standard model. *Annual Review of Psychology,* 53, 53–81.

Neath, I., & Nairne, J. S. (1995). Word-length effects in immediate memory: Overwriting trace decay. *Psychonomic Bulletin & Review,* 2, 429–41.

Oberauer, K., & Lewandowsky, S. (2008). Forgetting in immediate serial recall: decay, temporal distinctiveness, or interference? *Psychological Review,* 115, 544–76.

Ornstein, P. A., & Naus, M. J. (1978). Rehearsal processes in children's memory. In P. A. Ornstein (Ed.), *Memory development in children* (pp. 69–99). Hillsdale, NJ: Erlbaum.

Pascual-Leone, J. (2005). A neoPiagetian view of developmental intelligence. In O. Wilhelm & R. W. Engle (Eds.), *Understanding and measuring intelligence* (pp. 177–201). London: Sage.

Pascual-Leone, J., & Smith, J. (1969). The encoding and decoding of symbols by children: A new experimental paradigm and a neo-Piagetian model. *Journal of Experimental Child Psychology,* 8, 328–55.

Peterson, L. R. & Peterson, M. J. (1959). Short-term retention of individual verbal items. *Journal of Experimental Psychology,* 58, 193–98.

Portrat, S., Barrouillet, P., & Camos, V. (2008). Time-related decay or interference-based forgetting in working memory? *Journal of Experimental Psychology: Learning, Memory, and Cognition,* 34, 1561–64.

Raye, C. L., Johnson, M. K., Mitchell, K. J., Greene, E. J., & Johnson, M. R. (2007). Refreshing: A minimal executive function. *Cortex,* 43, 135–45.

Ricker, T. J., & Cowan, N. (2010). Loss of visual working memory within seconds: The combined use of refreshable and non-refreshable features. *Journal of Experimental Psychology: Learning, Memory, and Cognition,* 36(6), 1355–68.

Rundus, D. (1971). Analysis of rehearsal processes in free recall. *Journal of Experimental Psychology,* 89, 63–77.

Tan, L., & Ward, G. (2000). A recency-based account of the primacy effect in free recall. *Journal of Experimental Psychology: Learning, Memory, and Cognition,* 26, 1589–625.

Tang, Y. -Y., Ma, Y., Wang, J., Fan, Y., Feng, S., Lu, Q., Yu, Q., Sui, D., Rothbart, M. K., Fan, M., & Posner, M. I. (2007). Short-term meditation training improves attention and self-regulation. *Proceedings of the National Academy of Science USA,* 104, 17152–56.

Todd, J. J., & Marois, R. (2004). Capacity limit of visual short-term memory in human posterior parietal cortex. *Nature,* 428, 751–54.

Tulving, E., & Patkau, J. E. (1962). Concurrent effects of contextual constraint and word frequency on immediate recall and learning of verbal material. *Canadian Journal of Psychology,* 16, 83–95.

Xu, Y., & Chun, M. M. (2006). Dissociable neural mechanisms supporting visual short-term memory for objects. *Nature, 440,* 91–95.

Vogel, E. K., McCollough, A. W., & Machizawa, M. G. (2005). Neural measures reveal individual differences in controlling access to working memory. *Nature, 438,* 500–503.

Wason, P. C. & Shapiro, D. (1971). Natural and contrived experience in a reasoning problem. *Quarterly Journal of Experimental Psychology, 23,* 63–71.

Watson, J. B. (1913). Psychology as the behaviorist views it. *Psychological Review, 20,* 158–77.

Wilding, J. (2001). Over the top: Are there exceptions to the basic capacity limit? *Behavioral and Brain Sciences, 24,* 152–53.

Williams, B. R., Ponesse, J. S., Schachar, R. J., Logan, G. D., & Tannock, R. (1999). Development of inhibitory control across the life span. *Developmental Psychology, 35,* 205–213.

Zelazo, P. D. (2003). The development of executive function in early childhood. *Monographs of the Society for Research on Child Development, 68,* Serial No. 274.

Zhang, W. & Luck, S. J. (2009). Sudden death and gradual decay in visual working memory. *Psychological Science, 20,* 423–28.

Chapter 8

Applications of cognitive science to education

Henry L. Roediger, III, Bridgid Finn and
Yana Weinstein

Overview

We present five topics (sets of findings) from cognitive psychology that can be directly applied to the classroom and/or to students' study strategies outside of class. These include: (1) retrieval practice through testing, which enhances learning more than restudying does; (2) spaced periods of study of the same topic relative to massed study of that topic; (3) interleaving of different topics of study rather than blocked studying of the same topic; (4) improved metacognitive monitoring for students (that is, teaching students to assess accurately the state of their own knowledge); and (5) teaching in ways that will facilitate transfer of learning to novel situations. Current educational practice often uses the less beneficial of these strategies (e.g. repeated massed study of the same topic), and thus principles derived from cognitive psychology can help to encourage better practices in the classroom and in individual study periods outside of class. Although these five topics are only some of the ways cognitive psychologists have shown that student learning can be improved, we believe they represent techniques that are most applicable to improving education today. These bodies of knowledge have stood the test of time (some of the topics have been studied for over 100 years) and their validity is not in doubt. The trick for educators and students is to find ways to apply these strategies in beneficial ways.

The goal of this volume is to provide insights from neuroscience and cognitive science aimed at improving educational practice. Our chapter falls squarely into the latter camp, that of cognitive science (although we do point to some neuroscientific findings in places). We extrapolate research from purely behavioural cognitive psychology and make recommendations for educational practice. In a few cases, our suggestions have already been implemented in the classroom, at least on an experimental basis. Of course, neural processes underlie any cognitive processes, so in a broad (and vacuous) sense our chapter can be conceived as being about neuroscience and education. However, we certainly do not make this claim. We seek to generalize findings from behavioural studies to possible educational practice.

8.1 **Cognition and its relation to neuroscience and education**

Because the orientation of our chapter differs from some other chapters in the book, we pause to consider and defend our perspective. In 1997, John Bruer wrote an article entitled 'Education and neuroscience: A bridge too far'. He reviewed the enthusiasm sweeping educational circles in the mid 1990s for applying neuroscience to education but argued that it was misplaced, often based on simplistic interpretations or even misunderstandings of the claims of neuroscience. He wrote: 'Educational applications of brain science may come eventually, but as of now neuroscience has little to offer teachers in terms of informing classroom practice. There is, however, an applied science of mind, cognitive science, which can serve as a basic science of the development of an applied science of learning and instruction' (p. 4). Bruer argued that the link between neuroscience and education represented a 'bridge too far', and he suggested that two shorter spans must be constructed before researchers could traverse the divide between neuroscience and education. The first bridge that must be constructed is from cognition to instruction: What are the cognitive underpinnings of instructional practice and how can these be made more effective? Once a good start has been made at this task, the second bridge must link from cognition to neural circuitry: what neural circuits underlie the cognitive processes that are linked to instructional practice? In the past 15 years, good progress has been made on this latter front in some areas (e.g. brain mechanisms for reading and for straightforward mathematical computation), but certainly much more work remains to be done. Our chapter aims mostly at building the first of Bruer's bridges, the link between cognition and instruction.

Other chapters in this volume build a strong case for neuroscience and education, or what Carew and Magsamen (2010) have called neuroeducation. They defined it as 'a nascent discipline that seeks to blend the collective fields of neuroscience, psychology, cognitive science and education to create a better understanding of how we learn and how this information can be used to create more effective teaching methods, curricula and educational policy' (p. 685). By this very broad definition, our chapter does fit under the umbrella of neuroeducation.

8.2 **Organisation of the chapter**

We have selected five topics from cognitive psychology of learning and memory (see further discussion in Chapter 7, this volume) that we believe have special import for education: retrieval practice (testing) of information as a mnemonic enhancer; spacing of study episodes to improve performance relative to massed (back-to-back) study episodes; interleaving of various topics of study rather than blocking them; metacognition (the knowledge of one's own cognitive processes and how they affect learning); and transfer of knowledge (applying learned information in a new domain). These topics have been thoroughly studied in laboratory settings (some more than others, of course), and we believe all five are ripe for application in classroom settings.

8.3 **Retrieval practice via testing**

Educators generally give tests to assess students' learning and to assign grades. In addition, administrators give standardized tests to assess student learning on a common measurement scale (say, within a state or country or even across countries). Controversies abound in the use of testing, especially for this latter purpose. In this section, we argue that tests can serve another important purpose, too—they can foster better learning and retention as well as assessing it. The act of taking a test can greatly enhance knowledge of the tested material. Experimental psychologists have known this phenomenon, called the testing effect, for over 100 years; Abbott (1909) seems to have been the first to note it through experimentation, although thinkers back to Aristotle have supposed that the principle is true. In his essay on 'Memory and reminiscence', Aristotle wrote: 'Exercise in repeatedly retrieving a thing strengthens the memory'. It only took another 2500 years or so for his claim to receive empirical confirmation. However, Aristotle did not get everything right. He thought the heart to be the repository of memories, and the phrase of 'learning by heart' provides a vestige of this claim.

A straightforward experiment by Roediger and Karpicke (2006) shows the power of testing with only a single test. They had students read fact-filled passages about a variety of topics (the sun, sea otters). Shortly after reading a passage, students either took an initial test on the passage (lasting 7 minutes) or they read it again for 7 minutes. For the test, they were given the title of the passage (sea otters) and asked to recall as much as they could. The passage was divided into idea units (fundamental ideas or concepts in each sentence) so that it could be reliably scored. On this first test, students in the relevant condition recalled 70% of the idea units. Of course, in the condition that involved restudying the whole passage, students were exposed to 100% of the units. This re-exposure condition would seem to be at an advantage over the testing condition because students did not perfectly recall the passage during the test.

The critical part of the experiment occurred later, when students were given a final criterial test. This test was given to different groups of subjects either shortly after the manipulation (about 5 minutes later), after 2 days, or after 7 days. The results are shown in Figure 8.1. At the short retention interval, the repeated study condition produced greater recall than the study/test condition. This outcome shows what students have long known—cramming (repeated study) just before a test elevates performance on that test. However, performance on the two delayed tests showed just the opposite, with the study/test condition outperforming the study/study condition by a wide margin. This outcome represents the testing effect: taking a test generally produces better performance on a later test than does repeated study. Notice that the increased delay caused greater forgetting, but the forgetting was much smaller for the condition that took an initial test than for the condition in which students repeatedly studied the information. Testing seems to insulate against forgetting (see, too, Carpenter, Pashler, Wixted & Vul, 2008; Wheeler, Ewers &

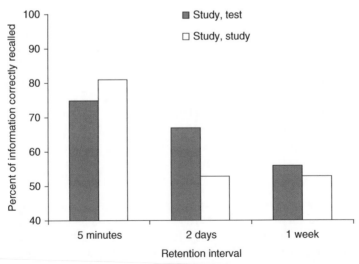

Fig. 8.1 Data from Roediger and Karpicke (2006a, experiment 1). On the test taken 5 minutes after study, material that students repeatedly studied was recalled better than material that was studied once and then tested. On the two delayed tests, however, the pattern reversed: studying and taking a test led to better performance than did repeated studying.

Buonanno, 2003). One idea to explain the benefit of testing is that it permits students to practice the skill they will need later, a form of transfer appropriate processing (e.g. Roediger, Gallo & Geraci 2002). More specifically, research by Zaromb and Roediger (2010) supports the idea that in practising retrieval of large sets of information with few cues, students organise the information more effectively (create good retrieval plans) than they do when they repeatedly study the information, and this retrieval plan aids later organised recall.

Another fact to glean from Figure 8.1 is the nature of the interaction between the study/study and study/test conditions and retention interval, viz., a crossover interaction. As noted previously, the interaction arises (descriptively) because repeated study leads to better initial performance, but the initial test slows forgetting, thus leading to the crossover interaction. Forgetting does occur in the study/test condition, of course, but it is more gradual. In this experiment a testing effect occurred only on a delayed test, but in many other experiments testing effects are found in the same experimental session (e.g. Carrier & Pashler, 1992 among many others).

One boundary condition for the testing effect is the level of initial performance on the test. In the Roediger and Karpicke (2006) experiment just described, students recalled 70% of the idea units on the first test. But what if the first test were delayed 2 weeks and performance had dropped to, say, 30%? Then the rereading condition would have been exposed to 100% of the idea units 2 weeks later and the testing group to only 30%. Would the testing effect still occur? The answer is no, or not usually. Initial level of performance is one of the boundary conditions, as Kang, McDermott and Roediger (2007) showed. However, the testing effect does emerge even with low initial test performance if feedback is given. That is, even if students are tested at a delay and recall only 30% of the material, if they receive feedback a testing effect will still occur. This fact may seem odd; after all, giving feedback simply re-exposes students to the material as in the comparison rereading condition. However, the requirement for students to expend effort in trying to retrieve seems to exert a positive effect when they receive feedback. Izawa (1970) referred to this as the potentiating effect of testing, arguing that taking a test enhances the information encoded and stored from the next study opportunity relative to a pure restudy condition (see Arnold & McDermott, in preparation). This test-potentiating effect can even occur when a test is given before any material is studied (Kornell, Hays & Bjork, 2009; Richland, Kornell & Kao, 2009).

In all the cases just cited, there seems to be something about making an effort to retrieve that facilitates performance, and Pyc and Rawson (2009) have produced evidence consistent with the idea that expending retrieval effort may underlie the testing effect in certain situations. Nonetheless, the important point for practical purposes is that feedback should be given whenever possible after tests. Feedback corrects errors students make (Pashler, Cepeda, Wixted & Rohrer, 2005) and can benefit even correct answers if they are ones made with low confidence (Butler, Karpicke & Roediger, 2008). The usual advice is that feedback should be given as soon as possible after a test, preferably immediately. However, in experiments directly comparing immediate to (somewhat) delayed feedback, delayed feedback produces larger benefits on later tests (Butler, Karpicke & Roediger, 2007). The reason may be that delayed feedback permits spaced presentation of material and, as we shall see in the next section of the chapter, spaced practice benefits retention relative to massed practice. Providing immediate feedback amounts to massed presentation and thus produces less long-term benefit.

A large amount of research on the benefits of testing has been published in recent years (see Roediger & Butler, 2011 and Roediger, Agarwal, Kang & Marsh, 2010 for short and long reviews, respectively). The important point for present purposes is that testing provides a robust benefit in later recall relative to restudying. Further, testing effects have been obtained in simulated classroom settings (e.g. Butler & Roediger, 2007). The power of testing effects has led researchers to bring testing into actual classrooms to see if their use can improve student achievement (e.g. Carpenter, Pashler & Cepeda, 2009; McDaniel, Agarwal, Huelser, McDermott & Roediger, 2011; Roediger, Agarwal, McDaniel & McDermott, 2011).

We will use one experiment from Roediger et al. (2011) to illustrate the possible effectiveness of retrieval practice via testing as a tool for improving learning in a middle school classroom. The research was fully integrated into a social studies class for 11–12-year-olds during the first term of an academic year. Topics covered included major civilizations (e.g. ancient Egypt, India, China and so on). The experiments were conducted within-students, with some facts about each topic randomly assigned for quizzing and another set of facts assigned to the no-quiz condition. The teacher had six classes, and there was a different random assignment of materials to condition for each class; the teacher was unaware of which facts were assigned to which condition (she was outside the classroom when the materials were quizzed). A research assistant performed the quizzing using a student response (clicker) system that permitted immediate feedback (but also permitted the assistant to obtain scores). The quizzes and final criterial tests were all multiple choice, because that is the evaluation mode the teacher preferred.

The quizzing was a common part of students' daily routine. From their point of view, they read (or not) the assigned material, the teacher covered it in class and they reviewed (or not) the material before the chapter exam. The quizzes given in class were viewed as practice, and students seemed to enjoy them (they often complained on the days when quizzes were not given—'why can't we use the clickers today?'). Facts that were tested were quizzed three times, once before the teacher presented the material in class, once the day after she presented it and once just before the chapter exam.

The scores of interest occurred during three later tests: a chapter exam, an end-of-the-term exam and a final exam at the end of the school year in May (approximately 7–8 months after the manipulation). In the case of chapter and term exams, the grades counted for the student grades; that is, our dependent measures were collected as part of the exams the students were taking for grades. At the end of the year this feature was not included; the students received a surprise test on material covered many months earlier.

The results are shown in Figure 8.2, where it can be seen that a testing effect occurred on all three tests. On the chapter exams, items that had been quizzed were correctly answered 91% of the time relative to 74% of the time for non-quizzed items. The teacher reported that her classes

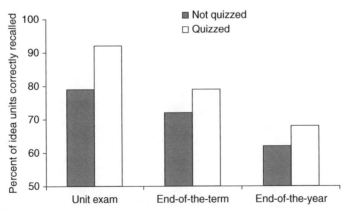

Fig. 8.2 Performance of middle school students at three different time-points on science material that was previously quizzed in class, or was not (from McDaniel, Agarwal, Huelser, McDermott & Roediger, 2011). Material that was previously quizzed produced consistently better performance on all three tests.

usually scored in the mid 70s on her tests, so the fact that we could raise performance from this level to the low 90s means that (on the grading scale used in class) students rose from a C+ to an A– average. Obviously, on the delayed tests given at the end of the semester and the end of the year, performance dropped (forgetting occurred), but the testing advantage remained. Even at the end of the school year items quizzed in October and November were better remembered than items not quizzed. We suspect that if we had been able to use production (recall tests) such as short-answer tests, we might have found even larger effects due to the greater retrieval effort they engender (see Kang et al., 2007).

In sum, retrieval practice via testing leads material to be better retained than does either no practice or repeated study of material. This finding has important implications for study strategies of individual students (e.g. Karpicke, 2009) and for classroom practice (see, too, McDaniel, Agarwal, Huelser, McDermott & Roediger, 2011; McDaniel, Thomas, Agarwal, McDermott & Roediger, in preparation). In addition, material learned via retrieval practice also seems to transfer to novel situations (when the material is tested in other ways) and so represents more than 'teaching to the test' (Butler, 2010). Retrieval practice serves as a critical mnemonic booster. We turn now to other techniques to improve learning and memory.

8.4 **Spaced practice**

The spacing effect is the robust finding that distributing practice, by spacing out several study episodes over time rather than massing them all at once, can substantially boost long-term learning. Ebbinghaus (1885/1964) noted that, 'with any considerable number of repetitions a suitable distribution of them over a space of time is decidedly more advantageous than the massing of them at a single time' (p. 89). Since this early observation, the benefits of spaced practice on long-term retention have been established across a wide range of populations and domains. Besides hundreds of positive findings with college students, spacing effects have been shown in studies with children (Son, 2010; Toppino, 1991), older adults (Balota, Duchek & Paullin, 1989), non-human animals (Davis, 1970) and even amnesic patients (Cermak, Varfallie, Lanzoni, Mather & Chase, 1996). The advantages of spacing have also been shown to occur across a vast range of domains and tasks, including motor learning, classical and operant conditioning, implicit measures of memory, recognition memory, paired associate learning, free recall, text processing, statistics learning and vocabulary acquisition (see, e.g. Cepeda, Vul, Pashler, Wixted & Rohrer, 2006; Dempster, 1988, 1996; Glenberg, 1979; Greene, 1990 and Hintzman,1974, provided reviews). Finally, the spacing effect occurs whether the presentation modality is visual or auditory (Melton, 1970), and even when the two presentations occur in distinct modalities (Hintzman, Block & Summers, 1973).

In general, the benefits of spacing grow with practice and increase with the length of the lag between presentations (Pavlik & Anderson, 2005). In free recall experiments, the number of items that occur between presentations shows a systematic relationship with later recall, with greater lags leading to better recall (e.g. Glenberg, 1976; Madigan, 1969; Melton, 1967). Melton presented subjects with words two times. Some of the words were massed—in other words, presented twice in a row. Other repetitions were spaced by 2, 4, 6, 20, or 40 different interpolated words. Results showed that the probability of recall on a later test steadily increased with the lag between presentations. It is important to note, however, that there may be some differences in the temporal parameters of the effect among different populations and experimental conditions (Hintzman, 1974; Toppino & DeMesquita, 1984; Wilson, 1976).

Spacing of learning has been shown to be an effective technique in improving the cognitive function of patients with various types of neurological dysfunction (Schacter, Rich & Stampp, 1985).

Spacing provides an advantage relative to other techniques in that it does not require effort on the part of the patients. Judged against other techniques that have been shown to aid learning in patient groups, such as imagery, organization, or verbal labelling (e.g. Cermak, 1975, Cermak, Reale & DeLuca, 1977; Gianutsos & Gianutsos, 1979), spacing is relatively undemanding on the patient's cognitive resources (Bjork, 1979; Schacter et al., 1985). After exposure to spaced training, some patients show spontaneous use of spaced retrieval (Schacter et al., 1985), demonstrating that spaced training can be implemented and retained by patient populations.

In the classroom, spaced practice has long been regarded as a way to enhance learning. In William James's (1899/1958, pp. 93–94) essays to teachers and students he advises,

> Cramming seeks to stamp things in by intense application immediately before the ordeal. But a thing thus learned can form but few associations. On the other hand, the same thing recurring on different days, in different contexts, read, recited on, referred to again and again, related to other things and reviewed, gets well wrought into the mental structure…There is no moral turpitude in cramming. It would be the best, because the most economical, mode of study if it led to the results desired. But it does not…

Some of the earliest studies of distributed practice were done in educational settings. Pyle (1913) drilled 8–9-year-olds on math problems once a day for 10 days, or once in the morning and once in the afternoon for 5 days. When practice was extended over more days, learning was more effective. Similarly, Smith and Rothkopf (1984) found that distributing a lesson over 4 days was more effective than presenting it for the same amount of total time on one day.

The spacing effect has been demonstrated with a variety of educationally relevant materials. Rea and Modigliani (1985) showed better spelling and mathematics performance following spaced practice. Vocabulary learning is facilitated by spaced relative to massed practice (Bahrick & Phelps, 1987; Dempster, 1987), with the gains shown to be quite long lasting. For example, Bahrick and Phelps (1987) tested people's retention of Spanish–English vocabulary words that they had learned 8 years earlier. They found that people who had initially learned the vocabulary pairs in spaced training sessions showed much better retention than the group who had received massed training.

In a recent study, Kornell and Bjork (2008) tested the effectiveness of massed versus spaced practice on students' ability to learn concepts and categories. Successive presentations of category exemplars might result in superior learning by allowing people to notice similarities amongst the exemplars, whereas targeting within-category similarities might be more difficult if exemplars are spaced out. Participants studied six paintings by each of 12 artists, such as Braque and Seurat. Six of the artists' paintings were studied in massed presentation, and six of the artists' paintings were studied with spaced presentations. On the final test, participants were shown new paintings from each of the 12 artists and were required to indicate whom they thought was the artist for each painting. Classification of the new paintings to artists was superior following spaced practice. Interestingly, despite the large advantage for the spaced items, when students were asked to judge which type of presentation they thought was most effective, they reported that massed study was more effective than spaced study. We will return to this point shortly.

Spaced practice is equally, if not more effective when the second presentation is a test trial or a trial involving a test plus a restudy opportunity for feedback (Glover, 1989; Landauer & Bjork, 1978; Whitten & Bjork, 1977). So, for instance, students study an item (say, a face–name pair) once, and on the second presentation the item is tested (e.g. the face is given and subjects try to produce the name) before the whole pair is presented again. As we have discussed in the previous section, testing provides a considerable boost to long-term retention. It is possible to further enhance the testing effect by spacing test trials. Glover (1989) examined the extent to which taking massed and spaced tests improved performance on a later test. Seventh graders studied materials that were part of their normal science course. Before the final test, the students received intervening tests according to one of four conditions: a single test, two massed tests, two spaced tests, or no intervening test. As can be seen in Figure 8.3, reconstructed from the data reported in Glover (1989), final test performance was worst in the no intervening test condition and best in the two spaced tests condition. Importantly, performance did not differ between the single test and two massed tests conditions. Additional tests were more effective than a single test only when they were spaced apart, possibly because of the additional processing involved in the second spaced test relative to the second massed test.

In another demonstration of the benefits of spaced testing, Landauer and Bjork (1978) asked participants to learn face–name pairs. The pairs were repeated according to different rehearsal (retrieval) patterns. The initial presentation was the face and a first and last name. On subsequent presentations, the pairs received a test-type practice in which the face was shown with the first name and a blank for participants to attempt recall of the last name. These presentations were either massed or spaced. In contrast to the massed condition, the items that had received spaced test-type practice showed much better recall on a final test (see Roediger & Karpicke, 2011 for a review).

Over 20 years ago, Dempster (1988) described the irregular use of the spacing effect in education as 'a failure to apply the results of psychological research'. In spite of the many demonstrations of enhanced learning of educationally relevant materials as a result of spacing, students are reluctant to space their learning. Why? For one reason, massed practice may just feel better. With massed study, processing each repetition is fluent whereas processing during a spaced practice can feel more challenging. Similarly, when tests are massed, recall attempts are easier and so success rates are high.

Metacognitive assessments about whether something will be remembered in the future are often based on feelings of fluency, even though it can be a misleading index of later retention

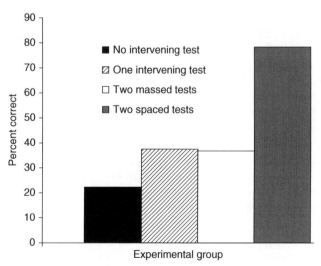

Fig. 8.3 Percent correct in a flower labelling task as a function of the number of intervening tests (0, 1 or 2) and the distribution of those tests (massed or spaced). One test was better than 0 tests, and a second test was only more beneficial than a single test if the two tests were spaced rather than massed. Reported in Glover (1989; Experiment 3).

(Benjamin, Bjork & Schwartz, 1998). Because learning with massed practice feels faster and easier than when practice is spaced (even when massing produces inferior performance), people rate massing as the more effective learning strategy, as they did in the Kornell and Bjork (2008) study cited earlier (see, too, Baddeley & Longman, 1978 and Zechmeister & Shaughnessy, 1980). This confidence is misplaced, because often the easier something is to process, the worse recall will be later (Bjork, 1994; cf. Koriat, 2008). Many studies have shown that when given the opportunity to control their own learning, people will usually choose to mass rather than space their study (e.g. Landauer & Ross, 1977), and this is especially so when learning feels more difficult to the subjects (Toppino, Cohen, Davis & Moors, 2009; but see Benjamin & Bird, 2006). People will space items that they judge to be well learned (Son, 2004, Toppino et al., 2009), although young children tend to mass their practice (Son, 2005).

In sum, most studies show that spaced practice provides benefits relative to massed practice, yet too often teachers provide information in massed fashion in class (one topic, then another topic, then a third and so on) rather than trying to space out reviews of previously covered topics. In addition, students tend to study the same way, concentrating or massing their efforts on one topic after another. As Dempster (1988) noted, spacing of information would doubtless be a boon to education if it could be more widely adopted in schools and if students could be taught to study using this principle.

8.5 Interleaving of topics

Just as spacing is more beneficial to learning than massing, interleaving different sets of materials during study can promote learning as compared to blocking materials together. The interleaving of materials always involves their spacing, but interleaving goes beyond spacing by having students study one skill, or type of problem, or set of material, and then cycle through the various skills, etc., in various orders. (Spacing can be achieved in other ways besides interleaving, as discussed later.) This is opposed to practising the same skill or set of material repeatedly and then moving on to another. For example, if children are learning the skills of multiplication and division, one strategy for teaching (and a common one) is to have them do 25 multiplication problems and then 25 division problems. This blocked presentation often promotes accurate and efficient performance after a short period of practice. The interleaving alternative is to randomly alternate multiplication and division problems. Under these conditions, accuracy and speed grow more slowly over trials, but the benefits are more long lasting. That is, in most studies retention is much better following interleaved practice than following massed practice. However, as with spacing, students and teachers generally prefer massed training because students seem to learn so quickly under these conditions.

The benefits of interleaving have been demonstrated with physics problems (Rohrer & Taylor, 2007), mathematics problems (Le Blanc and Simon, 2008) and even learning to identify artists (Kornell & Bjork, 2008—as described in the previous section). As an example, Rohrer and Taylor had students learn how to find the volume of four geometric solids. Students then practised solving problems for each solid, in one of two conditions: either the problems for one solid were all solved before moving on to the next solid or the problems for all four solids were randomly intermixed. On a follow-up test, students in the intermixed conditions performed over three times better than students who had practised the problems blocked by each type of solid. Despite these benefits, an overwhelming majority of mathematics (and other) textbooks block their practice problems.

Before any strong conclusions can be drawn, however, it is important to distinguish between the benefit arising from spacing described in detail previously, and the benefit derived from interleaving specifically, because (as just noted) interleaving by definition involves spacing. Taylor and Rohrer (2010) designed a study to do just that by creating blocked tasks that involved spacing and comparing performance on interleaved problems (with spacing). The basic question they addressed is 'Does interleaving problems benefit their later retention more than spacing the problems?'

Rohrer and Taylor (2010) had students aged 9–10 years solve four different types of mathematics problems. In the task, students were shown a picture of a prism and told how many base sides it had. They then had to use one of four formulas to calculate either the number of faces (number of base sides + 2), the number of corners (number of base sides × 2), the number of edges (number of base sides × 3), or the number of angles (number of base sides × 6). In the learning phase, students either solved blocks of the same problem (e.g. they repeatedly calculated the number of corners in different prisms before moving on to calculate the number of angles in a different set of prisms), or they practised all four types of problems in an interleaved fashion. This is the typical comparison of blocking and interleaving, but in their experiment there was a twist imposed on the blocked condition because it included a 30-second filler task (puzzles unrelated to the mathematics problems of interest) between sets of the same type of problem. The reason for inserting these filler tasks was to match the gap between problems of the same kind with that of the interleaved group. In other words, because there were four different problem types, if a student in the interleaved condition attempted a problem that involved calculating the number of faces first, they would attempt the problems involving corners, edges and angles before being given a further problem where they had to calculate the number of faces. At 10 seconds per problem, the gap between the first and second problems involving calculating the number of faces was 30 seconds—exactly the same length of time that students in the blocked group performed a filler task. Thus the blocked condition was changed to be the same as a typical spaced condition in a spacing effect experiment; therefore, any improved learning in the interleaved condition can be attributed to interleaving specifically as opposed to spacing more generally.

Figure 8.4 shows average performance during learning in the two conditions as well as performance on a delayed test the next day. Despite the fact that the blocked condition involved spacing in

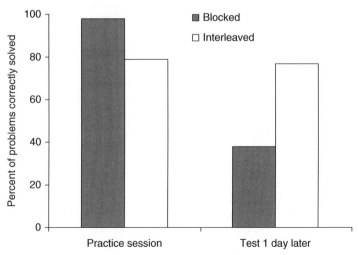

Fig. 8.4 Performance of students aged 9–10 years on four different types of mathematical problems during a practice session and on a test a day later, from an experiment by Taylor and Rohrer (2010). In the blocked condition, each type of problem was practised separately. In the interleaved condition, all four problem types were randomly intermixed during practice. Reproduced from Kelli Taylor and Doug Rohrer, The effects of interleaved practice, *Applied Cognitive Psychology*, 24(6), 837–48, © 2010, John Wiley and Sons.

this experiment, interleaving had its usual effects on performance. The bars on the left of Figure 8.4 show that interleaving slowed learning relative to blocking, as it usually does. That is partly why both students and teachers do not like to interleave practice—it slows initial learning. Yet performance on a test a day after learning was far better in the interleaved compared with the blocked condition, as shown by the bars on the right side of Figure 8.4. Comparing the two sides of Figure 8.4, it is apparent that there was great forgetting in the blocked condition and none in the interleaved condition. Thus, interleaving has a benefit on learning (as measured on delayed tests) over and above spacing. Presumably there is something about the effort of switching between tasks and reinstating the mental procedures necessary for solving them that creates better retention for the long term in the interleaved condition.

Although interleaving has not been studied from a neuroscientific perspective (yet), a similar concept has been studied extensively under a different name. Multitasking—defined by Delbridge (2001) as 'engaging in frequent switches between individual tasks'—has been studied extensively by neuroscientists, with the focus largely on identifying and locating the costs of switching between tasks. This line of research paints a negative picture of interleaving, because performance and brain activity are measured during the learning phase and never during later recall, when the benefits of interleaving appear. Neurological theories of why interleaving works would be purely speculative at this stage, but it is reasonable to propose that the extra cognitive load caused by switching between tasks or types of problems in an interleaved study phase actually benefits retrieval.

The take away point for the classroom is that the customary practice of blocking should be diminished in favour of more interleaved and spaced practice. Students learning to print letters should not necessarily do A 25 times, then B and so on. Massed practice creates quick learning but also produces quicker forgetting relative to interleaved (or spaced) practice.

8.6 Metacognition

Surveys of university students' study practices show that they differ greatly from those that cognitive psychologists recommend (Karpicke, Butler & Roediger, 2009; Kornell & Bjork, 2007; McCabe, 2011). Why might this be so? One fact might be that many good learning strategies proposed by cognitive psychologists tend to hurt performance in the short term while improving it in the long term. Students may be more focused on short-term processes—ones that they can readily judge—and so be guided by their immediate intuitions and beliefs. Central to this paradox is the issue of whether students are able to monitor and regulate their own learning in an effective manner.

The primary model of metacognition (Nelson & Narens, 1990) proposes that people evaluate the progress of their learning and use this evaluation to regulate study strategies. Students use their evaluations to plan and control their activities such as allocating additional study time to material that seems to be learned less well. Research has shown that metacognitive monitoring plays a key role in how people control their study (e.g. Metcalfe & Finn, 2008; Thiede, 1999; but see Weinstein, Finn, McDermott & Roediger, in preparation). The efficacy of this process of course hinges on the accuracy of the metacognitive assessments as well as the integrity of the strategies that are implemented.

In the sections above we touched on how students' metacognitive evaluations can be insensitive to the beneficial effects that a number of learning strategies (retrieval practice, spacing and interleaving) have on learning. In all three cases, the 'easy' version of the task that seems most natural—repeated massed study—leads to poorer long-term retention than the alternative strategy (testing, spaced or interleaved study). In some cases, metacognitive inaccuracies may occur due to reliance on heuristics that would be predictive in other instances, as Kahneman (2003) has argued.

For example, massed study works fine for an immediate test so students use it; however, massed practice is poor for performance on delayed tests, but students may not realize that fact and use the strategy even when it is inappropriate.

Metacognitive inaccuracies can also arise from reliance on cues like familiarity or fluency that may not be diagnostic of whether or not something is learned. For example, rereading may lead to fluent processing of the material and this fluency can lead to overconfidence that one knows the material that seems so fluent and familiar (e.g. Benjamin et al., 1998; Glenberg, Sanocki, Epstein & Morris, 1987; Metcalfe, Schwartz & Joaquim, 1993; Reder & Ritter, 1992). Thus, overconfident students who reread the material repeatedly may stop studying before they have truly mastered material because of faulty monitoring (e.g. Metcalfe & Finn, 2008; Zechmeister & Shaughnessy, 1980). Because monitoring is tightly linked to the control of study, students need to learn how to evaluate their current state of knowledge by using cues or tests that are diagnostic of their state of learning; otherwise they can fall prey to illusions of knowing. For example, if students test themselves at a reasonable delay from learning, they can discover whether or not they can produce the studied material when it is needed. If they do not quiz themselves but simply continue to restudy, the material may seem fluent and easy, but they may not be able to retrieve it after a delay (e.g. Roediger & Karpicke, 2006).

One technique used to study the development of metacognitive knowledge during learning is called the judgement of learning (JOL) paradigm. The JOL paradigm simply asks students to predict whether they will later be able to recall or recognize a fact that has been learned. If you learn today that Jefferson City is the capital of Missouri, you might be asked to judge how well you could recall this fact in a month. If people are asked to perform JOLs while studying material, the JOLs are usually not very accurate (though usually not at chance, either); however, if the JOLs are made after a short delay when attempting to retrieve the correct target information, they are much more predictive of later success (Nelson & Dunlosky, 1994). During study, while students are examining the material, the facts being learned may seem easy and fluent; yet when students judge learning after even a brief delay, they may appreciate how difficult the material is to retrieve and thus not be fooled. As noted earlier, metacognitive accuracy can also be improved following a test (Finn & Metcalfe, 2007; 2008; Koriat & Bjork, 2006) by asking students to think of reasons why their answer might be wrong (Koriat, Litchtenstein & Fischhoff, 1980), by asking students to summarize what they have learned before making their judgements (Thiede & Anderson, 2003), and by warning them about the possibility of bias (Jacoby & Whitehouse, 1989).

When metacognitive accuracy is good, control of study efforts is effective (Nelson, 1992). When people control their own study time, their performance is better than when they are given a random amount of time (Atkinson, 1972; Mazzoni & Cornoldi, 1993) or when they are given the opposite value of the amount of time they chose for study (Kornell & Metcalfe, 2006). Nelson (1993) had students study vocabulary pairs and then make JOLs. After making the JOLs, students were given additional study trials for some items. One group of participants received additional study time for the items that they had given the lowest JOLs, which indicated that they were sure that they would not remember them on a later test. Another group restudied the items that they had given the highest JOLs, items that they were sure they would remember. The control group restudied items that were made up of half high JOL and half low JOL items. The group that had restudied the low JOL items performed the best on the follow-up test, indicating that people were sensitive to what they did and did not know and that studying the items that they did not know benefited learning.

A large body of literature has shown that people often choose spend the most time studying the items that they have given the lowest JOLs (see Son & Metcalfe, 2000 for a review). This pattern fits with the idea that people will study until there is no longer a discrepancy between their judgement of learning and the learning objective. This idea is called the discrepancy reduction model of study time allocation and was proposed by Dunlosky and Thiede (1998) to explain why students tend to study material they judge as more difficult for longer periods of time. However, a discrepancy reduction strategy does not always work. For some (by definition, difficult) material, students can be given a large amount of time to study and still not master the material. Nonetheless, when permitted to control their study strategies, students persist in processing this difficult material even when studying has diminishingly small effects, a process Nelson and Leonesio (1988) called labouring in vain. People often do not know when to stop studying (or perhaps to change strategies and study in a different way).

When study time is limited, as it usually is, students may not always study the most difficult material. Work by Metcalfe and her collaborators has shown that people do not always choose to study the most difficult items when faced with learning a list of pairs that vary in difficulty (Metcalfe, 2002; Metcalfe & Kornell, 2003, 2005; Kornell & Metcalfe, 2006). Under time pressure, study strategies shift. With time constraints, items that are of moderate difficulty are prioritized over items that are already known or items that are too difficult. These moderately difficult items are in what Metcalfe and her collaborators have called the 'region of proximal learning.' That is, students will go for the low-hanging fruit—the material that has not been learned but that seems to be learnable in a short amount of time. In support of this model, Metcalfe and Kornell (2005) showed that when given the option, students prioritize easier unknown items in preference to extremely difficult items.

Additional support for this region of proximal learning idea comes from research by Kornell and Metcalfe (2006). In one series of experiments, students were given an initial test on Spanish–English vocabulary pairs. From the set of incorrect items participants selected half for restudy. Results showed that performance on the final test was better when people's study choices were honoured and they were given the items that they had selected for restudy relative to when their choices were dishonoured and they were given the items that they had not selected. Importantly, the items that they chose were the easiest items (according to norms) but ones as yet not learned.

Some neuroscientific evidence does bear on the issue of metacognition when learning. Converging evidence from patient and neuroimaging studies suggest that a number of regions in the prefrontal cortex are activated in metacognitive monitoring, and that the areas involved in metamemory do not completely overlap with those areas involved in remembering (see, e.g. Pannu & Kaszniak, 2005; Schwartz & Bacon, 2008; and Shimamura [2008] for a review). Using functional magnetic resonance imaging (fMRI), Kao, Davis and Gabrieli (2005) investigated the neural circuitry that mediates JOLs. Participants studied pictures and predicted how well they would later remember each scene. Results showed that activation in the medial temporal lobe was associated with actual, but not predicted, recall success. Activation in the left ventromedial prefrontal cortex was associated with predictive accuracy, but not with actual performance. Areas in the lateral and dorsomedial prefrontal cortex were associated with both successful recall and JOLs. The results point to at least somewhat distinct processes involved in remembering and judgements about remembering.

Research with patients with damage to their frontal lobes provides further support for the idea that the prefrontal cortex is important for metacognitive monitoring and control. Lesions in the prefrontal cortex have been associated, in particular, with impairments in feeling of knowing (FOK) judgements (Janowsky, Shimamura & Squire, 1989; Pannu & Kaszniak, 2005; Schnyer et al., 2004). During a FOK task participants make a judgement about the likelihood that they will be able to recognize the answer to the question that they were not able to correctly answer at that time. An example is: 'What is the capital of Kentucky?'. If the answer is unknown, a FOK judgement is made. Later the participant is asked to recognize the answer from among several alternatives: Lexington, Louisville, Frankfort or Paducah? The FOK judgement taps into people's access to partial information about the target, which might be comprised of the first letter (Koriat & Leiblich, 1975), syntactic characteristics (Miozzo & Caramazza, 1997) or other kinds of knowledge (Metcalfe & Finn, 2011). Frontal patients have much worse accuracy in their predictions of future performance (e.g. Janowsky et al., 1989; Pinon, Allain, Kefi, Dubas & LeGall, 2005; Vilkki, Servo & Surma-aho, 1998).

The frontal lobes develop throughout childhood and are thought to mature in adolescence or even early adulthood (e.g. Welsh & Pennington, 1988; Welsh, Pennington & Groisser, 1991; and see Romine & Reynolds, 2005 for a review) and metacognitive capacities follow a similar developmental trajectory (e.g. Schneider, 2010). For example, while young children's metamemory judgements can be quite inaccurate, by the time they reach middle school their ability to monitor memory appropriately seems to be in place (e.g. Metcalfe & Finn, submitted; Schneider & Lockl, 2002). For learning to proceed successfully though, the output of metacognitive evaluations needs to be implemented into good study strategies. Research suggests that the ability to monitor accurately may appear before the student knows how to use that information effectively (Metcalfe & Finn, submitted; Schneider, 2010). Metacognitive control can lag behind accurate monitoring, but it does show age-related improvements (e.g. Dufresne & Kobasigawa, 1989; Lockl & Schneider, 2004). While adults and older children spend more time studying items that they do not know, younger children make less adaptive choices and may need assistance to choose optimally (Metcalfe & Finn, submitted).

Because metacognitive processes are used to regulate children's learning behaviours, it is crucial that their monitoring is accurate and their control decisions are adaptive. Several studies have shown that overconfidence is related to poor exam performance (Bol & Hacker, 2001; Kruger & Dunning, 1999). Moreover, overconfidence in a particular study strategy may encourage students to persist in using ineffective strategies (Hacker, Bol & Bahbahani, 2008). Low-achieving students face a double burden: Over a wide spectrum of academic domains, those students with the lowest levels of performance show the greatest overconfidence (e.g. Kruger & Dunning, 1999; and see Hacker et al., 2008 for a review). Fortunately, metacognitive strategies can be successfully trained (e.g. Brown & Campione, 1990), and research points toward the importance of classroom practices in teaching students how to regulate their own learning (see Schneider & Pressley, 1997 for a review). Specifically, instructing students as to why a particular study strategy is useful increases use of that strategy (O'Sullivan & Pressley, 1984). Pressley and collaborators have shown that the most effective teachers are those who incorporate instruction about how to use metacognition to select and modify study strategies (e.g. Pressley, Goodchild, Fleet & Zajchowski, 1989).

As Hacker and others (2008) have noted, however, the large bulk of research on metacognition has been conducted in laboratory settings with hopes that it can be generalized to more naturalistic educational contexts. The needed translational research is just now underway. Findings from the lab do not always transfer neatly into the classroom. In classroom settings, teachers may need to provide more information on how to use metacognition appropriately, especially with children and with low-achieving students. By explicitly training metacognitive monitoring and providing feedback about strategy selection, educators can improve how students use metacognition to learn (Hacker, 2004; Nietfeld, Cao & Osborne, 2005).

8.7 **Transfer of learning**

The techniques described so far in this chapter—testing, spacing, interleaving—will not be of much use unless the knowledge and techniques acquired during their implementation can be transferred to other materials and situations. Do these techniques lead to encapsulated knowledge or do they lead to learning that is flexible and can be transferred to new contexts? *Transfer* has been a buzzword in educational policy ever since Edward Thorndike and Robert Woodworth began studying the topic over 100 years ago (Thorndike & Woodworth, 1901a, 1901b, 1901c). Today many educational organizations would agree that 'A main reason for formal education is to facilitate learning in situations outside school' (Klausmeier, 1961, p. 352). The exact definition of transfer has been long debated and continues to be discussed (see, e.g. Barnett & Ceci, 2002; Beach, 1994). In the context of education, transfer can refer to the production of the same piece of information in response to two differently worded questions, but also to the application of problem-solving skills acquired in one context to another. The former sort of transfer is often called near transfer and the latter is termed far transfer, with the distance terms referring to how similar the new task is to one that was trained. If transfer is achieved when the new task is only a slight modification of the original one, then this is a case of near transfer. When the two tasks are conceptually and procedurally different and yet transfer is achieved, it is said to be far transfer. Not surprisingly, many studies find near transfer to be quite robust, but relatively few have reported far transfer (although such studies do exist, as we shall see). In light of the effective study techniques described earlier, transfer could also refer to the propensity to use these techniques with new materials once their efficiency has been experienced.

Despite educators' general agreement that transfer of learning is crucial, surprisingly little evidence of such transfer exists in the lab, much less in the schools. In part, this difficulty arises from differences in definitions of transfer (and in near and far transfer). In Thorndike and Wordsworth's (1901) original work on the topic entitled 'The influence of improvement in one mental function upon the efficiency of other functions', the authors were disappointed to find zero evidence of transfer. For example, subjects trained to estimate the areas of rectangles showed no transfer to estimating the areas of other shapes. This led the authors to conclude that 'Improvement in any single mental function rarely brings about equal involvement in any other function, no matter how similar . . .' (p. 247). More recently, this conclusion was echoed by Detterman (1993), who branded transfer an epiphenomenon and argued strongly that no real evidence of transfer had been produced by the literature. On the other side of the debate, Schwartz, Bransford and Sears (2005) argued that the classic definitions of far transfer are too narrow and lead to incorrect conclusions of failed transfer in many studies. According to their view, failure of transfer occurs because researchers are expecting their students to directly map a learned procedure onto an entirely new context rather than modifying learned procedures to suit the new situation. Schwartz et al. further suggested that a more important facet of transfer involves improvement in the ability to learn new ways of solving problems. With this revised definition, transfer can be identified in many situations, and the picture looks less bleak.

Nonetheless, the distinction between near and far transfer, referring to the degree to which the initial learning and transfer contexts differ, has allowed important advances in research on transfer. Educators are most concerned with far transfer (i.e. transfer to a context largely dissimilar to the initial learning context), but even the definition of far transfer has been problematic. For instance, Hamers, de Koning and Sijtsma (1998) claimed to have shown far transfer when the 8–9-year-olds trained on attribute classification transferred that skill to an intelligence quotient (IQ) test. However, Barnett and Ceci (2002) argued that since both the trained task and the transfer task were from the same *domain* (paper and pencil tests) the authors were really demonstrating near transfer. In an attempt to prevent what they considered to be misrepresentation of far transfer, Barnett and Ceci proposed a taxonomy of transfer, a framework for evaluating past experiments and designing new ones. They identified two major factors—content and context—with the latter consisting of a set of six dimensions that could be used to determine whether a given situation involves near or far transfer: *knowledge domain, physical context, temporal context, functional context, social context* and *modality*. For instance, the *functional context* dimension describes the motivational mind set involved in an activity. Answering a geography question in the context of a school quiz and then remembering the same fact during a conversation with friends would thus involve far transfer due to the different functions that the piece of information served in the two cases. Figure 8.5 provides examples of near and far transfer involving each of the six dimensions.

Barnett and Ceci (2002) did describe two studies that demonstrated far transfer by their criteria. Fong, Krantz and Nisbett (1986) showed that statistical training in college courses transferred to an unexpected phone survey that required use of statistical knowledge. Likewise, Ceci and Ruiz (1993) found that real-life racetrack handicapping skills transferred to a laboratory-based stock market experiment; better handicapping skills led to better performance in the laboratory experiment that required use of the skills in a totally different context. However, most dimensions outlined by Barnett and Ceci (e.g. *social context*, or learning in collaboration with others at school but then coming up with the information alone during a test), have not received as much attention in the literature. Whether such transfer can be demonstrated must await future research.

Working closely with the Barnett and Ceci (2002) framework, Butler (2010) designed a series of experiments to vary only the knowledge domain dimension. After studying passages on topics such as 'tropical cyclones' and 'vaccines', students answered inferential questions within the same knowledge domain (for near transfer) or a different domain (for far transfer). As an example of near transfer, after answering a question on the general uses of vaccines, students would attempt a question about the purpose of a particular vaccine. The far transfer question could be about the function of a vehicle, and students were expected to draw upon their understanding of the human body gained from the vaccine passage to correctly explain the processes occurring inside the

	Near ←				→ Far
Knowledge domain	Mouse vs. rat	Biology vs. botany	Biology vs. economics	Science vs. history	Science vs. art
Physical context	Same room at school	Different room at school	School vs. research lab	School vs. home	School vs. the beach
Temporal context	Same session	Next day	Weeks later	Months later	Years later
Functional context	Both clearly academic	Both academic but one nonevaluative	Academic vs. filling in tax forms	Academic vs. informal questionnaire	Academic vs. at play
Social context	Both individual	Individual vs. pair	Individual vs. small group	Individual vs. large group	Individual vs. society
Modality	Both written, same format	Both written, multiple choice vs. essay	Book learning vs. oral exam	Lecture vs. wine tasting	Lecture vs. wood carving

Fig. 8.5 Examples of near and far transfer for the six domains defined by Barnett and Ceci (2002; adapted from their figure 1B).

vehicle—clearly a very different domain to the one they had studied! Crucially, the design also involved a comparison between testing and restudying, to investigate which technique led to greater transfer (either near or far transfer). That is, students either repeatedly studied facts or they studied the facts and took tests on them; then they took a final test that assessed near or far transfer as described above.

Butler (2010) found that initial testing led to better transfer than repeated studying (a testing effect), and further that the testing effect occurred on both tests of near transfer and far transfer. This is an important finding for the testing effect, since it demonstrates that testing is not only beneficial for retaining the specific facts that are rehearsed at retrieval, but also improves transfer of knowledge to new domains.

In another study, Rohrer, Taylor and Sholar (2010) had 9–11-year-olds study or practise retrieving city locations. Afterwards, the students were given a task in which they had to determine which city they would drive through on a given route. Performance on the driving task was better after retrieval practice than after repeated study of the city locations, another indication of transfer. Evidence of improved transfer has also been demonstrated for other techniques described above such as spacing; that is, spaced study leads to greater transfer on a later test than does massed study (Helsdingen, van Gog & van Merriënboer, 2009).

Atherton (2007) proposed a neuroscientific theory of transfer. According to this theory, transfer will occur insofar as the brain regions activated in different contexts are interconnected. Atherton cites demonstrations of transfer from music ability to language processing (e.g. Anvari, Trainor, Woodside & Levy, 2002) along with findings of overlap between brain regions involved in music and language processing (e.g. Koelschet al., 2004) as indirect empirical evidence of this hypothesis. From an alternative perspective, Haskell (2001) proposed that the human brain evolved to support transfer. Clearly, these new theories need to be tested empirically before they can guide educational practice; for now, Barnett and Ceci's (2002) taxonomy of transfer and carefully specified behavioural work are leading the way.

Although studies of transfer of learning in educational contexts lag behind other realms of research reviewed in this chapter, this is a critical topic for future research. Most education aims to be useful, to provide students with skills and knowledge they will need throughout life. Thus educators will need to know how to teach (and have students study) to create flexible knowledge that can be retained and used over long periods of time.

8.8 Conclusion

This chapter has shown how five different topics studied by cognitive psychology can have implications for education: retrieval practice through testing; spaced periods of study of the same topic; interleaving of different domains of study; improving metacognitive monitoring of students; and teaching in ways that will facilitate transfer of learning to novel situations. These five topics are some of those that cognitive psychology can contribute to educators to help guide practice. However, we believe that these are only some of the tools that are needed. The strategies outlined here assume students will learn readily and will be motivated to learn, but sadly these assumptions are often not met. Indeed, among full-time university students only about 11% of first years spend as much time preparing for class as their professors expect (NSSE, 2008). Developmental and social psychologists, among others, must be enlisted to help understand how students can become motivated to learn and want to apply the strategies described here. The task for us all is long, and we can hope that neuroscientific approaches will increasingly shed light on these issues in the future.

References

Abbott, E. E. (1909). On the analysis of the factors of recall in the learning process. *Psychological Monographs*, 11, 159–77.

Anvari, S. H., Trainor, L. J., Woodside, J., & Levy, B. A. (2002). Relations among musical skills, phonological processing and early reading ability in preschool children. *Journal of Experimental Child Psychology*, 83, 111–30.

Arnold, K. M., & McDermott, K. B. (in preparation). Test potentiated learning. Manuscript in preparation.

Atherton, M. (2007). *A Proposed Theory of the Neurological Limitations of Cognitive Transfer*. Paper presented at the annual meeting of the American Educational Research Association, 9–13 April, Chicago, IL.

Atkinson, R. C. (1972). Optimizing the learning of a second language vocabulary. *Journal of Experimental Psychology*, 96, 124–29.

Baddeley, A. D., & Longman, D. J. A. (1978). The influence of length and frequency of training session on the rate of learning to type. *Ergonomics*, 21, 627–35.

Bahrick, H. P., & Phelps, E. (1987). Retention of Spanish vocabulary over 8 years. *Journal of Experimental Psychology: Learning, Memory, and Cognition*, 13, 344–49.

Balota, D. A., Duchek, J. M., & Paullin, R. (1989). Age related differences in the spacing of repetitions and retention interval. *Psychology & Aging*, 4, 3–9.

Barnett, S. M., & Ceci, S. J. (2002). When and where do we apply what we learn?: A taxonomy for far transfer. *Psychological-Bulletin*, 128(4), 612–37.

Beach, K. D. (1994). *A Sociohistorial Alternative to Economic and Cognitive Transfer Metaphors for Understanding the Transition from School to Work*. Paper presented at the annual meeting of the American Educational Research Association, April, New Orleans, LA.

Benjamin, A. S., & Bird, R. (2006). Metacognitive control of the spacing of study repetitions. *Journal of Memory & Language*, 55, 126–37.

Benjamin, A. S., Bjork, R. A., & Schwartz, B. L. (1998). The mismeasure of memory: When retrieval fluency is misleading as a metamnemonic index. *Journal of Experimental Psychology: General*, 127, 55–68.

Bjork, R. A. (1979). An information-processing analysis of college teaching. *Educational Psychologist*, 14, 15–23.

Bjork, R. A. (1994). Memory and metamemory considerations in the training of human beings. In J. Metcalfe and A. Shimamura (Eds.), *Metacognition: Knowing about Knowing* (pp. 185–205). Cambridge, MA: MIT Press.

Bol, L., & Hacker, D. (2001). The effect of practice tests on students' calibration and performance. *Journal of Experimental Education*, 69, 133–51.

Brown, A. L., & Campione, J. C. (1990). Communities of learning and thinking or a context by any other name. In D. Khun (Ed.), *Contributions to Human Development*, 21, 108–25.

Bruer, J. (1997). Education and the brain: A bridge too far. *Educational Researcher*, 26(8), 4–16.

Butler, A. C. (2010). Repeated testing produces superior transfer of learning relative to repeated studying. *Journal of Experimental Psychology: Learning, Memory, and Cognition*, 36, 1118–33.

Butler, A. C., Karpicke, J. D., & Roediger, H. L. (2007). The effect of type and timing of feedback on learning from multiple-choice tests. *Journal of Experimental Psychology: Applied*, 13, 273–81.

Butler, A. C., Karpicke, J. D., & Roediger, H. L. (2008). Correcting a metacognitive error: Feedback increases retention of low-confidence correct responses. *Journal of Experimental Psychology: Learning, Memory, and Cognition*, 34, 918–28.

Butler, A. C., & Roediger, H. L. (2007). Testing improves long-term retention in a simulated classroom setting. *European Journal of Cognitive Psychology*, 19, 514–27.

Carew, T. J., & Magsamen, S. (2010). Neuroscience and education: An ideal partnership for producing evidence based solutions to guide 21st century learning. *Neuron*, 67, 685–88.

Carpenter, S. K., Pashler, H., & Cepeda, N. J. (2009). Using tests to enhance 8th grade students' retention of U. S. history facts. *Applied Cognitive Psychology, 23,* 760–71.

Carpenter, S. K., Pashler, H., Wixted, J., & Vul, E. (2008). The effects of tests on learning and forgetting. *Memory & Cognition, 36,* 438–48.

Carrier, M., & Pashler, H. (1992). The influence of retrieval on retention. *Memory & Cognition, 20,* 632–42.

Ceci, S. J., & Ruiz, A. (1993). The role of context in everyday cognition. In M. Rabinowitz (Ed.), *Applied Cognition* (pp. 164–83). Hillsdale, NJ: Erlbaum.

Cepeda, N. J., Pashler, H., Vul, E., Wixted, J., & Rohrer, D. (2006). Distributed practice in verbal recall tasks: A review and quantitative synthesis. *Psychological Bulletin, 132,* 354–80.

Cermak, L. S. (1975). Imagery as an aid to retrieval for Korsakoff's patients. *Cortex, 77,* 163–69.

Cermak, L. S., Reale, L., & DeLuca, D. (1977). Korsakoff patient's nonverbal vs verbal memory: Effects of interference and mediation on rate of information loss. *Neuropsychologia, 15,* 303–310.

Cermak, L. S., Verfaellie, M., Lanzoni, S., Mather, M. & Chase, K. A. (1996). The spacing effect on the recall and recognition ability of amnesic patients. *Neuropsychology, 10,* 219–27.

Davis, M. (1970). Effects of interstimulus interval length and variability on startle-response habituation in the rat. *Journal of Comparative and Physiological Psychology, 78,* 260–67.

Delbridge, K. (2001). *Individual Differences in Multi-tasking Ability: Exploring a Nomological Network.* Unpublished Doctoral Dissertation, Michigan State University, East Lansing.

Dempster, F. N. (1987). Effects of variable encoding and spaced presentations on vocabulary learning. *Journal of Educational Psychology, 79,* 162–70.

Dempster, F. N. (1988). The spacing effect: A case study in the failure to apply the results of psychological research. *American Psychologist, 43,* 627–34.

Dempster, F. N. (1996). Distributing and managing the conditions of encoding and practice. In E. L. Bjork & R. A. Bjork (Eds.), *Handbook of Perception and Cognition, Vol.10: Memory,* (pp. 317–44). San Diego, CA: Academic Press.

Detterman, D. K. (1993). The case for prosecution: Transfer as an epiphenomenon. In D. K. Detterman, & R. J. Sternberg, *Transfer on Trial: Intelligence, Cognition, and Instruction,* (pp. 1–24). Norwood, NJ: Ablex Publishing Corporation.

Dufresne, A., & Kobasigawa, A. (1989). Children's spontaneous allocation of study time: Differential and sufficient aspects. *Journal of Experimental Child Psychology, 47,* 274–96.

Dunlosky, J., & Thiede, K. W. (1998). What makes people study more? An evaluation of factors that affect self-paced study. *Acta Psychologica, 98,* 37–56.

Ebbinghaus, H. (1885/1964). *Memory: A Contribution to Experimental Psychology.* Oxford: Dover.

Finn, B., & Metcalfe, J. (2007). The role of memory for past test in the underconfidence with practice effect. *Journal of Experimental Psychology: Learning, Memory, and Cognition, 33,* 238–44.

Finn, B., & Metcalfe, J. (2008). Judgments of learning are influenced by memory for past test. *Journal of Memory and Language, 58,* 19–34.

Fong, G. T., Krantz, D. H., & Nisbett, R. E. (1986). The effects of statistical training on thinking about everyday problems. *Cognitive Psychology, 18,* 253–92.

Gianutsos, R., & Gianutsos, J. (1979). Rehabilitating the verbal recall of brain-injured patients by mnemonic training: An experimental demonstration using single-case methodology. *Journal of Clinical Neuropsychology, 2,* 117–35.

Glenberg, A. M. (1976). Monotonic and nonmonotonic lag effects in paired-associate and recognition memory paradigms. *Journal of Verbal Learning and Verbal Behavior, 15,* 1–15.

Glenberg, A. M. (1979). Component-levels theory of the effects of spacing of repetitions on recall and recognition. *Memory & Cognition, 7,* 95–112.

Glenberg, A. M., Sanocki, T., Epstein, W., & Morris, C. (1987). Enhancing calibration of comprehension. *Journal of Experimental Psychology: General, 116,* 119–36.

Glover, J. A. (1989). The 'testing' phenomenon: Not gone but nearly forgotten. *Journal of Educational Psychology*, 81, 392–99.

Greene, R. L. (1990). Spacing effects on implicit memory tests. *Journal of Experimental Psychology: Learning, Memory, and Cognition*, 16, 1004–1011.

Hacker, D. J. (2004). Self-regulated comprehension during normal reading. In R. B. Ruddell & N. Unrau (Eds.), *Theoretical models and processes of reading* (5th ed., pp. 775–79). Newarek, DE: International Reading Association.

Hacker, D. J., Bol, L., & Bahbahani, K. (2008). Explaining calibration in classroom contexts: The effects of incentives, reflection, and attributional style. *Metacognition and Learning*, 3, 101–21.

Hamers, J. H. M., De Koning, E., & Sijtsma, K. (1998). Inductive reasoning in the third grade: Intervention promises and constraints. *Contemporary Educational Psychology*, 23, 132–48.

Haskell, R. E. (2001). *Transfer of Learning: Cognition, Instruction and Reasoning*. New York: Academic Press.

Helsdingen, A. S., van Gog, T., & van Merriënboer, J. J. G. (2009). The effects of practice schedule on learning a complex judgment task. *Learning and Instruction*, 21(1), 126–36.

Hintzman, D. L. (1974). Theoretical implications of the spacing effect. In R. L. Solso (Ed.), *Theories in Cognitive Psychology: The Loyola Symposium*. Hillsdale, NJ: Erlbaum.

Hintzman, D. L., Block, R. A., & Summers, J. J. (1973). Modality tags and memory for repetitions: Locus of the spacing effect. *Journal of Verbal Learning and Verbal Behavior*, 12, 229–38.

Izawa, C. (1970). Optimal potentiating effects and forgetting-prevention effects of tests in paired-associate learning. *Journal of Experimental Psychology*, 83, 340–44.

Jacoby, L. L., & Whitehouse, K. (1989). An illusion of memory: False recognition influenced by unconscious perception. *Journal of Experimental Psychology: General*, 118, 126–35.

James, W. (1899/1958). *Talks to Teachers on Psychology: And to Students on Some of Life's Ideals*. New York: Holt.

Janowsky, J. S., Shimamura, A. P., & Squire, L. R. (1989). Source memory impairment in patients with frontal lobe lesions. *Neuropsychologia*, 27, 1043–56.

Kahneman, D. (2003). A perspective on judgment and choice: mapping bounded rationality. *American Psychologist*, 58, 697–720.

Kang, S. H. K., McDermott, K. B., & Roediger, H. L. (2007). Test format and corrective feedback modulate the effect of testing on memory retention. *European Journal of Cognitive Psychology*, 19, 528–58.

Kao, Y. C., Davis, E. S., & Gabrieli, J. D. E. (2005). Neural correlates of actual and predicted memory formation. *Nature Neuroscience*, 8(12), 1776–783.

Karpicke, J. D. (2009). Metacognitive control and strategy selection: Deciding to practice retrieval during learning. *Journal of Experimental Psychology: General*, 138, 469–86.

Karpicke, J. D., Butler, A.C., & Roediger, H. L. (2009). Metacognitive strategies in student learning: Do students practice retrieval when they study on their own? *Memory*, 17, 471–79.

Klausmeier, H. J. (1961). *Educational Psychology: Learning and Human Abilities*. New York: Harper.

Koelsch, S., Kasper, E., Sammler, D., Schulze, K., Gunter, T., & Friederici, A. D. (2004). Music, language and meaning: brain signatures of semantic processing. *Nature Neuroscience*, 7(3), 302–307.

Koriat, A. (2008). Easy comes, easy goes? The link between learning and remembering and its exploitation in metacognition. *Memory & Cognition*, 36, 416–28.

Koriat, A., & Bjork, R. A. (2006). Illusions of competence during study can be remedied by manipulations that enhance learners' sensitivity to retrieval conditions at test. *Memory & Cognition*, 34, 959–72.

Koriat, A., & Lieblich, I. (1975). Examination of the letter serial position effect in the 'TOT' and the 'Don't Know' states. *Bulletin of the Psychonomic Society*, 6, 539–41.

Koriat, A., Lichtenstein, S., & Fischhoff, B. (1980). Reasons for confidence. *Journal of Experimental Psychology: Human Learning and Memory*, 6, 107–118.

Kornell, N., & Bjork, R. A. (2007). The promise and perils of self-regulated study. *Psychonomic Bulletin & Review*, 14, 219–24.

Kornell, N., & Bjork, R. A. (2008). Learning concepts and categories: Is spacing the 'enemy of induction'? *Psychological Science, 19,* 585–92.

Kornell, N., Hays, M. J., & Bjork, R. A. (2009). Unsuccessful retrieval attempts enhance subsequent learning. *Journal of Experimental Psychology: Learning, Memory, & Cognition,* 35, 989–98.

Kornell, N., & Metcalfe, J. (2006). Study efficacy and the region of proximal learning framework. *Journal of Experimental Psychology: Learning, Memory, & Cognition,* 32, 609–22.

Kruger, J., & Dunning, D. (1999). Unskilled and unaware of it: How difficulties in recognizing one's own incompetence lead to inflated self-assessments. *Journal of Personality and Social Psychology,* 77, 1121–34.

Landauer, T. K., & Bjork, R. A. (1978). Optimal rehearsal patterns and name learning. In M. M. Gruneberg, P. E. Harris, & R. N. Sykes (Eds.), *Practical Aspects of Memory* (pp. 625–32). New York: Academic Press.

Landauer, T. K., & Ross, B. H. (1977). Can simple instruction to space practice improve ability to remember a fact? An experimental test using telephone numbers. *Bulletin of the Psychonomic Society,* 10, 215–218.

Le Blanc, K., & Simon, D. (2008). *Mixed Practice Enhances Retention and Jol Accuracy for Mathematical Skills.* Paper presented at the 49th Annual Meeting of the Psychonomic Society, Chicago, IL.

Lockl, K., & Schneider, W. (2004). The effects of incentives and instructions on children's allocation of study time. *European Journal of Developmental Psychology,* 1, 153–69.

Madigan, S. A. (1969). Intraserial repetition and coding processes in free recall. *Journal of Verbal Learning and Verbal Behavior,* 8, 828–35.

Mazzoni, G., & Cornoldi, C. (1993). Strategies in study-time allocation: Why is study time sometimes not effective? *Journal of Experimental Psychology: General,* 122, 47–60.

McCabe, J. (2011). Metacognitive awareness of learning strategies in undergraduates. *Memory & Cognition,* 39, 462–76.

McDaniel, M. A., Agarwal, P. K., Huelser, B. J., McDermott, K. B., & Roediger, H. L. (2011). Test-enhanced learning in a middle school science classroom: The effects of quiz frequency and placement. *Journal of Educational Psychology,* 103, 399–414.

McDaniel, M. A., Thomas, R. C., Agarwal, P. K., McDermott, K. B., & Roediger, H. L. (in preparation). Quizzing promotes transfer of target principles in middle school science: Benefits on summative exams. Manuscript in preparation.

Melton, A. W. (1967). Repetition and retrieval from memory. *Science,* 158, 532.

Melton, A. W. (1970). The situation with respect to the spacing of repetitions and memory. *Journal of Verbal Learning and Verbal Behavior,* 9, 596–606.

Metcalfe, J. (2002). Is study time allocated selectively to a region of proximal learning? *Journal of Experimental Psychology: General,* 131, 349–63.

Metcalfe, J., & Finn, B. (submitted). Metacognition and control of study choice in children. Manuscript submitted for publication.

Metcalfe, J., & Finn, B. (2008). Evidence that judgments of learning are causally related to study choice. *Psychonomic Bulletin and Review,* 15, 174–79.

Metcalfe, J., & Finn, B. (2011). People's hypercorrection of high confidence errors: Did they know it all along? *Journal of Experimental Psychology: Learning, Memory, and Cognition.*

Metcalfe, J., & Kornell, N. (2003). The dynamics of learning and allocation of study time to a region of proximal learning. *Journal of Experimental Psychology: General,* 132, 530–42.

Metcalfe, J., & Kornell, N. (2005). A region of proximal learning model of study time allocation. *Journal of Memory and Language,* 52, 463–77.

Metcalfe., J., Schwartz, B. L., & Joaquim, S. G. (1993). The cue-familiarity heuristic in metacognition. *Journal of Experimental Psychology: Learning, Memory, and Cognition,* 19, 851–61.

Miozzo, M., & Caramazza, A. (1997). Retrieval of lexical-syntactic features in tip-of-the-tongue states. *Journal of Experimental Psychology: Learning Memory and Cognition,* 23, 1410–23.

National Survey of Student Engagement (NSSE) (2008). *Promoting engagement for all students: The imperative to look within (2008 results)*. Bloomington: Indiana University Center for Postsecondary Research.

Nelson, T. O. (1992). *Metacognition: Core Readings*. Needham Heights, MA: Allyn & Bacon.

Nelson, T. O. (1993). Judgments of learning and the allocation of study time. *Journal of Experimental Psychology: General, 122, 269–73.*

Nelson, T. O., & Dunlosky, J. (1994). Norms of paired-associate recall during multitrial learning of Swahili-English translation equivalents. *Memory, 2, 325–35.*

Nelson, T. O., & Leonesio, R. J. (1988). Allocation of self-paced study time and the 'labor-in-vain effect'. *Journal of Experimental Psychology: Learning, Memory, and Cognition, 14, 47–86.*

Nelson, T. O., & Narens, L. (1990). Metamemory: A theoretical framework and new findings. *The Psychology of Learning and Motivation, 26, 125–41.*

Nietfeld, J. L., Cao, L., & Osborne, J. W. (2005). Metacognitive monitoring accuracy and student performance in the classroom. *Journal of Experimental Education, 74, 7–28.*

O'Sullivan, J., & Pressley, M. (1984). The completeness of instruction and strategy transfer. *Journal of Experimental Child Psychology, 38, 275–88.*

Pannu, J. K., & Kaszniak, A. W. (2005). Metamemory experiments in neurological populations: a review. *Neuropsychology Review, 15, 105–30.*

Pashler, H., Cepeda, N., Wixted, J., & Rohrer, D. (2005). When does feedback facilitate learning of words? *Journal of Experimental Psychology: Learning, Memory, and Cognition, 31, 3–8.*

Pavlik, P. I. Jr., & Anderson, J. R. (2005). Practice and forgetting effects on vocabulary memory: An activation-based model of the spacing effect. *Cognitive Science, 29, 559–86.*

Pinon, K., Allain, P., Kefi, M. Z., Dubas, F., & Le Gall, D. (2005). Monitoring processes and metamemory experience in patients with dysexecutive syndrome. *Brain Cognition, 57, 185–88.*

Pressley, M., Goodchild, F., Fleet, J., & Zajchowski, R. (1989). The challenges of classroom strategy instruction. *Elementary School Journal, 89(3), 301–42.*

Pyc, M. A., & Rawson, K. A. (2009). Testing the retrieval effort hypothesis: Does greater difficulty correctly recalling information lead to higher levels of memory? *Journal of Memory and Language, 60, 437–47.*

Pyle, W. H. (1913). Economical learning. *Journal of Educational Psychology, 4, 148–58.*

Rea, C. P., & Modigliani, V. (1985). The effect of expanded versus massed practice on the retention of multiplication facts and spelling lists. *Human Learning, 4, 11–18.*

Reder, L. M., & Ritter, F. (1992). What determines initial feeling of knowing? Familiarity with question terms, not with the answer. *Journal of Experimental Psychology: Learning, Memory, and Cognition, 18, 435–51.*

Richland, L. E., Kornell, N., & Kao, L. S. (2009). The pretesting effect: Do unsuccessful retrieval attempts enhance learning? *Journal of Experimental Psychology: Applied, 15, 243–57.*

Roediger, H. L., Agarwal, P. K., Kang, S. H. K., & Marsh, E. J. (2010). Benefits of testing memory: Best practices and boundary conditions. In G. M. Davies & D. B. Wright (Eds.), *New Frontiers in Applied Memory* (pp. 13–49). Brighton: Psychology Press.

Roediger, H. L., Agarwal, P. K., McDaniel, M. A., & McDermott, K. B. (2011). Test-enhanced learning in the classroom: Long-term improvements from quizzing. *Journal of Experimental Psychology: Applied, 103, 299–414.*

Roediger, H. L. & Butler, A.C. (2011). The critical role of retrieval practice in long-term retention. *Trends in Cognitive Sciences, 15, 20–27.*

Roediger, H. L., & Karpicke, J. D. (2006). Test-enhanced learning: Taking memory tests improves long-term retention. *Psychological Science, 17, 249–55.*

Roediger, H. L. & Karpicke, J. D. (2011). Intricacies of spaced retrieval: A resolution. In A.S. Benjamin (Ed.), *Successful Remembering and Successful Forgetting: Essays in Honor of Robert A. Bjork*. New York: Psychology Press.

Roediger, H. L., Gallo, D. A., & Geraci, L. (2002). Processing approaches to cognition: The impetus from the levels of processing framework. *Memory, 10*, 319–32.

Rohrer, D., & Taylor, K. (2007). The shuffling of mathematics practice problems improves learning. *Instructional Science, 35*, 481–98.

Rohrer, D., Taylor, K., & Sholar, B. (2010). Tests enhance the transfer of learning. *Journal of Experimental Psychology: Learning, Memory, and Cognition, 36*, 233–39.

Romine, C. B. & Reynolds, C. R. (2005). A model of the development of frontal lobe functioning: findings from a meta-analysis. *Applied Neuropsychology, 12*, 190–201.

Schacter, D. L., Rich, S. A., & Stampp, M. S. (1985). Remediation of memory disorders: Experimental evaluation of the spaced retrieval technique. *Journal of Clinical and Experimental Neuropsychology, 7*, 79–96.

Schneider, W. (2010). Metacognition and memory development in childhood and adolescence. In H. Slaatas Waters & W. Schneider (Eds.), *Metacognition, Strategy Use, and Instruction* (pp. 54–81). Guilford Press: New York.

Schneider, W., & Lockl, K. (2002). The development of metacognitive knowledge in children and adolescents. In T. J. Perfect & B. L. Schwartz (Eds.), *Applied Metacognition* (pp. 224–57). Cambridge: Cambridge University Press.

Schneider, W., & Pressley, M. (1997). *Memory Development Between 2 and 20*. Hillsdale, NJ: Erlbaum.

Schnyer, D. M., Verfaellie, M., Alexander, M. P., Lafleche, G., Nicholls, L., & Kaszniak, A. W. (2004). A role for right medial prefrontal cortex in accurate feeling of knowing judgments: Evidence from patients with lesions to frontal cortex. *Neuropsychologia, 42*, 957–66.

Schwartz, B. L., & Bacon, E. (2008). Metacognitive neuroscience. In J. Dunlosky, & R. A. Bjork (Eds.), *Handbook of Memory and Metamemory: Essays in Honor of Thomas O. Nelson*, (pp. 355–71). Psychology Press: New York.

Schwartz, D. L., Bransford, & J. D., Sears, D. L. (2005). Efficiency and innovation in transfer. In J. Mestre (Ed.), *Transfer of Learning from a Modern Multidisciplinary Perspective* (pp. 1–51). Greenwich, CT: Information Age Publishing.

Shimamura, A. P. (2008). A neurocognitive approach to metacognitive monitoring and control. In J. Dunlosky & R. A. Bjork (Eds.), *Handbook of Metamemory and Memory*, (pp. 373–90). New York Psychology Press.

Smith, S. M., & Rothkopf, E. Z. (1984). Contextual enrichment and distribution of practice in the classroom. *Cognition and Instruction, 1*, 341–58.

Son, L. K. (2004). Spacing one's study: Evidence for a metacognitive control strategy. *Journal of Experimental Psychology: Learning, Memory, and Cognition, 30*, 601–604.

Son, L. K. (2005). Metacognitive control: Children's short-term versus long-term study strategies. *Journal of General Psychology, 132*, 347–63.

Son, L. K. (2010). Metacognitive control and the spacing effect. *Journal of Experimental Psychology: Learning, Memory, and Cognition, 36*, 255–62.

Son, L. K., & Metcalfe, J. (2000). Metacognitive and control strategies in study-time allocation. *Journal of Experimental Psychology: Learning, Memory, and Cognition, 26*, 204–21.

Taylor, K., & Rohrer, D. (2010). The effects of interleaving practice. *Applied Cognitive Psychology, 24*, 837–48.

Thiede, K. W. (1999). The importance of accurate monitoring and effective self-regulation during multitrial learning. *Psychonomic Bulletin & Review, 6*, 662–67.

Thiede, K. W., & Anderson, M. C. M. (2003). Summarizing can improve metacomprehension accuracy. *Contemporary Educational Psychology, 28*, 129–60.

Thorndike, E. L., & Woodworth, R. S. (1901a). The influence of improvement in one mental function upon the efficiency of other functions. *Psychological Review, 8*, 247–61.

Thorndike, E. L., & Woodworth, R. S. (1901b). The influence of improvement in one mental function upon the efficiency of other functions: II. The estimation of magnitudes. *Psychological Review, 8*, 384–95.

Thorndike, E. L., & Woodworth, R. S. (1901c). The influence of improvement in one mental function upon the efficiency of other functions: III. Functions involving attention, observation, and discrimination. *Psychological Review*, 8, 553–64.

Toppino, T. C. (1991). The spacing effect in young children's free recall: Support for automatic-process explanations. *Memory & Cognition*, 19, 159–67.

Toppino, T. C., Cohen, M. S., Davis, M., & Moors, A. (2009). Metacognitive control over the distribution of practice: When is spacing preferred? *Journal of Experimental Psychology: Learning, Memory, and Cognition*, 35, 1352–58.

Toppino, T. C., & DeMesquita, M. (1984). Effects of spacing repetitions on children's memory. *Journal of Experimental Child Psychology*, 37, 637–48.

Varma, S., McCandliss, B. D., & Schwartz, D. L. (2008). Scientific and pragmatic challenges for bridging education and neuroscience. *Educational Researcher*, 37(3), 140–52.

Vilkki, J., Servo, A., & Surma-aho, O. (1998). Word list learning and prediction of recall after frontal lobe lesions. *Neuropsychology*, 12(2), 268–77.

Weinstein, Y., Finn, B., Roediger, H. L., & McDermott, K. B. (in preparation). The effect of retention interval expectancy on study time and accuracy. Manuscript in preparation.

Welsh, M. C., & Pennington, B. F. (1988). Assessing frontal lobe functioning in children: Views from developmental psychology. *Developmental Neuropsychology*, 4(3), 199–230.

Welsh, M. C., Pennington, B. F., & Groisser, D. B. (1991). A normative-developmental study of executive function: A window of prefrontal function in children. *Developmental Neuropsychology*, 7(2), 131–49.

Wheeler, M.A., Ewers, M., & Buonanno, J. (2003). Different rates of forgetting following study versus test trials. *Memory*, 11, 571–80.

Whitten, W. B., & Bjork, R. A. (1977). Learning from tests: The effects of spacing. *Journal of Verbal Learning and Verbal Behavior*, 16, 465–78.

Wilson, W. P., (1976). Developmental changes in the lag effect: An encoding hypothesis for repeated word recall. *Journal of Experimental Child Psychology*, 22, 113–22.

Zaromb, F. M., & Roediger, H. L. (2010). The testing effect in free recall is associated with enhanced organizational processes. *Memory & Cognition*, 38, 995–1008.

Zechmeister, E. B., & Shaughnessy, J. J. (1980). When you know that you know and when you think that you know but you don't. *Bulletin of the Psychonomic Society*, 15, 41–44.

The influence of neurogenetics on education

Chapter 9

Genetics and genomics: good, bad and ugly

Yulia Kovas and Robert Plomin

Overview

In this chapter we discuss how genetic research can inform neuroscience, education and, crucially, the links between them to facilitate progress in learning and teaching. Throughout the chapter we highlight the most important recent behavioural genetic findings using research on individual differences in mathematics as an example. The same issues apply to other areas of cognition and learning. In terms of genetic influences, genetic research suggests that each individual is likely to have their own unique combination of genetic variants, each having only a small effect, contributing to their abilities and performance. In parallel, a unique set of environmental experiences, each having only a small effect, is likely to explain the rest of the variation. Moreover, each genetic variant is likely to contribute to many different traits. Contrary to common opinion, these genetic effects are not static or deterministic, but change throughout life and in different educational and cultural contexts. We stress the importance of quantitative genetic methods and the unique contribution of twin studies to our understanding of educationally relevant phenomena. One of our examples describes the recent finding that studying in the same classroom contributes very little (if anything) to the similarity between the two children in academic achievement beyond their genetic similarity. We provide potential explanations and implications of this and other findings, concluding that 'teaching equally by teaching differently' might be truly necessary in the future to optimize learning for all students. In addition, we discuss how recent and future advances in educationally-relevant molecular genetic and genomic research might help explain, understand, predict and use individual differences in learning.

The chapter considers the importance and potential contributions of genetics to education and to neuroscience in education (good), the general view about genetics in education (bad) and attempts to date to identify specific genes throughout the genome responsible for ubiquitous genetic influence (ugly). We will use as an example of research one topic of great importance to education—variation in mathematical ability and achievement—to illustrate the main points.

9.1 **Introduction**

Neuroscience is in fashion. People at dinner parties talk about creative brains, language and spatial hemispheres, brain training and brain areas for this and that. It is likely that much of these conversations are like those inspired by phrenology of the 19th century. Indeed, the word phrenology means 'knowledge of mind' in Greek. Attempts to apply the knowledge of the mind to education are also not new. Active attempts to use phrenological analysis to define an individual pedagogy were made by Paul Bouts (e.g. 1986) and his followers. Two decades later, Bouts's individual pedagogy remains a mirage. Current attempts to integrate neuroscience into the classroom do not go much further than adding a brain-related qualifier to the name: brain-based, brain-compatible, brain-friendly and brain-targeted instructional approaches are largely based on over-interpretation of existing data (Alferink and Farmer-Dougan, 2010). Although today's educationally relevant neuroscience has begun to offer the first tentative explanations for how existing educational practices might be supported by the developing brain (e.g. McCandliss, 2010), this is only the beginning of the long path towards a truly educationally relevant neuroscience. In this chapter we argue that adding genetics to both neuroscientific and educational research will help to bridge neurosciences and education and lead to improved education.

In the first part of the chapter we consider the contributions that quantitative and molecular genetic research has already made to the field of learning abilities and disabilities. We will argue that recent twin research provides many important insights into the origins of individual differences in learning ability and achievement. We will also argue that although we would use the word 'ugly' to describe attempts to date to find the actual DNA (deoxyribonucleic acid) polymorphisms that are involved in genetically-driven variation, we believe that greater progress will be achieved in the near future.

In the second half we discuss the implications of these findings to educational and neuroscientific research. In terms of education, one of the major goals of this chapter is to reverse the generally 'bad' view about genetics in education. This view is reflected in apathy and even antipathy about genetics. We believe that this view comes in part from misconceptions about genetics. Education is pragmatic and education of the future will use methods that can be shown to work best for children with particular cognitive, perceptual and motivational strengths and weaknesses. We are a long way away from such a personalized education, but, just as with personalized medicine (Collins, 2010), such education is possible, and genetic understanding will be a major part of it. Future progress in identifying genetic polymorphisms associated with variation in cognitive skills and complex patterns of covariation among these skills might bring the tools for early screening to education. Genetics might ultimately help with decisions on whether direct remediation of some impairments will be possible (such as reversing or even preventing the development of a particular perceptual or cognitive weakness), or whether providing compensatory approaches will be required (such as developing strategies that attenuate the negative effects of having a particular perceptual or cognitive weakness). The need for such knowledge is well recognized in the field of education (Krasa & Shunkwiler, 2009).

In terms of neuroscience, we discuss how recent genetic findings are inconsistent with some of the current paradigms and views in neuroscientific research into cognition and learning (see Chapter 11, this volume). We conclude the chapter with ways in which new insights from genetic research can inform and contribute to educationally-relevant neuroscience and education ('good').

Throughout this chapter we use research on individual differences in mathematics as an example. The same issues apply to other areas of cognition and learning. We chose to focus on mathematics because it is an area of great societal importance yet one that has only recently been studied from a genetic perspective. Adequate mathematical skills are necessary in today's technologically-driven societies. Moreover, high levels of mathematical achievement are required for continued technological advances, innovations and applications, and for all areas of sciences, technology, engineering and mathematics (STEM). Despite the increasing demand for STEM expertise and the focus of the UK National School Curriculum in maths and science, for the past two decades, the number of students in Britain opting for maths and science careers has been in decline. Moreover, the rates of mathematical underachievement remain consistently high. Individual differences in mathematical abilities as well as in motivation and interest in STEM subjects develop through a complex process of gene–environment interplay. Understanding the key determinants of variation in ability and achievement as well as interest, motivation, engagement in mathematics is necessary in order to take a crucial step towards successful interventions aimed at reversing the lack of interest in mathematics, improving mathematical achievement and decreasing mathematical disability. We believe that genetic research can inform neuroscience, education—and, crucially, the links between them—to facilitate progress in learning and teaching maths.

9.2 Insights from twin research

A general consensus in genetics exists today—that variation in all complex traits, including all psychological and behavioural traits, is partly explained by DNA differences (polymorphisms) among people, that many DNA polymorphisms are involved in each trait, and that each DNA polymorphism explains very little of the variation in any given trait (Plomin, Haworth & Davis, 2009). This quantitative trait locus (QTL) model of genetic involvement in quantitative traits has also been applied to common disorders and disabilities (Plomin & Kovas, 2005). In other words, learning disabilities such as dyslexia, dyscalculia or ADHD have been conceptualized as quantitative cut-offs on one or several continuous dimensions. Another consequence of the QTL model is that each individual is likely to have their own unique combination of genetic variants—contributing to their abilities and performance. In parallel, a unique set of environmental experiences, each having only a small effect on a trait (quantitative trait environment, QTE), is likely to explain the rest of the variation. In addition, complex interactions between different QTLs and between different QTEs may be taking place, as well as interactions and correlations between QTLs and QTEs. Finally, each QTL is likely to contribute to many different traits, a phenomenon known as pleiotropy, and the situation is likely to be the same for each QTE.

Recent quantitative genetic research has already provided important insights into the origins of the individual differences in mathematical ability and achievement. Using twin methodology, this research has led to the following five important conclusions:

Individual variation in mathematics develops under the influence of both genetic and environmental factors

Recent research into sources of individual differences in mathematical ability has led to the undisputed conclusion that both genes and environments shape people's individual profiles of strengths and weaknesses in this trait. In the school years, approximately 50% of the between-individual variation in mathematical ability is explained by genetic factors. The rest of the variation is largely driven by individual-specific (rather than family-wide or school-wide) factors (Kovas, Haworth, Dale & Plomin, 2007). A particularly interesting finding is that studying mathematics in the same classroom does not increase the similarity between the two children beyond their genetic similarity (Kovas, Haworth et al., 2007), at least in the UK where the curriculum is standardized. We found no statistical differences between twin correlations in mathematics (and other academic subjects) for pairs of twins in our sample who were taught by the same vs. different teacher.

This might mean that the child's cognitive/motivational profile interacts with the learning situation, so that the same classroom, teacher, or teaching method has a significantly different effect on different individuals. Alternatively, it might mean that under a highly structured curriculum, teachers contribute less to variance—whereas genes and non-shared environments (including children's individual perceptions of the teachers and classrooms) contribute to the variation.

Different aspects of mathematics are influenced by mostly the same QTLs

Multivariate genetic analysis investigates not only the variance of traits considered one at a time but also the covariance among traits. It yields the genetic correlation which can be roughly interpreted as the likelihood that genes found to be associated with one trait will also be associated with the other trait. Recent multivariate research has shown that different aspects of mathematics, such as computation, knowledge of mathematical procedures and operations, interpreting graphs and diagrams, problem solving and non-numerical operations are largely influenced by the same set of genes, at least in the early school years (Kovas, Haworth et al., 2007; Kovas, Petrill & Plomin, 2007; Plomin & Kovas, 2005). One major implication of this finding is that if a child underperforms selectively in some areas of mathematics, this discrepant performance is likely to stem from an environmental source. These discrepancies in performance must also mean that the same person may perform at, below, or above their genetic propensities. What follows is that not only different teaching methods will be required for different aspects of maths, but that any one way of teaching a particular aspect of maths is unlikely to be the best way for all students—'teaching equally by teaching differently' is necessary.

Many of the same QTLs are involved in mathematical and other learning abilities

In a review of a dozen multivariate genetic studies of learning abilities and difficulties, the average genetic correlation was 0.70, between reading and mathematics, between language and mathematics, and between reading and language (Plomin & Kovas, 2005). A recent multivariate genetic analysis based on web-based testing yields even higher genetic correlations between mathematics, reading and language (Davis, Haworth & Plomin, 2009).

Moreover, the general effects of genes appear to extend beyond specific learning abilities and disabilities such as reading and mathematics to other cognitive abilities such as verbal abilities (e.g. vocabulary and word fluency) and non-verbal abilities (e.g. spatial and memory). The average genetic correlation is about 0.60 between specific learning abilities and general cognitive ability (g), which encompasses these verbal and non-verbal cognitive abilities (Plomin & Kovas, 2005); the recent study mentioned earlier yielded genetic correlations greater than 0.85 between mathematics, reading and language versus general cognitive ability (Davis et al., 2009).

This genetic overlap among traits has been referred to as the 'generalist genes hypothesis'. Figure 9.1 illustrates several models through which genetic pleiotropy may lead to the observed effects of the 'generalist genes' using learning disabilities as an example. One possibility is that a generalist gene affects a single mechanism (for example, a brain area or function) that is pleiotropically involved in several cognitive processes (Figure 9.1 model 1). In this case, the brain structures and functions are uncorrelated genetically because they are influenced by different genes, even though at the level of learning disabilities the effect of these mechanism-specific genes appears to be pleiotropic. We believe that this possibility is unlikely because gene expression profiles in the brain suggest that any gene is likely to be expressed in more than one structure or function.

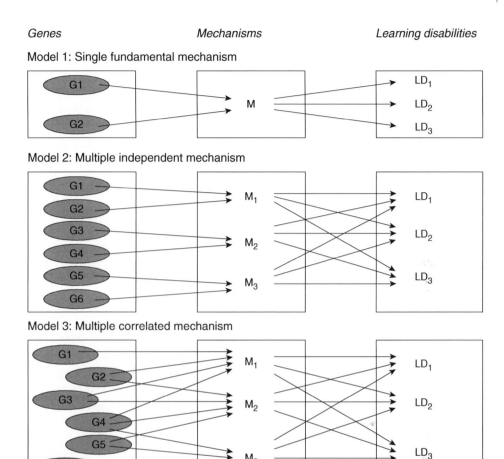

Fig. 9.1 Pleiotropy. Three models of effects of genes on learning disabilities as mediated by brain or cognitive mechanisms.

A second possibility is that multiple mechanisms are involved but each mechanism is influenced by its own independent set of genes (Figure 9.1 model 2). The third possibility, which we favour, is that generalist genes affect multiple mechanisms and that each of these affect multiple learning disabilities (Figure 9.1 model 3). This mechanism would lead to genetic correlations in the brain as well as in the mind.

The concept of generalist genes has far-reaching implications for understanding the genetic links between brain, mind and education (Plomin et al., 2007). For mathematics, more quantitative genetic research is needed to characterize the contribution of generalist and specialist genes to different mathematically relevant abilities and skills. Moreover, finding the actual DNA sequences responsible for these generalist genetic effects, as well specialist effects, will have important practical benefits, as discussed in the following section.

Genetic and environmental influences are not static, but change across age and across cultures.

Recent research suggests that genes contribute to both change and continuity in mathematical performance (Kovas, Haworth et al., 2007). In other words, although some of the same genes continue to influence mathematics across development, new genetic effects also come on line at different ages. This seemingly paradoxical finding is not that surprising. What we call mathematics in the early school years is very different operationally and conceptually from the complex set of knowledge and procedures that we call mathematics in later school years.

Another important finding is that genetic effects may be stronger or weaker depending on the environmental situation. An area of research with huge potential that has only begun to be explored in relation to learning abilities and difficulties is the developmental interplay between genes and environment. Several studies, conducted in different countries, have suggested that the effects of genes may be smaller in countries that do not use a centralized National Curriculum (e.g. Samuelsson, Byrne, Olson et al., 2008; Samuelsson, Olson, Wadsworth et al, 2007). In other words, if the curriculum and teaching methods do not differ from school to school, most of the variation in mathematics stems from genetic and individual-specific environments. Much more research involving cross-cultural comparisons is needed to help identify the relevant environments and their impact.

Aetiological continuity exists between low, normal and high performance in mathematics

The use of very large representative samples of twins in the community has made it possible to investigate the aetiology of development of difficulties in the context of the normal distribution (Kovas, Petrill, Haworth & Plomin, 2007; Kovas, Haworth et al., 2007; Petrill et al., 2009). These twin studies show that what we call 'learning difficulties' are largely the quantitative extreme of the same genetic and environmental factors responsible for normal variation in learning abilities (Plomin & Kovas, 2005). Stated more provocatively, these results suggest that there are no aetiologically distinct difficulties, only the low end of the normal bell-shaped distribution of abilities. In other words, when genes are found 'for' maths difficulty, these genes will not be limited to maths difficulty. Rather they will be associated with maths ability throughout the distribution, including high maths ability (Plomin, Haworth & Davis, 2009).

This finding has profound implications for the diagnosis of learning abilities, because it suggests that we should think in terms of quantitative dimensions rather than qualitative diagnoses. As discussed in the following section on molecular genetics, there are many chromosomal and single-gene causes of learning difficulties. However, these are rare and often severe forms of learning difficulties, whereas the quantitative genetic data are telling us that the vast majority of common learning difficulties are the quantitative extreme of the same genetic and environmental factors responsible for normal variation in these learning abilities. Properly understood, these results should help to avoid negative 'us versus them' stereotypes about learning difficulties.

Even before finding the genes, putting together the two genetic findings—that children with learning difficulties differ quantitatively not qualitatively and that genetic effects are general—suggests that genetic nosology differs from current diagnoses based on symptoms. First, the genetic data suggest that common learning difficulties are only the low end of the normal bell-shaped distribution of abilities. Second, because genetic effects are largely general, they blur distinctions between ostensibly different problems such as reading and maths difficulties. That is, most of what is going on genetically has broad general effects rather than specific effects on just one difficulty.

All of these implications will remain at a conceptual level until genes are found that are responsible for these genetic effects. When these genes are found, their implications for prediction and intervention may be even greater than their effect on diagnosis.

9.3 **Insights from molecular genetics**

Identifying the genes responsible for the genetic effects on variation in learning will provide the ultimate early diagnostic indicators of learning difficulties, because a DNA sequence does not change as the result of development, behaviour or experience. (The expression of DNA, which involves transcription of the DNA code into RNA, does change, but this is another matter.) Finding these genes may facilitate matching the most suitable teaching methods and learning environments to individual cognitive and motivational profiles. However, progress towards identifying the responsible genes has been slower than expected because, as is now widely recognized, genetic influence on common disorders like learning difficulties involves many genes of small effect.

The big questions are 'How many?' and 'How small?', because it will be extremely difficult to detect reliable effects if the influence of each gene is very weak. If a single gene were responsible, it would be easy to identify its chromosomal neighbourhood (linkage) and then its specific address (association), as has happened for thousands of single-gene disorders, which are typically severe and rare, often one in 10,000 rather than one in 10 or one in 20, as is the case for most learning difficulties. More than 250 single-gene disorders include cognitive difficulty among their symptoms (Inlow & Restifo, 2004) and many more are likely to be discovered (Raymond & Tarpey, 2006). However, together these single-gene disorders, as well as numerous chromosomal abnormalities, account for fewer than 1% of children with learning difficulties (Plomin, DeFries, McClearn & McGuffin, 2008). In contrast, the frequency of learning difficulties is often said to be as high as 5–10%. Although rare disorders are dramatically important for affected individuals, the vast majority of common learning difficulties are not explained by these and other known single-gene and chromosomal problems. It is possible theoretically that other, as yet unidentified, single-gene and chromosomal problems are responsible for some unexplained learning difficulties, but the quantitative genetic evidence reviewed earlier suggests that this is not the case.

As mentioned earlier, most geneticists have come around to the view that common disorders— medical as well as behavioural—are caused by multiple genes of small effect sizes. These genes are often called *quantitative trait loci* (QTLs) because, if a trait is influenced by many genes, the genetic effects will be distributed quantitatively as a normal bell-shaped distribution, regardless of whether we impose a diagnostic cut-off on the distribution (Plomin, Haworth & Davis, 2009). Although QTLs are genes like any others, the notion of QTLs is important conceptually. Single-gene disorders are deterministic. For example, if you have the gene for Huntington disease, you will die from the disease regardless of your other genes or your environments. In contrast, QTLs refer to probabilistic propensities rather than predetermined programming, and this warrants greater optimism for effective intervention.

New molecular genetic techniques have been developed to identify QTLs, and their first success was for reading difficulty—chromosomal linkages were identified in 1994 (Cardon et al., 1994). However, it has proved difficult to pinpoint the actual genes responsible for these linkages, although several genes involved in neuronal migration during development appear to be likely candidates (Scerri & Schulte-Koene, 2010). The latest development is called genome-wide association, which examines hundreds of thousands of DNA markers simultaneously for their association (correlation) with disorders (Hirschhorn & Daly, 2005). In the past 3 years, genome-wide association has transformed genetics research in the life sciences and has identified 779 significant ($p < 5 \times 10^{-8}$) associations between common genetic variants and 148 traits (Hindorff et al., 2010). Combined with large samples (in the thousands), it is possible to identify many genes of small effect. Genome-wide association is made possible by DNA microarrays (often called 'gene chips') the size of a postage stamp, that can genotype as many as a million DNA markers inexpensively (Plomin & Schalkwyk, 2007). Moreover, rather than genotyping individuals for particular sets of DNA markers, we will soon be able to sequence each individual's entire genome. The race is on to sequence all three billion bases of DNA of an individual for less than \$1000/£620 (Service, 2006), although the cost is currently about \$10,000/£6200.

A genome-wide view of QTLs is necessary because there may be hundreds of DNA markers that predict different components or constellations of learning difficulties, at different ages, in different environments, and in response to different intervention programmes. With today's microarray technology, thousands of such QTLs could easily be incorporated on an inexpensive 'learning difficulties' gene chip, even before complete DNA sequencing is available for each individual.

The first association scan of this type for reading difficulty identified 10 DNA markers associated with reading (Meaburn et al., 2007), although replication in other independent studies is necessary. These DNA markers or loci, called single nucleotide polymorphisms (SNPs) are points in the DNA sequence where not all people have the same letter (as is the case with most of the DNA), but every individual has one of two possible letters—also called variants or polymorphisms. If having one allele of a gene increases one's ability and having the other allele decreases it, the frequency of the two variants will differ for low- and high-ability groups. Although each of the 10 DNA markers in this study had a very small effect, by aggregating dozens or hundreds of such markers in a cumulative genetic risk index, it will be possible in the future to predict genetic risk for reading difficulty.

Only one molecular genetic investigation into mathematical ability has been reported so far (Docherty et al., 2010). In this first large genome-wide association study, 10 single nucleotide polymorphisms (SNPs) were nominated as associated with mathematical variation in 10-year-old children, participants in the Twins' Early Development Study (TEDS). Consistent with the results from the twin studies which suggest genes work additively, when these 10 SNPs were combined into a set, they accounted for 2.9% (p = 7.277e–14) of the phenotypic variance in mathematics. The association was linear across the distribution; the third of children in the sample who harbor 10 or more of the 20 risk alleles identified are nearly twice as likely (p = 3.696e–07) to be in the lowest performing 15% of the distribution. These results correspond with the QTL model implying genetic effects across the entire spectrum of ability.

It is important to note that attempts to replicate many of the first QTL associations of learning abilities have failed, as has been the case for all complex traits and common disorders (McCarthy et al., 2008). More highly powered studies will be needed to detect and replicate these QTL associations; other strategies are also under consideration (Manolio et al., 2009). Moreover, we are a long way from understanding the path from each of these (and other) SNPs to mathematical differences. One of the SNPs is in a *NRCAM* gene which encodes a neuronal cell adhesion molecule, potentially opening a window into one of the general brain mechanisms. The mechanisms through which other SNPs are involved in mathematics are yet to be studied. However, even before such mechanisms are understood, useful research can be carried out to further understand the role of these QTLs in mathematical learning.

One such investigation used the 10 SNPs described earlier and the Twins Early Development Study's longitudinal multivariate dataset to test the Generalist Genes Hypothesis which posits that many of the same genes influence diverse cognitive abilities and disabilities across age (Docherty, Kovas, Petrill & Plomin, 2011). 4927 children in this study were genotyped on these SNPs and had data available on measures of mathematical ability, as well as on other cognitive and learning abilities at 7, 9, 10 and 12 years of age. Using these data, the authors assessed the association of the available measures of ability at age 10 and other ages with the composite 'SNP-set' scores. The 'SNP-set' score is calculated by adding up a score for each of the 10 SNPs. For each SNP, an individual can have a score of 0 (no maths-increasing variant, allele), 1 (one maths-increasing allele) or 2 (two maths-increasing alleles). Because there are 10 SNPs, SNP-set scores can range from 0 (no maths-increasing alleles) to 20 (both alleles at all 10 loci are maths increasing). The results of this study supported the findings from the quantitative genetic studies described earlier in several ways: first, the SNP set associated with overall mathematics score at 10 was also similarly associated with each of the three mathematical components (mathematical computations, non-numerical knowledge and understanding number)—supporting the generalist genes hypothesis. Second, the SNP set yielded significant (although even smaller) associations with mathematical ability at ages 7, 9 and 12, suggesting that some of the same genetic effects contribute to mathematics throughout development, but that some new genetic effects might come online at each developmental stage. Third, the same SNP set was also significantly (and to a similar magnitude) associated with reading and general cognitive ability at age 10, again supporting the generalist genes hypothesis. Some genetic age or trait specificity was also found. For example, the SNP set was significantly associated with spatial ability at age 12, but the magnitude of this association was very small. With small effect sizes expected in such complex traits, future studies may be able to capitalize on power by searching for 'generalist and specialist genes' using longitudinal and multivariate approaches.

Discovery of the generalist and specialist genes involved in mathematical variation together with a list of candidate environments hypothesized as important for mathematical development open a new opportunity to study potential gene–environment interface. The first study to look at the maths QTL by QTE interactions has recently been reported (Docherty et al., 2011), examining whether the association between the 10-SNP set and mathematical ability differs as a function of 10 environmental measures in the home and school in a sample of 1888 children. Two significant GE interactions were found for environmental measures in the home: The associations between the 10-SNP set and mathematical ability differed as a function of level of chaos in their home and of their parents' negativity. Both QTL by QTE interactions were in the direction of the diathesis-stress type of GE interaction: The 10-SNP set was more strongly associated with mathematical ability in chaotic homes and when parents are negative.

These quantitative genetic (quantifying the genetic and environmental influences), molecular genetic (finding the relevant DNA polymorphisms) and molecular genomic (understanding each gene–behaviour mechanism) investigations of mathematical ability are just the beginning of a long path towards adequate explanations of individual variation in this trait. However, some important implications for neuroscience and education can already be identified based on these early findings.

9.4 **Implications for neuroscience and education**

Despite the universal ability to comprehend numerical information, learning mathematics is different for different people. Some develop a strong interest in the subject, comprehend mathematical concepts with ease and learn mathematical skills to the highest proficiency level. Others dislike mathematics, are afraid of it or struggle with understanding and applying mathematical concepts. As reviewed previously, genetics is a major part of the explanation of these differences.

Educational research shies away from genetic explanations of individual variation. Even when complex multilevel models of development are proposed, incorporating different levels of research and theory (e.g. Tommerdahl, 2010)—the basic pervasive level of genetics is strangely ignored. One reason is that genetics in education is mistakenly associated with inequality (Pinker, 2002). Moreover, much of the educational research on mathematics is devoted to finding the best methods for teaching mathematics for all children, a one-size-fits-all approach. This seems paradoxical because teachers, more than anyone, know that each child has a unique cognitive profile and even students who perform adequately may learn in diverse ways. No matter what instructional material a school has chosen, the teacher on the front line must adapt it to the needs of specific students (Krasa & Shunkwiler, 2009). Teachers are expected to obtain a 'portrait' of a student's reasoning, perceptual, cognitive and executive functioning to facilitate successful teaching. Indeed, in the US, the *Principles and Standards for School Mathematics of the National Council of Teachers of Mathematics* (NCTM, 2008; in Krasa & Shunkwiler, 2009) requires that teachers have the knowledge and ability to analyse students' thinking and to be knowledgeable about both internal factors (the child's developmental readiness, cognitive strengths and weaknesses, and interests) and external factors (the subject matter and teaching methods) affecting learning experiences (Krasa & Shunkwiler, 2009).

The existence of large individual differences in learning means that determining the appropriate pedagogical approaches for each maths skill will not be enough to optimize education for each learner (Krasa & Shunkwiler, 2009). Many examples demonstrate how the same method can be good for some and completely unsuitable for others. For example, one area of lively debate is whether the use of concrete story problems might facilitate mathematical learning better for some children rather than using abstract equations (Krasa & Shunkwiler, 2009). However, questions remain as to whether these different methods are suitable for different aspects of maths learning, as well as for children with different mathematics abilities, or different cognitive or perceptual impairments or styles. In the future, we might be able to identify children with aetiological weaknesses in imagination or abstraction—for them the use of analogies, suppositions and reflection on their own assumptions would not be the best method of learning mathematics.

Many other alternative learning and teaching paths may be available to different individuals and for different aspects of learning mathematics (see Chapter 6, this volume). Future education, using the power of computers to personalize learning, as well as differentiated teaching methods and learning paths, may achieve previously unimaginably optimized outcomes. This might sound far-fetched and even frightening for some people. However, we contend that such individually-tailored leaning is already the goal of education. Despite this, mathematical pedagogy informed by research is in its infancy. In the US, with the absence of the UK's National Curriculum, schools and districts try to adopt the best curricula for their students, but, with current lack of deeper understanding of the origins of the individual variation, the mathematics curriculum remains 'in chaos' (Krasa & Shunkwiler, 2009, p. 185). In countries like the UK with a centralized curriculum, the 'chaos' remains in the effect of the centralized curriculum—because one method cannot be suitable for all. In 2008, the US National Mathematics Advisory Panel (NMAP) reviewed a large number of educationally relevant studies, and concluded that there is no one ideal approach that is appropriate for all students (Krasa & Shunkwiler, 2009). The panel also recommended that the students' background, abilities and effort; the teachers' background and strengths; and the instructional context, approach and curriculum should all factor into pedagogy. However, before this can happen, much research—going beyond crude categories of aptitudes by intervention studies—is needed to predict which instructional materials and methods are effective for which particular teachers, students, topics and learning conditions.

To summarize, the most effective instruction will adapt to each individual, particularly with struggling students (Krasa & Shunkwiler, 2009). However, we are a long way from understanding individual variation and from being able to adapt to it successfully. We believe that an important step towards understanding individual differences and utilizing this understanding for pedagogy is incorporating genetics into the explanation and the research. Currently, when individual-specific variation is considered, the word 'genetic' is generally avoided and reference is made to child characteristics, internal factors, innate cognitive capacities, abilities, cognitive strengths and weaknesses, motivations, interests, attitudes and personality. Educationalists and educational researchers should not shy away from acknowledging that genetics plays a large part in shaping these child-specific factors.

Even when these 'child characteristics' are rightfully understood in terms of genetic and environmental aetiology, they are usually hypothesized to have causal influence on mathematical learning and achievement. For example, a concept of 'cognitive readiness' is often discussed in terms of competence in early cognitive skills that children require for successful acquisition of later skills. For mathematics, it is assumed that good numerosity discrimination abilities (approximate number system), working memory, intact phonological functions, receptive and expressive language, body (particularly finger) and other spatial awareness, and the ability to integrate all of these might be necessary. In addition, 'school readiness' for mathematics has been hypothesized to include some basic early number skills, such as symbolic number knowledge, counting and understanding of cardinality principle. Different interventions for mathematical problems have been suggested and are currently the target of many investigations that attempt to improve the level of mathematical achievement by raising the level of one or more of these cognitive abilities or early skills. However, quantitative genetic research suggests that the link between different cognitive skills might be pleiotropic, in that many of the same genes affect diverse learning skills. This means that improving one ability or skill (e.g. numerosity awareness) might not necessarily lead to an improvement in a correlated ability or skill (e.g. counting) because the common aetiology that contributes to their correlation is genetic in origin. In this sense, improving one skill could be expected to lead to an environmentally induced discrepancy between the two skills.

We believe that research into what skills and skill levels are required for successful early mathematics learning will progress faster by incorporating the findings from recent twin research. For example, quantitative genetic research can explore the extent to which the link between early number sense and later maths achievement (Halberda, Mazzocco & Feigenson, 2008) is due to common genetic or common environmental aetiology. Although this issue is still under investigation, such findings already exist for different traits, documenting the extent to which genes and different types of environments contribute to covariation between different skills at different ages. As discussed in an earlier section, this research (e.g. Plomin & Kovas, 2005) suggests that much of the covariation among different cognitive traits stems from pleiotropic effects of genes. This might mean that a useful future direction is understanding the processes through which resilience operates—understanding the sources of high mathematical performance, despite poor performance on early 'predictor' abilities. Previous research suggests that non-shared environments play a large role in discrepancies in an individual's abilities. For example, if one's mathematical ability is greater than one's reading ability—this is largely explained by the fact that different non-shared environments are important to the development of the two traits (e.g. Plomin, Kovas & Haworth, 2007).

Many of the mechanisms through which individual differences in learning emerge may be unexpected and even counter-intuitive. An example of such an unexpected mechanism, is a recent finding that the prediction of maths achievement from self-evaluation of mathematics (self-perceived ability and 'liking' of math), after accounting for any association with IQ, is genetic rather than environmental (Luo, Kovas, Haworth & Plomin, 2011). This finding is not unique to mathematics; other studies have demonstrated that variation in motivation and self-perceived abilities are partly genetic and the link between these traits and achievement is explained by shared genes rather than environmental links (e.g. Greven, Harlaar, Kovas, Chamorro-Premuzic & Plomin, 2009). This might mean that improving self-perceived ability will actually not lead to higher levels of mathematical performance, but rather to an environmentally influenced discrepancy between the two traits. Having higher self-perceived ability might have other positive psychological effects, however, other factors will need to be targeted in order to raise the level of mathematics performance.

Initial genetic differences might interact with environments, so that the same educational input leads to drastically different outcomes for different children. Behavioural genomic research will ultimately indicate which genes and which environments are involved. Accounting for, evaluating, and interpreting these genetic findings by educationalists is necessary to develop new, complex explanations of individual variation and associated teaching and learning methods. Research exploring the aetiology of the hypothesized links among different abilities and between early and later abilities, guided by questions from educationalists and psychologists, will help to avoid erroneous conclusions and policies and might ultimately help with decisions on whether direct remediation of some impairments will be possible or whether providing compensatory approaches will be required.

Another, related, important interface of genetics and education is the issue of screening and prediction of problems. Although most students are remarkably adaptable and do well with common teaching practices, many students do not. These students miss out on some or all of the mathematics curriculum, with enormous cost to their education, self-confidence and job prospects, as well as to the nation's standing in the global economy (Krasa & Shunkwiler, 2009). In the area of mathematics, several early screening tests have been devised to predict future performance. For example, early competence in number knowledge and number sense is the most powerful predictor for later learning in maths. However, research has also shown that even very early maths skill is composed of many partially independent abilities, where almost no component is an absolute prerequisite for any other. As discussed earlier, any prediction of one competence from another may actually reflect the shared genetic aetiology, and the dissociations among competencies may be environmental in origin. The prediction is further complicated by the fact that different cognitive skills may come into play later in the curriculum. Clearly, a complex pattern of cognitive strength and weakness needs to be taken into account to predict later maths performance. One problem is that by the time such strengths and weaknesses become apparent or testable the child might already be well on the way along a deviant developmental path. In the future, it is possible that an individual's DNA sequence can be easily examined for educationally-relevant variation in order to suggest the best methods, compensatory strategies, and learning approaches for each individual, as has been proposed in relation to personalized medicine (Collins, 2010). One can imagine a future where, for example, a child might be identified as having poor spatial but stronger linguistic and logical potential. A curriculum that utilizes this alternative path to successful mathematical learning might be suggested for this learner.

What about the links between educationally relevant neuroscience and genetics? Currently, neuroimaging research focuses on group analyses, rather than individual differences analyses. For example, several brain regions show increased brain activation on average during numerical tasks when compared to control tasks. These areas of increased activation include intraparietal sulcus, inferior and superior frontal gyri, as well as other co-ordinates within the precentral, dorsolateral and superior prefrontal regions (e.g. Stanesku-Cosson et al., 2000; Venkatraman, Ansari & Chee, 2005). According to one influential hypothesis, these regions, and in particular the horizontal segment of the intraparietal sulcus (hIPS), are the loci of a dedicated, domain-specific number system, subserving operations with both symbolic and non-symbolic stimuli (e.g. Dehaene, Piazza, Pinel & Cohen, 2003; Piazza, Pinel, LeBihan & Dehaene, 2007; Cantlon et al., 2006). It is important to remember that we must be careful with interpretation of these findings in causal ways to avoid slipping into neo-phrenology. For example, a common extension of these findings is the hypothesis that low mathematical ability stems from some damage or dysfunction in these brain areas. However, any 'abnormal' brain activation in these areas may be the result rather than the cause of the low mathematical ability. Such inappropriate interpretations and oversimplification of complex neuroscience research is widespread, and much of today's attempts at brain-based curricula and teaching approaches rely on these inappropriate interpretations (see Alferink & Farmer-Dougan, 2010 for examples of such bad practices). If the aim of neuroscientific research is to uncover the brain mechanisms causally involved in mathematical variation, the research needs to move towards individual differences and to consider the genetic contribution to variation as it is the variation that is of particular interest to education. The complexity of genetic and environmental aetiologies of individual differences in learning means that to the extent that normal variation in learning is driven by genetic factors, many neural processes of small effect mediate the effects of genes on cognition (Kovas & Plomin, 2006). Figure 9.1 could be used to illustrate these complex gene–brain–cognition links, by substituting 'genes' with 'QTLs' to reflect the polygenicity; 'mechanisms' with 'QTNs' to refer to the quantitative trait neural processes; and 'LD' (learning disabilities) with different quantitative cognitive traits. These QTNs of small effect are reflected in the widely distributed brain–behaviour associations consistently found in neuroimaging studies (Kovas & Plomin, 2006).

To date, limited research has addressed the issue of individual variation in relation to mathematical ability and associated brain activity. Most of this research involves case–control comparisons rather than individual differences throughout the distribution. For example, functional magnetic resonance imaging (fMRI) studies with patients with developmental dyscalculia (DD) have shown decreased or abnormally modulated activity or decreases in the grey matter density in parietal cortices in people with dyscalculia as compared to a control group (reviewed in Kucian et al., 2006). Other brain areas have also been implicated in the differences between typically developing children and children with DD (Kucian et al., 2006). One recent study explored the extent to which variation in fMRI response as a function of number task difficulty relates to high versus low mathematical ability (Kovas, Viding, Ng, Giampietro, Brammer, Happé & Plomin, 2009). This study found that high versus low numerosity skills in 10-year-old children were subserved by a widely distributed brain network. Some parts of this network appear to support numerical judgements in general (as shown by activation in both low- and high-ability group), whereas others may subserve individual differences in numerical ability, as manifested by magnitude differences in brain activation between low- and high-mathematical ability children. The causal direction of the association, as is the case for all non-genetically sensitive neuroimaging studies, remains unknown. Either small differences across a wide brain network lead to the individual differences in mathematical performance or differences in mathematical performance (caused by multiple genetic and environmental factors) cause the observed differences in activation during the approximate judgement task. The results of this study also suggest that different neural mechanisms may be involved in approximation per se and in individual differences in mathematical ability. This is suggested by the non-overlapping brain areas active in approximation versus baseline and low- versus high-ability comparisons. This finding could reflect a dissociation between the brain processes subserving invariable (species universal) ability to use approximate judgement, and the brain processes involved in individual variation in this ability.

Clearly, the complexity of the aetiology of the variation in complex traits calls for new neuroscientific approaches that view the brain as a functional system, rather than independent pieces of the puzzle. New methods and analyses of this type have begun to emerge. For example, network analysis of structural and functional connectivity aims to characterize the organization of brain networks—path lengths, clustering, hierarchy and regional interconnectivity. One very recent study (Rykhlevskaia, Uddin, Kondos & Menon, 2009) combined morphometry and tractography analyses of the whole brain to look at the structural network differences between typical and low mathematical performance groups of 7–9-year-old children, matched for gender, intelligence quotient (IQ), reading and working memory. The results of this study showed highly distributed differences between the two groups, involving both white and gray matter in many brain areas. The results suggest that multiple functional circuits in the brain are involved in mathematical ability. The authors hypothesized that a core white matter deficit might be involved that leads to a disconnection among different circuits. An alternative explanation is that individuals possess unique patterns of activation and connectivity among all of the circuits involved in a given complex trait. However, the mechanisms underlying individual differences are currently described in the literature are poorly understood, with only very general mechanisms being discussed. This is not surprising as most current research is based on groups, or, at best, compares ability groups.

It is likely that in 50 years current neuroscientific explanations will look naïvely crude, talking about large chunks of the brain (e.g. the prefrontal cortex) working 'in concert' with another (e.g. medial temporal lobe) as well as other memory 'systems' (e.g. dorsal basal ganglia memory systems' (Menon, 2011). The picture is likely to be infinitely more complex. For example, for mathematical reasoning, it is already clear that integration across a distributed brain network is involved in higher-order visuospatial processing, memory and cognitive control, as well as more specific numerical networks (Menon, 2011). The complexity of these inter-relationships as well as the degree of individual differences in these networks is yet to be described and understood. We believe that the future of neuroscience lies in moving away from group comparisons to applying new exciting methods to tracing neural profiles of individuals. Until neuroscience can fully embrace individual variation, the hypothesized brain mechanisms underlying cognition and learning will remain unknown and neuroscience will continue to have limited impact on education. Much more research is needed that involves large samples in order to gain enough statistical power to detect processes of small effects in multiple brain areas at the whole brain level of analysis, and to identify and replicate the complex neural networks suggested (but not established) by the existing literature.

Not surprisingly, although educationalists have initially embraced neuroscience, with calls for 'brain-based' educational programmes, the initial hype has now subsided. The US NMAP has recently cautioned that attempts to connect research in the brain sciences to classroom teaching and student learning in mathematics are premature and that instructional programmes that claim to be based on brain sciences research remain to be validated. In a recent book about mathematical learning, written by a clinical psychologist and mathematical educationalist the authors state: 'One looks forward to the day when brain research can inform pedagogy in the trenches. In the meantime, educators should interpret all "brain-based claims with scepticism"' (Krasa & Shunkwiler, 2009, p. 185).

Teachers want more studies that focus on individual differences—studies that explore how different methods, teaching and learning styles are reflected in the brain, and what it means for education. To be relevant to education, neuroscience will need to conduct research into appropriate pedagogical approaches for each maths skill as well as methods that work best for any given set of cognitive and perceptual strengths and weaknesses. Indeed, understanding of the complex patterns of connectivity among different brain circuits involved in mathematics may in the future contribute to the development of remediation training programmes to strengthen particular connections. One recent example of such research has shown that specialized reading instruction can normalize the reading circuit, demonstrating that experience can alter brain functions (Meyler et al, 2008; Shaywitz et al., 2004). More research into interventions and into differential impact of interventions on different people is required to inform scientifically based pedagogy. Do different people code arithmetic facts differently—for example phonologically versus visually (but also many other ways, beyond this crude dichotomy)? Do children with particular cognitive styles or impairments show different pattern of mathematical representations? How does explicit instruction, exposure, drill, and practice change brain function and brain structure? Do children in different educational cultures differ in their mathematical representations? Research comparing brain function across modalities and strategies as well as focusing on maths task-related brain functioning of students across the ability spectrum in language, reading and maths will have significant implications for the classroom (Krasa & Shunkwiler, 2009). When genes involved in variation in these traits are found, a field of educational imaging genomics will emerge that explores these issues focusing on children with different genetic profiles.

9.5 **Conclusion**

It is not yet possible to predict the strength of any direct impact that identifying sets of genes associated with learning difficulties will have on teachers in the classroom confronted with a particular child with a learning difficulty. However, the capability of predicting genetic risk from DNA is likely to have far-reaching implications in terms of diagnosis, treatment and intervention (Plomin & Walker, 2003). Gene-based diagnoses of learning difficulties are likely to be very different from current diagnoses. For example, many of the same genes (the 'generalist genes' mentioned earlier) that predict reading difficulty will also predict maths difficulty, although some genes will be specific. That is, a learning difficulties gene chip in the future would mostly contain genes that can predict which children are likely to have general problems with reading *and* maths, but it would also contain some genes that can predict specific problems with reading *or* maths. Moreover, genes on the learning difficulties gene chip that predict learning difficulties will also predict normal variation in learning abilities as well as high ability, which means that these genes will be useful for predicting the educational progress of all children, not just children at the low end of the normal distribution. Identifying these genes will lead to dimensional rather than diagnostic systems of classification of learning abilities and difficulties, based on aetiology rather than symptomatology. It will also lead to research on the brain and mind pathways between genes and behaviour that can account for these general as well as specific effects (Kovas & Plomin, 2006).

A learning difficulties gene chip could be even more important for treatment and intervention than for diagnosis. In terms of treatment, an untapped opportunity for genetic research is to identify genes that predict, not disorders themselves, but response to treatment. This goal is part of a 'personalized treatment' movement rather than imposing one-size-fits-all treatments.

However, the most important benefit of identifying genes associated with learning difficulties is the power to predict problems very early in life, which will not only serve as the earliest possible warning system but also facilitate research on interventions that prevent learning difficulties from developing, rather than waiting until problems are so severe that they can no longer be ignored. Genetic prediction will complement any prediction that might in the future be available from neural data (e.g. the hypothesized link between the differential organization of white matter and later variation in reading; Gabrieli, 2009). In addition, genetic prediction will avoid the problem of disentangling the direction of effect associated with many brain measurements, and may offer more precision than behavioural measures. The goal of early intervention fits with a general trend toward preventative medicine which is much more cost-effective for children as well as for society. Because vulnerability to learning difficulties involves many genes of small effect size, genetic engineering is unimaginable for learning difficulties; interventions will rely on environmental engineering, not genetic engineering.

It could be argued that genetics is unimportant because we need to provide resources to prevent children from falling off the low end of the bell curve, regardless of the causes of their poor performance. However, genetics is likely to facilitate the development of successful preventative interventions that can focus on diagnoses based on aetiology rather than symptomatology.

Genetics can also help to target children most likely to profit from interventions. Targeting is likely to be important because successful prevention programmes usually require extensive and intensive, and thus expensive, interventions (Alexander & Slinger-Constant, 2004; Hindson et al., 2005; Horowitz, 2004).

What about the ethical issues raised by finding genes associated with learning abilities and difficulties? For example, will gene chips justify social inequality? Knowledge alone does not account for societal and political decisions. Values are just as important in the decision-making process. Decisions both good and bad can be made with or without knowledge. Finding genetic influence on mathematics ability does not mean that we ought to put all our resources into educating the best mathematicians and forgetting the rest. Depending on our values, genetics could be used to argue for devoting more resources to help disadvantaged children. Indeed, genetics makes this view more palatable because it avoids assigning blame for poor performance solely to environmental failures of the school and family.

We are a long way away from identifying and understanding the mechanisms through which QTLs, QTNs and QTEs are involved in shaping individual differences in learning and achievement. However, the new understanding of the complexity of these mechanisms highlights the importance of genetics to neuroscience in education (*good*) and cautions against the simplistic and deterministic *bad* view about genetics in education. Continuous technological and conceptual advances offer promise of changing the '*ugly*' state of knowledge to date, through further progress in attempts to identify specific genes throughout the genome responsible for ubiquitous genetic influence. We believe that only with better integration between educational, genetic and neuroscience research we will achieve the optimized education of the future.

Acknowledgements

We gratefully acknowledge the ongoing contribution of the parents and children in the Twins Early Development Study (TEDS). TEDS is supported by a programme grant (G0500079) from the UK Medical Research Council; our work on environments and academic achievement is also supported by grants from the US National Institutes of Health (HD44454 and HD46167); and from the Government of the Russian Federation (11.G34.31.003).

References

Alexander, A. W., & Slinger-Constant, A. M. (2004). Current status of treatments for dyslexia: critical review. *Journal of Child Neurology*, 19, 744–58.

Alferink, L.A., & Farmer-Dougan, V. (2010). Brain-(not) based education: Dangers of misunderstanding and misapplication of neuroscience research. *Exceptionality*, 18, 42–52.

Bouts, P. (1986). *La Psychognomie*. Paris: Dervy-Livres.

Cantlon, J. F., Brannon, E. M., Cater, E. J., & Pelphrey, K. A. (2006). Functional imaging of numerical processing in adults and 4-y-old children. *PLoS Biology*, 4, 844–54.

Cardon, L. R., & Fulker, D. W. (1994). The power of interval mapping of quantitative trait loci using selected sib pairs. *American Journal of Human Genetics*, 55, 825–33.

Collins, F. (2010). *The Language of Life*. New York: Harper Collins.

Dehaene, S., Piazza, M., Pinel, P., & Cohen, L. (2003). Three parietal circuits for number processing. *Cognitive Neuropsychology*, 20, 487–506.

Docherty, S. J., Davis, O. S. P., Kovas, Y., Meaburn, E. L., Dale, P. S., Petrill, S. A., Schalkwyk, L.C., & Plomin, R. (2010). A genome-wide association study identifies multiple loci associated with mathematics ability and disability. *Genes, Brain, and Behavior*, 9, 234–47.

Docherty, S. J., Kovas, Y., Petrill, S. A., & Plomin, R. (2010). Generalist genes analysis of DNA markers associated with mathematical ability and disability. *BMC Genetics*, 11, 61.

Docherty, S. J., Kovas, Y., & Plomin, R. (2011). Gene-environment interaction in the etiology of mathematical ability using SNP-sets. *Behavioral Genetics*, *41*(1), 141–54.

Fisher, S. E., and Francks, C. (2006). Genes, cognition and dyslexia: learning to read the genome. *Trends in Cognitive Science*, 10, 250–57.

Gabrieli, J. D. E. (2009). Dyslexia: A new synergy between education and cognitive neuroscience. *Science*, 325, 280–83.

Greven, C. U., Harlaar, N., Kovas, Y., Chamorro-Premuzic, T., & Plomin, R. (2009). More than just IQ: School achievement is predicted by self-perceived abilities for genetic reasons. *Psychological Science*, 20, 753–62.

Halberda, J., Mazzocco, M. M. M., & Feigenson, L. (2008). Individual differences in non-verbal number acuity correlate with maths achievement. *Nature*, 455, 665–68.

Hindorff, L.A., Junkins, H.A., Hall, P.N., Mehta, J.P., & Manolio, T.A. (2010). *A Catalog of Published Genome-Wide Association Studies*. Available at: http://www.genome.gov/gwastudies.

Hindson, B., Byrne, B., Fielding-Barnsley, R., Newman, C., Hine, D. W., & Shankweiler, D. (2005). Assessment and early instruction of preschool children at risk for reading disability. *Journal of Educational Psychology*, 97, 687–704.

Hirschhorn, J. N., & Daly, M. J. (2005). Genome-wide association studies for common diseases and complex traits. *Nature Review Genetics*, 6, 95–108.

Horowitz, S. H. (2004). From research to policy to practice: prescription for success for students with learning disabilities. *Journal of Child Neurology*, 19, 836–39.

Inlow, J. K., & Restifo, L. L. (2004). Molecular and comparative genetics of mental retardation. *Genetics*, 166, 835–81.

Kovas, Y., Haworth, C. M. A., Dale, P. S., & Plomin, R. (2007). The genetic and environmental origins of learning abilities and disabilities in the early school years. *Monographs of the Society for Research in Child Development*, 72, 1–144.

Kovas, Y., Petrill, S. A., Haworth, C., & Plomin, R. (2007). Mathematical ability of 10-year-old boys and girls: Genetic and environmental etiology of typical and low performance. *Journal of Learning Disabilities*, 40, 554–67.

Kovas, Y., Petrill, S. A., & Plomin, R. (2007). The origins of diverse domains of mathematics: Generalist genes but specialist environments. *Journal of Educational Psychology*, 99, 128–39.

Kovas, Y., & Plomin, R. (2006). Generalist genes: Implications for cognitive sciences. *Trends in Cognitive Science*, 10, 198–203.

Kovas, Y., Viding, E., Ng, V., Giampietro, V., Brammer, M., Barker, G.J., Happé, F. G. E., & Plomin, R. (2009). Brain correlates of non-symbolic numerosity estimation in low and high mathematical ability children. *PloS One*, 4, e4587.

Krasa, N., & Shunkwiler, S. (2009). *Number sense and number nonsense. Understanding the challenges of learning math*. Baltimore, MD: Paul H Brookes Publishing Co.

Kucian, K., Loenneker, T., Dietrich, T., Dosch, M., Martin, E., & von Aster, M. (2006). Impaired neural networks for approximate calculation in dyscalculic children: a functional MRI study. *Behavioral and Brain Functions*, 2, 1–17.

Luo, Y., Kovas, Y., Haworth, C., & Plomin, R. (2011). The etiology of mathematical self-evaluation and mathematics achievement: understanding the relationship using a cross-lagged twin study from age 9 to 12. *Learning and Individual Differences*, 21(6), 710–18.

Manolio, T.A., Collins, F.S., Cox, N.J., Goldstein, D. B., Hindorff, L. A., Hunter, D. J., et al. (2009). Finding the missing heritability of complex diseases. *Nature*, 461(7265), 747–53.

McCandliss, B. D. (2010). Educational neuroscience: The early years. *Proceedings of the National Academy of Sciences of the United States of America*, 107, 8049–50.

McCarthy, M.I., Abecasis, G.R., Cardon, L.R., Goldstein, D. B., Little, J., Ioannidis, J. P. A., et al. (2008). Genome-wide association studies for complex traits: consensus, uncertainty and challenges. *Nature Reviews Genetics,* 9 (5), 356–69.

Meaburn, E. L., Harlaar, N., Craig, I. W., Schalkwyk, L. C., & Plomin, R. (2007). Quantitative trait locus association scan of early reading disability and ability using pooled DNA and 100K SNP microarrays in a sample of 5760 children. *Molecular Psychiatry,* 13, 729–40.

Menon, V. (2011). Developmental cognitive neuroscience of arithmetic: implications for learning and education. *ZDM: The International Journal on Mathematics Education,* 42(6), 515–25.

Meyler, A., Keller, T. A., Cherkassky, V. L., Gabrieli, J. D. E., & Just, M. A. (2008). Modifying the brain activation of poor readers during sentence comprehension with extended remedial instruction: A longitudinal study of neuroplasticity. *Neuropsychologia,* 46, 2580–92.

OECD. (2008). *The PISA 2006 Assessment Framework for Science, Reading and Mathematics.* Paris: OECD.

Paracchini, S., Scerri, T., & Monaco, A. P. (2007). The genetic lexicon of dyslexia. Annual Review of Genomics, *Human Genetics,* 8, 57–79.

Petrill, S.A., Kovas, Y., Hart, S.A., Thompson, L.A., & Plomin, R. (2009). The genetic and environmental etiology of high math performance in 10-year-old twins. *Behavioral Genetics,* 39, 371–79.

Piazza, M., Pinel, P., Le Bihan, D., & Dehaene, S. (2007). A magnitude code common to numerosities and number symbols in human intraparietal cortex. *Neuron,* 53, 293–305.

Pinker, S. (2002). *The blank slate: The modern denial of human nature.* New York: Penguin.

Plomin, R., DeFries, J. C., McClearn, G. E., & McGuffin, P. (2008). *Behavioural Genetics* (5th ed.). New York: Worth.

Plomin R., Haworth C. M. A., & Davis, O. S. P. (2009). Common disorders are quantitative traits. *Nature Reviews Genetics,* 10, 872–78.

Plomin, R., & Kovas, Y. (2005). Generalist genes and learning disabilities. *Psychological Bulletin,* 131, 592–617.

Plomin, R., Kovas, Y., & Haworth, C. M. A. (2007). Generalist genes: Genetic links between brain, mind and education. *Mind, Brain and Education,* 1, 11–19.

Plomin, R., Owen, M. J., & McGuffin, P. (1994). The genetic basis of complex human behaviors. *Science,* 264, 1733–39.

Plomin, R., & Schalkwyk, L. C. (2007). Microarrays. *Developmental Science,* 10, 19–23.

Plomin, R., & Walker, S. O. (2003). Genetics and educational psychology. *British Journal of Educational Psychology,* 73, 3–14.

Raymond, F. L., & Tarpey, P. (2006). The genetics of mental retardation. *Human Molecular Genetics,* 15, R110–R116.

Rykhlevskaia, E., Uddin, L. Q., Kondos, L., & Menon, V. (2009). Neuroanatomical correlates of developmental dyscalculia: combined evidence from morphometry and tractography. *Frontiers in Human Neuroscience,* 3, 1–13.

Samuelsson, S., Byrne, B., Olson, R. K., Hulslander, J., Wadsworth, S., Corley, R., Willcutt, E. G., & DeFries, J. C. (2008). Response to early literacy instruction in the United States, Australia, and Scandinavia: A behavioral-genetic analysis. *Learning and Individual Differences,* 18, 289–95.

Samuelsson, S., Olson, R. K., Wadsworth, S., Corley, R., DeFries, J. C., Willcutt, E. G., Hulslander, J., & Byrne, B. (2007). Genetic and environmental influences on prereading skills and early reading and spelling development in the United States, Australia, and Scandinavia. *Reading and Writing* 20, 51–75.

Scerri, T.S., & Schulte-Koene, G. (2010). Genetics of developmental dyslexia. *European Child & Adolescent Psychiatry,* 19, 3, 179–197.

Service, R. F. (2006). Gene sequencing. The race for the $1000 genome. *Science,* 311, 1544–46.

Shaywitz, B. A., Shaywitz, S. E., Blachman, B. A., Pugh, K. R., Fulbright, R. K., Skudlarski, P., et al. (2004). Development of left occipitotemporal systems for skilled reading in children after a phonologically-based intervention. *Biological Psychiatry,* 55, 926–33.

Stanesku-Cosson, R., Pinel, P., van de Moortele, P. F., Le Bihan, D., Cohen, L., & Dehaene, S. (2000). Understanding dissociations in dyscalculia. A brain imaging study of the impact of number size on the cerebral networks for exact and approximate calculation. *Brain,* 123, 2240–55.

Tommerdahl, J. (2010). A model for bridging the gap between neuroscience and education. *Oxford Review of Education,* 36, 97–109.

Venkatraman, V., Ansari, D., & Chee, M. W. (2005). Neural correlates of symbolic and nonsymbolic arithmetic. *Neuropsychologia,* 43, 744–53.

Wolfgang, C., Stannard, L., & Jones, I. (2001). Block play performance among preschoolers as a predictor of later school achievement in mathematics. *Journal of Research in Childhood Education,* 15, 173–81.

Chapter 10

Genetic sciences for developmentalists: an example of reading ability and disability

Elena L. Grigorenko

Overview

There is more and more traffic at the junction of genomic and social sciences, a complex pattern of intersecting questions, studies and research possibilities. And it is likely to become only heavier in the near future. What does this mean for such social sciences as psychology and education? In this chapter, these issues are explored using the research on the genetic aetiology of reading ability and disability as an illustration.

10.1 What is known about the genetic aetiology of reading ability and disability?

However referred to, defined, diagnosed or measured, reading disability (RD) has always been viewed as a condition whose pathogenesis involves hereditary factors. This idea was initially presented in the description of the first documented case of RD in the late 19th and early 20th century, and although it has been challenged, it has gradually won full (or near-full) acceptance (Fletcher, Lyon, Fuchs & Barnes, 2007). Thus, although the assumption that RD is, at least to a certain degree, a genetic disorder has been dominant for a while, the field's understanding of the impact of these heritable factors and the biological and genetic machinery behind them has changed over the century of scientific inquiry into RD.

These changes in understanding have paralleled the emergence of the view that reading is a complex system of cognitive processes supported by multiple areas of the brain (Pugh & McCardle, 2009). Reading engages different cognitive representations that are rooted in various anatomical areas of the brain, each characterized by particular architecture. In addition, the geography of reading in the brain assumes adequate connectivity between various reading-related areas. In short, the complex multiprocess cognitive system of reading is supported by a developmentally emergent, amalgamated, functional brain system (see Chapter 5, this volume). The current view asserts that this functional brain system is, in turn, established under the influence of and contributed to by complex genetic machinery. This brain system can be 'broken' or challenged in more than one way, causing RD (Pernet, Andersson, Paulesu & Demonet, 2009). Correspondingly, it is plausible to assume that the malfunctioning of the brain system that supports reading may be caused by multiple deficiencies in the corresponding genetic machinery (Grigorenko, 2009).

At this point, researchers have only a general sketch of the components of this machinery and how they operate. Yet, there are some clear elements of this sketch that seem quite permanent.

It is widely accepted that reading is a skill that requires socialization. In other words, for the overwhelming majority of people, the acquisition of reading as a skill requires teaching: the presence of social 'others'—either people who have already acquired that skill and can transmit the knowledge of it (e.g. teachers), or social tools that can aid the acquisition of reading by capitalizing on the crystallized knowledge of how this skill can be transmitted (e.g. computers). While attempting to understand the genetic machinery behind reading, researchers assume the presence of adequate schooling. In other words, inquiries into the genetic factors underlying reading are conducted under the assumption that deficient reading performance cannot be explained by the absence or the quality of teaching.

Yet, pretty much any characteristic of reading (e.g. reading speed and reading accuracy) or any reading-related process (e.g. phonemic awareness or lexical retrieval), even in the presence of adequate schooling, is continuously distributed in the general population or, in other words, is marked by a wide range of individual differences. Investigations into the sources of these individual differences indicate that a substantial portion of variation between people in both characteristics of reading and reading-related processes can be attributed to the variation in their genetic endowments (i.e. their genomes). Estimates of the magnitude of this portion of variance vary (1) across the life span; (2) for different languages; and (3) for different societal groups. These estimates are typically derived from studies of relatives and are called heritability estimates. To obtain heritability estimates, researchers recruit different types of genetic relatives, for example, identical (monozygotic) and fraternal (dizygotic) twins or other family members (e.g. parents and offspring, siblings or members of extended families). When the degree of genetic relatedness is known, participating relatives can be assessed with reading or related tasks and the degree of their genetic similarity can be compared to the degree of similarity with which they perform these tasks. Multiple statistical techniques have been developed to obtain heritability estimates from different types of relatives. There is no single perfect method for obtaining these estimates; therefore, researchers often use multiple types of relatives and multiple statistical approaches in order to minimize errors and maximize precision. In completing these studies, researchers have made a number of observations. First, although consistently attributing differences in performance to differences in the genome on an average of 40–60% (Grigorenko, 2004), they noticed some systematic fluctuations in these estimates. Specifically, these estimates are lower if they are obtained earlier in the development of the individual (e.g. among preschoolers or in early grades). These estimates also tend to vary depending on the language in which they are obtained and the specific characteristics of reading they are obtained for, suggesting that there is tremendous variation in how genetic factors manifest themselves in the different languages in which reading is acquired. Finally, these estimates tend to diverge depending on the characteristics of the sample in which they were obtained; these characteristics include socioeconomic status (SES), ethnic constellations and quality of schooling. In summary, when characteristics of reading (e.g. speed and accuracy) and performance on reading-related tasks (e.g. tasks of word segmentation and object naming) are considered in the general population, a substantial portion of related individual differences can be attributed to genetic differences between people. Thus, normal variation in reading is associated with genetic variation.

Genetic influences appear to be of even greater magnitude when challenged (or disabled) reading is considered. In a number of studies, researchers considered heritability estimates in samples selected through poor readers. Limiting variation (often relatives of poor readers demonstrate similarly depressed levels of performance) typically results in higher heritability estimates. Moreover, in these selected samples, researchers used different—in addition to heritability— statistics, such as relative risk estimates. These estimates typically provide an approximation of the likelihood that a relative of a poor reader (i.e. an individual affected with reading disability) will also be a poor reader. These estimates are compared to the general population risk. Specifically, it has been estimated that the prevalence of RD is estimated at 5–12% of school-aged children (Katusic, Colligan, Barbaresi, Schaid & Jacobsen, 2001). Relative risk statistics suggest that the prevalence of RD among relatives of individuals who suffer from this disorder is substantially higher than the general population estimates. These findings also attest to the important role of genes in the development and manifestation of RD. To obtain relative risk statistics, researchers recruit families of individuals with RD (so-called RD probands). Multiple types of family units can be utilized in this research: sibling units (i.e. pairs or larger sibships), nuclear families and extended families. Once again, there is no single method that is ideal; working with different constellations of relatives is associated with various strengths and weaknesses and, similar to heritability studies, to maximize the accuracy and precision of findings, researchers utilize multiple approaches.

Although the literature is replete with data supporting the hypothesis that genetic factors are important for understanding individual differences in reading acquisition and reading performance, a clear delineation of the specifics of these factors has been challenging. Currently, the literature contains references to about 20 (Schumacher, Hoffmann, Schmal, Schulte-Korne & Nothen, 2007) potential genetic susceptibility loci (i.e. regions of the genome that have demonstrated a statistically significant linkage to RD; typically these regions involve more than one and often hundreds of genes) and six (Grigorenko & Naples, 2009) candidate genes for RD (i.e. genes located within susceptibility loci that have been statistically associated with RD), but none of these loci or genes have been either fully accepted or fully rejected by the field. In addition, there is an ongoing debate regarding the specificity of the impact of RD-related genes; the issue here is whether the genes that are referred to as candidate genes for RD are sources of the specific genetic variation that accounts for individual differences in reading and reading-related processes only, or whether these genes have a broader impact on other types of learning (i.e. learning mathematics) and other cognitive processes. In an attempt to address this issue, a so called 'generalist gene hypothesis' was put forward (Plomin & Kovas, 2005). This hypothesis argues that genes that affect one area of learning, such as reading performance, are largely the same genes that affect other abilities (and disabilities). Thus, there are some promising developments in the field, but also numerous unanswered questions.

The information that has contributed to the identification of susceptibility loci and candidate genes for RD has been generated by so-called molecular studies of reading and reading-related processes. Unlike heritability and relative risk studies, where participants need to be characterized only behaviourally, that is, through their performance on reading and reading-related tasks, these studies assume the collection of genetic material, DNA (deoxyribonucleic acid). These studies can be subdivided into a number of major overlapping categories, by the type of samples they engage (i.e. genetically unrelated cases/probands and matched controls or family units such as siblings or nuclear and extended families) and by the type of genetic units they target (i.e. specific genes, specific genetic regions, or the whole genome). The first molecular-genetic study of RD was completed with a number of extended families of individuals with RD (Smith, Kimberling, Pennington & Lubs, 1983). In such studies, families of individuals with RD (typically, severe RD) are approached and their members are asked to donate both behaviour indicators of their performance on reading and reading-related tasks, as well as biological specimens. The task, once again, is to correlate the similarities in performance to similarities in genes, only now, the genetic similarities are not estimated, but measured using special molecular-genetic (i.e. genotyping and sequencing) and statistical (i.e. linkage and association analyses) techniques. Family units can vary from pairs of siblings to large extended families, impacting, correspondingly, the sample size that is needed for enough statistical power to distinguish a true genetic signal from noise.

Generally, extended families are harder to identify and harder to work with, but they tend to have more statistical power to identify the genetic source of RD (at least in these families). Smaller familial units such as nuclear families or sib-pairs are easier to identify and recruit, but the sample size requirements are much greater. The literature has illustrations of different types of samples used in molecular genetic studies of RD (Grigorenko, 2005). Similarly, the literature has examples of the utilization of different genetic units as targets. The very first molecular genetic study of RD was a whole-genome scan, in which the genome in its entirety was screened for linkage with RD, although that study had very few markers and they were protein markers (the technology then did not allow work with DNA markers). To date, nine genome-wide screens for RD (Brkanac et al., 2008; de Kovel et al., 2004; Fagerheim et al., 1999; Fisher et al., 2002; Igo et al., 2006; Kaminen et al., 2003; Meaburn, Harlaar, Craig, Schalkwyk & Plomin, 2008; Nopola-Hemmi et al., 2002; Raskind et al., 2005) have been reported. These studies utilized hundreds, thousands and hundreds of thousands of markers, as technology and cost permitted. There are also studies that focus on particular, selected regions of the genome. The selection of these regions is typically determined either by a previous whole-genome scan or by a theoretical hypothesis capitalizing on a particular aspect of RD[1] (Skiba, Landi, Wagner & Grigorenko, 2011). Yet, some of the studies settled on candidate regions through different means, such as through a known chromosomal aberration. Denmark, for example, has a health policy of screening all of its newborns for macro-chromosomal changes (e.g. large rearrangements). In these cases, researchers can screen individuals who have such rearrangements for the presence of RD (Buonincontri et al., 2011). The hypothesis then is that a gene that is affected by such an aberration is somehow related to RD. As indicated earlier, at the present time quite a few (±20) different genomic regions are entertained as harbouring candidate genes for RD, but this list is likely to continue to expand (Rubenstein, Matsushita, Berninger, Raskind & Wijsman, 2011). In addition, as the goal of this work is, ultimately, to identify specific genes whose function is related to the transformation of a brain into a reading brain (i.e. the establishment of brain networks supporting the different types of cognitive representations required for the acquisition of reading), there are studies of specific candidate genes. Currently, there are six candidate genes being evaluated as causal genes for RD (*DYX1C1, KIAA0319, DCDC2, ROBO1, MRPL2* and *C2orf3*), but more genes have been reported as putative additions to this list (Buonincontri et al., 2011; Newbury et al., 2011). At this point, the field contains both support and lack of support for the involvement of each of these genes; thus, the findings are somewhat difficult to interpret. More time and effort are needed to understand the involvement of each of these genes with reading and its related processes.

10.2 **How can these results be interpreted?**

In short, the accumulated findings, to date, can be separated into definite and tentative. One definitive conclusion that the field has made pertains to the fact that individual differences in the ways people acquire and practise reading are associated with differences in their genomes (see Chapter 11, this volume). In other words, individual differences in reading, in their typical and atypical forms, are, at least partially, genetic. The association between reading and the

---

[1] For example, one early study (Cardon et al., 1994, 1995) focused on the short arm of chromosome 6 (6p), specifically, on the HLA region, assuming that there were connections between RD and left-handedness, and left-handedness and the autoimmune function. Neither of the connections has been confirmed, but the region 6p21, identified as a result of this study, remains a prominent player in the field, harboring two candidate genes for RD, *DCDC2* and *KIAA0319*.

genome appears to be stronger for those individuals who experience difficulties attempting to acquire and exercise reading compared to those individuals who do it seamlessly (assuming, of course, the proper developmental stage and adequate learning and teaching conditions). This observation is widely accepted in the field and can be perceived as a fact. However, there are caveats. It appears that the strength of the connection between reading and the genome is variable, being weaker in early childhood and, gradually, getting stronger through the school years and into adolescence and adulthood. Virtually nothing is known, however, about reading in older adults—there simply have not been enough studies to investigate the dynamics of heritability estimates for reading and reading-related processes as they continue to develop during old and old-old ages.

From this ultimate statement (i.e. that reading is, at least partially, controlled by genes), the degree of certainty in interpreting the observations in the field drops off rather quickly. Yes, the heritability estimates are high, but what specific genetic mechanisms generate and substantiate them? The transition from statistical estimates of the role of genetic factors to the identification of these factors has proven to be difficult. This is not specific to studies of reading only; in fact, there is a coined phrase about 'missing heritability,' referring to the rather common situation of multiple not-so-fruitful attempts to translate the high heritability estimates obtained for a variety of complex human conditions (i.e. disorders such as diabetes, ADHD and autism) into their underlying genetic foundations (Avramopoulos, 2010). The field of understanding the genetic bases of reading has the imprint of the field of the genetics and genomics of complex disorders in general. Specifically, many initial positive findings are often followed by non-replications, suggesting either a high level of heterogeneity of the involved genetic mechanisms or a high level of false positive results. Either interpretation is difficult to grapple with. The former logically leads to a supposition of the field's inability to generalize effectively from specific deficiencies that might be characteristic of specific families, or from specific samples to the general population. The latter assumes that the initial samples were too small and lacking statistical power to differentiate true and false findings, and that much larger samples are needed to weed out the initial field of promising results by marking a number of them as 'false positives'.

Needless to say, neither of these perspectives appeals. Yet, although these two possibilities are the most obvious ones, they are not the only alternatives. The *first* possibility pertains to the fact that the last few years of studies in genetics and genomics have resulted in breath-taking discoveries that overshoot all of the expectations raised by the first generation of the human genome sequence in 2000. Among these discoveries, three, arguably, deserve the most attention: (1) other, in addition to previously known, sources of structural variation in the genome; (2) the role of epigenetic mechanisms in how the genome changes throughout development, and in different contexts and experiences (e.g. experiences ranging from diet to schooling); and (3) the role of unstable or movable elements in the genome. Although it is not possible to discuss these 'new' genetic mechanisms in detail in this overview, it is important to note that none of them have been considered in studies of the genetic bases of reading and RD. Clearly, there is much to do there! The *second* possibility is related to the attempt of the field to sift through the findings and, putting aside their inconsistent nature, try to hypothesize about the various underlying theoretical considerations that might bring these findings together. Thus, although, at first glance, the collection of the six candidate genes for RD seems quite disparate, they all appear to be contributing, to different degrees, to the process of brain maturation and neuronal migration (Galaburda, LoTurco, Ramus, Fitch & Rosen, 2006). Thus, through further investigations, it is possible that the emergent system will lead the field toward the involvement of particular genetic pathways, rather than specific genes.

10.3 Why might these findings be of interest to psychologists and educators?

Recent rapid progress in cellular and molecular technologies and the speedy application of these technologies in research have generated a lot of hype in the scientific community and numerous subsequent remarkable discoveries. These discoveries are fuelling hopes and expectations regarding radical changes in the ways prevention, treatment and remediation-recovery may be carried out in the fields of medical conditions and neuropsychiatric disorders. Some of these discoveries have already generated practical implications, mostly in the areas of diagnostic and pharmaceutical medicine. A new field has emerged called *public health genomics* (PHG)[2]—a rapidly developing multidisciplinary research and practice area whose objective is to bring genome-based knowledge responsibly and effectively into the domain of public health with the goal of improving population health (A. Brand, Schroder, Brand & Zimmern, 2006; H. Brand, 2007; Burke, Khoury, Stewart, Zimmern & Bellagio Group, 2006). This genome-based knowledge is referred to as the '-omics'—a family of sciences focusing on the structure and function of the genome: the examination of the genome (genomics), the messenger RNA (ribonucleic acid) transcribed from active genes (transcriptomics), the proteins coded for by this mRNA (proteomics) and the metabolites which are the end products of gene expression (metabolomics)—of today (Tan, Lim, Khan & Ranganathan, 2009) and is characterized by an overabundance of data and information, along with the enormous wealth of opportunities associated with the utilization of these data.

However, it is probably accurate to say that the hype is still mostly limited to the scientific community. Practitioners rarely use, still, these discoveries in their everyday operations and, moreover, often lack understanding of the current discoveries and developments in the field (Chen & Goodson, 2007). Similarly, interactions with the general public reveal a lack of depth in understanding the general scope of recent discoveries in genetics and genomics, prevalent misconceptions reflecting a deficiency of knowledge (Chapple, May & Campion, 1995; Condit, 2001; Richards & Ponder, 1996; Walter, Emery, Braithwaite & Marteau, 2004), negative biases toward particular terms (e.g. mutation; Condit, Dubriwny, Lynch & Parrott, 2004) and the presence of strong personal, moral and global concerns (Barns, Schibeci, Davison & Shaw, 2000; B. R. Bates, Lynch, Bevan & Condit, 2005; Henneman, Timmermans & van der Wal, 2004) about the use and abuse of genetic and genomic information (Hahn et al., 2010). Yet, the very success of the integration of genome-based knowledge into public health relies on the public's ability to understand the need to collect and apply family history that is relevant to diseases and disorders, and the importance of being able to consult with knowledgeable healthcare providers, and to self-trigger and self-monitor health-related behaviours (Charles, Gafni & Whelan, 1999; Sheridan, Harris & Woolf, 2004).

It is fair to say that very little (if any) of these discoveries, at least at this point, have made their impact on either behavioural therapy or education. Although there are no data that have been collected specifically with or from educators or other types of providers in the sphere of education (e.g. educational tutors or occupational and behaviour therapists), there is no reason to believe that their level of mastery of this knowledge is any different from that of the general public. Yet, although perhaps seemingly a bit more distant from recent genome-based discoveries, educators are only a step away from healthcare practitioners. Indeed, developmental disorders, especially

---

[2] PHG is closely related to *personalized genomic medicine*—a subfield of medicine where genetic information is used to improve health outcomes (Guttmacher & Collins, 2002; Holtzman, 2006; Khoury & Gwinn, 2006) via incorporating data on an individual's genotype, family medical history and/or expression analysis into disease risk assessment (Khoury, 2003).

common developmental disorders such as learning disabilities (LD) in general and RD in particular, are public health issues that are serviced, primarily, by educators.

If, indeed, RD is recognized as a public health condition, then current thinking with regard to PHG, as with other common conditions, can be applied to understanding the role of genetics and genomics in issues pertaining to RD. In this context, a number of other useful parallels can be drawn connecting the literature on PHG and RD.

The essence of PHG is to personalize healthcare to maximize the well-being and longevity of each individual. The essence of quality education is the individualization of teaching and learning to maximize productive accomplishment and life satisfaction. Both personalized medicine and individualized education show similar trends in their utilization of technology and multimedia, engaging multiple sources of information and knowledge, and capitalizing on the usage of profiles of information rather than single data points on an individual. Both personalized medicine and education demonstrate changes in their definition of 'problems.' In medicine, previously, disease was defined through the presence of symptoms. Now, having accumulated a substantial corpus of data on the genomic bases of many diseases, medicine defines diseases through the presence of a genotype or a genomic signature that confers a susceptibility to these clinical symptoms (Ford et al., 2008). Similarly, education is striving to find early precursors of various learning difficulties and focusing on prevention rather than on failure and subsequent remediation. Yet, although PHG has focused on the utilization of genomic information in medicine and public health since early in this century, education has largely ignored the advances in genome-based knowledge. Yet, similar to the situation concerning common diseases in medicine, educational difficulties appear to have genetic bases. In fact, heritability estimates for academic difficulties exceed those for such common medical conditions as obesity, diabetes and cardiovascular problems. So, if healthcare practitioners are applying genome-based knowledge to the understanding, prevention and treatment of such common conditions in medicine, then why should not educators apply this knowledge to dealing with common LD, such as RD? In fact, the 'preventive' logic of PHG seems totally reasonable when applied to common LD. If the genetic bases of LD are understood and there are tools and resources available for diagnostic purposes, educators might no longer have to wait for signs of academic failure. If indicators of genetic risks are identified, the specific measure aimed at preventing the onset of difficulty or failure, or at least minimizing their extent, will become increasingly powerful. And, although the initial costs of early diagnostic and preventive actions might not be negligible, subsequent remedial costs might be substantially diminished by these activities. Although this type of reasoning has not been practised in education on a large scale, it has been present in the public health literature for over a decade (e.g. Gibson, Martin & Singer, 2002; Zajtchuk, 1999), and there is much to learn from this literature.

One such useful line of discourse addresses the ways and stages of translating genetic and genomic discoveries into public health applications. Although many translation paradigms have been proposed, here only one is exemplified, developed by Khoury and colleagues (Agurs-Collins et al., 2008; Khoury et al., 2009; Khoury, Gwinn et al., 2007; Khoury, Valdez & Albright, 2008). Connecting '-omics' sciences with public health, four different stages are differentiated for the purposes of validating scientific knowledge and integrating it into disease control and prevention programs (Khoury et al., 2009).

At the first stage, the discovery stage, also referred to as the analytical validity stage, research attempts to connect a genetic or genomic mechanism with a particular public health condition. At the second stage, referred to as the clinical validity phase, research appraises the value of the connection established in the first stage for health practice, validating the first-stage observations in different settings and different samples to collect evidence and assess its replicability and robustness. At this stage, evidence-based guidelines connecting a genetic or genomic discovery and a public health issue are developed. Subsequently, at the third stage, referred to as the clinical utility phase, research attempts to apply these stage-two guidelines to health practice by mechanisms of knowledge transfer and delivery, as well as dissemination and diffusion practices. Only having planted these practices into every-day application on a large scale can research, at stage four, access health outcomes of a genomic discovery in practice. This stage engages multiple ethical and social considerations.

It is important to note that the literature today acknowledges that most current genome-based research is unfolding within stage one (Khoury et al., 2008). It has been estimated that only approximately 3% of published studies are conducted within stage two, and only a very few—at stage three or four (Khoury, Gwinn et al., 2007).

In the context of viewing LD (and, correspondingly, RD) as a public-health problem and capitalizing on the Khoury and colleagues translational paradigm, what is the current state of affairs with respect to the relevant genetic and genomic discoveries in the field of LD in general and RD in particular?

Currently, the majority of activity is still happening at stage one. This stage defines the analytical validity (Haddow & Palomaki, 2003) of whatever genetic or genomic test needs to be conducted so that the genotype of interest for the disorder (in this case, RD) may be measured accurately, reliably and at a minimum expense. Although a number of specific candidate genes for RD have been identified, the field is rather far from grasping the mechanism that unifies these genes in their RD-specific action. Moreover, the field is also rather far from verifying the numerous hypotheses regarding the specific genomic regions and other candidate genes that are currently under consideration. In other words, stage one of the related work in the field of RD is far from being completed. It is rapidly unfolding, but it is too early, at this stage, to ascertain the specific genetic tests that have diagnostic validity for RD.

Yet, the research on some of the candidate genes (e.g. *KIAA0319, DCDC2, ROBO1* and *DYX1C1*) has transitioned to stage two, where the degree of association between these genes and RD is being validated in a variety of samples and contexts, but the field has yet to formalize the various stage-two findings into comprehensive guidelines connecting the specific genetic variants in these genes or the specific mechanisms these genes support to genetic vulnerability for RD or its specific components.[3] These activities are crucially important for the clinical validity (Haddow & Palomaki, 2003) of a genetic or genomic finding, i.e. its capacity to detect or predict the condition of interest (e.g. RD). Moreover, establishing and explicating these connections is absolutely necessary for the possibility of transitioning to stage-three practices that should involve service providers for individuals with RD, ranging from educators to medical doctors and career counsellors. These guidelines need to be as clear and as robust as possible, presenting a list of vulnerability genetic mechanisms and identifying the risks associated with these mechanisms. As it appears now, it is quite likely, based on the indications from the RD literature so far (T. C. Bates, Luciano, Montgomery, Wright & Martin, 2011; Meaburn et al., 2008; Newbury et al., 2011; Paracchini et al.,

---

[3] To appreciate the magnitude of the task faced by researchers of the genetic bases of RD, it is important to consider comparable efforts with regard to other complex common disorders. The Centers for Disease Control and Prevention has developed two knowledge synthesis initiatives in genomics (for details, see Khoury, Valdez et al., 2008; Yu, Gwinn, Clyne, Yesupriya & Khoury, 2008): The Human Genome Epidemiology Network (HuGENet), (http://www.cdc.gov/genomics/hugenet/default.htm) and The Evaluation of Genomic Applications in Practice and Prevention (EGAPP) initiative (http://www.egappreviews.org). The first initiative is aimed at developing the knowledge base on genes and diseases. The second initiative is intended to comment on the clinical validity and utility of genetic tests in specific clinical and population health scenarios.

2008) and the evidence from other common disorders[4] (Khoury, Little et al., 2007), that the relative risks associated with each of these vulnerability genetic mechanisms is of small magnitude (e.g. as determined by genetic association studies, the estimated relative risk for common diseases is around 1.5 (Khoury, Little et al., 2007). Thus, the field should expect either a long list of genetic risk variants for RD or a discovery of some type of clustering for these variants that results in a substantial magnification of risk, when more than one variant is present in an individual. Once again, though, the evidence in the field of RD that has accumulated so far is rather far from being leveraged into policy recommendations. This, of course, assumes that all policy should be evidence-based and, in many cases, especially in education research, this is simply not the case. Yet, the field is working very hard both at stage one—generating new discoveries on the genetic bases of RD—and at stage two, testing the extent of the generalizability of these discoveries and their role in various samples and contexts. A fundamental question that needs to be answered at stage two is whether an understanding of the genetic mechanism for RD may be used for pharmacological treatment of these and related disabilities (i.e. so that pharmacological, but not educational, approaches can be individualized), for individualized education (i.e. so that educational approaches are matched to genetic mechanisms and pharmacological treatment is engaged minimally or not at all), or for both. There have been attempts to treat RD pharmacologically in Europe, but not in the USA. Whether the idea of treating RD pharmacologically will be even acceptable in the USA is not a trivial question.

At stage three, a cadre of qualified and knowledgeable providers should be able to implement the guidelines developed at stage two in their everyday practice. This phase of clinical utility is aimed at establishing the net health benefit resulting from the introduction of a particular diagnostic procedure related to the genetic or genomic findings and associated with relevant interventions (Haddow & Palomaki, 2003). This stage ideally requires well-designed studies in the community. As the ultimate goal of PHG is to generate public health policies and recommendations that can be easily translated in individualized approaches for the beneficiaries of these policies, and since the major 'battle' that engages individuals with RD unfolds in schooling, if there are ever PHG policies for RD, they will inevitably involve schools. At this point, it is unclear whether these policies will be delivered by educators themselves or by PHG providers placed in schools (similar to the medical outreach services delivered to schools through school nurses), but it is clear that academic activities will be a part of the PHG policies that address RD.

Finally, only when there are data collected at stage three with regard to short- and long-term outcomes of RD-related PHG will the field be able to qualify and quantify the importance and relevance of the findings described earlier (and the findings to be made) to RD-related outcomes. Research at this stage calls for joint efforts from public health providers, bench and education scientists, economists, policymakers, stakeholders and community. There is a tremendous body of literature indicating that reading failure and deficiencies in the mastery of reading are risk factors for many negative life outcomes. Correspondingly, an assessment of the impact of RD-related PHG needs to be coupled not only with short-term academic outcomes, but with long-term life outcomes.

It is likely that 'moving' through Khoury's stages in the field of LD is not going to be easy. Even if the first-stage findings are convincingly validated at the second stage, and the variability of findings we see now is interpreted in a systematic and interpretable manner such that the mechanism behind these findings is understood, the field faces the behemoth tasks of completing stages three

---

[4] As determined by genetic association studies, the estimated relative risk for common diseases is around 1.5 (Khoury, Little, Gwinn & Ioannidis, 2007).

and four before being able to arrive to a conclusion with regard to the ultimate value of genetic and genomic studies for the prevention and treatment of RD and associated deficits, and for the enhancement and promotion of literacy. Yet, many current public health policies are associated with breakthroughs and shortcuts; it has certainly been the case for many PHG applications and might well be the case for its RD-related applications. Thus, while the field is making its way through Khoury's stages, whether in a sequential or simultaneous manner, it is important to focus on tasks that connect ongoing genetic and genomic RD-associated research, at whatever stage, with the public's perception of it. There are at least three such tasks that appear to be of high priority.

The first task is that of educating the public. If traditional PHG for 'conventional' health conditions such as cancer and diabetes generate so much misunderstanding and misinterpretation by the general public, the task of communicating the idea of PHG for developmental disorders, especially such common disorders as LD, will be even more difficult. Yet, it has to be done because modern public health policies count on and are targeted at an educated consumer. The literature indicates that genome-based knowledge is associated with misconceptions and misinterpretations in the public as a whole (Burton & Adams, 2009), but appears to be particularly the case for various minority groups (Hahn et al., 2010). One of the major issues here is the removal of the negative connotation associated with the concepts of genome-based science and PHG.[5]

The second, even more mammoth task is educating the providers, the professionals of both the fields of medicine and education, who serve individuals with LD. Here of note is a tremendous discrepancy in the literature, where there is a large current body of work on the importance of including basic knowledge of the genome-based sciences in the education of medical practitioners and virtually no work advocating the importance of the mastery of these basics by educators.

The third and, perhaps, the most important task is to trigger and stimulate substantial discourse at the intersection of PHG and education. There is a large-scale highly engaging discussion in the literature on the development of national policies on PHG in the USA and elsewhere (Gonzalez-Andrade & Lopez-Pulles, 2010; Little et al., 2009; Metcalfe, Bittles, O'Leary & Emery, 2009), but the issues concerning where and how PHG and education should meet are not part of this discussion. Because PHG is aimed primarily at common conditions directly associated with lifestyle (Agurs-Collins et al., 2008; Boccia, Brand, Brand & Ricciardi, 2009; Khoury et al., 2008; Sanderson, Wardle & Humphries, 2008), it seems that such life-long conditions as RD, which are directly associated with life outcomes, should be a part of these discussions.

Completing these three tasks is crucial for the development, implementation and evaluation of PHG for LD. Thus, engaging with them early (now?) is highly important. The observed slow rate of translation of genome-based knowledge into public health policy and practice (Boccia et al., 2009) has been connected with numerous factors, such as the low relative risk for common disorders, the complexity of the four-stage translational process and the lack of critical knowledge in the public. An unfortunate mark of the rapid development of genome-based sciences and technologies has been a popularization—to the point of hyperbole—of the scale and immediacy of the application of these developments to health-related practices (Davey Smith et al., 2005; Kamerow, 2008). Such harmful popularization, coupled with commercial potential, has often resulted in ill-justified use of genetic and genomic testing for susceptibility to complex disorders, mostly with no clear application or guidance for proper utilization of this knowledge. It is inevitable, as any knowledge-based progress, that soon, perhaps even in the foreseeable future, the genome will become part of the common record for personalized medicine and individualized education.

---

[5] For example, it has been reported that the word 'variant' is preferred by the general public over the word of 'mutation'.

Thus, it is very important to prepare the general public and service providers for this transformation of the genome from something that once cost a tremendous amount of funds to sequence, in draft, only 10 years ago, but soon will become just a part of the birth record for each of us.

Acknowledgements

The preparation of this article was supported by funds from the USA National Academy of Sciences (NAS) and National Institutes of Health (NIHDC07665 and HD052120). Grantees undertaking such projects are encouraged to express freely their professional judgement. This article, therefore, does not necessarily represent the position or policies of the Division of Behavioral and Social Sciences and Education of the NAS and the NIH and no official endorsement should be inferred. I express my gratitude to Ms Mei Tan for her editorial assistance and to *Psykhe* for permission for this text to overlap with Grigorenko, E. L. (2011).

References

Agurs-Collins, T., Khoury, M. J., Simon-Morton, D., Olster, D. H., Harris, J. R., & Milner, J. A. (2008). Public health genomics: translating obesity genomics research into population health benefits. *Obesity, 16,* S85–94.

Avramopoulos, D. (2010). Genetics of psychiatric disorders methods: Molecular approaches. *Psychiatric Clinics of North America, 33,* 1–13.

Barns, I., Schibeci, R., Davison, A., & Shaw, R. (2000). 'What do you think about genetic medicine?' Facilitating sociable public discourse on developments in the new genetics. *Science Technology & Human Values, 25,* 283–308.

Bates, B. R., Lynch, J. A., Bevan, J. L., & Condit, C. M. (2005). Warranted concerns, warranted outlooks: a focus group study of public understandings of genetic research. *Social Science & Medicine, 60,* 331–44.

Bates, T. C., Luciano, M., Montgomery, G. W., Wright, M. J., & Martin, N. G. (2011). Genes for a component of the language acquisition mechanism: ROBO1 polymorphisms associated with phonological buffer deficit. *Behavior Genetics, 41,* 50–57.

Boccia, S., Brand, A., Brand, H., & Ricciardi, G. (2009). The integration of genome-based information for common diseases into health policy and healthcare as a major challenge for Public Health Genomics: the example of the methylenetetrahydrofolate reductase gene in non-cancer diseases. *Mutation Research, 667,* 27–34.

Brand, A., Schroder, P., Brand, H., & Zimmern, R. (2006). Getting ready for the future: integration of genomics into public health research, policy and practice in Europe and globally. *Community Genetics, 9,* 67–71.

Brand, H. (2007). Good governance for the public's health. *European Journal of Public Health, 17,* 541.

Brkanac, Z., Chapman, N. H., Igo, R. P. J., Matsushita, M. M., Nielsen, K., Berninger, V. W., et al. (2008). Genome scan of a nonword repetition phenotype in families with dyslexia: Evidence for multiple loci. *Behavior Genetics, 38,* 462–75.

Buonincontri, R., Bache, I., Silahtaroglu, A., Elbro, C., Veber Nielsen, A.-M., Ullmann, R., et al. (2011). A cohort of balanced reciprocal translocations associated with dyslexia: identification of two putative candidate genes at DYX1. *Behavior Genetics, 41,* 125–33.

Burke, W., Khoury, M. J., Stewart, A., Zimmern, R. L., & Bellagio Group. (2006). The path from genome-based research to population health: development of an international public health genomics network. *Genetics in Medicine, 8,* 451–58.

Burton, H., & Adams, M. (2009). Professional education and training in public health genomics: a working policy developed on behalf of the public health genomics European network. *Public Health Genomics, 12,* 216–24.

Cardon, L. R., Smith, S. D., Fulker, D. W., Kimberling, W. J., Pennington, B. F., & DeFries, J. C. (1994). Quantitative trait locus for reading disability on chromosome 6. *Science, 226,* 276–79.

Cardon, L. R., Smith, S. D., Fulker, D. W., Kimberling, W. J., Pennington, B. F., & DeFries, J. C. (1995). Quantitative trait locus for reading disability: correction. *Science*, 268, 1553.

Chapple, A., May, C., & Campion, P. (1995). Lay understanding of genetic disease: a British study of families attending a genetic counseling service. *Journal of Genetic Counseling*, 4, 281–300.

Charles, C., Gafni, A., & Whelan, T. (1999). Decision-making in the physician-patient encounter: revisiting the shared treatment decision-making model. *Social Science & Medicine*, 49, 651–61.

Chen, L. S., & Goodson, P. (2007). Public health genomics knowledge and attitudes: a survey of public health educators in the United States. *Genetics in Medicine*, 9, 496–503.

Condit, C. M. (2001). What is 'public opinion' about genetics? *Nature Reviews Genetics*, 2, 811–815.

Condit, C. M., Dubriwny, T., Lynch, J., & Parrott, R. (2004). Lay people's understanding of and preference against the word 'mutation'. *American Journal of Medical Genetics*, 130A, 245–50.

Davey Smith, G., Ebrahim, S., Lewis, S., Hansell, A. L., Palmer, L. J., & Burton, P. R. (2005). Genetic epidemiology and public health: hope, hype, and future prospects. *Lancet*, 366, 1484–98.

de Kovel, C. G. F., Hol, F. A., Heister, J., Willemen, J., Sandkuijl, L. A., Franke, B., et al. (2004). Genomewide scan identifies susceptibility locus for dyslexia on Xq27 in an extended Dutch family. *Journal of Medical Genetics*, 41, 652–57.

Fagerheim, T., Raeymaekers, P., Tonnessen, F. E., Pedersen, M., Tranebjaerg, L., & Lubs, H. A. (1999). A new gene (DYX3) for dyslexia is located on chromosome 2. *Journal of Medical Genetics*, 35, 664–69.

Fisher, S. E., Francks, C., Marlow, A. J., MacPhie, I. L., Newbury, D. F., Cardon, L. R., et al. (2002). Independent genome-wide scans identify a chromosome 18 quantitative-trait locus influencing dyslexia. *Nature Genetics*, 30, 86–91.

Fletcher, J. M., Lyon, G. R., Fuchs, L. S., & Barnes, M. A. (2007). *Learning disabilities.* New York: Guilford.

Ford, P., Seymour, G., Beeley, J. A., Curro, F., Depaola, D., Ferguson, D., et al. (2008). Adapting to changes in molecular biosciences and technologies. *European Journal of Dental Education*, 1, 40–47.

Galaburda, A. M., LoTurco, J. J., Ramus, F., Fitch, R. H., & Rosen, G. D. (2006). From genes to behavior in developmental dyslexia. *Nature Neuroscience*, 9, 1213–1217.

Gibson, J. L., Martin, D. K., & Singer, P. A. (2002). Priority setting for new technologies in medicine: a transdisciplinary study. *BMC Health Services Research*, 2, 14.

Gonzalez-Andrade, F., & Lopez-Pulles, R. (2010). Ecuador: public health genomics. *Public Health Genomics*, 13, 171–80.

Grigorenko, E. L. (2004). Genetic bases of developmental dyslexia: A capsule review of heritability estimates. *Enfance*, 3, 273–87.

Grigorenko, E. L. (2005). A conservative meta-analysis of linkage and linkage-association studies of developmental dyslexia. *Scientific Studies of Reading*, 9, 285–316.

Grigorenko, E. L. (2009). At the height of fashion: what genetics can teach us about neurodevelopmental disabilities. *Current Opinion in Neurology*, 22, 126–30.

Grigorenko, E. L. (2011). At the junction of genomic and social sciences: An example of reading ability and disability. *Psykhe*, 20, 81–92.

Grigorenko, E. L., & Naples, A. J. (2009). The devil is in the details: Decoding the genetics of reading. In P. McCardle & K. Pugh (Eds.), *Helping children learn to read: Current issues and new directions in the integration of cognition, neurobiology and genetics of reading and dyslexia* (pp. 133–48). New York: Psychological Press.

Guttmacher, A. E., & Collins, F. S. (2002). Genomic medicine—a primer. *New England Journal of Medicine*, 347, 1512–20.

Haddow, J. E., & Palomaki, G. E. (2003). ACCE: a model process for evaluating data on emerging genetic tests. In M. Khoury, J. Little & W. Burke (Eds.), *Human genome epidemiology: A scientific foundation for using genetic information to improve health and prevent disease* (pp. 217–33). New York: Oxford University Press.

Hahn, S., Letvak, S., Powell, K., Christianson, C., Wallace, D., Speer, M., et al. (2010). A community's awareness and perceptions of genomic medicine. *Public Health Genomics*, 13, 63–71.

Henneman, L., Timmermans, D. R., & van der Wal, G. (2004). Public experiences, knowledge and expectations about medical genetics and the use of genetic information. *Community Genetics,* 7, 33–43.

Holtzman, N. A. (2006). What role for public health in genetics and vice versa? *Community Genetics,* 9, 8–20.

Igo, R. P. J., Chapman, N. H., Berninger, V. W., Matsushita, M., Brkanac, Z., Rothstein, J. H., et al. (2006). Genomewide scan for real-word reading subphenotypes of dyslexia: novel chromosome 13 locus and genetic complexity. *American Journal of Medical Genetics (Neuropsychiatric Genetics),* 141, 15–27.

Kamerow, D. (2008). Waiting for the genetic revolution. *British Medical Journal,* 336, 22.

Kaminen, N., Hannula-Jouppi, K., Kestila, M., Lahermo, P., Muller, K., Kaaranen, M., et al. (2003). A genome scan for developmental dyslexia confirms linkage to chromosome 2p11 and suggests a new locus on 7q32. *Journal of Medical Genetics,* 40, 340–45.

Katusic, S. K., Colligan, R. C., Barbaresi, W. J., Schaid, D. J., & Jacobsen, S. J. (2001). Incidence of reading disability in a population-based birth cohort, 1976–1982, Rochester, Minnesota. *Mayo Clinic Proceedings,* 76, 1081–92.

Khoury, M. J. (2003). Genetics and genomics in practice: the continuum from genetic disease to genetic information in health and disease. *Genetics in Medicine,* 5, 261–68.

Khoury, M. J., Bowen, S., Bradley, L. A., Coates, R., Dowling, N. F., Gwinn, M., et al. (2009). A decade of public health genomics in the United States: Centers for Disease Control and Prevention 1997–2007. *Public Health Genomics,* 12, 20–29.

Khoury, M., & Gwinn, M. (2006). What role for public health in genetics and vice versa? *Community Genetics,* 9, 282.

Khoury, M. J., Gwinn, M., Yoon, P. W., Dowling, N., Moore, C. A., & Bradley, L. (2007). The continuum of translation research in genomic medicine: how can we accelerate the appropriate integration of human genome discoveries into health care and disease prevention? *Genetics in Medicine,* 9, 665–74.

Khoury, M. J., Little, J., Gwinn, M., & Ioannidis, J. P. (2007). On the synthesis and interpretation of consistent but weak gene-disease associations in the era of genome-wide association studies. *International Journal of Epidemiology,* 36, 439–45.

Khoury, M. J., Valdez, R., & Albright, A. (2008). Public health genomics approach to type 2 diabetes. *Diabetes,* 57, 2911–2914.

Little, J., Potter, B., Allanson, J., Caulfield, T., Carroll, J. C., & Wilson, B. (2009). Canada: public health genomics. *Public Health Genomics,* 12, 112–20.

Meaburn, E., Harlaar, N., Craig, I., Schalkwyk, L., & Plomin, R. (2008). Quantitative trait locus association scan of early reading disability and ability using pooled DNA and 100K SNP microarrays in a sample of 5760 children. *Molecular Psychiatry,* 13, 729–40.

Metcalfe, S. A., Bittles, A. H., O'Leary, P., & Emery, J. (2009). Australia: public health genomics. *Public Health Genomics,* 12, 121–28.

Newbury, D. F., Paracchini, S., Scerri, T. S., Winchester, L., L., A., Walter, J., et al. (2011). Investigation of dyslexia and SLI risk-variants in reading- and language-impaired subjects. *Behavior Genetics,* 41, 90–104.

Nopola-Hemmi, J., Myllyluoma, B., Voutilainen, A., Leinonen, S., Kere, J., & Ahonen, T. (2002). Familial dyslexia: neurocognitive and genetic correlation in a large Finnish family. *Developmental Medicine and Child Neurology,* 44, 580–86.

Paracchini, S., Steer, C. D., Buckingham, L. L., Morris, A. P., Ring, S., Scerri, T., et al. (2008). Association of the KIAA0319 dyslexia susceptibility gene with reading skills in the general population. *The American Journal of Psychiatry,* 165, 1576–84.

Pernet, C. R., Andersson, J., Paulesu, E., & Demonet, J. F. (2009). When all hypotheses are right: A multifocal account of dyslexia. *Human Brain Mapping,* 30, 2278–92.

Plomin, R., & Kovas, Y. (2005). Generalist genes and learning disabilities. *Psychological Bulletin,* 131, 592–617.

Pugh, K., & McCardle, P. (Eds.). (2009). *How children learn to read: Current issues and new directions in the integration of cognition, neurobiology and genetics of reading and dyslexia research and practice.* New York: Psychology Press.

Raskind, W. H., Igo, R. P. J., Chapman, N. H., Berninger, V. W., Thomson, J. B., Matsushita, M., et al. (2005). A genome scan in multigenerational families with dyslexia: Identification of a novel locus on chromosome 2q that contributes to phonological decoding efficiency. *Molecular Psychiatry,* 10, 699–711.

Richards, M., & Ponder, M. (1996). Lay understanding of genetics: a test of a hypothesis. *Journal of Medical Genetics,* 33, 1032–36.

Rubenstein, K., Matsushita, M., Berninger, V. W., Raskind, W. H., & Wijsman, E. M. (2011). Genome scan for spelling deficits: effects of verbal IQ on models of transmission and trait gene localization. *Behavior Genetics,* 41, 31–42.

Sanderson, S. C., Wardle, J., & Humphries, S. E. (2008). Public health genomics and genetic test evaluation: the challenge of conducting behavioural research on the utility of lifestyle-genetic tests. *Journal of Nutrigenetics & Nutrigenomics,* 1, 224–31.

Schumacher, J., Hoffmann, P., Schmal, C., Schulte-Korne, G., & Nothen, M. M. (2007). Genetics of dyslexia: the evolving landscape. *Journal of Medical Genetics,* 44, 289–97.

Sheridan, S. L., Harris, R. P., & Woolf, S. H. (2004). Shared decision making about screening and chemoprevention. *a suggested approach from the U.S. Preventive Services Task Force. American Journal of Preventive Medicine,* 26, 56–66.

Skiba, T., Landi, N., Wagner, R., & Grigorenko, E. L. (2011). In search of the perfect phenotype: An analysis of linkage and association studies of reading and reading-related processes. *Behavior Genetics,* 41, 6–30.

Smith, S. D., Kimberling, W. J., Pennington, B. F., & Lubs, H. A. (1983). Specific reading disability: identification of an inherited form through linkage analyses. *Science,* 219, 1345–47.

Tan, T. W., Lim, S. J., Khan, A. M., & Ranganathan, S. (2009). A proposed minimum skill set for university graduates to meet the informatics needs and challenges of the '-omics' era. *BMC Genomics,* 10, S36.

Walter, F. M., Emery, J., Braithwaite, D., & Marteau, T. M. (2004). Lay understanding of familial risk of common chronic diseases: a systematic review and synthesis of qualitative research. *Annals of Family Medicine,* 2, 583–94.

Yu, W., Gwinn, M., Clyne, M., Yesupriya, A., & Khoury, M. J. (2008). A navigator for human genome epidemiology. *Nature Genetics,* 40, 124–25.

Zajtchuk, R. (1999). New technologies in medicine: biotechnology and nanotechnology. *Disease A Month,* 45, 449–95.

Chapter 11

Education 2.0: genetically-informed models for school and teaching

Timothy C. Bates

Overview

School is intended to raise life-course outcomes for children and society. What we do in school to achieve this necessarily depends intimately on our beliefs about how children's minds work. Key questions include the following: do children differ in their ability to learn, or are they all identical? Is the mind a blank slate or a complex set of specialized cognitive mechanisms? Do differences in learning simply reflect prior experience? Are these differences remediable through experience and neuronal plasticity? Would successful teaching minimize or increase differences in capability between children? Are differences in children largely due to gene-by-environment (G×E) interactions, such that all children can achieve equal outcomes given the correct special environments or teaching styles? This article uses behaviour and molecular genetic examples to address these questions. Of course, not all answers are clear, and many are surprisingly preliminary given the importance of education to society. It is suggested that neuroscience indicates that children's minds are made up of multiple genetically and psychologically distinct mechanisms underlying learning; that children differ in the functioning of these systems; and that much of this difference is genetic is origin and general in nature, with limits to brain plasticity. A systems framework is presented linking these basic biological bases to the lifespan thriving which school is intended to foster. This systems model suggests that the principal outcome of neuroscience research will be to focus attention on how non-neuroscientific choices impact learning: teacher selection and training, teaching methods and curriculum. The chapter concludes by suggesting that much of education is based on different answers to the research questions posed here than those flowing from current neuroscience. If we are to maximize the capability of all children, it is suggested that it will be important to achieve consilience between what we know about the mind and what educationalists do.

11.1 Introduction

To discuss even a fraction of the information flowing in a single year from biological and genetic approaches relevant to understanding children's learning would take an encyclopaedia. As noted in other chapters in this volume, our understanding of disorders such as dyslexia, language impairment, autism, etc. is voluminous. Rather than summarizing the information of relevance for teaching or curriculum regarding research on some fraction of this information, here I focus on describing how genetic and biological research on individual differences provides frameworks that help us understand what education can do for children, how education may work and what it is reasonable to expect from education. The central emphasis of this chapter is on a framework linking basic biological factors to the outcomes which education fosters, such as prosperity and independence, via an intervening layer of cognitive systems. It is these cognitive systems that interact with education and which are the proximal causes of behaviour that education is designed to enhance.

Education appears to be important for a range of reasons, not least of which is its strong association with social position (Gottfredson, 2004). At an international level too, education is highly correlated with technological prowess and with GNP (gross national product) per capita. This seems to be due to the association of education with the skills and adaptability of employees (Rindermann & Ceci, 2009). Finally, in societies with a broad electoral franchise, the task of ensuring that citizens are culturally literate was deemed important enough that education is free and compulsory until late adolescence. Doing a good job at education, then, seems critical, and for that reason around 5–7% of the GNP of developed nations is devoted to education, provided by most states at low or zero direct cost.

But what does education do—what happens at school, and can behaviour genetic research tell us anything about this critical social enterprise? What would be seen as a mainstream view from the genetics of behaviour? Most important among the results from research, I suggest, are the following three. First, the finding that the mind is not a holistic instrument of infinite capacity, but rather a collection of dissociable specialized systems (see, for instance articles in Rapp, 2002), underpinned by distinct genetic mechanisms (e.g. Bates, Castles et al., 2007). Second is that much of this genetic variance is shared across the distinct components of the mind, creating a general ability factor explaining much of the differences between children (Deary, Spinath & Bates, 2006; Kovas & Plomin, 2006). Third and arguably most important is that much of these differences between children in mental capacity originate in genetic differences (Royal Society Working Group, 2011).

Some key 'explanatory gaps' remain, however, between educators' understandings of what children's biology brings to the classroom compared to what neurobiological and genetic research indicates. Recent commentaries have documented a desire to 'leave out the brain' in education (Blake & Gardner, 2007). This view that brains don't matter is in contradiction to the three claims noted earlier, and also represents a strong thread of discourse in education rejecting a role of biology in cognition and learning, and therefore in education (Hirsh-Pasek & Bruer, 2007). This view is similar to that expressed humorously in Garrison Keillor's radio broadcasts when he described the fictional US Midwestern town of Lake Wobegon where 'all the women are strong, all the men are good looking, and all the children are above average'. From 'no child left behind' to anguished newspaper headlines regarding differential success at entering university predicted from class background, this idea that teaching should be able to achieve equal outcomes for all children irrespective of the child distracts attention from understanding effective strategies for educating. Indeed, as argued later in this chapter, those practices most successful at increasing the capabilities of children may also magnify these differences (e.g. see Figure 11.2). Because evidence that the mind is componential, and with large genetic individual differences in the performance on these components is basic to answering this question of what education should be capable of, it forms the topic for this first section in the present chapter.[1]

---

[1] Much of this argument depends on information regarding the different perspectives on learning and development emerging from the individual differences, genetics and neuropsychology perspectives. In order to avoid distracting from the central argument presented here, a discussion of the way in which behaviour–genetic models of cognition can be integrated with those flowing from neuropsychology and differential psychology is presented in appendix 1.

11.2 **Differences on educationally-relevant traits have a genetic component**

Nearly all researchers have long since agreed that nearly all human behaviours have a substantial heritability (Turkheimer, Haley, Waldron, D'Onofrio & Gottesman, 2003). All educationally relevant traits from reading ability (Bates, Luciano et al., 2007; Haworth et al., 2010; Olson & Byrne, 2005; Paracchini, Scerri & Monaco, 2007) to autism (Ronald et al., 2006b) to intelligence (Deary et al., 2006) are heritable. Indeed, these three are among the most heritable. The same is true for non-cognitive resources such as conscientiousness, and curiosity (Luciano, Wainwright, Wright & Martin, 2006).

Most of our information about the genetics of educationally relevant traits comes from twin studies. Perhaps the single most challenging, but important finding from twin research is that fraternal twins' scores on ability measures and educational outcomes correlate much less than do those of identical twins, despite shared social status, neighbourhoods, parents, diet, housing and, often, schools and teachers (see Chapter 9, this volume). This sharp drop in correlation is as expected for traits that are highly heritable, and suggests that family environments have modest effects (Plomin, 1991). A caveat is that some (Turkheimer et al., 2003) but not all (Asbury, Wachs & Plomin, 2005; van der Sluis, Willemsen, de Geus, Boomsma & Posthuma, 2008) studies in populations with extremely low social position show large effects of family environment on cognition which suppress genetic variance (Tucker-Drob, Rhemtulla, Harden, Turkheimer & Fask, 2011). The general lack of shared environment effects is a success story for education (Bates, 2008a). They mean that our school systems, carefully crafted and optimized over the last century, largely but not entirely, equalize differences between family environments and the ability to profit from education they provide (Haworth, Asbury, Dale & Plomin, 2011). I think we should take pride, therefore, in having reduced shared-environment as close to zero as we have, while focusing on effective school interventions which address residual shared environmental effects in low-socioeconomic status groups. Teacher quality appears to be a critical factor in this respect (Taylor, Roehrig, Soden Hensler, Connor & Schatschneider, 2010), suggesting that more rigorous teacher selection and training may be the single most significant factor in addressing social status effects on the cognitive underpinnings of capability.

Molecular genetics: learning has mechanisms

If minds and brains are complex cognitive mechanisms as indicated by differential, genetic and neuropsychological research (see Appendix 1), then education is the process of working with these mechanisms to maximize the capability of the developing child. Molecular genetics is in its relative infancy compared to psychology, and currently few, if any, biological mechanisms are well understood, especially for complex cognitive systems such as memory and reasoning. However, the last decade has seen a dramatic increase in genetic knowledge, mostly based on increases in the power of molecular genetics, available now at relatively very low cost. Beyond the ultimate benefit of showing the mechanisms underlying learning, concrete genetic associations give a sense of reality to difficulties in learning which are often claimed to be malingering or an unwarranted call for special attention. A second finding of direct relevance to educators is that molecular research has confirmed that the same genes are responsible for the full range of normal variation, for instance, in reading ability and intelligence seen in the classroom as are involved in severe clinical reading disorder or dyslexia (Bates, Luciano et al., 2007) and learning disability (Kovas & Plomin, 2006) respectively (see Chapter 10, this volume). This tells us that biology does not apply to only special cases, but that all children face the limits and potentials of their brain mechanisms for learning. Concrete genetic associations thus provide evidence for the reality of difficulties in learning and for continuity between normal and abnormal.

Moving to candidate genes, researchers have begun to find the pathways in which these effects operate. In the case of reading, all of the genes discovered to date appear to play a role in fetal neuronal migration. *DYX1C1*, for instance, affects the short-range laminar migration of axons, resulting in both inappropriately increased and foreshortened intracortical connectivity (Rosen et al., 2007). These specific genes alter neuronal migration in ways that are believed to underlie the cognitive difficulties encountered in dyslexia. At this gene-specific level, the same variants involved in clinical dyslexia are reliably found in normal samples, unselected for dyslexia. Thus *KIAA0139* (Luciano et al., 2007), *DCDC2* (Lind et al., 2009) and *DYX1C1* (Bates et al., 2010) have all been linked to normal variation in reading ability. The study of other traits relevant for education such as language and long-term storage and learning is in its infancy, but already gene effects are being found which mediate differences in learning mechanisms in spoken language (Bates et al., 2011; Hannula-Jouppi et al., 2005) and for memory or the retention and recall of information over time (Papassotiropoulos et al., 2006). Both these traits are, of course, of direct relevance to understanding the mechanisms of learning in the classroom.

Understanding the underlying biochemistry of these mechanisms casts a powerful light on the origins of differences in learning in the classroom. The new genetics opens possibilities for earlier and more targeted diagnosis (these opportunities for translational research are outlined briefly in Appendix 2). The nature of the structural biological differences coded for by these genes may also carry some implications for the kinds of plasticity teachers should expect in children. These factors are discussed next.

Plasticity and limits to plasticity in complex systems

Breakthroughs allowing researchers to selectively knockdown genes of interest during development (in animal models, of course) have recently allowed the functional effect of genes implicated in dyslexia to be visualized directly. This work is ongoing, but strongly suggests that the biological basis of dyslexia lies in failures of neuronal migration (Gabel, Gibson, Gruen & LoTurco, 2010). These knockdown mice have psychological disorders plausibly related (although distantly, of course) to those found in dyslexia—for instance, reduced working memory in *DYX1C1* knockdown models (Szalkowski et al., 2011), a trait also altered by polymorphisms in this gene in human subjects (Bates et al., 2010). For education, perhaps the most important generalizable conclusion flowing out of this research lies in the timing and complexity of the processes involved in brain development. This in turn affects the kind of change (plasticity) that we will observe in education.

The brain is 'plastic' in the psychological sense of supporting learning, and in the biological sense underpinning this, with connection weights and even connections being almost continuously remodelled throughout the brain and underlying learning. However, the term 'brain plasticity' is often wielded like a light-sabre in education—implying that anything is possible (Bates, 2008a). It is near certain that education cannot replace a temperature-unstable version of the catechol-*O*-methyltransferase molecule with a more stable version, or a *KIBRA* allele enabling better retention, anymore than we should expect Lamarckian processes to know to look on chromosome 22 for the *COMT* gene and within this for the guanine/adenine substitution that underlies a change in thermo-stability of the catechol-*O*-methyltransferase enzyme important in the functional availability of dopamine and other catecholamines. The effect of genetic differences such as these on educational outcomes is a matter of empirical study. It is logically possible that selective application of an educational remedy might obviate all effects of some genetic differences, but this must be empirically shown. It is also plausible that differences in hippocampal function coded by *KIBRA* and frontal functions affected by genes such as *COMT* affect the relative capability of children in ways that education cannot remove but must simply build upon.

Genetic differences are not restricted to neurotransmitter or membrane plasticity effects, but also underlie the differentiation and migration of neurons into the developing brain-space, unfolding a complex, ordered, pattern of compartmentalized connectivity. As the individual is developing their psychological characteristics, these genes along with thousands of others then play repeated roles in learning. An example of the effects of knocking down the current best candidate genes for dyslexia is shown in Figure 11.1. This figure shows (in the leftmost pane) a normal cortex with its typical complex multilayered structure, laid down *in utero*. In each of the panels to the right, the effect of knocking down a dyslexia-linked gene is shown. In each case, hundreds of millions of neuronal cell bodies fail to migrate out of the ventricular zone to construct the complex vertically organized laminar cortex (Meng et al., 2005).

This complex matrix of neuronal migration, much of it occurring during gestation, has implications for expectations about cognitive plasticity in response to interventions. Beyond the remodelling of connectivity among existing cells and systems, the evidence for new functional neurons in the neocortex, as it were 'on demand' throughout life, is almost completely negative. Studies of neuronal plasticity have highlighted the general lack of neuronal cell replication in the brain outside a single layer of a small region of the hippocampus and (possibly) the olfactory bulb where neurons are exposed to the environment (Bartlett et al., 1998). Production of new neocortical cells, while technically feasible, has not to my knowledge been observed in humans. Bhardwaj et al. (2006) addressed this question in a highly innovative and insightful way, utilizing variation in atmospheric radiation following nuclear testing to 'carbon date' the timing of neuron formation based on isotope levels in cells. Postmortem studies conducted by Bhardwaj et al. indicated clearly that all neocortical neuronal cells were tagged with the radiation level present during gestation: compelling evidence against the routine generation of new neocortical neurons post-birth in humans. While behaviour can change, learning is unlikely to be able to roll-back development and re-engineer the billions of cells and their complex migrations and connections involved in neocortex, compensating altered DNA code using experience. We might hope to develop the technology to grow new functional cortical neurons, and thus remediate the effects of structural alterations by remodelling neuronal structure. This would have to activate developmental migration processes normally terminated years before training in reading begins. These reactivated processes

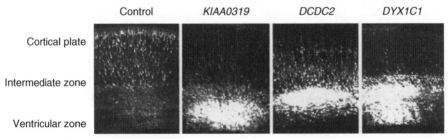

Fig. 11.1 Normally-occurring neocortical neuronal migration ('Control') and the effects of RNA interference (RNAi) gene-knockout for three candidate dyslexia genes: *KIAA0319* and *DCDC2* on chromosome 6, and *DYX1C1* on chromosome 15. Reproduced from *Cerebral Cortex*, *17*, Disruption of neuronal migration by RNAi of Dyx1c1 results in neocortical and hippocampal malformations, Glen D. Rosen, Jilin Bai, Yu Wang, Christopher G. Fiondella, Steven W. Threlkeld, Joseph J. LoTurco, and Albert M. Galaburda, Copyright (2007) with permission from Oxford University Press.

would then have to operate within a large already-developed brain; organizing cell differentiation, cell body, axonal and dendritic migration, in an environment quite distinct from that of the relatively 'empty' (albeit scaffolded) space in which normal fetal migration occurs.

Tabula rasa: genes, modules, mechanisms and innateness

A final education-related outcome of behaviour and molecular genetic research (along with neuropsychological patients, and normal and abnormal developmental psychology) is the light it casts on the centuries-old debate about the organization of the mind caricatured as the 'tabula rasa' versus ideas of innate representation associated with Locke (1690/2009) and Kant (1788/1996) respectively. Research indicates compellingly that we are not blank slates, but rather are collections of mechanisms.

The research articulated earlier on the complex mechanistic systems of gene effects unfolding in the womb and beyond, along with the observation of large genetic differences in this developmental programme and limited evidence for neuron-level neocortical plasticity after birth is of relevance for the tabula rasa view of the child's mind. Biological research suggests that education builds on an underlying biology determining the nature and complexity of primitive concepts in the young learner's mind. Consilience between different fields is often an indication of an important insight into nature's organization. It is valuable then to see that this view from behaviour and molecular genetics as specifying the neuronal architecture of the brain is compatible with a burgeoning body of work in developmental psychology supporting the view of the mind as composed of a number of primitives (including numerosity) out of which more complex concepts are assembled (Bloom, 2004; Carey, 1992, 2000, 2004, 2009).

Our own work in dyslexia is framed within a modular view of mind in which distinct mechanisms not only receive different inputs, but also run different types of algorithm on those inputs (Coltheart, Rastle, Perry, Langdon & Ziegler, 2001; Pinker, 1991). The finding that different subtypes of dyslexia appear to have different genetic bases is compatible with the view that the box-and-arrow diagrams of neuropsychology (see Appendix 1) are reflected in genes programming for development (Bates, Castles et al., 2007). A range of other models, however, continue to be supported as compatible to varying degrees with adult cognitive data. Some implement just one kind of computation (e.g. a neural network) and place a deeper burden on the stimuli to which people are exposed (e.g. written language) as the basis for the emergence of differential sensitivity to different features in different brain regions (e.g. regions of cortex specialized for letter-groups, words and meanings: Harm & Seidenberg, 2001). Variations on this model suggest that the brain ends up being modular, but that this emerges from a uniform substrate with only weak constraints on innate capability present (Karmiloff-Smith, Scerif & Ansari, 2003; Scerif & Karmiloff-Smith, 2005). Thus while most mainstream experts agree that the brain ends up being specialized, there are divergent views on the role of the environment in development. This in turn raises the 'nature nurture' debate: a discussion that has been central to discussions of mind and education.

The preceding sections suggest that there are significant differences between children in their ability to reason and learn and that specialized skills and talents such as language, spatial skills and memory are also scripted in our genome in important ways. It may be, however, that these apparent differences do not reflect main effects of genes raising or lowering the capacity to learn or reason, but rather that they generate a set of trade-offs such that all children have similar net capacities, but which must be accessed in different ways. A similar though less well-specified position is that genes act *only* in interaction with the environment and that psychological development is so complex and interactive that main effects are unlikely to cause differences between children.

As Pinker (2002) has noted, this 'no-debate; it's all a transaction' position meets little resistance as it appears sophisticated and reasonable. In particular it appears to offer hope for change: perhaps learning outcomes *are* merely complex, and better diet, smaller classes, brain gym, or attention to learning styles could make differences in outcomes go away. Research highlighted in several chapters of the present book suggests, however, that this is not the case to more than a limited degree. Despite the apparent sophistication of the 'there is no-debate' position on nature and nurture, 'Nature and nurture won't go away' for very good reasons (Pinker, 2004). Understanding nature is critical in understanding what is common to human beings and for understanding the implications of children differing in nature as well as in nurture.

Nature, nurture and differences between children in an effective school

The kinds of relationship of genes to environment are diverse, and often shaped by natural selection. As might be expected for important traits, evolution and selection has often tightly controlled many aspects of development: what Waddington called highly 'canalized' traits. This term refers to traits where redundant genetics systems are designed to robustly achieve a particular phenotype in almost any obtaining environment. That is why all humans have blood that carries oxygen, two more-or-less functioning eyes, ears etc. and a very precise set of many dozens of cognitive (cortical) maps and brain sub-systems to parse these inputs in robustly human manners: working memory, phonological loops, visual motion areas, social values and even mechanisms for evaluating group membership and other core evolved needs (Tooby, Cosmides & Barrett, 2003). The innate equipment of the brain is eager for certain very particular kinds of environmental input, but individuals differ in what they cognitively can do with a given input, and also, non-cognitively, in which input they prefer (Heckman, 2007) and evoke in their environment (Haworth et al., 2011). From these effects flow the large (often very large) main effects of genes.

Genes also interact with each other, and, because of these 'G×G' interactions, effects of even small changes can be extremely wide-ranging and powerful; the gene *MECP2*, for example, regulates over 1000 other genes and has widespread effects on brain function, cognition and disease (Joyner et al., 2009). Indeed while the great majority of DNA is not expressed as protein, significant sections of this non-expressed DNA underlie the regulation of gene expression itself, forming a complex operating system for the body (Qureshi, Mattick & Mehler, 2010).

Finally, genes interact with the environment: G×E. There are again diverse forms of G×E effect, including covariation, evocative and mechanistic interactions (Plomin, 1994). Interactions with the environment are of particular relevance to education, as schooling involves providing environments to children without which they would not develop the capabilities they acquire in interaction with education. In this case gene alleles are usually expected to multiply the effect of environments, reflecting the observed differences in children in their rates of learning (steeper or shallower slopes). Thus alleles affecting rate of learning will shift a child with a higher learning rate even further forward relative to her peers when the learning environment is optimized. Therefore while it might be expected that improvements in education would ameliorate prior environmental insufficiencies, leading to reduced differences between children, better teaching may act to magnify differences between children.

In their paper 'The rhetoric and reality of gap closing' Ceci and Papierno (2005) point to a large body of evidence showing that 'when the "have-nots" gain the "haves" gain even more'. This situation is illustrated graphically in Figure 11.2. Strong empirical evidence for this multiplier model comes from the decades-long programme of research into the lifespan developmental of talent of Baltes group, in his 'testing-the-limits' paradigm. Baltes concluded that intensive and organized

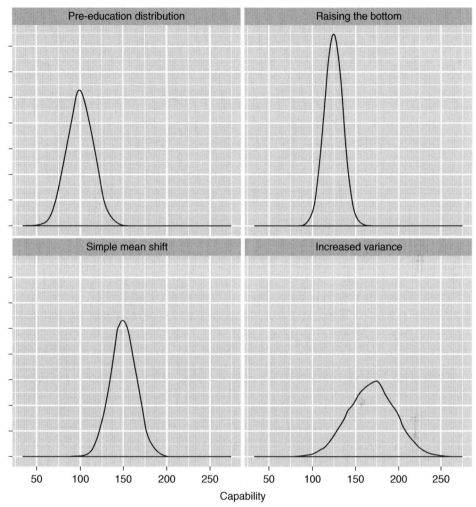

Fig. 11.2 Rhetoric and reality of gap closing: four ways in which education might impact differences between students. Note: the top-left panel depicts a normal distribution of ability prior to education beginning. The three additional plots reflect three possible effects of education on student capability. The top-right panel shows the expectation if children do not differ in their ability to gain from education: Education will reduce variance, with the initially less-capable students being remediated, falling closer to the more capable student's performance. The lower-left panel represents a simple shift, in which education preserves rank-order differences and variance, raising the average scores of all students equally. The final lower-right hand shows the empirically observed effect described by Ceci & Papierno (2005) regarding the 'rhetoric and reality of gap closing': all students gain capability, but the initially most capable students gain even more, increasing variance.

training in memory benefited all subjects, and lead to exceptionally high levels of performance. However, after intensive training, subjects' maximum performance asymptoted. Moreover existing differences were not washed out with practice, but rather the exact opposite result was obtained: original differences were magnified by massed practice (Baltes & Kliegl, 1992). Similar to Baltes, Detterman and Ruthsatz (1999) suggest that exceptional achievement is the product of (at least) intelligence, domain specific talent or skill and practice. This formula has been supported in case studies of musical expertise (Ruthsatz, Detterman, Griscom & Cirullo, 2008) musical prodigy (Ruthsatz & Detterman, 2003), and even domain-specific skill in savants (Ruthsatz et al., 2008). These empirical results contrast sharply with the widely cited model of talent proposed by Howe, Davidson and Sloboda (1998). Howe et al. argued for a version of the blank slate model, denying the existence of talent as anything other than early environmental choices, exposures and opportunities. In this model all variance in talented performance is allocated to practice. Detterman, Gabriel & Ruthsatz (1998) call this claim 'absurd environmentalism' for ignoring any role of individual differences. The results of Detterman, and, especially compelling due to their magnitude and duration, of Baltes's group suggest that this model is wrong and that the observation of Ceci and Papierno (2005) is the norm.

A second form of interaction that has been proposed to explain differences in children are so-called 'cross-over' gene–environment interactions. This kind of interaction maps neatly onto the learning-styles literature (see Chapter 13, this volume), and the related suggestion that all children could learn equally well, but that they will respond to different environments. In order to sustain the notion that all children performing equally well when given the learning environment that suits their style, these models need to propose that environments which are good for some children are bad for others (otherwise some children would do better than others across the board, reducing the model to the 'general ability + specific talents' compelling model (Spearman, 1927; Vernon, 1950)). While intuitively appealing, placed into a concrete context this model leads to paradoxical expectations. For instance, to take the example of reading, the cross-over G×E model would predict that on beginning school lessons based on phonics (or whole-language models), some children would learn to read, while others (with different alleles) would not only fail to learn as rapidly or even fail to learn at all, but may unlearn what reading skills they had with each exposure to the (for them) toxic school-based reading method. In practice, except in cases of disease or injury, reading improves with exposure to teaching for all children. Study of these models in psychiatry suggests that this particular form of interaction of exactly counterbalanced benefits and costs in different environments are highly unlikely to be detectable, other than as false-positives, a position supported both empirically (Munafo & Flint, 2009; Risch et al., 2009) and theoretically (Maes et al., 2006).[2]

Strong multiplicative G×E effects with weak (if any) cross-over interactions mean that good schools will increase rather than decrease differences between children; by building on and multiplying psychological capabilities, school will act to exaggerate even small underlying differences in biological capacity. A critical consequence of differences between children is that while we can set standards for performance, and can raise the performance of all children within practical limits, we cannot hope to meet targets defined in terms such that no child can be left behind any other. This suggests that education policy ought, therefore, to focus on maximizing individual children's capabilities rather than reducing differences between children. Of course this is not a carte blanche for ignoring differences. One statistic, for example, which does suggest a current failing in education is the finding that children equated for intelligence quotient (IQ) at age 11 do better at age 17 according to their social class. This is prima facie evidence that schools taking able children from lower social class backgrounds are failing to provide them with the opportunities to

---

[2] A lack of interaction effects is also relevant for claims about selective dietary requirements for adequate brain development also claimed to be relevant for differences in educational outcomes, e.g. the effects of breastfeeding on intelligence e (Caspi et al., 2007) where replications now have found either a opposite (Steer, Davey Smith, Emmett, Hibbeln & Golding, 2010), or null effects (Martin et al., 2011).

fulfil their potential—a finding supported by a growing literature (Taylor et al., 2010; Tucker-Drob et al., 2011).

11.3 A three-level cognitive systems framework: educating for capability

Understanding that already in the brain, billions of genetic decisions have led to the migration and differentiation of stem cells into precisely located cell bodies with their myriad connections is a perspective-changing event. Education is not working with a blank slate, nor with a magic space of infinite capacity, but with a delicate organ comprised of numerous specialized, limited capacity mechanisms. A brain in which some skills are innate: not just perception with its dozens of cortical maps in various tonotopic, retinotopic and other representational frames, but also behaviours as diverse as number skill and social relationships (Carey, 2009). Despite these specialized mechanisms, many tasks critical for success in society are effortful, require practice, and are subject to forgetting. This highlights limitations on our cognitive skill, but also focuses our attention on the role of education in achieving effortful, difficult learning. In this section I wish to bring together earlier material into a framework capable of representing biological capacities and differences, psychological content and processes, and the desired objective biographical goals of education, such as productive employment, and better health and social relationships. The section first introduces such a model. Using this framework, the mechanism of education in driving these goals is discussed focusing on three factors: what kinds of teaching lead to effective learning, which methods can effectively deliver this teaching, and what the outcomes of effective education will be in terms of between student differences.

If models that treat biology as uniform, or subsume it in a transaction where biology places no limits on performance, are unhelpful in understanding the mind, what framework can contain biological elements (such as G-coupled proteins and axonal guidance mechanisms), psychological mechanisms (such as memory and reasoning), as well as content (such as knowledge of procedures, memory for facts, or storage of phonemes in a buffer), and also valued outcomes such as being offered a job, keeping a stable relationship, or starting a successful company which motivate our interest in education? Here I suggest a framework in which these three factors are viewed as three distinct layers in a bio-psycho-social framework. This model is presented in Figure 11.3.

This bio-psycho-social model proposes that individual differences in behaviour span three domains: (1) *biological capacities*—which have no psychological properties; (2) *psychological adaptations*—psychological constructs such as values, memory, vocabulary, reasoning skills, which are formed under the joint influences of basic tendencies and external factors such as life-events and cultural norms; and (3) *objective biography*—observable outcomes explained by the interplay of external stimuli (also observable) with psychological adaptations. This model specifies the proximal cause of objective biography as the psychological layer, with influences from basic tendencies being fully mediated by psychological adaptations—with no direct links to behaviour. Education is explicitly represented as an external stimulus interacting with psychological adaptations (the decision to form and fund education is, of course, part of the objective biography of citizens and politicians).

In this model, we no longer expect social environment and genes to form an inseparable or inscrutable nexus, but rather to function as distinct layers of a bio-psycho-social system, with cognition in the middle layer. Cognition is dependent on underlying biological capacities, but it is this psychological layer with which education interacts and which generates observed behaviours. In this model, teachers and schools are elements of the external environment, interacting with the current mental state to generate new mental capacities. Elements of the left hand pane are tightly linked to genetic code. Measured in this way, they will often (but not always) have close to 1.0 heritabilities—as has been observed for components of reading (Olson, Forsberg & Wise, 1994).

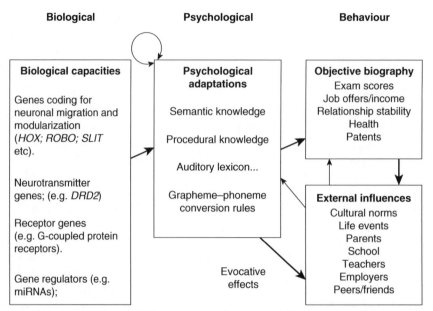

Fig. 11.3 A bio-psycho-social model for education. This model defines the observable outcomes for the individual student as their 'objective biography'. This biography of achievements is viewed as resulting from their psychological adaptations and external influences. These in turn interact, allowing for mental adaptations to evoke different environments and vice versa. At the left is the basic biological.

A key benefit of this three-layer model is the affordance of a conceptual framework for separating basic genetic and biological factors from the proximal causes of behaviour in the form of psychological content and skills. This has the potential to defuse misleading debates about 'nature' and 'nurture' which often tend to lose both in a 'complex transaction', suggesting that development is so exceptionally complicated and intrinsically interactive that nothing true can be said about how nature (genes and their programmes) unfolds (Pinker, 2002). It is of course important that scientists retain a respect for the complexity of their subject, moderating extreme claims so as not to suppress new approaches unnecessarily and not to simplify that which is abstract and complex. This can lead, however, to a failure to progress scientifically. There is a very good reason that biologists do not think it sophisticated to say when comparing a tall and short variant of plant differing in a gene allele that it would be necessarily simplistic to account for changed height in terms of the gene due to this invoking nature. Instead they would examine and report the mechanism by which the allele's expression leads to growth. They would trace out the pathways, promoters and regulators of the allele searching for additional genes affecting plant height. This is, of course, exactly what occurs in human medicine to great benefit, and is occurring for the trait of height in humans (Lango Allen et al., 2010; Yang et al., 2010). If we are to understand variation in normal development, as we do in disease it is important that we avoid reflexively retreating to concepts such as pervasive interaction at the expense of concentrating on the mechanisms of nature.

This perspective suggests answers to several important questions about the role of education in helping children absorb the content and the tools of culture. To show the utility of the framework, I focus briefly on two: what teaching methods work? And what impact do differences between teachers have in education?

How does school raise capabilities?

Parents, psychologists and even moral philosophers widely agree on things they would like their children and fellow citizens to have, for instance, the ability to gain and keep well-paid work, to have fulfilling social relations, to obey reasonable laws (Sen, 2009), to cope with challenges and master new skills (Ryff & Keyes, 1995), and to share a common cultural vocabulary allowing efficient communication and cooperation (Hirsch, 1999). Views on how school might achieve this vary dramatically. On the one hand school is viewed as a place where children construct their own reality with as little structure of guidance being imposed from teachers as possible. A key role for teachers in this model is simply to encourage the child in whatever they are doing. Sternberg (2000) in his article 'In search of the zipperump-a-zoo' suggests, perhaps somewhat tongue in cheek, a model of school performance in which pupils (using himself as an example) simply produce what is expected of them based on largely capricious expectations. Rather opposed to this view, is the idea that school is a place for explicit teaching of widely agreed facts about the world, theories of why these facts are so, and skills in how to do things, from painting to playing an instrument, to writing, and designing and constructing objects. This latter approach favours explicit instruction over personal construction. Rather than ask children to construct reality, it seeks to expand their capabilities by direct instruction: 'When you greet someone, shake their hand. Here is how: look the other person in the eye, look at their hand: grasp it firmly but not harshly, and, again, make eye contact'. Rather than presenting people and events value and date free and leaving children to extract what they can from these, events are used as examples of concepts and their development over time: 'At Trafalgar in 1805, Britain was at war with Napoleon's France and faced the combined fleets of France and Spain. Lord Nelson showed exemplary valour.
. .' It favours also using practical tests to ingrain knowledge: for instance, using an abstract concept in practical situations with repeated practice.

Somewhat aligned with this dimension running from modern structured learning to post-modern constructivism, are questions of whether school should teach knowledge, or raise basic abilities such as reasoning and memory, and in a related theme, whether school should focus on creativity rather than knowledge and skill acquisition. Experiments that can be interpreted as raising basic abilities such as intelligence are likely to be widely reported (Jaeggi, Buschkuehl, Jonides & Perrig, 2008; Rauscher, Shaw & Ky, 1993) and even sold commercially, but we are less likely to read about significant caveats (Sternberg, 2008), or critiques (Moody, 2009) and failures of replication (Chabris, 1999; Steele, Bass & Crook, 1999; Stough, Kerkin, Bates & Mangan, 1994). The bio-psycho-social model is of value here in distinguishing between two very distinct meanings attributable to words such as 'memory' which have both biological and psychological interpretations. To the extent that 'memory' is understood as a basic biological mechanism, we cannot expect education to give a person a better memory any more than we could expect education to rewrite a person's DNA to give the more effective version of the *KIBRA* gene. If 'memory' is, however, considered psychologically as the ability to store and recall knowledge, then memory can plainly be improved through practice and testing. The job of effective education then, becomes not equalizing children through neuronal plasticity or other mechanisms, but enhancing psychological capability by creating normative expectations and teaching situations and activities that increase desired objective biographical outcomes.

A similar approach can be taken in understanding the choice to teach creativity or understanding of existing knowledge. To give a recent example, Sternberg has argued for the former, stating that we should be 'teaching creativity not memorization' (Sternberg, 2010). This option (if choice is required) perhaps sounds attractive: after all who would not rather have the founder of Nokia, the inventor of a high capacity battery, or the creators at Apple as fellow citizens than someone who memorized the periodic table but did nothing new? However the statement masks several assumptions. The first is that creativity can be taught and that it is independent of intelligence and openness. Both these propositions have been powerfully critiqued (Brody, 2003a, 2003b). A second lies in the use of the term 'memorization'. Why not 'learning', or 'understanding'? A suggestion that 'school should teach creativity instead of understanding' perhaps does not sound as compelling. In the bio-psycho-social model, effective education is about increasing capability by doing what works. The question then becomes empirical: what works to maximize children's lifespan capability? This places a large focus for research and practice on understanding and implementing effective teaching that transfers to improvements in average measured outcomes. Paradoxically then, neuroscience tell us that education is about *teaching*.

The bio-psycho-social model suggests that a core difference between an educated and an uneducated mind consists in acquiring a large store of knowledge and skills. If this is correct, then a key question is to resolve the debate as to whether skills and knowledge require systematic practice and work in committing them to memory, or if learning should be similar to play, in that it is discovery-based and self-directed. In this latter view, rather than teachers teaching the facts that others agree on and giving practice with this knowledge, children should construct reality for themselves (Mayer, 2004). Likewise, testing is seen as having little or no role in learning, serving no positive purpose and acting to divide children, either along class lines reflecting coaching at home, or by giving damaging and self-fulfilling feedback to those unlucky enough to score badly on the test day. Any apparent learning as a result of study for tests is seen as fragile and quickly forgotten due to its rote nature.

While widely supported amongst educationists (Mayer, 2004), this view is quite opposite to traditional models of learning. To quote one of the oldest educational theorists 'Exercise in repeatedly recalling a thing strengthens the memory' (Aristotle, 350 BCE/2007). What do the data say (see Chapter 7, this volume)? Testing dramatically increases learning, *especially* over longer time periods (Roediger & Karpicke, 2006). Testing provides opportunities for retrieval, and retrieving memories appears to enhance them (Karpicke & Roediger, 2008). Speaking informally, the brain gets the message that retrieved memories are in use and valuable, and enhances their durability. We often hear, too, that feedback on poor performance must be damaging (via lowering self-esteem, it is suggested). Again, however, evidence suggests that feedback after testing further enhances learning over the testing effect (Butler, Karpicke & Roediger, 2008). Even the idea that rote learning only 'teaches the test' rather than building generalizable knowledge is false (McDaniel, Howard & Einstein, 2009). Why does structured learning and testing seem to work so well? Verbatim knowledge is often valuable in its own right—either in creating a social currency for compact communication ('There is a tide in the affairs of men. . .') or for precise recipes where quantities matter (seven eights are 56, not 57). This is true not only for learning one's times-tables, but of all facts and axioms: these building blocks of knowledge cannot be derived, but rather form the basis of derived knowledge, and the basis for augmenting our ability to manipulate information. In the bio-psycho-social model, this augmentation takes place in the psychological layer and is what raises capability by multiplying the effectiveness of biological ability.

Rote as one end of a dimension of depth of knowledge

Rote learning where meaning was denied would be reprehensible. But learning the skills that multiply capability nearly always involves typically rote-learning and practice. Take an example from mathematics in the form of matrix algebra. Matrix algebra is ubiquitous in engineering, mathematics and statistics. Like the card game Bridge, it consists of abstract and arbitrary rules which meet a goal: in this case for making thinking about complex problems very much easier by abstracting away potentially enormous sets of arithmetic operations into single symbols. The pay off is immense: $A \cdot B^{-1}$ can express in five characters what might be several million underlying mathematical operations on many thousands of numbers in large matrices. It is used in dozens of fields from accounting and forecasting to statistical modelling the behaviour of complex systems and materials in science and technology. Learning of matrix algebra involves some insight, but developing a facility in doing matrix algebra requires dozens of hours of practice. While the goal of matrix algebra is not arbitrary, the rules and symbolisms of matrix algebra are: indeed, as in calculus, alternative formulations competed early on. Any person wanting to use this technology must learn the symbolism and rules of transformation by rote so they become habitual.

The opposite of rote learning might be characterized by conceptual knowledge in which the student can recognize a concept in various forms and media, manipulate it comfortably, and relate it to other knowledge. We can, however, often be surprised by how shallow our own knowledge is. We see this in children who can draw a square but not a diamond, though neither has been taught by rote. In what sense does the child understand the concept 'square' when they cannot also draw a diamond? Or the concept diamond, when they cannot draw one? As adults we may feel beyond this stage, but it appears we are not. Imagine if you will a simple object: say a cube. Most readers will think they have deep knowledge of a cube: You could probably define it ('a symmetrical three-dimensional shape, either solid or hollow, contained by six equal squares'); recognize it in a sentence, and visually, recognize it translated in different orientations, distances or materials. You can relate its properties in mathematical terms of angle (90°), orientation of sides (orthogonal), equality (of edge lengths). You can also compute characteristics such as area or volume. Imagine a cube in front of you. How many sides does it have? A moment's inspection says six. Now, is our knowledge of this simplest three-dimensional figure 'rote' or deep? Refraining from verbal or mathematical solutions, imagine placing a finger on one corner, rocking the cube over and picking the cube up by placing a second finger on the diametrically opposite corner. Can you see how the cube can spin on the axis you have created? Now: how many corners are there around the middle of the cube? Few people answer this question correctly without turning to verbal knowledge. Clearly a continuum of representation exists, with much of it shallower than we think, and elaborated mostly by extensive use and testing.

Neuroscience in the classroom

In the bio-psycho-social model, school multiplies capability by providing new knowledge and methods of manipulating this knowledge. Differences in educational technique and their effects appear in this model as transactions between the child's layer two, and teachers and teaching systems in layer three (see Figure 11.3). Variables in this transaction include teacher differences, class sizes and other factors. The model focuses attention on the optimal trade-offs of factors such as teacher quality and class size (given the costs of each). Other variables include setting of children by ability and non-cognitive factors such as classroom discipline. Each of these factors can be examined empirically. Here I mention briefly two recurrent issues: class size and teacher quality.

Class-size reductions have been argued to be important (Krueger, 1999) and having large effects on learning. This is important, because in a system in which salary and pension costs dominate expenditure, class size is the single most expensive variable to change in education. It therefore has to have a very large effect to warrant the opportunity cost of decreasing class size. Effects of class size, however, are typically found to be small, often approaching zero (Hanushek, 1998; Hoxby, 2000). Moreover, because reductions in class size lead to reductions in teacher pay over the medium term, such interventions may further reduce teacher quality where funds are limited (Jepsen & Rivkin, 2009).

This latter conclusion by Jepsen and Rivkin (2009) focuses attention on the concept of teacher quality. Views on the impact of teacher ability and subject matter knowledge on educational outcomes vary. At one extreme, it is suggested to be largely irrelevant (Hanushek, 1998). Others conclude that teacher's cognitive ability is an important factor in raising student capabilities. Some data are available. Teacher effects have been examined in large samples (Leigh, 2010). This study controlled for between-student differences, looking at the effect of teachers on the children they taught, controlling for the previous year results of those children. This was a large study with data on 10,000 teachers and over 90,000 pupils. There was evidence for a general ability to teach, with significant between-teacher effects on student literacy and on numeracy. Between-teacher differences also influence the likelihood of individual children maximizing the expression of genetic differences in ability, with good evidence for a strong specific role of teachers on effective teaching of reading (Taylor et al., 2010). Similar data suggest significant lifetime health and income benefits of teacher quality early as kindergarten (Chetty et al., 2011). The nature of teacher quality is, however, unclear at present and warrants more research.

The impact of reduced prestige and pay relative to other occupational choices in lowering the ability of entrants to teacher training is often highlighted. Objectively, teacher cognitive scores appear to have been falling. For instance, over the past decades the ratio of trainee teachers with average IQ (100) versus high ability (IQs of 130 or above) appears to have fallen from approximately equal numbers in both groups to current ratios of 4:1 (Murnane, 1992). Studies of teacher-trainees also indicate that some teachers lack a full understanding of the material that they will be required to teach as assessed by an ability to gain good passes in the tests that their pupils will sit, for instance submissions to the Donaldson report on education indicated that teachers often need multiple attempts to gain good passes in primary school mathematics tests, and say that they often learn the material they will teach one lesson ahead of the pupils. This has been attributed to a cultural norm in schools that 'undermines the need to have a sound understanding of what is being taught' (Donaldson, 2011). Links of teacher cognition and knowledge to student improvements in student capabilities may repay research investment.

11.4 **Conclusion**

Genetics research and research on the brain suggests that children bring to school very large and robust differences in their learning capacity, with large proportions of these differences reflecting heritable differences (Royal Society Working Group, 2011), but also that it is decisions about how we teach that will have the biggest impact on raising children's capabilities. This highlights two implications of neuroscience research. The first is that because education works via biological systems that differ between children, differences between children in turn are magnified by effective education. The second is that improving education can be aided by reliable answers to practical questions about effective teaching techniques: Under what circumstances does learning best occur? How are skills like painting best taught? How is historical knowledge best taught? Do dates help or hinder children's understanding and recall? Should classes be small? Can a mix of large instructional classes and small tutorials and practice sessions deliver better educational outcomes? Does teacher ability and knowledge matter? Should children be 'set' in classes with others of like-ability? Solid, well-referenced and readily accessible answers to these questions are basic to optimizing the capabilities of children, and yet surprisingly hard to find. Debates in the news are often fuelled by apparent disagreement over core facts: are exams easier today than in the past? Work by the Royal Society of Chemistry (Royal Society of Chemistry, 2008) appears to demonstrate compellingly that this is the case. But this information is often available only in a scientific report rather than in open, well-referenced, live and contestable summaries, easily accessible to journalists, employers and parents alike. While there are numerous special committee reports and documents into education, these date rapidly and necessarily have a limited authorship. Often too, there simply do not seem to be robust reliable studies to address particular questions. It would be hoped, then, that the coming years will see a much tighter integration across academic departments of education, psychology and other disciplines basic to learning, as well as with policymakers, parents and curriculum decision-making. My vote would be to invest modest sums in answering these questions in ways that confront the educational debate with facts about how education works.

11.5 Appendix 1: describing and explaining the mind

At least three branches of psychology have emerged to explain differences between individuals: neuropsychology, individual differences research and behaviour genetics. The three have distinct perspectives on the mind, but can be integrated under an umbrella of cognitive genetics, providing a framework for understanding the biological bases of learning relevant for education (Bates, 2008b). These three approaches to cognition are shown in Figure 11.4 (differential psychology model of intelligence); Figure 11.5 (neuropsychological model of specialized cognitive mechanisms and their connections); and Figure 11.6, showing a behaviour genetic model of reading and elements of language and working memory.

The three figures have quite different forms, reflecting the focus of researchers in the three domains. The results of all three of these measures are of course reconcilable, and highlight two facts of broad importance for education.

The differential psychology approach is shown in Figure 11.4. This depicts the relationships amongst 13 cognitive ability measures, focusing on the covariance among these measures. Of relevance to education, this covariation approach revealed that a single general factor account for around 40–50% of the variance in any diverse and arbitrarily large battery of tests (Carroll, 1993).

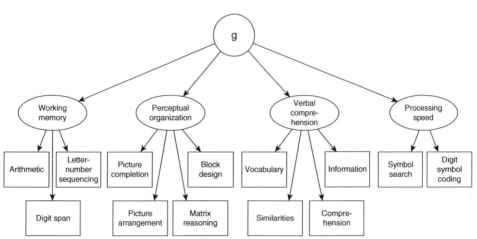

Fig. 11.4 Differential psychology model of cognition: The factor structure of ability as measured in the Wechsler Adult Intelligence Scale III. General ability accounts for around half the variance in performance, with domains of ability also exhibiting some structure (for instance, verbal and perceptual factors) and substantial variance among the specific abilities. It is these specific abilities, or even more fractionated forms thereof, which appear in neuropsychological models, such as that of Figure 11.5.

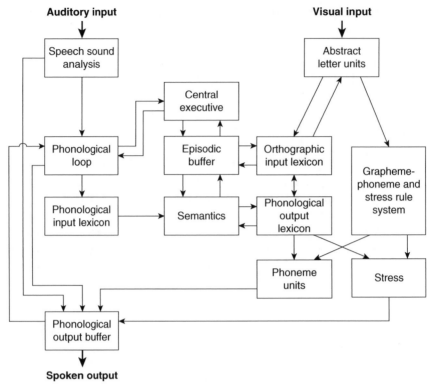

Fig. 11.5 Reading, language and working memory in cognitive neuropsychology. On the right-hand side are shown components of the dual-route cascaded system for reading (Coltheart et al., 2001). Data from patients supports dissociations between the ability to read aloud pseudo-words such as 'SLINT' and the ability to read-aloud irregular words such as 'YACHT' (Castles, Bates, Luciano, Martin & Coltheart, 2005; Castles & Holmes, 1996). In the centre we see semantics and the executive components of working memory (Baddeley, 2007). These are represented as separate modules because dissociations again suggest that there are patients with acquired brain damage who retain the ability to read but no longer access meaning from written language, and other patients who retain meaning despite losing the ability to read. Finally on the left hand side of the model are components of mind implicated in specific language disorder, specifically the systems for speech sound analysis, for brief storage of phonological strings, independent of meaning, and connections supporting rehearsal in short-term memory. Adapted from Baddeley, A. D., Working memory: Multiple models, multiple mechanisms. In H. L. Roediger, Y. Dudai & S. M. Fitzpatrick (Eds.), *Science of Memory: Concepts*, 2007, with permission from Oxford University Press.

This general or *g* factor is heritable (Deary et al., 2006) and highly stable across the lifespan (Deary, Whalley, Lemmon, Crawford & Starr, 2000). General ability measures are highly correlated with a range of life-course outcomes, and account for much of the variance in educational outcomes. For instance, in a study of some 70,000 children followed from age 11 to their GCSE exams, IQ measures at age 11 accounted for 81% of school grade variance at age 16 (Deary, Strand, Smith & Fernandes, 2007). This suggests that children differ not merely on what they know at a given moment, but on their ability to learn. Thus even lengthy exposure to education tends to be highly rank-preserving. Figure 11.4 also highlights how hierarchical modelling within individual differences captures clusters of variance beyond *g*. Shown here are representing visuospatial and

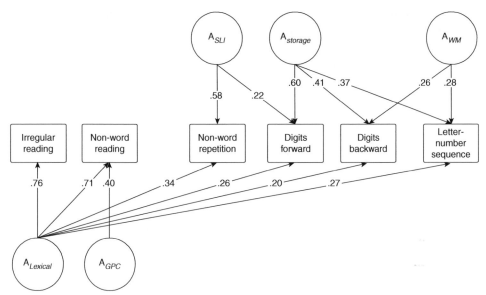

Fig. 11.6 Behaviour genetics model of the relationships between components of reading ability, phonological buffering important in language acquisition, forward and backward digit span and working memory (letter number sequencing). Multivariate behaviour–genetic models reveal both the correlations which dominate the differential psychology model (Figure 11.4), and the distinctions which are articulated in the cognitive model (Figure 11.5).

verbal ability intermediate between the single tests themselves and *g* or general ability. These two facts about cognition: that it contains both a substantial general component and a substructure of specialized faculties emerged in the 1920s (Spearman, 1927) and have been widely agreed since the 1950s (Vernon, 1950; Neisser et al., 1996).

Figure 11.5 shows the second powerful organizing framework in psychology: that of cognitive neuropsychology (Rapp, 2002). By focusing on double dissociations in patients, cognitive neuropsychology has built models of cognition based on the cascaded flow of information through distinct processing systems revealed in cases of developmental and acquired cognitive deficit (Shallice, 1988). A critical piece of information from this model is the suggestion that the mind has a complex internal structure, with concepts and processes of the mind being basic building blocks of its function, rather than arbitrary arrangements imposed by external processes such as education.

The individual differences and neuropsychological models appear, on the face of it, incompatible: one contains a latent, unobserved *g* factor and de-emphasizes specialized mechanisms; the other contains explicit boxes and arrows and focuses almost entirely on mechanisms and information transfer, rather than correlated abilities (which are uninformative about the dissociability of components of ability Shallice, 1988). Elsewhere, I argued that this apparent discrepancy is, however, illusory: both sets of information can be represented in a common framework (Bates, 2008b). An example of such a common framework is that of behaviour genetics as described below.

The behaviour genetics model (Figure 11.6) closely maps onto the model from neuropsychology. Both models capture irreducible components of the mind. The difference is that for behaviour genetics, evidence for these modules is based on developmental genetic dissociations flowing from Mendel's laws, rather than on acquired differences in patients. Research on the genetics of autism, for instance, recently demonstrated that autism fractionates into at least three distinct and independent components involving affective processing, language and flexibility of thinking: each with distinct biological bases and therapeutic needs (Ronald et al., 2006a). It is interesting to note that behaviour-genetic research (Bates, Castles et al., 2007) also replicates the dissociation of surface and phonological dyslexia found in acquired dyslexia. This genetic linkage suggests that the origin of the dissociable mechanisms found in the adult mind, and exposed by damage to the adult brain, do not arise simply from environmental regularities promoting specialization (Karmiloff-Smith, 1998), but that this specialization is itself programmed (Pinker, 2002).

As well as articulating the separately heritable mechanisms of the mind found in neuropsychology, behaviour genetics also highlights an aspect of the mind that does not emerge naturally from neuropsychological models (because of their focus on double dissociation): namely the general and group ability components of cognition captured in individual differences research. General ability or 'g' is not a mental mechanism per se, hence there is no 'g' box in neuropsychological models (though g is more tightly linked to brain volumes and functional differences in parieto-frontal and infero-temporal regions of the brain supporting attention, language, and working memory than with other regions of the brain (Jung & Haier, 2007)). Nevertheless, g accounts for around half the variance in cognitive ability and this is an important finding for education: it implies that children performing poorly (or well) on one element of their schooling are likely to be performing similarly poorly (or well) on other domains of their learning. Genetic studies (reviewed in Deary, Spinath & Bates, 2006) indicate that not only is a large portion of stable cognitive differences heritable, but that almost all of the correlations between different mental abilities are accounted for by genetic influences on multiple cognitive traits. Our own research into the genetics of reading and language supports this idea. The genetics of reading is mostly overlapping: not only at the level of comprehension where general ability is predominant, but in the very specific subtasks of reading: building a sight vocabulary of whole words and building a facility for decoding letters into sound. Genes then as often have very general effects (Kovas, Harlaar, Petrill & Plomin, 2005) just as they also have very specific effects (Bates, Castles et al., 2007; Wilmer et al., 2010).

11.6 **Appendix 2: risks and future developments**

Gene discovery creates the possibility of early diagnostic testing based on large numbers of aggregated genetic risk factors. Because genes associated with reading are present at birth, genetic knowledge allows an assessment of specific risk years before signs of reading delay could ever be detected by conceivable behavioural testing. This affords both much more effective allocation of this expensive resource to at-risk individuals, much earlier intervention and, possibly, interventions tailored to particular risk factors. As the discovery of specific genes for cognitive development continues, affordable effective prognostic testing is now emerging. As a result of genome-wide association studies (GWAS) assessing around one million single nucleotide polymorphisms (SNPs) in each individual for around £200, we can also soon expect dramatic improvements in early (pre-symptomatic) diagnosis of risk based on prediction functions. For many complex disorders, thousands of genes seem likely to each play very small, but cumulative roles. Prediction functions aggregate these thousands of tiny effects distributed across the genome and impacting on a complex disorder like schizophrenia or autism. In the case of height, this has been effective, suggesting that the present methods of genomic testing capture around 40% of the genetic basis of height (Lango Allen et al., 2010; Yang et al., 2010). These levels of prediction afford great opportunity. When genetic tests become available for disorders such as dyslexia and autism, it seems likely that their effects could be beneficial. Genetic tests can assess risk well before any behavioural signs, allowing focused intervention much earlier than ever before.

More speculatively, understanding the molecular genetic pathways of disorders such as autism may lead to entirely novel biologically informed therapies. While this seems remote given our lack of understanding of complex disorders at a neurological level, the dramatic response of apparently inevitable disorders such as phenylketonuria (PKU) and insulin-dependent diabetes to very simple biochemical interventions, suggests that substantial improvements, or even elimination may be possible when the triggers of gene expression in pathways to dyslexia, autism, attention deficit hyperactivity disorder and even general intelligence (as in the case of PKU) are understood. Secondly, and as seen in several other disorders already, understanding the genetic basis of differences in educational aptitude is likely to transform these into facts that we can acknowledge and address, rather than unexplained outcomes for which teachers are blamed as 'leaving children behind'. This new knowledge seems likely to promote not only acceptance and additional resources for tailored programmes (as is beginning to occur for autism and dyslexia) but also earlier and more effective intervention.

While early genetic testing offers the hope that we might put in place ameliorative cognitive interventions well before any negative social consequences of falling behind in school can ever emerge, this power, of course, enables various forms of GATTACA-like society. On the hopeful side, many important changes in education were accompanied by research results from individual differences and genetics: from socially inclusive education, where Binet and others provided evidence for ability in disadvantaged groups, to equal provision for education for both sexes, which was preceded by research from Burt and Moore (1912) who concluded in their paper in the *Journal of Experimental Pedagogy* that 'with few exceptions innate sex differences in mental constitution are astonishingly small—far smaller than common belief and common practice would lead us to expect' (Burt & Moore, 1912), a result supported by more sophisticated contemporary studies also showing almost equal average ability across the sexes (Deary, Irwing, Der & Bates, 2007). More, not less opportunity and resources for the less able, has likewise accompanied research on the biology of differences. Thus compelling evidence for high heritability of dyslexia did not result in abandonment or ostracism of children at risk for reading disability. Instead research on behavioural therapies and teaching methods continued to occur, with the goal of enabling all children to access written language. A similar pattern of positive change occurred in autism. Previously believed due to 'refrigerator mothers', evidence for high heritability (Folstein & Rutter, 1977) helped focus work on the cognitive basis of autism (Baron-Cohen, Wheelwright, Skinner, Martin & Clubley, 2001), the fractionation of its genetic components (Ronald et al., 2006a), and therapeutic innovations, as well as a public acceptance of the disorder, and a dramatic increase in therapeutic resources consequent on a spectrum diagnosis.

References

Aristotle. (350 BCE/2007). *On Memory and Reminiscence* (J. I. Beare, Trans.). Adelaide: eBooks@Adelaide, University of Adelaide.

Asbury, K., Wachs, T. D., & Plomin, R. (2005). Environmental moderators of genetic influence on verbal and nonverbal abilities in early childhood. *Intelligence,* 33(6), 643–61.

Baddeley, A. D. (2007). Working memory: Multiple models, multiple mechanisms. In H. L. Roediger, Y. Dudai, & S. M. Fitzpatrick (Eds.), *Science of Memory: Concepts.* (pp. 151–53). Oxford: Oxford University Press.

Baltes, P. B., & Kliegl, R. (1992). Further testing of limits of cognitive plasticity: Negative age differences in a mnemonic skill are robust. *Developmental Psychology – a Quarterly Journal of Human Behavior,* 28, 121–25.

Baron-Cohen, S., Wheelwright, S., Skinner, R., Martin, J., & Clubley, E. (2001). The autism-spectrum quotient (AQ): evidence from Asperger syndrome/high-functioning autism, males and females, scientists and mathematicians. *Journal of Autism & Developmental Disorders,* 31(1), 5–17.

Bartlett, P. F., Brooker, G. J., Faux, C. H., Dutton, R., Murphy, M., Turnley, A., et al. (1998). Regulation of neural stem cell differentiation in the forebrain. *Immunology and Cell Biology,* 76, 414–418.

Bates, T. C. (2008a). Current genetic discoveries and education: Strengths, opportunities, and limitations. *Mind Brain and Education,* 2(2), 74–79.

Bates, T. C. (2008b). Intelligence: An overview. In A. Verri (Ed.), *The Life Span Development in Genetic Disorders* (pp. 1–8). New York: Nova Publishers.

Bates, T. C., Castles, A., Luciano, M., Wright, M., Coltheart, M., & Martin, N. (2007). Genetic and environmental bases of reading and spelling: A unified genetic dual route model. *Reading and Writing,* 20(1–2), 147–71.

Bates, T. C., Lind, P. A., Luciano, M., Montgomery, G. W., Martin, N. G., & Wright, M. J. (2010). Dyslexia and DYX1C1: deficits in reading and spelling associated with a missense mutation. *Molecular Psychiatry,* 15(12), 1190–96.

Bates, T. C., Luciano, M., Castles, A., Coltheart, M., Wright, M. J., & Martin, N. G. (2007). Replication of reported linkages for dyslexia and spelling and suggestive evidence for novel regions on chromosomes 4 and 17. *European Journal of Human Genetics,* 15(2), 194–203.

Bates, T. C., Luciano, M., Medland, S. E., Montgomery, G. W., Wright, M. J., & Martin, N. G. (2011). Genetic variance in a component of the language acquisition device: ROBO1 polymorphisms associated with phonological buffer deficits. *Behavior Genetics,* 41, 50–57.

Bhardwaj, R. D., Curtis, M. A., Spalding, K. L., Buchholz, B. A., Fink, D., Bjork-Eriksson, T., et al. (2006). Neocortical neurogenesis in humans is restricted to development. *Proceedings of the National Academy of Sciences of the United States of America,* 103(33), 12564–68.

Blake, P., & Gardner, H. (2007). A first course in mind, brain, and education. *Mind Brain and Education,* 1(2), 61–65.

Bloom, P. (2004). *Descartes' Baby: How the science of child development explains what makes us human.* New York: Basic Books.

Brody, N. (2003a). Construct validation of the Sternberg Triarchic Abilities Test: Comment and reanalysis. *Intelligence,* 31, 319–30.

Brody, N. (2003b). What Sternberg should have concluded. *Intelligence,* 31, 339–42.

Burt, C. L., & Moore, R. C. (1912). The mental differences between the sexes. *Journal of Experimental Pedagogy,* 1, 273–84, 355–88.

Butler, A. C., Karpicke, J. D., & Roediger, H. L. (2008). Correcting a metacognitive error: feedback increases retention of low-confidence correct responses. *Journal of Experimental Psychology. Learning Memory and Cognition,* 34(4), 918–28.

Carey, S. (1992). The origin and evolution of everyday concepts. *Minnesota Studies in the Philosophy of Science,* 15, 89–128.

Carey, S. (2000). The origin of concepts. *Journal of Cognition and Development,* 1(1), 37–41.

Carey, S. (2004). Bootstrapping & the origin of concepts. *Daedalus,* 133(1), 59–68.

Carey, S. (2009). *The Origin of Concepts.* Oxford: Oxford University Press.

Carroll, J. B. (1993). *Human cognitive abilities: A survey of factor-analytic studies.* New York: Cambridge University Press.

Caspi, A., Williams, B., Kim-Cohen, J., Craig, I. W., Milne, B. J., Poulton, R., et al. (2007). Moderation of breastfeeding effects on the IQ by genetic variation in fatty acid metabolism. *Proceedings of the National Academy of Sciences of the United States of America,* 104(47), 18860–65.

Castles, A., Bates, T. C., Luciano, M., Martin, N. G., & Coltheart, M. (2005). Two different sets of genes for reading (and spelling). *Australian Journal of Psychology,* 57, 47–47.

Castles, A., & Holmes, V. M. (1996). Subtypes of developmental dyslexia and lexical acquisition. *Australian Journal of Psychology,* 48(3), 130–35.

Ceci, S. J., & Papierno, P. B. (2005). The rhetoric and reality of gap closing: when the 'have-nots' gain but the 'haves' gain even more. *American Psychologist,* 60(2), 149–60.

Chabris, C. F. (1999). Prelude or requiem for the 'Mozart effect'? *Nature,* 400(6747), 826–27; author reply 827–828.

Chetty, R., Friedman, J. N., Hilger, N., Saez, E., Schanzenbach, D. W., & Yagan, D. (2011). How Does Your Kindergarten Classroom Affect Your Earnings? Evidence from Project Star. *The Quarterly Journal of Economics* 126(4), 1593–660; doi:10.1093/qje/qjr041.

Coltheart, M., Rastle, K., Perry, C., Langdon, R., & Ziegler, J. (2001). DRC: A dual route cascaded model of visual word recognition and reading aloud. *Psychological Review,* 108(1), 204–56.

Deary, I. J., Irwing, P., Der, G., & Bates, T. C. (2007). Brother-sister differences in the g factor in intelligence: analysis of full, opposite-sex siblings from the NLSY1979. *Intelligence,* 35, 451–56.

Deary, I. J., Spinath, F. M., & Bates, T. C. (2006). Genetics of intelligence. *European Journal of Human Genetics,* 14(6), 690–700.

Deary, I. J., Strand, S., Smith, P., & Fernandes, C. (2007). Intelligence and educational achievement. *Intelligence,* 35(1), 13–21.

Deary, I. J., Whalley, L. J., Lemmon, H., Crawford, J. R., & Starr, J. M. (2000). The stability of individual differences in mental ability from childhood to old age: follow-up of the 1932 Scottish Mental Survey. *Intelligence,* 28, 49–55.

Detterman, D. K., Gabriel, L. T., & Ruthsatz, J. M. (1998). Absurd environmentalism. *Behavioral and Brain Sciences,* 21(3), 411–12.

Detterman, D. K., & Ruthsatz, J. (1999). Toward a more comprehensive theory of exceptional abilities. *Journal for the Education of the Gifted,* 22(2), 148–58.

Donaldson, G. (2011). *Review of Teacher Education in Scotland.* Available at: http://www. reviewofteachereducationinscotland.org.uk/

Folstein, S., & Rutter, M. (1977). Infantile autism: a genetic study of 21 twin pairs. *Journal of Child Psychology and Psychiatry,* 18(4), 297–321.

Gabel, L. A., Gibson, C. J., Gruen, J. R., & LoTurco, J. J. (2010). Progress towards a cellular neurobiology of reading disability. *Neurobiology of Disease,* 38(2), 173–80.

Gottfredson, L. S. (2004). Intelligence: is it the epidemiologists' elusive 'fundamental cause' of social class inequalities in health? *Journal of Personality and Social Psychology,* 86(1), 174–99.

Hannula-Jouppi, K., Kaminen-Ahola, N., Taipale, M., Eklund, R., Nopola-Hemmi, J., Kaariainen, H., et al. (2005). The axon guidance receptor gene ROBO1 is a candidate gene for developmental dyslexia. *PLoS Genetics,* 1(4), e50.

Hanushek, E. A. (1998). *The evidence on class size.* Rochester, NY: University of Rochester, W. Allen Wallis Institute of Political Economy.

Harm, M. W., & Seidenberg, M. S. (2001). Are there orthographic impairments in phonological dyslexia? *Cognitive Neuropsychology,* 18(1), 71–92.

Haworth, C. M., Asbury, K., Dale, P. S., & Plomin, R. (2011). Added value measures in education show genetic as well as environmental influence. *PLoS One,* 6(2), e16006.

Haworth, C. M., Wright, M. J., Luciano, M., Martin, N. G., de Geus, E. J., van Beijsterveldt, C. E., et al. (2010). The heritability of general cognitive ability increases linearly from childhood to young adulthood. *Molecular Psychiatry,* 15(11), 1112–20.

Heckman, J. J. (2007). The economics, technology, and neuroscience of human capability formation. *Proceedings of the National Academy of Sciences of the United States of America,* 104(33), 13250–55.

Hirsch, E. D., Jr. (1999). *The Schools We Need: And Why We Don't Have Them.* New York: Anchor Books.

Hirsh-Pasek, K., & Bruer, J. T. (2007). The brain/education barrier. *Science,* 317, 1293.

Howe, M. J., Davidson, J. W., & Sloboda, J. A. (1998). Innate talents: reality or myth? *Behavioural & Brain Sciences,* 21(3), 399–407; discussion 407–42.

Hoxby, C. M. (2000). The effects of class size on student achievement: New evidence from population variation. *The Quarterly Journal of Economics,* 115(4), 1239–85.

Jaeggi, S. M., Buschkuehl, M., Jonides, J., & Perrig, W. J. (2008). Improving fluid intelligence with training on working memory. *Proceedings of the National Academy of Sciences of the United States of America*, 105(19), 6829–33.

Jepsen, C., & Rivkin, S. (2009). Class size reduction and student achievement. The potential tradeoff between teacher quality and class size. *Journal of Human Resources*, 44(1), 223–50.

Joyner, A. H., J., C. R., Bloss, C. S., Bakken, T. E., Rimol, L. M., Melle, I., et al. (2009). A common MECP2 haplotype associates with reduced cortical surface area in humans in two independent populations. *Proceedings of the National Academy of Sciences of the United States of America*, 106(36), 15483–88.

Jung, R. E., & Haier, R. J. (2007). The Parieto-Frontal Integration Theory (P-FIT) of intelligence: converging neuroimaging evidence. *Behavioural & Brain Sciences*, 30(2), 135–54; discussion 54–87.

Kant, I. (1788/1996). *Critique of Practical Reason* (T. K. Abbott, Trans.). New York: Prometheus Books.

Karmiloff-Smith, A. (1998). Development itself is the key to understanding developmental disorders. *Trends in Cognitive Sciences*, 2(10), 389–98.

Karmiloff-Smith, A., Scerif, G., & Ansari, D. (2003). Double dissociations in developmental disorders? Theoretically misconceived, empirically dubious. *Cortex*, 39(1), 161–63.

Karpicke, J. D., & Roediger, H. L., 3rd. (2008). The critical importance of retrieval for learning. *Science*, 319(5865), 966–68.

Kovas, Y., Harlaar, N., Petrill, S. A., & Plomin, R. (2005). 'Generalist genes' and mathematics in 7-year-old twins. *Intelligence*, 33(5), 473–89.

Kovas, Y., & Plomin, R. (2006). Generalist genes: implications for the cognitive sciences. *Trends in Cognitive Sciences*, 10(5), 198–203.

Krueger, A. B. (1999). Experimental estimates of education production functions. *The Quarterly Journal of Economics*, 114(2), 497–532.

Lango Allen, H., Estrada, K., Lettre, G., Berndt, S. I., Weedon, M. N., Rivadeneira, F., et al. (2010). Hundreds of variants clustered in genomic loci and biological pathways affect human height. *Nature*, 467(7317), 832–38.

Leigh, A. (2010). Estimating teacher effectiveness from two-year changes in students' test scores. *Economics of Education Review*, 29(3), 480–88.

Lind, P., Luciano, M., Duffy, D., Castles, A., Wright, M. J., Martin, N. G., et al. (2009). Dyslexia and DCDC2: normal variation in reading and spelling is associated with DCDC2 polymorphisms in an Australian population sample. *European Journal of Human Genetics*, 18(6), 668–73.

Locke, J. (1690/2009). *An Essay Concerning Human Understanding*. New York: BiblioLife.

Luciano, M., Lind, P. A., Duffy, D. L., Castles, A., Wright, M. J., Montgomery, G. W., et al. (2007). A haplotype spanning KIAA0319 and TTRAP is associated with normal variation in reading and spelling ability. *Biological Psychiatry*, 62(7), 811–817.

Luciano, M., Wainwright, M. A., Wright, M. J., & Martin, N. G. (2006). The heritability of conscientiousness facets and their relationship to IQ and academic achievement. *Personality and Individual Differences*, 40(6), 1189–99.

Maes, H. H., Neale, M. C., Kendler, K. S., Martin, N. G., Heath, A. C., & Eaves, L. J. (2006). Genetic and cultural transmission of smoking initiation: an extended twin kinship model. *Behavior Genetics*, 36(6), 795–808.

Martin, N., Benyamin, B., Hansell, N., Martin, N. G., Wright, M. J., & Bates, T. C. (2011). Cognitive function in adolescence: Testing for interactions between breastfeeding & FADS2 polymorphisms. *Journal of the American Academy of Child and Adolescent Psychiatry*, 50(1), 55–62.e4.

Mayer, R. E. (2004). Should there be a three-strikes rule against pure discovery learning? The case for guided methods of instruction. *American Psychologist*, 59(1), 14–19.

McDaniel, M. A., Howard, D. C., & Einstein, G. O. (2009). The read-recite-review study strategy: effective and portable. *Psychological Science*, 20(4), 516–22.

Meng, H., Smith, S. D., Hager, K., Held, M., Liu, J., Olson, R. K., et al. (2005). DCDC2 is associated with reading disability and modulates neuronal development in the brain. *Proceedings of the National Academy of Science of the United States of America*, 102(47), 17053–58.

Moody, D. E. (2009). Can intelligence be increased by training on a task of working memory? *Intelligence*, 37(4), 327–28.

Munafo, M. R., & Flint, J. (2009). Replication and heterogeneity in genexenvironment interaction studies. *International Journal of Neuropsychopharmacology*, 1–3.

Murnane, R. J. (1992). *Who Will Teach?: Policies That Matter*. Boston, MA: Harvard University Press.

Neisser, U., Boodoo, G., Bouchard, T. J., Jr., Boykin, A. W., Brody, N., Ceci, S. J., et al. (1996). Intelligence: Knowns and unknowns. *American Psychologist*, 51(2), 77–101.

Olson, R. K., & Byrne, B. (2005). Genetic and environmental influences on reading and language ability and disability. In H. W. Catts & A. G. Kamhi (Eds.), *The connections between language and reading disabilities* (pp. 173–200). Mahwah, NJ: Lawrence Erlbaum Associates.

Olson, R. K., Forsberg, H., & Wise, B. (1994). Genes, environment, and the development of orthographic skills. In V. W. Berninger (Ed.), *The varieties of orthographic knowledge, 1: Theoretical and developmental issues* (Vol. 8, pp. 27–71). Dordrecht: Kluwer.

Papassotiropoulos, A., Stephan, D. A., Huentelman, M. J., Hoerndli, F. J., Craig, D. W., Pearson, J. V., et al. (2006). Common Kibra alleles are associated with human memory performance. *Science*, 314(5798), 475–78.

Paracchini, S., Scerri, T., & Monaco, A. P. (2007). The genetic lexicon of dyslexia. *Annual Review of Genomics and Human Genetics*, 8, 57–79.

Pinker, S. (1991). Rules of language. *Science*, 253(5019), 530–35.

Pinker, S. (2002). *The Blank Slate: The Modern Denial of Human Nature*. Boston, MA: Allen Lane Science.

Pinker, S. (2004). Why nature & nurture won't go away. *Daedalus*, 133(4), 5–13.

Plomin, R. (1991). Why children in the same family are so different from one another. *Behavioral and Brain Sciences*, 14(2), 336–38.

Plomin, R. (1994). *Genetics and experience: The interplay between nature and nurture*. Thousand Oaks, CA: Sage.

Qureshi, I. A., Mattick, J. S., & Mehler, M. F. (2010). Long non-coding RNAs in nervous system function and disease. *Brain Research*, 1338, 20–35.

Rapp, B. (Ed.). (2002). *The Handbook of Cognitive Neuropsychology: What Deficits Reveal About the Human Mind*. New York: Psychology Press.

Rauscher, F. H., Shaw, G. L., & Ky, K. N. (1993). Music and spatial task performance. *Nature*, 365(6447), 611.

Rindermann, H., & Ceci, S. J. (2009). Educational policy and country outcomes in international cognitive competence studies. *Perspectives on Psychological Science*, 4(6), 551–77.

Risch, N., Herrell, R., Lehner, T., Liang, K. Y., Eaves, L., Hoh, J., et al. (2009). Interaction between the serotonin transporter gene (5-HTTLPR), stressful life events, and risk of depression: a meta-analysis. *Journal of the American Medical Association*, 301(23), 2462–71.

Roediger, H. L., & Karpicke, J. D. (2006). Test-enhanced learning: taking memory tests improves long-term retention. *Psychological Science*, 17(3), 249–55.

Ronald, A., Happe, F., Bolton, P., Butcher, L. M., Price, T. S., Wheelwright, S., et al. (2006a). Genetic heterogeneity between the three components of the autism spectrum: a twin study. *Journal of the American Academy of Child and Adolescent Psychiatry*, 45(6), 691–99.

Ronald, A., Happe, F., Bolton, P., Butcher, L. M., Price, T. S., Wheelwright, S., et al. (2006b). Genetic heterogeneity between the three components of the autism spectrum: A twin study. *Journal of the American Academy of Child and Adolescent Psychiatry*, 45(6), 691–99.

Rosen, G. D., Bai, J., Wang, Y., Fiondella, C. G., Threlkeld, S. W., LoTurco, J. J., et al. (2007). Disruption of neuronal migration by RNAi of DYX1C results in neocortical and hippocampal malformations. *Cerebral Cortex*, 17(11), 2562–72.

Royal Society of Chemistry (Producer). (2008). *The Five Decade Challenge: A wake-up call for UK science education?* Available at: http://www.rsc.org/images/ExamReport_tcm18-139067.pdf (accessed 16 February 2011).

Royal Society Working Group. (2011). *Brain Waves Module 2: Neuroscience: implications for education and lifelong learning.* London: Royal Society.

Ruthsatz, J., & Detterman, D. K. (2003). An extraordinary memory: The case study of a musical prodigy. *Intelligence,* 31(6), 509–518.

Ruthsatz, J., Detterman, D., Griscom, W. S., & Cirullo, B. A. (2008). Becoming an expert in the musical domain: It takes more than just practice. *Intelligence,* 36(4), 330–38.

Ryff, C. D., & Keyes, C. L. M. (1995). The structure of psychological well-being revisited. *Journal of Personality and Social Psychology,* 69(4), 719–27.

Scerif, G., & Karmiloff-Smith, A. (2005). The dawn of cognitive genetics? Crucial developmental caveats. *Trends in Cognitive Sciences,* 9(3), 126–35.

Sen, A. (2009). *The Idea of Justice.* New York: Allen Lane.

Shallice, T. (1988). *From Neuropsychology to Mental Structure.* Cambridge: Cambridge University Press.

Spearman, C. (1927). *The abilities of man.* New York: Macmillan.

Steele, K. M., Bass, K. E., & Crook, M. D. (1999). The mystery of the Mozart effect: Failure to replicate. *Psychological Science,* 10(4), 366–69.

Steer, C. D., Davey Smith, G., Emmett, P. M., Hibbeln, J. R., & Golding, J. (2010). FADS2 polymorphisms modify the effect of breastfeeding on child IQ. *PLoS One,* 5(7), e11570.

Sternberg, R. J. (2000). In search of the zipperump-a-zoo: Half a career spent trying to find the right questions to ask about the nature of human intelligence. *The Psychologist,* 13, 250–55.

Sternberg, R. J. (2008). Increasing fluid intelligence is possible after all. *Proceedings of the National Academy of Sciences of the United States of America,* 105(19), 6791–92.

Sternberg, R. J. (2010). *College Admissions for the 21st Century.* Boston, MA: Harvard University Press.

Stough, C., Kerkin, B., Bates, T. C., & Mangan, G. (1994). Music and spatial IQ. *Personality & Individual Differences,* 17(5), 695.

Szalkowski, C. E., Hinman, J. R., Threlkeld, S. W., Wang, Y., Lepack, A., Rosen, G. D., et al. (2011). Persistent spatial working memory deficits in rats following in utero RNAi of Dyx1c1. *Genes, brain, and behaviour,* 10(2), 244–52.

Taylor, J., Roehrig, A. D., Soden Hensler, B., Connor, C. M., & Schatschneider, C. (2010). Teacher quality moderates the genetic effects on early reading. *Science,* 328(5977), 512–514.

Tooby, J., Cosmides, L., & Barrett, H. C. (2003). The second law of thermodynamics is the first law of psychology: evolutionary developmental psychology and the theory of tandem, coordinated inheritances: comment on Lickliter and Honeycutt (2003). *Psychological Bulletin,* 129(6), 858–65.

Tucker-Drob, E. M., Rhemtulla, M., Harden, K. P., Turkheimer, E., & Fask, D. (2011). Emergence of a gene x socioeconomic status interaction on infant mental ability between 10 months and 2 years. *Psychol Sci,* 22(1), 125–33.

Turkheimer, E., Haley, A., Waldron, M., D'Onofrio, B., & Gottesman, I. (2003). Socioeconomic status modifies heritability of IQ in young children. *Psychological Science,* 14, 623–28.

van der Sluis, S., Willemsen, G., de Geus, E. J., Boomsma, D. I., & Posthuma, D. (2008). Gene-environment interaction in adults' IQ scores: measures of past and present environment. *Behavior Genetics,* 38(4), 348–60.

Vernon, P. E. (1950). *The structure of human abilities.* London: Methuen.

Wilmer, J. B., Germine, L., Chabris, C. F., Chatterjee, G., Williams, M., Loken, E., et al. (2010). Human face recognition ability is specific and highly heritable. *Proceedings of the National Academy of Sciences of the United States of America,* 107(11), 5238–41.

Yang, J., Benyamin, B., McEvoy, B. P., Gordon, S., Henders, A. K., Nyholt, D. R., et al. (2010). Common SNPs explain a large proportion of the heritability for human height. *Nature Genetics,* 42(7), 565–69.

Misuse of neuroscience in the classroom

Chapter 12

Neuroscience, education and educational efficacy research

Max Coltheart and Genevieve McArthur

Overview

There are many different kinds of educational/cognitive difficulties from which children can suffer as they develop, and there are many different commercially-available treatment programmes which are advertised as able to assist such children. However, it is rare for any evidence to be supplied that demonstrates the efficacy of such programmes. We show how neuroscience cannot help education here. Studies which collect neuroscientific data from children before and after a treatment has been applied cannot tell us whether or not the treatment has ameliorated the children's educational/cognitive difficulty. Whether a treatment has worked can only be assessed by collecting cognitive-testing data before and after treatment. Even when this is done and test performance is better after treatment than before, there are several potential confounding factors which need to be ruled out before a justified claim for efficacy of the treatment can be made. We describe two treatment research designs which allow one to rule out these confounding factors.

12.1 Introduction

There exists a plethora of commercial products that are claimed to be of assistance for children with educational difficulties, especially 'learning disabilities' (see Chapters 13–15 this volume). Typically, these products are claimed by their originators to be based on something neuroscientific, or at least to be 'inspired' by neuroscience. For many of these programmes, any true link with neuroscience is non-existent, or tenuous at best. But parents and even teachers are not well-placed to evaluate the relative merits of such programmes, or to decide whether the strong claims for efficacy made by the programme vendors are justified or not.

Consider, for example, the Miracle Belt™, a weighted belt that was invented by a professional baseball player in the USA in 2004. According to the website for this product (http://www.miraclebelt.com/), wearing this belt for periods of time will 'benefit children diagnosed with Autism, Attention Deficit Disorder (ADD), Attention Deficit Hyperactivity Disorder (ADHD), Angelman syndrome, Apraxia, Aspergers syndrome (AS), Ataxia, Cerebral Palsy, Down syndrome, Dyslexia, Fetal Alcohol syndrome (FAS), Hypotonia, Pervasive Developmental Disorder (PDD-NOS), Prader–Willi syndrome (PWS), Rett syndrome, Sensory Integration Disorder (SID) and Sensory Processing Disorder (SPD)'. How could a parent of a child suffering from any of these 17 disorders, or a clinician concerned with the treatment of these disorders, decide whether the Miracle Belt™ is in fact beneficial for such children?

The website offers a neuroscientific rationale for the product:

> Weighted therapy products have been found to help stimulate interaction between the brain and the senses which is crucial for healthy neurological function. Miscommunication between the brain and the senses can be regulated with the use of sensory therapy products, weighted products, such as weighted belts, weighted blankets and/or weighted lap pads. By stimulating the proprioceptive system these weighted products help to increase neurological feedback to the brain. Seeking to encourage the nervous system to integrate sensory information through the use of weighted sensory products can make a dramatic impact in the development of your child.

The site also quotes approving testimonials from 52 people: these consist of 24 from family members of affected children and 28 from professionals of various kinds (including 14 from occupational therapists). But what the site does not provide is any evidence that the treatment is efficacious for any of the 17 disorders mentioned.

12.2 Using anecdotes and testimonials to assess an educational treatment

It is very common for educational product websites to include anecdotes or testimonials praising the product. No matter how glowing these are, the reasons for ignoring them are numerous. These include:

1. The sample is selected by an interested party: there is no way of knowing whether a poll of a random sample of users of the product would return an overall positive or overall negative verdict.

2. The data are subjective, reflecting merely the opinions of those providing the anecdotes or testimonials rather than objective measurements of any aspects of behaviour of the treated children.

3. Commercially available products claiming to assist children with any kind of educational or cognitive difficulty, such as a reading difficulty, are often expensive. It is not easy for parents or teachers who have bought and used such products to accept that a large financial outlay has not after all assisted the children with whom the product was used.

These issues are well-illustrated in a study of Sunflower Therapy carried out by Bull (2007). According to the Sunflower Therapy website at http://www.sunflowertrust.com/, this therapy has 'helped thousands of children. Many have obvious learning difficulties: dyslexia, dyspraxia, ADD, ADHD, Aspergers Syndrome, Autism.' The website includes some positive anecdotal case studies and some claims about the neuroscientific basis of the therapy, with references to kinesiology and neurolinguistic programming.

Bull (2007) carried out a randomized control trial with 70 dyslexic children aged 6–13 years. After treatment with Sunflower Therapy, children in the treated group scored significantly higher than children in the untreated group on two subjective measures: a questionnaire on academic self-esteem and a questionnaire on reading self-esteem. The majority of parents (57.13%) of the treated children subjectively judged that the treatment had been effective. However, the treatment had no effect on objective measures of reading comprehension, word recognition or spelling. This is a very clear illustration of why subjective measures such as anecdotes and testimonials should be ignored when one is seeking to evaluate the effectiveness of any method claiming to ameliorate any kind of educational or cognitive difficulty, such as reading difficulties, spelling disorders or spoken language impairments.

12.3 **Using controlled trials to assess educational treatments**

The websites for the Miracle Belt™ and Sunflower Therapy treatments do not report any research aimed at investigating the efficacy of these treatment methods, and hence do not provide any objective evidence that these treatments are beneficial for any conditions. The term 'Miracle Belt' yielded no hits on the PsycInfo and Medline publication databases, and the only hit for 'Sunflower Therapy' on either database was the Bull study described in the previous section.

However, some websites for commercially available treatment methods do provide references to such research. One example is the Davis Dyslexia Correction method (see http://www.dyslexia.com). At http://www.dyslexia.com/science/research.htm, one can find a section entitled 'Published Peer-Reviewed Research' which cites four publications. The first of these is authored by people from the Davis Dyslexia organization, and all four are in somewhat exotic journals (*Reading Improvement*, a magazine published by Project Innovation, Alabama; *New Thoughts on Education*, which is published from the Faculty of Education and Psychology of the University of Iran; *Africa Education Review*, published from the Faculty of Education of the University of South Africa; and *the Pertanika Journal of Social Sciences and Humanities,* which is the official journal of Universiti Putra Malaysia) but of course it does not follow from any of this that all four of these studies were methodologically inadequate.

But what do we mean here by 'methodologically inadequate'? What methodological criteria must be met by a treatment study in order for it to qualify as methodologically adequate?

If it is claimed that any cognitive ability, such as reading, has been improved by a treatment, then that cognitive ability must be tested both before treatment begins (pre-test) and after the treatment has finished (post-test). If statistics reveal that performance on a cognitive test is not significantly higher at post-test than pre-test, then no claim for efficacy of the treatment can be made.

But suppose that an appropriate statistical analysis has been done and performance on the cognitive test after treatment was significantly better than before treatment. Is this sufficient to claim that the treatment worked? No, because several other confounding factors could explain that improvement. These confounding factors—which must be ruled out for a claim of efficacy to be justifiable—are:

1. A test–retest or practice effect: performance was better in the post-test because children had practice doing the tests in the pre-test.

2. Maturation effect: the children performed better on the cognitive tests in the post-test than in the pre-test because they were older and had received more education in the interim.

3. Regression to the mean: if a treatment group is selected on the basis of having particularly poor performance on a cognitive test at pre-test, then their extreme scores are expected to be a bit closer to the population mean at post-test for purely statistical reasons. The lower the psychometric reliability of the cognitive test being used, the larger the regression-to-the-mean effect will be.

4. A placebo effect: simply receiving treatment, regardless of the nature of that treatment, may improve performance through increased self-confidence or motivation arising from being singled out for any kind of help.

Treatment studies need to be designed so that they can demonstrate that none of these confounding factors can explain the improved performance on a cognitive test after treatment. One way to do this is to conduct a randomized controlled trial. Here one first identifies a group of children in need of treatment. Half of the children are randomly assigned to the treatment group; the other half are assigned to the untreated (control) group. Since the group assignment is done randomly, at pre-test one would expect that the two groups do not differ on the cognitive ability that is to be treated (e.g. a reading difficulty). Suppose that at post-test, the treated group's scores on a test of this cognitive ability are significantly higher than the untreated group. This cannot be due to a practice effect because the two groups have had equal amounts of practice on the cognitive test at pre-test; nor can it be due to the treated group being older or more educationally experienced because the untreated control group increased in age and education to the same degree as the treated group over the same period of time; nor can it be due to regression to the mean, because random assignment ensured that the two groups started with the same level of performance on a cognitive task at pre-test.

Thus listed explanations (1) through (3) can be safely ruled out. If one is worried about a placebo effect, that can be addressed by giving the untreated group an equal amount of attention and help by providing them with exposure to a different kind of treatment programme, one that has nothing to do with the ability which the study itself is investigating.

One drawback of a traditional randomized control trial is that the control group either misses out on treatment, or is given a treatment that is not expected to work. This is less than ideal for children with cognitive difficulties who desperately need treatment. One solution is to use a different design: the double-baseline controlled trial.

Suppose the treatment is going to take two months. This means that the pre-test and post-test will be separated by two months. In this case, a double-baseline controlled trial adds a 'pre-' pre-test (pre-test 1). After two months of no treatment, there is a second pre-test (pre-test 2). Then, after 2 months of training, there is a post-test.

If test performance on a cognitive test in pre-test 2 is no different to pre-test 1 (i.e. after 2 months of no training), then (1) the cognitive test being used is not subject to practice effects; (2) the subject's performance is not changing due to increasing age and educational experience; and (3) there is no statistically significant regression effect. Given this, if the subject's scores are then significantly better at post-test than at pre-test 2, then one can rule out these three confounding factors. If one is also concerned about a placebo effect, one can add to pre-test 1, pre-test 2 and post-test, a test for an ability that is unrelated to the ability being treated. If there is no post-test improvement in performance on tests of this unrelated ability, then there is no placebo effect, and so any superiority at post-test of the treated ability can be ascribed to a specific effect of the particular treatment effect being used.

If treatment effects are to be of practical significance, they must be lasting. The durability of any treatment effect can be assessed by adding a second, delayed, post-test (i.e. post-test 2). That delay—a period in which there is no further treatment—might be two months, or it might be longer—say, 6 months. Comparison between scores from the delayed post-test 2 and scores from the pre-test 2 will help determine how long the treatment effect lasts. Desirable as a delayed post-test is, practical reasons can make this difficult or even impossible; and the occurrence of an improvement in an immediate post-test remains important even if a delayed post-test cannot be carried out.

In sum, randomized controlled trials and double-baseline controlled trials—with or without a delayed post-test—are two methodologically sound treatment study designs that are not difficult to implement. Furthermore, the latter can be used with single cases. So the failure of most commercial programmes that claim to help children with some kind of educational difficulty to provide any evidence that their products actually work is not because there are no sound ways of assessing the efficacy of such treatments.

12.4 **Using neuroscientific data to assess educational treatments**

We have described ways in which methodologically rigorous controlled studies of educational treatment programmes can be designed and carried out. The reader may notice that all the examples of controlled studies that we used measured the effect of treatment using performance on cognitive educational tests. None of our examples measured the effect of treatment using measures of the brain (neuroscientific measures). There is a good reason for this: neuroscientific data cannot tell us if an educational treatment is effective or not.

For example, suppose a study tests the brains of children with some form of educational difficulty immediately before and immediately after an educational treatment (but does not collect behavioural test data). And suppose this study finds a significant effect of treatment on these children's brains that is not explained by any of the confounding factors outlined previously. This means that there is a genuine treatment effect on the brain. Does this mean that the treatment has been effective with respect to the children's educational difficulty? We cannot tell. We know that the educational treatment changed the children's brains, but we do not know if these brain changes have fixed their educational difficulty. The only way to find out if the treatment really worked is to use a relevant educational cognitive test to measure the children's educational performance before and after treatment. The neuroscientific data are irrelevant for the purpose of treatment evaluation.

Now, suppose that a second study tests children with an educational difficulty on a relevant educational cognitive test before and after a treatment (and their brain responses are also measured before and after treatment). And suppose that this study finds a genuine treatment effect on the children's performance on the educational test, but no effect on their brain responses. Does this mean that the treatment was not effective? Clearly not, since the children's performance on the educational cognitive test improved. This improvement must stem from a change in the brain somewhere, but the study failed to detect the neural effect. So the neuroscientific data are again irrelevant. It is the educational cognitive data that tells us that the educational treatment worked.

Suppose instead that in this second study a genuine treatment effect on the brain was found, but there was no effect on the educational cognitive test. Do the changes in the brain mean that the treatment worked? No. The treatment did not alleviate the educational difficulty at which it was aimed and so it was not effective—even though there were changes in the brain. Yet again the neuroscientific data are irrelevant.

As far as assessing the effectiveness of educational treatments is concerned, then, neuroscience does not speak to education. Efficacy studies of treatments for educational difficulties must use cognitive tests of the relevant educational abilities before and after treatment. Neuroscientific measurements before and after treatment cannot tell us anything about efficacy of educational treatments.

If our conclusion that neuroscience does not speak to education is correct, then we should be alarmed at a number of aspects of teacher training, because teacher trainees are given a very different impression. In the UK, Howard-Jones (2011) reports a study he and colleagues did of 158 teacher-training graduates about to enter secondary schools. Of this group, 20% believed that their brains would shrink if they drank less than six to eight glasses of water per day; 65% believed that physical coordination exercises could improve the integration of left-hemisphere functions with right-hemisphere functions; and 82% believed that studies of brain function justify the conclusion that teaching children in their preferred learning style could improve learning outcomes. These are, of course, all neuromyths.

The situation is no better in Australia. Brain Gym® is an educational programme based upon a model of brain functioning that involves three important factors: laterality, focusing and centring.

This model is rejected by neuroscientists (Geake, 2008); furthermore, there is no evidence that the Brain Gym® programme conveys any educational benefits (Hyatt, 2007). Yet the programme is widely used in Australian schools, and has official approval on the websites of the Departments of Education in every Australian State and Territory; and the Departments of Education of New South Wales, Victoria, Tasmania, Queensland and South Australia have all provided funding for teachers to attend Brain Gym® training classes as professional development (Stephenson, 2009).

12.5 **Conclusion**

We began our chapter by asking, in relation to the Miracle Belt™, 'How could a parent of a child suffering from any of these 17 disorders, or a clinician concerned with the treatment of these disorders, decide whether the Miracle Belt™ is in fact beneficial for such children?'. We'll conclude with a discussion of this question as it applies to any form of educational treatment.

Suppose a parent or clinical professional is considering the use of treatment X (e.g. weighted vests) with a child who has an educational impairment in ability A (e.g. reading). We suggest that here are four scenarios under which it is a reasonable course of action to decide to embark upon treatment X. They are, ordered from the most to least powerful, as follows:

1. There is at least one publication in the peer-reviewed scientific literature that describes a methodologically sound treatment study that found that treatment X improves cognitive ability A in children with an impairment in that ability.

2. The scientific literature provides good reason to believe that treatment X should improve cognitive ability A, though there are no direct demonstrations that this is so.

3. The scientific literature provides good reason to believe that proficiency in ability B (e.g. spelling) improves the acquisition of cognitive ability A (e.g. reading), and there is at least one publication in the peer-reviewed literature which has shown in a methodologically adequate way that treatment X improves ability B.

4. The scientific literature provides good reason to believe that proficiency in ability B (e.g. spelling) improves the acquisition of cognitive ability A (e.g. reading), and also provides good reason to believe that treatment X should improve the acquisition of ability B (e.g. spelling), though there have not been any direct demonstrations that this is so.

Thus one can consider, in relation to any potential treatment, whether any of these four scenarios hold: if any of them do, then there is at least some evidence for the efficacy of the treatment, and hence some sound justification for embarking on the treatment. In this case, a parent or educator might rationally decide—depending on cost, and on how well the treatment matches their child's difficulties—to try the treatment with their child. However, if none of these four scenarios hold, then the treatment has not been properly assessed by a methodologically sound study (despite any claims on a company's website), and there is currently no scientific basis for that treatment.

We argue that only programmes that fit one or other of these four scenarios can claim to be efficacious. So it is a sobering thought that almost all of the commercially available programmes advertised as capable of assisting children with educational/cognitive difficulties fail to fit any of the four scenarios at all.

In contemplating the use of such programmes, parents or educators might make one of two decisions. They might decide to stick with the scientific evidence and so not use the treatment. Or they might try the treatment anyway; if so, they need to be fully aware that this decision is a leap in the dark and so the chances of success are limited at best.

References

Bull, L. (2007). Sunflower therapy for children with specific learning difficulties (dyslexia): a randomised, controlled trial. *Complementary Therapies in Clinical Practice*, 13, 15–24.

Geake, J. (2008). Neuromythologies in education. *Educational Research*, 50, 123–33.

Howard-Jones, P. A. (2011). From brain scan to lesson plan. *The Psychologist*, 24, 110–113

Hyatt, K. (2007). Building stronger brains or wishful thinking? *Remedial and Special Education*, 28, 117–24.

Stephenson, J. (2009). Best practice? Advice provided to teachers about the use of Brain Gym® in Australian schools. *Australian Journal of Education*, 53, 108–24.

Chapter 13

Educational double-think

Michael C. Corballis

Overview

Is your child a Dickens or a Degas? A Plato or a Picasso? The idea that people are either verbal or visual thinkers is current not only in popular mythology, but also in the educational concept of 'learning styles'. Teaching methods, it is claimed, should be tailored to accommodate either the verbal or visual proclivities of the students. Careful examination of the evidence indicates that the distinction is without foundation, at least with respect to the effectiveness of educational practice. It may be true that some people are good at verbal skills, some at visual skills, but just as many are good at both or good at neither—and there are, of course, many shades in between. An associated but equally suspect idea is that people are either left-brained or right-brained. These distinctions may reflect a natural human tendency to see the world and its inhabitants in terms of dichotomies, thereby overlooking the very complexities that make us human. They may also reflect a politically correct but often misguided (and indeed dangerous) attempt to define talent in those who fail to achieve academically. The natural disposition to dichotomize is exploited by individuals and organizations only too eager to profit from human folly.

A while ago, as one does, I submitted a paper to a journal in the hope of having it published. One of the referees, an esteemed visual scientist, took exception to the fact that I had presented the data in tables. He asked that I resubmit them as graphs, so he could see what was going on. He was, he said, a visual thinker rather than a verbal one.

Then there was Nobel Prize winner Albert Einstein, reputedly a slow learner and even a poor mathematician, who revolutionized modern physics with the theory of relativity. He is said to have developed relativity theory through visual imagery, at the age of 16 imagining himself travelling beside a light beam and considering the consequences. You might therefore think that Einstein, too, was a visual thinker rather than a verbal one. Hope springs eternal that any child not performing well in school may turn out to be another Einstein (see also Chapters 12, 14 and 15, this volume).

13.1 Learning styles

The idea that people can be divided into visual and verbal thinkers underlies the notion of *learning styles*, which has wide currency in folklore, pop psychology and educational practice. Type 'learning styles' into Google, and you will be immediately rewarded with any number of online tests to tell you what particular style suits you. Much of this is gimmickry, but learning styles have also found their way into respectable educational theory and practice. According to one influential textbook, 'Some students seem to learn better when information is presented through words (*verbal* learners), whereas others seem to learn better when it's presented through pictures (*visual* learners)' (Ormrod, 2008, p. 160, italics in original). In the US, the National Association of Secondary School Principals established a test of learning styles that is widely distributed (Keefe, 1988). The idea has permeated the noble halls of academe. The Yale University Graduate School of Arts and Sciences (2009) has a website offering advice for instructors, which includes the recommendation that instructors determine their own learning style as well as that of their students.

There is money to be made. One popular organization is the International Learning Styles Network (http://www.learningstyles.net) which markets four different assessment tests, ELSA for ages 7–9, LSCY for ages 10–13, LIVES for ages 14–18 and BE for ages 17 and older. Each costs $5 per student. A programme offering certification costs $1225 per trainee, and for a further $1000 you can learn how to conduct research on learning styles. The tests are based on work by Rita and Kenneth Dunn (1992, 1993), and those attending the inauguration of the Dr Rita Dunn Educational Foundation on 14 November, 2009 were invited to donate $35 to the foundation.

Are learning styles meaningful?

To their credit, the Dunns have made at least some attempt to verify their methods. In a meta-analysis of 42 experimental studies, Dunn, Griggs, Olsen, Beasley and Gorman (1995) claimed that taking learning styles into account could be expected to raise children's learning by 75% of a standard deviation. For example, on a test on which the average score is 100 and the standard deviation 20, accommodating individuals' learning styles should raise the average score to 115.

Nevertheless one should perhaps be sceptical of studies in which the investigators have a financial interest in an affirmative outcome. More recent examinations of the evidence have not proven so positive. Several authors have examined the evidence and found it wanting (Coffield, Moseley, Hall & Eccloestone, 2004; Curry, 1990; Willingham, 2005). The most recent critique is that of Pashler, McDaniel, Rohrer and Bjork (2008), who lay down the conditions that must be met to empirically demonstrate the influence of learning styles. Basically, you need a means of classifying individuals into verbal and visual learners, and you need two methods of instruction, one designed to teach by verbal methods and the other designed to teach by visual methods. Then you need a test score to measure the outcome of the learning under each method. To show that learning is most efficient when the method matches the learning style, you need results such as those shown in Figure 13.1. In statistical terms, this is known as an interaction between teaching method and learning styles, and the two curves should cross, as in the hypothetical examples shown in Figure 13.1.

Pashler et al. scoured the vast literature on learning styles, and found only one that 'could be described as even potentially meeting the criteria' (p. 111). This was a study by Sternberg and colleagues (1999), but according to Pashler et al. even that study was suspect. Only about a third of the participants were actually included in the analysis that produced the desired interaction, and the scores themselves were 'highly derived', rather than directly reflecting the actual test scores. Pashler et al. conclude their detailed analysis as follows:

> The contrast between the enormous popularity of the learning-styles approach within education and the lack of credible evidence for its utility is, in our opinion, striking and disturbing. If classification of students' learning styles has practical utility, it remains to be demonstrated (p. 117).

This does not mean that people don't differ in verbal or visual aptitudes. Tests of verbal ability will assuredly yield a range of scores, as will tests of visuospatial ability. Indeed, tests of verbal ability have low correlations with tests involving visualization or spatial orientation, and a technique

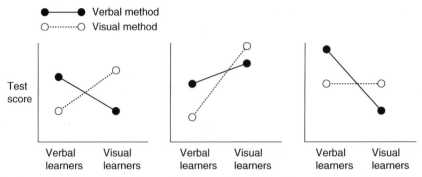

Fig. 13.1 Examples of test results that would support the notion of learning styles. The critical requirement is that the two curves plotted this way should cross, so that verbal learners score better with a verbal method, and visual learners better with a visual method. There is no good evidence from the literature that results actually conform to this pattern (after Pashler et al., 2008).

called factor analysis typically extracts separate factors that correspond to these basic dimensions (e.g. Guilford, 1967—see his Chapter 3 for more detail). Given that these factors are uncorrelated, some individuals will tend to score high on both, some high on one but low on the other, and some low on both. To suppose that there are verbal learners who score high on verbal but low on visual, and visual learners who score high on visual but low on verbal, is to ignore half the population. Of course those with good verbal skills may well gravitate toward careers in journalism, law or (heaven help them) academia, while those with good visual skills may well become artists, architects or professional tennis players. What is unclear from the literature is whether the learning of any particular accomplishment, be it algebra, art or critical writing, can be tuned to people's different aptitudes, and so taught in different ways. The evidence remains stubbornly negative.

13.2 **Hemisphericity**

Lurking beneath the learning-style movement is the dual spectre of the left and right brains. This features in some of the writing of Rita and Kenneth Dunn, with the supposition that verbal learners are characteristically left-brained and visual learners right-brained. The idea that people differ in which side of the brain is dominant has been termed *hemisphericity*, a term first coined by Joseph E. Bogen (Bogen, DeZure, R, TenHouten & Marsh, 1969), one of the surgeons who performed the famous split-brain operations in the 1960s as a treatment for intractable epilepsy. The rationale for the operation was that cutting the forebrain commissures connecting the two sides of the brain would prevent seizure activity from spreading over the whole brain. In this respect the surgery was more successful than expected, since with the help of medication seizures could be largely prevented altogether.

The surgery also meant that each side of the brain was effectively disconnected from the other, so that each side could be tested independently. Research soon showed that the left side of the brain was dominant for speech and the right for certain non-verbal functions, including visuospatial ones (e.g. Sperry, 1982). This led to the idea that even neurologically intact individuals would tend to habitually favour one or other hemisphere. Left-brained people, it was claimed, tend to be verbally oriented, logical and systematic, while right-brained people would be spatially oriented, intuitive and creative. Bogen himself was primarily responsible for initiating this idea, and characterized the left brain as *propositional* and the right side as *appositional*, leading to corresponding differences in cognitive style (Bogen, 1969).

Bogen's terminology has not gained widespread support, but the notion that people could be either left-brained or right-brained gained huge popularity in the 1970s, led by Robert Ornstein's best-selling book *The Psychology of Consciousness* (1972). Explicit tests of hemisphericity were devised by a number of authors (e.g. Gordon, 1986; Zenhausern, 1978) and the terms 'left-brained' and 'right-brained' became part of common usage. Even the staid *American Heritage® Dictionary of the English Language* (2008) offered the following definitions:

Left-brained *adj*: 1. Having the left brain dominant. 2. Of or relating to the though processes, such as logic and calculation, generally associated with the left brain. 3. Of or relating to a person whose behavior is dominated by logic, analytical thinking and verbal communication, rather than emotion and creativity.

Right-brained *adj*: 1. Having the right brain dominant. 2. Of or relating to the thought processes involved in creativity and imagination, generally associated with the right brain. 3. Of or relating to a person whose behavior is dominated by emotion, creativity, intuition, nonverbal communication and global reasoning rather than logic and analysis.

By the 1980s, though, the concept of hemisphericity began to wear thin, especially among academics and brain scientists. In what was then relatively youthful exuberance, I offered a critique (Corballis, 1980), and shortly afterwards a more trenchant review of evidence came to the following conclusion:

On the basis of the review presented, it would seem prudent to abandon the notion of hemisphericity, at least in so far as it claims to make any reference to the lateral function of the cerebral hemispheres. Such a claim cannot be supported by current scientific studies of the cognitive functions of the cerebral hemispheres, and it is most unlikely that more thorough understanding of the relations between cognitive function and cerebral structural systems will lead to any change in this state of affairs (Beaumont, Young & McManus, 1984, p. 206).

But popular opinion seldom heeds academic opinion, especially if there is profit to be made, and a search in Google still scores over 25,000 hits, including online tests to measure your own hemisphericity. The notion of hemisphericity is also incorporated into such cult activities as Neuro Linguistic Programming (NLP) or a more recent fad, Brain Gym®. At its origin, at least, NLP was based on the idea that you can tell which part of the brain is active in terms of where a person looks when asked simple questions. Looking up or down, left or right, reveals the secrets of your mind, so that any inconsistencies between mind and behaviour are revealed and can be fixed (Bandler & Grinder, 1979). Watch where you look, Big Brother may be watching—and your brain may not be safe.[1]

The notion of eye-movements as the key to mental activity was something of a fad in the 1970s, and based in part on the idea that looking to the right was a door into the left hemisphere, and looking to the left a door into the right hemisphere. This was largely discredited in an article that appeared the year before Bandler and Grinder's (1979) *Frogs into Princes,* one of the NLP manifestos (Ehrlichman & Weinberger, 1978). In any event, NLP is a movement that is still going strong, but has little scientific credibility. The term 'neurolinguistics' is itself a respectable term, and if you

---

[1] Educationalists beware, too. One John O. Stevens endorsed Bandler and Grinder's *Frogs into Princes* as follows: 'NLP represents a huge quantum leap in our understanding of human behavior and communication. It makes most current therapy and education totally obsolete.' Huge leaps seem better suited to the turning of princes into frogs.

search for it on the Web of Science—the compendium of respectable science—you discover some 320 articles over the past 46 years using the term, mostly in good journals of linguistics or neuroscience, or both. These articles have been cited over 2,100 times. The term 'Neuro Linguistic Programming' yields a mere 50 entries in a comparatively motley collection of journals over the past 31 years, cited 42 times. Many of these entries are critical of NLP. One of them bears the title 'Growing anti-intellectualism in Europe: A menace to science' (Drenth, 2003).

You do better with snake oil, actually. Search for that in the Web of Science and you get 249 entries, cited 782 times.[2]

Scepticism over hemisphericity and left-brain/right-brain folklore does not mean that the two cerebral hemispheres do not differ. Cerebral asymmetry of function has been amply confirmed by brain-imaging studies (e.g. Corballis, 2008), and was in fact documented well before the split-brain studies of the 1960s. The left-brain dominance for speech was first demonstrated in the 1860s by Broca (1861), and as early as 1864 the eminent neurologist John Hughlings Jackson wrote that if 'expression' resided in one side of the brain, then we might suppose that 'perception—its corresponding opposite' might reside in the other. The point at issue is whether individuals differ with respect to the extent to which one or other hemisphere dominates. Certain tasks may exercise one hemisphere more than the other, but there is no good reason to suppose that individuals differ markedly in which side of the brain is the more active during any given task or learning assignment.

Handedness

Hemisphericity is sometimes confused with handedness. A once-ubiquitous t-shirt carried the slogan 'The right hemisphere of the brain controls the left side of the body, therefore only left-handers are in their right minds'. This may have led to the myth that left-handers are somehow more creative and intuitive than right-handers, and that such luminaries as Albert Einstein, Pablo Picasso, Benjamin Franklin and Bob Dylan were (or are) left-handed. At one point, Wikipedia listed all four in the inventory of left-handers, but McManus (2002) pointed out that each of them was almost certainly right-handed—and Wikipedia appears to have been amended accordingly. In any event, some 70% of left-handers have language represented predominantly in the *left* side of the brain (e.g. Corballis, 2009), so there is no reason to suppose that they differ from right-handers in terms of hemispheric specialization. Of those 30% with language represented in the right side of the brain the conventions we might expect the functions normally associated with the left side to be in the right brain, and vice versa. In short, there is little reason to suppose that left- and right-handers would differ in their learning styles.

There is nevertheless some evidence that scholastic success does vary with handedness, but more with respect to degree of handedness than with its direction. A large sample of 12,770 11-year-olds in the UK were given tests of verbal ability, visuospatial ability, reading comprehension and mathematical ability, as well as a test of handedness, and on all four tests there was a dip in performance at the point at which left- and right-hand performance was equal. That is, ambidextrous children were somewhat worse than either left- or right-handers (Crow, Crow, Done & Leask, 1998). The same may apply to adults. Analysis of results from a television show in which people were given various tests of intellectual ability showed that those who described themselves as ambidextrous scored lower than those who were either left- or right-handed (Corballis et al., 2008). In both studies, the deficit applied to abilities traditionally associated with both the left brain (verbal, mathematical, reasoning) and with the right brain (visuospatial, spatial), so there

---

[2] All these counts are as of 1 April, 2010, and I'm not fooling you.

was no sense that the ambidextrous represent one or other learning style. They were simply worse.

If there is a typology to be derived from handedness, and by implication with brain asymmetry, it may have to do with the degree to which the brain is lateralized, and not with the direction. Although brain symmetry may lead to difficulties in some intellectual abilities, there may be compensations. Indeed, there may be an evolutionary trade-off between symmetry and asymmetry. Like the vast majority of animals, we belong to the order of species known as the Bilateria, dating from some 600 million years ago (Chen et al., 2004). Bilateral symmetry is generally adaptive in a world without systematic left–right bias, preserving balance, equal awareness and sensitivity to events on the left and right sides of the body and equal competence of the limbs on each side. These advantages may persist in athletic activities, hand–eye coordination, and perhaps activities such as music and art, but they may be forfeited to the extent that the brain has evolved left–right asymmetries, with corresponding advantages in more complex processing and hemispheric specialization. The genetics and evolution of cerebral asymmetry can be understood in terms of a balance between these competing tendencies (Corballis, 2006). Those ambidextrous kids who don't do so well in school may be the heroes of the sports field, or even the concert chamber.

Here, though, I am in danger of perpetrating a new mythology, so it is best perhaps to move on.

13.3 **Dichotomania**

Why, then, are we so insistent on categorizing people as left- or right-hemispheric, or as verbal or visual learners? I think there may be at least two reasons. First, there seems to be an overwhelming imperative to dichotomize. We think of people as fat or thin, tall or short, smart or dumb, happy or sad, good or bad, extravert or introvert. Learning styles slot easily into this compendium. It was also once said that there are two kinds of people, those who think there are two kinds of people, and those who don't. Reducing human variation to dichotomies is sometimes a useful shorthand, but ignores the texture of human individuality.

Second, dichotomous typology can serve as an excuse for low performance, especially in an age when the very notion of failure is politically incorrect. A child scoring poorly in academic subjects with a heavy verbal or computational emphasis may be excused on the grounds that she thinks differently; every parent whose child is not doing well at school may be bolstered by the possibility that they have produced another Einstein, and her talents have not yet been revealed. This may conceivably be true, but the detection of varied talents is best revealed by offering a wide range of subjects in the curriculum. The evidence suggests this to be a more appropriate strategy than one of attempting to adapt the teaching of individual subjects to each child's putative learning style.

What, then, of Einstein himself? The idea that he was a slow learner seems to be largely a myth, perhaps designed to give hope to the low-achieving child. His alleged deficiency in mathematics was famously mentioned in Ripley's *Believe it or Not* column, which trumpeted 'Greatest living mathematician failed in mathematics'. When shown this in 1935, Einstein is said to have laughed and said 'I never failed in mathematics. Before I was fifteen I had mastered differential and integral calculus' (Isaacson, 2007, p. 16). He was in fact top of his class in primary school, and continued to achieve top grades at the gymnasium. He was apparently slow to speak as a child, but this was probably attributable more to his then inward-looking personality than to any intellectual deficiency. It may not even be true, as letters from his grandparents record that he was just as clever and endearing as every grandchild is (Isaacson, 2007), and in adult life he was well-known as a fluent and charming speaker. The best guess is that Einstein was simply a genius, able to think verbally, visually, mathematically and above all creatively.

And he was right-handed.

References

American Heritage Dictionary of the English Language, Fourth Edition (2008). New York: Houghton Mifflin.

Bandler, R. & Grinder, J. (1979). *Frogs into princes*. Moab, UT: Real People Press.

Beaumont, J. G., Young, A. W. & McManus, I. C. (1984). Hemisphericity: A critical review. *Cognitive Neuropsychology*, 1, 191–212.

Bogen, J. E. (1969). The other side of the brain. II. An appositional mind. *Bulletin of the Los Angeles Neurological Society*, 34, 135–62.

Bogen, J. E., DeZure, R., TenHouten, W. D., & Marsh, J. F. (1972). The other side of the brain. III. The corpus callosum and creativity. *Bulletin of the Los Angeles Neurological Society*, 37, 49–61.

Broca, P. (1861). Remarques sur la siège de la faculté du langage articulé, suivies d'une observation d'aphémie. *Bulletin de la Société Anatomique de Paris*, 2, 330–57.

Chen, J. -Y., Bottjer, D. J., Oliveri, P., Dornbos, S. Q., Gao, F., Ruffins, S., Chi, H., Li, C. W., & Davidson, E. H. (2004). Small bilaterian fossils from 40 to 55 million years before the Cambrian. *Science*, 305, 218–22.

Coffield, F., Moseley, D., Hall, E., & Eccloestone, K. (2004). *Learning styles and pedagogy in post-16 learning. A systematic and critical review*. London: Learning and Skills Research Centre.

Corballis, M. C. (1980). Laterality and myth. *American Psychologist*, 35, 284–95.

Corballis, M. C. (2006). Cerebral asymmetry: A question of balance. *Cortex*, 42, 117–118

Corballis, M. C. (2009). The evolution and genetics of cerebral asymmetry. *Philosophical Transactions of the Royal Society of London B: Biological Sciences*, 364, 867–79.

Corballis, M. C., Hattie, J., & Fletcher, R. (2008). Handedness and intellectual achievement: An even-handed look. *Neuropsychologia*, 46, 374–78.

Crow, T. J., Crow, L. R., Done, D. J., & Leask, S. (1998). Relative hand skill predicts academic ability: Global deficits at the point of hemispheric indecision. *Neuropsychologia*, 36, 1275–82.

Curry, L. (1990). One critique of the research on learning styles. *Educational Leadership*, 48, 50–56.

Drenth, P. J. D. (2003). Growing anti-intellectualism in Europe: A menace to science. *Studia Psychologica*, 45, 5–13.

Dunn, R. & Dunn, K. (1992). *Teaching elementary students through their individual learning styles*. Boston, MA: Allyn & Bacon.

Dunn, R. & Dunn, K. (1993). *Teaching secondary students through their individual learning styles*. Boston, MA: Allyn & Bacon.

Dunn, R., Griggs, S. A., Olsen, J., Beasley, M. & Gorman, B. S. (1995). A meta-analytic validation of the Dunn and Dunn model of learning-style preferences. *Journal of Educational Research*, 88, 353–62.

Ehrlichman, H. & Weinberger, A. (1978). Lateral eye movements and hemispheric asymmetry: A critical review. *Psychological Bulletin*, 85, 1080–1101.

Gordon, H. W. (1986). The Cognitive Laterality Battery: Tests of specialized cognitive function. *International Journal of Neuroscience*, 29, 223–44.

Guilford, J. P. (1967). *The nature of human intelligence*. New York: McGraw-Hill.

Isaacson, W. (2007). *Einstein: His life and universe*. New York: Simon & Schuster.

Jackson, J. H. (1864). Clinical remarks on cases of defects of expression (by words, writing, signs, etc) in diseases of the nervous system. *Lancet*, 2, 604.

Keefe, J. W. (1988). *Profiling and using learning style. NASSP learning style series*. Reston, VA: National Association of Secondary School Principals.

Ormrod, J. E. (2008). *Educational psychology: Developing learners* (6th ed.). Upper Saddle River, NJ: Pearson.

Ornstein, R. E. (1972). *The psychology of consciousness*. San Francisco, CA: Freeman.

McManus, C. (2002). *Right hand, left hand: The origins of asymmetry in brains, bodies, atoms, and cultures*. London: Weidenfeld & Nicolson.

Pashler, H., McDaniel, M., Rohrer, D. & Bjork, R. (2008). Learning styles: Concepts and evidence. *Psychological Science in the Public Interest*, 9, 106–119.

Sperry, R. W. (1982). Some effects of disconnecting the cerebral hemispheres. *Science*, 217, 1223–27.

Sternberg, R. J., Grigorenko, E. L., Ferrari, M., & Clinkenbeard, P. (1999). A triarchic analysis of an aptitude-treatment interaction. *European Journal of Psychological Assessment*, 15, 1–11.

Willingham, D. T. (2005). Do visual, auditory, and kinaesthetic learners need visual, auditory, and kinaesthetic instruction? *American Educator*, 29(2), 31–35.

Yale University Graduate School of Arts and Sciences (2009). *Graduate Teaching Center: Teaching students with different learning styles and levels of preparation* [online] http://www.yale.edu/graduateschool/teaching/learningstyles.html.

Zenhausern, R. (1978) Imagery, cerebral dominance, and style of thinking—unified field model. *Bulletin of the Psychonomic Society*, 12, 381–84.

Chapter 14

Rose-tinted? The use of coloured filters to treat reading difficulties

Robert D. McIntosh and Stuart J. Ritchie

Overview

In this chapter, we discuss the use of coloured filters to treat reading difficulties, and the theoretical and practical claims that underpin it. We review evidence for the efficacy of coloured filters, and report the results of a new trial in schoolchildren with reading difficulties. We conclude that there is a chasm between the dramatic claims often made for this treatment, and the small and inconsistent effects that have been demonstrated experimentally. Indeed, we suggest that coloured filters have no proven efficacy, beyond some probable placebo effect, and that their use should not be recommended to private individuals, or supported by public bodies. Resources should instead be directed towards better-proven remedial interventions.

14.1 Introduction

'Wearing rose-tinted glasses' is not merely a metaphor for optimism. Around the world, millions of children and adults are using coloured filters, in the form of tinted spectacles or coloured plastic overlays, hoping or believing that this will improve their reading. The coloured filters have been prescribed to treat a perceptual condition that goes by various names, including, but not limited to: Irlen syndrome, Meares–Irlen syndrome, scotopic sensitivity syndrome and visual stress (which we will use here). Commonly-listed symptoms of visual stress are: light sensitivity, difficulty reading high-contrast text, perceptual distortions and illusory movements on the page, poor recognition of text outside a narrow line of sight and a lack of sustained attention (Irlen, 1991). According to Irlen, visual stress causes eye-strain, restlessness, fatigue and headaches, and is a major cause of reading disability; but it can be alleviated simply by viewing the printed page, or the whole world, through filters of the right colour. However, visual stress is a controversial diagnostic entity that is not recognized by the American Optometric Association, or the Royal College of Ophthalmologists in the UK. And, we will argue, despite the widespread use of coloured filters, there is no good neuroscientific reason to expect them to aid reading, nor compelling evidence that they do. We critically review the most relevant studies, and report our own trial of coloured filters in schoolchildren, after briefly introducing the major proponents of this treatment.

14.2 The Irlen method: a colourful history

The use of colour for its apparent healing properties dates back at least to ancient Egypt (Gottlieb & Wallace, 2001), and has recurred in diverse forms up to the present day. Collins (2002) recounts the story of A. J. Pleasonton, who, in the mid 19th century, was celebrated briefly for his claims to treat illness, and boost plant growth, using light filtered through blue glass. Babbitt (1878) also advocated colour, though not only blue, as a panacea, a tradition alive in various complementary therapies, such as the use of homeopathic remedies created by shining coloured light through water (see Wauters, 1999). Similarly without scientific basis, though with greater face plausibility, is the field of 'syntonic optometry', founded by Spitler (1941), which uses exposure to coloured light specifically to treat disorders of the visual system (see Barrett, 2009). The more recent, and prevalent, strategy of using coloured filters for reading difficulties has the further plausibility of being an in situ intervention, in which colour is used for its immediate effects on the visual system *during* reading.

In a 1964 volume on dyslexia, Critchley noted that a child who had difficulty reading normal text appeared to read better from coloured card. Nothing further of this nature was reported until 1980, when Meares described several children with visual discomfort and perceptual distortions during reading, which were ameliorated using coloured card or filters. Helen Irlen subsequently proposed that such visually-based reading difficulties represent a distinct, previously unrecognized syndrome. She originally suggested scotopic sensitivity syndrome as a name (Irlen, 1983), though she now refers to it exclusively as Irlen syndrome (Irlen, 2010; as noted earlier, we shall use the more neutral term, visual stress). Irlen patented her treatment system of coloured overlays and lenses, formalized the procedures for diagnosing visual stress and matching filter colours to individuals, and founded an eponymous Institute in California. The Irlen Institute has since enjoyed mass media exposure, and developed a worldwide influence as the leading proponent, and marketer, of coloured filters for reading.

A brief visit to the main Irlen Institute website, viewable in various colours, will enable you to complete a checklist to see whether you may have visual stress, and if so to locate your nearest Irlen diagnostician for a formal assessment. In the first session, you will be asked about adverse experiences during reading, such as distortions and movement of the text, light sensitivity, headaches, eyestrain, tiredness, loss of concentration and so on. You will then be shown a series of high-contrast pictures composed of closely-packed lines and symbols, and required to perform visual tasks such as counting the number of lines or symbols in parts of the picture (see Figure 14.1). For each task, you will rate your impression of different visual distortions (e.g. blurring, movement, flickering, glowing) and your overall visual discomfort. If your responses suggest visual stress, then the screener will try to identify the best colour of overlay for you (from a set of ten). You may thus be shown a page of dense text, with each half covered by a different overlay, and asked to decide which side is more comfortable to look at, the non-preferred overlay being replaced by another in a process of elimination until a winning colour emerges. This session may be followed by a more detailed assessment to determine whether you would benefit from Irlen lenses and to define your optimal lens colour (which may not be the same as your optimal overlay colour). You are unlikely to be referred for a professional eye examination, even though some symptoms of visual stress are common in people with well-understood visual problems, especially binocular and accommodative abnormalities (Blaskey et al., 1990; Scheiman, 2004; Singleton & Trotter 2005). In the UK, the cost of your Irlen Institute diagnosis and treatment would be around £50–100 for coloured overlays, or £350–450 for coloured lenses.

If you do visit an Irlen diagnostician, especially if you experience reading difficulties, be prepared for a positive diagnosis.[1] According to Irlen's (1991) book, *Reading by the Colours*, visual

---

[1] (Helveston (2001) reports that two of his colleagues who underwent training as Irlen screeners were explicitly instructed to aim for a 50% positive diagnosis rate.

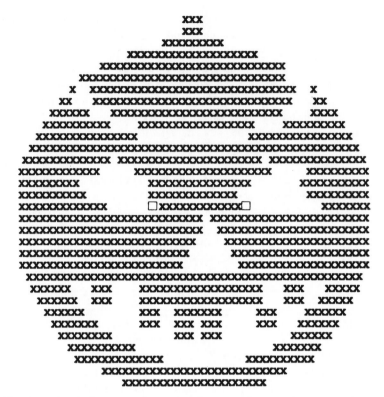

Fig. 14.1 The viewer is asked to count the symbols between the central square markers, and to report on visual distortions and discomfort. Figure from the Irlen Reading Perceptual Scale (IRPS) Manual (Irlen, 1987). © PDC/Irlen 1987–2010.

stress is present in 65% of people diagnosed with dyslexia, and in 12% of the non-dyslexic population. Moreover, reading is by no means the only activity implicated: visual stress is also claimed to affect writing, depth perception, coordination and motivation. In Irlen's most recent book (2010, *The Irlen Revolution*), the list of conditions related to visual stress has lengthened to include: attention deficit hyperactivity disorder (ADHD), autism, chronic fatigue syndrome, epilepsy, Tourette's syndrome, head injuries, agoraphobia, anxiety attacks, depression and conduct disorder. Like all of Irlen's specific claims regarding symptoms, prevalence and treatment efficacy, these statements are made without supporting evidence other than the author's experience. Crucially, Irlen (1991, 2010) also asserts that only filters provided by the Irlen Institute are effective in treating visual stress, as 'the Irlen method has been developed, refined and researched over many years' (Irlen, 1991, p. 194). No details of this research process are provided, so it could be argued that other systems of colour-treatment, such as Wilkins' intuitive system (see section 14.3) have been researched far more rigorously, or at least more openly. Finally, Irlen's writings offer no ideas about the nature of visual stress, beyond vacuous statements that it may reflect 'a structural brain deficit involving the nervous system' (Irlen, 1991, p. 57).

In an opinion piece for the *International Dyslexic Association*, Helveston (2001) summarized the Irlen Institute's approach as displaying:

> ... classic group behavior. The concept has a strong charismatic personality as originator and sustained leader. The supporting evidence is almost entirely anecdotal. The syndrome is becoming associated with an even more diverse array of maladies, tinted lenses now being offered for relief of problems far

removed from reading difficulty. The procedure for determining the specific tint has not been divulged and remains a type of 'trade secret'. Finally, a financially rewarding franchise activity is at the basis of the Irlen Institute activity (p. 13).

Even if this assessment is justified, the credentials of the Irlen Institute must be kept separate from the more important practical question of whether coloured filters can benefit reading, and if so for whom. Despite its obvious lack of scientific focus, Irlen's work has provided a spur for subsequent researchers, who have attempted to answer this question within better-specified theoretical frameworks.

14.3 Wilkins' intuitive colour system

The scientist most closely linked with coloured filters is Arnold Wilkins. His initial studies in this field were performed in the early 1990s, in collaboration with the Irlen Institute, but he subsequently developed an independent research programme, and devised his own system for diagnosing visual stress and determining optimal filter colours. Wilkins' approach shares some core features with the Irlen method: each person is said to have a precise colour that will reduce their visual stress symptoms and facilitate their reading optimally, and this optimal colour may differ between overlays and lenses (Wilkins, 2003; Wilkins, Sihra & Smith, 2005; though see Simmers, Gray & Wilkins, 2001). Unlike Irlen, Wilkins has proposed a broader theory of visual stress, tested his hypotheses and published his research.

The basic set of overlay colours in Wilkins' system are broadly similar to those provided by the Irlen Institute, and the optimal overlay is chosen by a similar process of comparison and elimination (see Wilkins, 1994). To confirm that the selected overlay makes a difference, the Wilkins Rate of Reading Test is applied, consisting of simple common words (e.g. my, play, see, cat, dog) presented in random order as lines of text, allowing the visual aspects of reading to be assessed independently of vocabulary and comprehension. An overlay is considered effective if reading rate increases by more than 5% when it is used (Wilkins, Jeanes, Pumfrey & Laskier, 1996). The prescription of coloured lenses requires a more precise examination, and Wilkins invented the 'intuitive colorimeter' for this purpose (Wilkins, 2003). This is a light-sealed box, within which the person views a page of text, being able to adjust the hue, saturation and luminance of the lighting for maximum comfort; tinted lenses can then be created to match the preferred colour. Moreover, Wilkins has founded a Society for Coloured Lens Prescribers to govern those using his system of treatment, with a code of conduct that binds members to evidence-based practice, and stipulates that visual stress should be assessed only as part of a wider package of orthoptic and optometric assessments.

Wilkins (1995) proposed that visual stress, where it cannot be attributed to orthoptic or optometric conditions, may result from cortical hyperexcitability: an over-sensitivity of certain cortical areas to particular visual stimuli. This idea relates visual stress theoretically to other conditions in which cortical hyperexcitability has been proposed, most prominently migraine (Wilkins, 1995; Wilkins, Huang & Cao, 2007), but also photosensitive epilepsy (Wilkins et al., 1999), multiple sclerosis (Newman-Wright, Wilkins & Zoukos, 2007), and autism (Ludlow, Wilkins and Heaton, 2006). The link was first suggested by the observation that children who benefit from Wilkins' intuitive overlays (i.e. who have visual stress) were almost twice as likely to have a family history of migraine as those without visual stress. One prediction that follows is that coloured filters might reduce the symptoms of migraine. Wilkins, Patel, Adjamian and Evans (2002) tested this prediction in 17 adults with visually-sensitive migraine, finding a marginal reduction in headaches during periods of wearing 'optimal' coloured lenses, as compared with a non-optimal tint. This result is encouraging enough to merit further investigation, but the cortical hyperexcitability theory of visual stress remains highly speculative, especially given controversy over whether the theory is valid even for migraine (Ambrosini, De Noordhout, Sandor & Schoenen, 2003; cf. Aurora & Wilkinson, 2007).

14.4 Oxford blues, and yellows

A third group has now hit the scene, with prime-time coverage on television (BBC1, *The One Show*) and radio (BBC Radio 4, *All in the Mind*). An award-winning garden was even created at the 2010 Chelsea Flower Show to publicize their ideas (Figure 14.2). John Stein, chair of the Dyslexia Research Trust at the University of Oxford, has claimed that around 25% of dyslexics will be helped by blue filters, and 25% by yellow.[2] He explains this by reference to the theory that dyslexia is often caused by abnormalities of the visual magnocellular system, which is essential for fine eye movement control and stable fixations (e.g. Stein, 2001). According to the theory, weakened magnocellular function leads to unstable fixations, causing visual distortions and confusions during reading. Yellow filters may boost magnocellular function because visual magnocells are most sensitive to yellow light (Ray, Fowler & Stein, 2005; Stockman, MacLeod & DePriest, 1991). By this token, one might expect blue filters to exacerbate the problem, as blue light is opponent to yellow, and inhibits magnocells (Stockman et al., 1991). However, Stein thinks that blue filters work by a different mechanism; specifically, by mimicking the light of early morning, and 'tricking' the brain's circadian circuitry to respond as if it were morning, thus boosting arousal, with benefits

Fig. 14.2 'Dyslexia: A Barrier to Education', an award-winning garden, designed by Tim Fowler to promote the theories of the Dyslexia Research Trust at the Chelsea Flower Show, 2010.

---

[2] Stein has elsewhere suggested that dyslexics may also be helped to read by dietary supplements of fish oils (Stein, 2003, 2010). The suggestion derives from a theorized lack in dyslexics of docosahexaenoic acid (DHA), important for magnocellular function, available in fish oils (Stein, 2010). For discussions of the equivocal educational benefits of fish oils, see Chapters 4 and 15, this volume.

for magnocellular function. These ideas are clearly speculative. Indeed, the magnocellular theory of dyslexia itself is highly controversial, though a detailed discussion is beyond the scope of this chapter (see Skoyles & Skottun, 2004; White et al., 2006).

More important for our purpose is the practical question of whether blue and yellow filters do aid reading. Here, despite the media fanfare and floral tributes, convincing evidence seems to be lacking. The claimed benefit of yellow filters rests largely on one study by the Dyslexia Research Trust (Ray et al., 2005) in which 38 schoolchildren with reading difficulties were encouraged to use yellow lenses or a 'placebo': a piece of card with a 'letterbox' space cut out, so that only one line of text could be read at a time. After 3 months, reading ability had improved slightly but significantly more in the yellow-lens group than in the placebo group. Unfortunately, the oddly-matched nature of the placebo treatment makes this result hard to interpret. Without a control group wearing plain lenses, it is difficult to judge whether the outcome reflected accelerated progress in the yellow-lens group, or retarded progress in the placebo group. The influence of *blue* overlays on reading is similarly hard to interpret. They have been reported to improve reading comprehension whilst reducing reading rate, which suggests a change of strategy rather than ability, and this effect was not specific to reading disordered groups (Iovino, Fletcher, Breitmeyer & Foorman, 1998). For blue *lenses*, the evidence is thus far limited to a single study, entitled, 'Failure of blue-tinted lenses to change reading scores of dyslexic individuals' (Christenson, Griffin & Taylor, 2001).

Given the paucity of published evidence for specific influences of blue and yellow filters, this treatment system will not be considered much further here, but it is worth commenting briefly on its peculiar disconnection from other literature in this field. For more than 20 years, Irlen and Wilkins have strongly advocated the use of coloured filters, framing their treatment in terms of visual stress, but the key study by the Dyslexia Research Trust (Ray et al., 2005) makes no reference at all to Irlen, Wilkins or to visual stress. Moreover, Ray et al. (2005) state that all 38 of their reading-disabled children reported improved clarity of text when viewing through yellow filters, compared with viewing through blue, pink, green, red, or grey filters, or no filter. This is flatly inconsistent with Irlen and Wilkins's insistence on the individuality of the optimal tint, and similarly hard to square with the fact that no other trial of coloured filters has noted an overall preference for yellow (though Kriss & Evans, 2005, did comment on the popularity of mint-green overlays). Finally, it even seems inconsistent with the recent statements of the same research group, who now claim that 25% of dyslexics benefit from yellow filters, and 25% from blue.

14.5 **From theory to practice**

The major proponents of coloured filters disagree over the mechanisms of the filters' effects, and even over fundamental issues such as the colour-specificity of the treatment, and the existence of visual stress as distinct from dyslexia. The sense of confusion arising from the literature is mirrored by inconsistent attitudes to coloured filters amongst educational and health professionals. In June 2010, we surveyed Principal and Deputy-Principal Educational Psychologists throughout Scotland about their experience and use of coloured overlays. All of our respondents were aware of the treatment, and their responses ranged from cautious endorsement:

> Have seen tinted glasses and overlays used. Young person reported an improvement.
> I would refer a child who was having trouble reading [for assessment of visual] after checking that this seemed more a visual problem… than a phonic problem.

To outright scepticism:

> …I'm reluctant to refer families for what seems to me little more than an expensive placebo.
> I have not been persuaded that diagnosis or interventions prescribed in terms of overlays etc. are anything other than a placebo. It is the teaching of reading and support that make the difference.

This range of responses sums up the wider situation. Some Local Councils support visual stress screening in schools, and provide coloured overlays and jotters to children, but others make no such provision. Some Health Boards, such as NHS Lothian and NHS Fife, do not offer the treatment; but, NHS Ayrshire & Arran, for example, spend an estimated £33,650 per year on coloured overlays, lenses and associated diagnosis. This is a mere drop in the ocean of an NHS budget; but the variability of expenditure across regions shows a deep lack of consensus over the treatment. In the rest of this chapter, we will leave aside the contested *theoretical* basis for using coloured filters, and focus on the *practical* issue at stake, whether coloured filters really do improve reading.

14.6 Do coloured filters improve reading?

Anecdotal evidence for the benefits of coloured filters is plentiful, as the 'testimonial' pages of Irlen websites show. Personal stories and case histories constitute much of the motivating material in books that recommend the use of coloured filters (Irlen, 1991, 2010; Stone, 2003; Wilkins, 2003), and many practitioners in the field have been convinced of the efficacy of the filters by witnessing their dramatic effects first hand; some even wear coloured lenses themselves. One understandable inference from such anecdotal evidence is that a true therapeutic effect must be at work: there is no smoke without fire. The persistence of the idea that colour can aid reading, which has recurred in various forms across at least half a century, would tend to support this view. On the other hand, if the dramatic benefits claimed are real, one would expect them to have been demonstrated many times over, and for coloured filters to have been embraced by health and education services worldwide. As we shall see, not only does the treatment remain controversial, but there is arguably *no high-quality evidence that coloured filters benefit reading at all*, far less at the robust practical level that would make this a viable approach to the remediation of reading difficulties.

High-quality evidence for any treatment must do more than just show an improvement when the treatment is applied; it must establish that the improvement is a *specific* consequence of the treatment (see Chapter 12, this volume). If a person with reading problems views text through a carefully-selected coloured filter, which they believe will improve their visual comfort and enhance their reading, then heightened motivation and expectations may be sufficient to bring about some or all of those changes, at least in the short term. Children may be especially susceptible to such placebo effects. Knowledge of placebo effects is as old as medicine itself, and the basic concept is now familiar to all educated people, but there is still an under-appreciation of how pervasive and powerful such effects can be (Goldacre, 2008). As our responses from Educational Psychologists made clear, the real question is not whether coloured filters can improve reading, but whether they do so more than a well-matched placebo. If good things happen to the optimist in rose-tinted glasses, is the source of the benefit in the glasses, or only in the optimism?

Answering this kind of question places stringent demands upon study design, and few studies in the literature have come close to meeting these demands. A systematic review was undertaken by Albon, Adi and Hyde (2008), to assess 'the effectiveness and cost-effectiveness of coloured filters for reading difficulty'. They identified 23 studies of direct relevance to this question.[3] Fifteen of these were evaluated, according to widely-accepted criteria (Jadad et al., 1996), as providing very poor quality evidence, because they were inadequately designed or reported. Eight other studies contained at least the essential elements of a randomized controlled trial, in that the effects of coloured filters on some measures of reading were compared against those of another

---

[3] A further 67 articles on the topic were rejected for reasons such as measuring inappropriate outcomes, testing participants without reading problems or being non-empirical articles (e.g. review articles).

intervention (either a placebo, or an alternative treatment), with participants randomly allocated to treatment conditions. The most notable of these studies will be described briefly.

14.7 Randomized controlled trials

The first randomized controlled trial (Blaskey et al., 1990) was conducted in the US, where people, aged from 9–51 years, reporting poor reading due to visual stress symptoms, were recruited following a feature about the Irlen method on the CBS news show *Sixty Minutes*. All recruits had their diagnosis confirmed and their optimal lens colour determined by an Irlen practitioner. On optometric assessment, 95% were found to have a clinically significant visual problem, most often binocular or accommodative disorder, and all 30 people that participated in the full study had such problems. Participants were randomized to three groups, being treated for 2 weeks by Irlen lenses, by a vision-therapy treatment based on eye exercises, or receiving no treatment. Subjective visual discomfort declined in both treatment groups, but only the vision-therapy group showed improved reading. Following treatment, people in the vision-therapy group no longer had sufficient symptoms to be candidates for the Irlen method, suggesting that visual stress is often caused by optometric disorders, and can be alleviated by therapies targeting these. This particular study found few cases of visual stress without optometric disorders, but other investigations have reported such cases, and some have even excluded participants with optometric problems (e.g. Bouldoukian, Wilkins & Evans, 2002, discussed later in this section).

Wilkins et al. (1994) performed a trial for which they selected 68 children (11–12 years old) with visual stress symptoms, who had been using a coloured overlay without prompting for at least 3 weeks. Children had their optimal *lens* colour determined using Wilkins' intuitive colorimeter, and two pairs of lenses were produced, one pair of the optimal tint, the other of a subtly-different control tint. Each pair was used for 1 month, with visual stress symptom diaries being kept, though only 36 of the children completed their diaries. As in the study of Blaskey et al. (1990), reports of visual stress symptoms were reduced with the optimal tint relative to the control, but there were no significant differences in reading outcomes. Children could not reliably distinguish the optimal from the control lenses in direct comparisons, suggesting that the subjective benefit of the optimal tint could not be ascribed to a placebo effect.[4] However, the reduction in reported symptoms was slight, even though the study included only children who were already using coloured filters to alleviate visual stress. This selection bias will inevitably have increased the chances of finding a positive effect of coloured filters.

The same problem of selection bias applies to a later study of 33 children and adults (7–44 years) who used coloured overlays (Bouldoukian et al., 2002). This study found that reading speed on the Wilkins Rate of Reading Test was higher with the optimal overlay than with a pale yellow placebo overlay. The experimenters were at pains to promote the placebo as a plausible treatment by describing it as 'a wonderful discovery to help patients with reading difficulties', and attaching a sticker to it stating 'Research Model A16 Anti UV/IR Filter—Made in USA'. This may nonetheless have been less compelling to participants than their familiar filter, previously selected by them in a process of elimination. Since participants and experimenter would have known which was the usual filter, placebo effects cannot be excluded. The study thus shows that poor readers who habitually use overlays may read better with them, but it does not establish whether this is due to the overlays themselves, or to the expectations of the users.

---

[4] There is some uncertainty here, because the ability to discriminate the lenses was not reported separately for those 36 children on whom the analysis of visual stress symptoms was based.

These three studies give a flavour of the higher-quality end of the evidence, but other studies are similarly inconsistent, and have similar limitations, many of which are discussed in the systematic review by Albon et al. (2008). These authors concluded that there is yet 'no convincing evidence to suggest that coloured filters can successfully improve reading ability in subjects with reading disability or dyslexia when compared to placebo, or other types of control' (p. 93). They repeatedly emphasize the poor quality of evidence, which may be less a criticism of the work reviewed than an acknowledgement of the difficulty of conducting research in the area. Finally, they highlight the need for further well-designed studies, identifying the design of the control treatment as a key issue. Highest-quality evidence requires 'double-masking' (also known as 'double-blinding', though the latter is considered less appropriate terminology for studies involving vision; see Morris, Fraser & Wormald, 2007), so that neither the experimenter nor the participant knows which is the treatment condition and which the control. Only then can we be sure that the outcomes are not skewed by expectations.

14.8 **The Port Glasgow study**

Our interest in this topic was first stirred when we learned that Inverclyde Council in Scotland was running a 'dyslexia-friendly schools' initiative, following a similar scheme in neighbouring East Renfrewshire. The steps for a school to acquire dyslexia-friendly status include the screening of its children for Irlen syndrome (which we refer to here as 'visual stress'), with the council providing Irlen products (coloured overlays and jotters) for those diagnosed. We approached Liz McKelvie of Irlen Scotland to ask about this aspect of the scheme, and were encouraged by her eagerness to help us conduct a fair trial of its effectiveness. Indeed, she modified her normal practice in order to facilitate a double-masked design. Between January and June of 2010, our study (Ritchie, Della Sala & McIntosh, 2011) ran at Newark Primary School, Port Glasgow, where teachers identified 75 children aged 7–12 years whom they considered to have problems reading. Sixty-one of these children participated in the study, with full data collected for 60. All were assessed by Liz McKelvie using the standard Irlen procedures, and an optimal overlay from the Irlen set was prescribed for each child diagnosed with visual stress. At the end of the session, when these children would normally have been given their overlay and instructed in its use, they were instead told that the testing was not complete and would continue on another day. At the next session, we measured performance on the Wilkins Rate of Reading Test (Wilkins et al. 1996) with the prescribed overlay, an Irlen overlay of a non-prescribed colour and a colourless overlay (for children not diagnosed with visual stress, the 'prescribed' overlay was assigned arbitrarily). Neither the child nor the experimenter knew whether the child had been diagnosed with visual stress, or which was the prescribed overlay. In a third session, the child was assessed by an independent orthoptist, from whom the diagnosis was also masked.

The first notable outcome was that 47 (77%) of the 61 children were diagnosed with visual stress, suggesting a liberal approach to diagnosis. This would seem to be deliberate, as Liz McKelvie expressed to us that it is better to receive an overlay that is not needed, than not to receive one that is needed.[5] Orthoptic abnormalities were found in more than half of the visual stress group (54.3%), but also in 38.5% of the non-visual-stress group; the difference was not statistically reliable, though the power to detect a difference was limited by the low numbers in this group. The most important outcome was the absence of any immediate effects of overlays on reading rate in either group of children, as illustrated in Figure 14.3. A 5% increase in reading speed has previously

---

[5] Your own view will depend on your evaluation of coloured overlays, as well as of the possible harms of inappropriately labelling a child with a 'syndrome'.

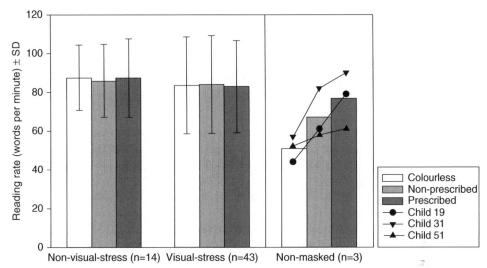

Fig. 14.3 Bars show mean Wilkins Rate of Reading Test scores for each overlay condition, for the visual stress and non-visual stress groups. Error bars indicate one standard deviation either side of the mean. The rightmost bars show average performance for the three children from whom the treatment condition was not successfully masked; individual results for each child are also plotted.

been suggested as a suitable threshold by which to define a positive response to overlays (Wilkins et al. 1996). In our study, 28% of children in the visual stress group read more than 5% (and up to 25%) faster with the prescribed than with the colourless overlay, but 32% read more than 5% (and up to 28%) more slowly (Ritchie, Della Sala & McIntosh, 2011). Nothing in the data suggests that these numbers are anything other than random variation around a true overlay effect of zero. This lack of effect is not limited to reading speed; the prescribed overlays had equally little impact on a test of reading comprehension.

A further interesting result arose by serendipity. Three children had to be excluded from the visual stress group because they stated that they knew their overlay colour (two of them were already using overlays). These three children, from whom the treatment condition was not successfully masked, showed a clear and consistent effect of the prescribed overlay (78%, 58% and 15% increases in reading speed), illustrated in Figure 14.3. Although it cannot be proven, it is tempting to view these patterns as placebo effects. This is supported by the fact that reading was also boosted, but less powerfully, when using the non-prescribed overlay. A colour-specific overlay effect would not predict a positive response to a non-prescribed colour. In fact, in our study, it might predict the opposite, because we selected the non-prescribed overlays, where possible, from the opposing part of the colour spectrum to the prescribed overlay. According to the colour-specificity theory, the optimal overlay will filter the wavelengths of light most troublesome for an individual. Since a filter of an opponent colour will preferentially transmit these very wavelengths, it should, if anything, make visual stress worse. The behaviour of our non-masked children thus suggests strong placebo effects for coloured overlays, and vindicates our double-masked design.

We believe that the effective double-masking, and the size and homogeneity of our sample of poor readers, make this the highest-quality evidence yet on the efficacy of coloured filters. We do not of course imagine that the study is definitive. Like any single piece of research, it is limited in scope: we tested the effects of overlays and not lenses; we evaluated one practitioner's application of the Irlen method only; and we assessed only immediate and not long-term effects on reading speed and comprehension (though we will follow-up these children after 1 year). On the other hand, a great strength of the work is its real-world relevance, since this was an evaluation of a pre-existing, government-funded initiative to tackle reading problems in schools. The intention is laudable; but our data demonstrate the potential dangers of investing in interventions for which no firm evidence exists. However, if unproven interventions are ever to be tried, then the opportunity to collect data on their effectiveness should not be wasted.

14.9 **Conclusions**

John Stein, being interviewed for BBC Radio 4, suggested that coloured filter treatments remain controversial because people are resistant to the idea that an answer for dyslexia could be so simple. We would argue that the treatments are controversial because of the lack of supporting evidence, but that it is precisely their 'magic-bullet' simplicity that makes them so attractive and tenacious. For a person restricted by reading difficulties, or watching a child struggle in school, miraculous tales of coloured filters may seem too good to be true, but it would take an arch cynic to reject them out of hand. When the treatment comes couched in the language of neuroscience, it gains further credibility, at least in the eyes of non-experts (Weisberg et al., 2008). And if faith, motivation or sheer novelty induce positive effects, then an underappreciation of these factors will tend to ensure that the credit is placed with the coloured-filter, so swelling the tide of anecdote. With such factors at play, it is nigh-impossible for any individual to know for sure the cause of a short-term reading boost. Long-term improvements are hardly more transparent, especially in children, who tend to improve over time with or without intervention. Wider public awareness of the nature and value of scientific evidence could help individuals make informed judgements in these murky areas. But this does not lessen the onus on public institutions to provide informed guidance, and to base health and education policies on a balanced appraisal of the scientific evidence, as well as on economic and political considerations.

For coloured filters, our own appraisal should now be clear: there is no good neuroscientific reason to expect coloured filters to aid reading, nor compelling evidence that they do (see Hyatt, Stephenson & Carter, 2009, for a similar evaluation). Future evidence may always change this assessment; but, at present, coloured filters should neither be recommended to private individuals, nor provided by public bodies. We cannot exclude the possibility that a specific colour-based mechanism provokes positive changes in some people, but such people must at least be rare. On the other hand, there can be little doubt that filters can enhance subjective visual comfort and boost reading performance in many poor readers via novelty and placebo effects. One might ask whether it is self-defeating to dismiss such benefits, especially since our own data suggest how dramatic they may be: we saw a 78% increase in reading speed in one child from whom the treatment was not successfully masked. Should we not instead harness this potential, and advocate filters for all poor readers? We would say not, though we should certainly heed the lesson that extra motivation and attention can give a temporary boost. Placebo effects are likely to accompany all interventions to some degree, but a pure placebo intervention could only be countenanced if there were no *effective* treatments; and even then it would raise a host of ethical dilemmas.

Happily, effective treatments do exist. None are as simple as a coloured overlay, but then dyslexia is a complex disorder. Sensory problems may contribute in some cases, and we have already highlighted the importance of visual assessment, especially if visual distortions are experienced (Blaskey et al., 1990). However, the consensus is that the majority of dyslexics have a cognitive language processing problem (Hulme & Snowling, 2008; White et al., 2006), and the most promising interventions target this core problem. Torgesen (2002, 2005) has provided authoritative overviews, suggesting that teaching the recognition of individual phonemes produces reliable long-term improvements in literacy (see also Hulme & Snowling, 2008). These interventions are often based on intensive one-to-one instruction; there are no easy short-cuts and no magic bullets. Nonetheless, our understanding of dyslexia is progressing on several fronts, from cognitive factors affecting reading, to predictive diagnostic measures based on brain function and genetic risk factors (Gabrieli, 2009). This expanding knowledge-base can only aid the refinement of effective treatments, and allow them to be targeted to vulnerable children at earlier stages of development. Surely that is cause for optimism.

Acknowledgements

The Port Glasgow study was supported by the Moray Endowment Fund of the University of Edinburgh. It was made possible by the collaboration of Liz McKelvie of Irlen Scotland, the orthoptic services of Nadia Northway and Jillian Campbell (Glasgow Caledonian University), and the generous assistance of pupils, teachers and support staff at Newark Primary School. Stuart Ritchie completed this work as part of an MSc by Research, and is now supported by an ESRC +3 Studentship for a PhD programme which will include a follow-up of the Port Glasgow study cohort.

References

Albon E., Adi Y., & Hyde C. (2008). *The Effectiveness and Cost-effectiveness of Coloured Filters for Reading Disability: A Systematic Review.* Birmingham: University of Birmingham Department of Public Health and Epidemiology.

Ambrosini, A., De Noordhout, A. M., Sandor, P. S., & Schoenen, J. (2003). Electrophysiological studies in migraine: a comprehensive review of their interest and limitations. *Cephalalgia,* 23(1), 13–31.

Aurora, S. & Wilkinson, F. (2007). The brain is hyperexcitable in migraine. *Cephalalgia,* 27(12), 1442–53.

Babbitt, E. D. (1878) *The Principles of Light and Colour.* London: Babbitt & Co.

Barrett, B. T. (2009) A critical evaluation of the evidence supporting the practice of behavioural vision therapy. *Ophthalmic and Physiological Optics,* 29, 4–25.

Blaskey, P., Scheiman, M., Parisi, M., Ciner, E. B., Gallaway, M., & Selznick, R. (1990) The effectiveness of Irlen filters for improving reading performance: a pilot study. *Journal of Learning Disabilities,* 23(10), 604–612.

Bouldoukian, J., Wilkins, A. J., & Evans, B. J. W. (2002). Randomised controlled trial of the effect of coloured overlays on the rate of reading of people with specific learning difficulties. *Ophthalmic and Physiological Optics,* 22(1), 55–60.

Christenson, G. N., Griffin, J. R., & Taylor, M. (2001). Failure of blue-tinted lenses to change reading scores of dyslexic individuals. *Optometry,* 72(10), 627–33.

Collins, M. (2002) *Banvard's Folly: Thirteen Tales of People Who Didn't Change the World.* New York: Picador.

Critchley, M. (1964) *Developmental Dyslexia.* London: Whitefriars Press.

Gabrieli, J. D. E. (2009). Dyslexia: a new synergy between education and cognitive neuroscience. *Science,* 325(5938), 280–83.

Goldacre, B. (2008) *Bad Science.* London: Harper Perennial.

Gottlieb, R. & Wallace, L. (2001) Syntonic phototherapy. *Journal of Behavioural Optometry,* 12(2), 31–38.

Helveston, E. M. (1990) Scotopic sensitivity syndrome. *Archives of Ophthalmology,* 108(9), 1232–33.

Helveston, E. M. (2001). Tinted lenses. *Perspectives: The International Dyslexic Association,* 27(3) 12–13.

Hulme, C. & Snowling, M. J. (2008). *Developmental Disorders of Language Learning and Cognition.* Oxford: Wiley-Blackwell.

Hyatt, K. J., Stephenson, J., & Carter, M. (2009). A review of three controversial educational practices: Perceptual motor programs, sensory integration, and tinted lenses. *Education and Treatment of Children,* 32(2), 313–42.

Iovino, I., Fletcher, J. M., Breitmeyer, B. G., & Foorman, B. R. (1998). Colored overlays for visual perceptual deficits in children with reading disability and attention deficit/hyperactivity disorder: are they differentially effective? *Journal of Clinical & Experimental Neuropsychology,* 20(6), 791–806.

Irlen, H. (1983) Successful treatment of learning difficulties. Paper presented at the 91st Annual Convention of the American Psychological Association, August, Anaheim, CA.

Irlen, H. (1987). *Irlen Differential Perceptual Schedule.* Long Beach: Perceptual Development Corporation.

Irlen, H. (1991). *Reading by the Colors: Overcoming Dyslexia and Other Reading Disabilities Through the Irlen Method.* New York: Avery Publishing Group.

Irlen, H. (2010). *The Irlen Revolution: A Guide to Changing Your Perception and Your Life.* New York: Square One Publishers.

Jadad, A. R., Moore, R. A., Carroll, D., Jenkinson, C., Reynolds, D. J., & Gavaghan, D. J. (1996). Assessing the quality of reports of randomised clinical trials: is blinding necessary? *Controlled Clinical Trials,* 17, 1–12.

Kriss, I. & Evans, B. J. W. (2005). The relationship between dyslexia and Meares-Irlen syndrome. *Journal of Research in Reading,* 28(3), 350–64.

Ludlow, A. K., Wilkins, A. J., & Heaton, P. (2006). The effect of coloured overlays on reading ability in children with autism. *Journal of Autism and Developmental Disorders,* 36(4), 507–516.

Meares, O. (1980) Figure/background, brightness/contrast and reading disabilities. *Visible Language,* 14, 13–29.

Morris, D., Fraser, S., & Wormald, R. (2007). Masking is better than blinding. *British Medical Journal,* 334(7597), 799.

Newman Wright, B., Wilkins, A. J., & Zoukos, Y. (2007). Spectral filters can improve reading and visual search in patients with multiple sclerosis. *Journal of Neurology,* 254(12), 1729–35.

Ray, N. J., Fowler, S., & Stein, J. F. (2005). Yellow filters can improve magnocellular function: motion sensitivity, convergence, accommodation, and reading. *Annals of the New York Academy of Sciences,* 1039, 283–93.

Ritchie, S. J., Della Sala, S., & McIntosh, R. D. (2011). Irlen colored overlays do not alleviate reading difficulties. *Pediatrics,* 128(4), e932–e938.

Scheiman, M. (2004). Coloured lenses to improve reading comfort and performance: are underlying vision problems being missed? *Journal of Optometric Vision Development,* 35, 37–41.

Simmers, A. J., Gray, L. S., & Wilkins, A. J. (2001). The influence of tinted lenses upon ocular accommodation. *Vision Research,* 41(9), 1229–38.

Singleton, C. & Trotter, S. (2005). Visual stress in adults with and without dyslexia. *Journal of Research in Reading,* 28(3), 365–78.

Skoyles, J. & Skottun, B. C. (2004). On the prevalence of magnocellular deficits in the visual system of non-dyslexic individuals. *Brain and Language,* 88(1), 79–82.

Spitler, H. R. (1941). *The Syntonic Principle.* Eaton, OH: College of Syntonic Optometry.

Stein, J. F. (2001). The magnocellular theory of developmental dyslexia. *Dyslexia,* 7(1), 12–36.

Stein, J. F. (2003) Visual motion sensitivity and reading. *Neuropsychologia,* 41, 1785–93.

Stein, J. F. (2010) Simple ways to treat dyslexia. *SEN Magazine,* 46, 28–30.

Stockman, A., MacLeod, D. I. A., & DePriest, D. D. (1991). The temporal properties of the human short-wave photoreceptors and their associated pathways. *Vision Research,* 31(2), 189–208.

Stone, R. (2003). *The Light Barrier: Understanding the mystery of Irlen syndrome and light-based reading difficulties.* New York: St. Martin's Griffin.

Torgesen, J. K. (2002). Lessons learned from intervention research in reading: a way to go before we rest. In R. Stainthorp & P. Tomlinson (Eds.), *Learning and teaching reading* (pp. 89–103). London: British Psychological Society.

Torgesen, J. K. (2005). Recent discoveries on remedial interventions for children with dyslexia. In M. J. Snowling & C. Hulme (Eds.), *The science of reading: A handbook* (pp. 521–537). Oxford: Blackwell.

Wauters, A. (1999). *Homeopathic Colour Remedies.* Freedom, CA: Crossing Press.

Weisberg, D. S., Keil, F. C., Goodstein, J., Rawson, E., & Gray, J. R. (2008). The seductive allure of neuroscience explanations. *Journal of Cognitive Neuroscience*, 20(3), 470–77.

White, S., Milne, E., Rosen, S., Hansen, P., Swettenham, J., Frith, U., & Ramus, F. (2006). The role of sensorimotor impairments in dyslexia: a multiple case study of dyslexic children. *Developmental Science*, 9(3), 237–69.

Wilkins, A. J. (1994). Overlays for classroom and optometric use. *Ophthalmic & Physiological Optics,* 14, 97–99.

Wilkins, A. J. (1995). *Visual Stress.* Oxford: Oxford University Press.

Wilkins, A. J. (2003). *Reading Through Colour.* Chichester: John Wiley and Sons.

Wilkins, A. J., Baker, A., Amin, D., Smith, S., Bradford, J., Zaiwalla, Z., Besag, F. M. C., Binnie, C. D., & Fish, D. (1999). Treatment of photosensitive epilepsy using coloured glasses. *Seizure: European Journal of Epilepsy*, 8(8), 444–49.

Wilkins, A. J., Evans, B. J. W., Brown, J. A., Busby, A. E., Wingfield, A. E., Jeanes, R., & Bald, J. (1994). Double-masked placebo-controlled trial of precision spectral filters in children who use coloured overlays. *Ophthalmic and Physiological Optics,* 14(4), 365–70.

Wilkins, A. J., Huang, J., & Cao, Y. (2007). Prevention of visual stress and migraine with precision spectral filters. *Drug Development Research*, 68(7), 469–75.

Wilkins, A.J., Jeanes, R.J., Pumfrey, P.D., & Laskier, M. (1996). Rate of Reading Test: its reliability, and its validity in the assessment of the effects of coloured overlays. *Ophthalmic and Physiological Optics*, 16(6), 491–97.

Wilkins, A. J., Patel, R., Adjamian, P., & Evans, B. J. W. (2002). Tinted spectacles and visually sensitive migraine. *Cephalalgia,* 22(9), 711–719.

Wilkins, A. J., Sihra, N., & Smith, I. N. (2005). How precise do precision tints have to be and how many are necessary? *Ophthalmic and Physiological Optics*, 25(3), 269–76.

Chapter 15

Don't try this at school: the attraction of 'alternative' educational techniques

Stuart J. Ritchie, Eric H. Chudler and Sergio Della Sala

Overview

A variety of 'alternative' educational techniques are being used in classrooms worldwide, and it is often unclear whether or not they are supported by anything other than vaguely 'neuroscientific-sounding' rationales. Here, the evidence is reviewed for a sample of these techniques—the classroom use of Brain Gym®, drinking water, brain training games, fish oil supplements and chewing gum—a selection of 'alternative' educational techniques that are either already popular, or becoming ever-more popular in schools worldwide. The evidence is mainly found wanting, and we suggest that what links these techniques, and to some extent explains their popularity, is a common flaw in reasoning which we name the *ferrous fallacy*.

Before critically reviewing the evidence for the efficacy of each of these techniques, we will first describe a particular logical fallacy that is commonly used in support of them. We will then describe the characteristics of reliable evidence, giving advice about how to evaluate controversial claims. After our reviews, we will attempt to explain the attraction of these techniques, and provide some recommendations for what should be done about them.

15.1 **The ferrous fallacy**

Consider the following argument, used by many advocates of magnet therapy for pain reduction:

1. Blood contains iron.

2. Magnets can be used to manipulate iron.

3. Therefore, magnets can be used to change aspects of blood flow, potentially influencing pain.

The premises are both true, but the conclusion is still false, for one devastating reason: the iron in blood haemoglobin is not ferromagnetic (that is, magnets do not attract it). The seductive-sounding analogy between established facts about the human body and a potentially life-changing treatment breaks down completely under the slightest scrutiny (see Figure 15.1). However, even if the magnets-attracting-blood reasoning were sound, it would still not prove that magnet therapy is efficacious. The only way to establish *that* is to examine rigorous scientific studies of the therapy. Upon doing this, we find that the evidence suggests magnet therapy does not work and cannot be recommended for pain relief (Pittler, Brown & Ernst, 2007).

Unfortunately, the making of such tenuous analogies from one context to another is pervasive in the world of 'alternative' educational techniques. In honour of the ill-fated magnetic theories just discussed, we will dub this line of reasoning the *ferrous fallacy,* defined as:

> Claiming a technique, intervention or treatment will work based not on evidence, but on analogy to related effects in a different context.

Throughout our discussion of controversial educational techniques in this chapter, we will point out specific instances of the ferrous fallacy, ranging from analogies based on perfectly good scientific data to analogies based on mere figments. Plainly, the only acceptable way one can justify educational techniques is with reference to rigorous evidence. Unfortunately, this is not the way that many 'alternative' methods have been promoted.

Rather like an 'alternative' medicine, there is no clear-cut definition of an 'alternative' educational technique. Indeed, 'complementary'—another word used by advocates of herbalism, homeopathy and the like—may seem more appropriate for the techniques discussed in this chapter, as they are not seriously entertained as an *alternative* to mainstream education. However, in this chapter we use 'alternative' to mean 'unconventional' or 'non-traditional'.

15.2 **Assessing the evidence**

The definition of the ferrous fallacy mentions evidence, which is—naturally—of critical importance in deciding whether or not to use a particular educational technique. However, not all forms

Fig. 15.1 The ferrous fallacy: accepting claims on the basis of analogy and not evidence can lead to some rather strange beliefs . . . © Dario Battisti, 2011.

Box 15.1 A checklist for evidence

A good piece of evidence need not check *all* of these boxes, but the more that are checked, the more reliable the evidence:

✓ Is the evidence from an appropriate scientific study?

Anecdotes, while often interesting, cannot be scientifically assessed.

✓ Is the evidence published in a peer-reviewed journal?

While not perfect, peer review is the best way we have to separate good from poor science.

✓ Is the sample appropriate?

The sample should be representative of the studied population, large enough, and randomized to avoid biasing the results

✓ Was a control used?

This is essential so we can see which effects would have occurred anyway. Control participants should be as similar as possible to experimental participants.

✓ Is the experiment blinded?

This reduces placebo effects. Both experimenters and participants should be blinded ('double-blinding').

✓ Is the statistical analysis appropriate?

Statistics such as gain scores and one-tailed t-tests, as well as multiple comparisons, can often confuse or inflate results—see Hyatt, Stephenson, and Carter (2009).

✓ Is the effect statistically and/or clinically significant?

Effect size is often of great interest—see Ziliak and McCloskey (2008).

✓ Is the experiment based on established theory?

If well-known parts of science are contradicted by the evidence, one should be sceptical.

✓ Are there any conflicts of interest?

What do the authors, or their funders stand to gain if the technique becomes widely accepted?

✓ Has the experiment been replicated?

Can the same effects be found by independent researchers?

For a more detailed discussion, see Chapter 12, this volume.

of evidence are equal. Before we examine the techniques themselves, we will briefly discuss the hallmarks of high-quality evidence.

We can use historical experiments on a different kind of magnetism—'animal magnetism'—to illustrate one important aspect of good evidence. In 18th-century France, Franz Mesmer—whose name is the root of the verb 'mesmerize'—became famous amongst the public for his miraculous healing powers, based on a new force he claimed to have discovered—'animal magnetism'. The voluminous eyewitness accounts of this force, which could apparently be directed at patients to relieve medical symptoms, aroused the interest of no less than Louis XVI, who commissioned an investigation by members of his Faculty of Medicine and the Royal Academy of Sciences, headed by the American ambassador of the time, Benjamin Franklin.

Since patients claimed to feel the effects of this magnetism on their bodies, an experiment was devised to control for the fact that the patients could see the mesmerist in the room with them. This resulted, in 1784, in history's first *blinded* experiment. When blindfolded, participants made wildly inaccurate guesses as to where a mesmerist was directing his magnetic energy; when sat behind a paper curtain, they claimed to feel the energy when in fact there was no mesmerist present (Kaptchuk, 1998). The commission concluded that there was no such thing as animal magnetism; we would say today that any successful healing achieved by the mesmerists worked via the *placebo effect*, due to the expectations of the patient. While it was a long time before the bogus healing technique died away, blinding was a vital contribution to scientific methodology.

Proper blinding, then, earns experiments a tick on our checklist of good scientific evidence, which can be found in Box 15.1, and includes other aspects such as suitable controls, peer-review and appropriate statistical testing. In addition to these, there are challenges which are peculiar to the field of education, such as the fact that children are constantly maturing—their performance on various measures (for example, arithmetic-based tasks), will naturally improve as they develop. Researchers must control for this if they are to make accurate comparisons over periods of time.

Sadly, this chapter will illustrate that the evidence for several techniques being used in classrooms worldwide gets a very poor mark indeed on our checklist. It will show that the attention-grabbing testimonials offered in support of 'alternative' educational techniques are the equivalent of advertisements for fast food which lure us in with pictures of succulent burgers, dripping with delicious sauces. Usually, the grey, shrivelled, cardboard-like articles we *actually* purchase at fast food restaurants fail to live up to the overblown claims. The same goes—as we will see—for these 'alternative' educational techniques.

Now, armed with our understanding of the ferrous fallacy and our evidence checklist, we can evaluate the techniques, starting with arguably the most deplorable example, Brain Gym®.

15.3 Brain Gym®: the ferrous fallacy goes feral

It would be of incredible significance if a few brief, simple exercises and movements, performed several times during a school day, could significantly improve children's academic performance. It would be of even greater significance if there were a variety of specific actions that improved particular skills, so each child could have a set of movements tailored to their learning needs. This may sound too good to be true, but proponents of Brain Gym®, a programme of movements which apparently promote 'whole-brain learning' (Dennison & Dennison, 2010), say they have found just such a thing.[1]

Many educators rate the Brain Gym® activities extremely favourably—indeed, after denigrating the 'data-driven' approach to education as unhelpful, McChesney (in a foreword to Dennison & Dennison 2010) calls Brain Gym® 'the triumphant uplifting and transforming of the human condition' (p. xiii). Cecilia Koester, a proponent, states on her website: 'I have seen miraculous improvement in both children and adults who have used Brain Gym®. In fact, three children with whom I've worked have gone from blindness to sight. One child began to walk independently at age five and, now seven and a half, has never returned to his wheelchair' (Koester, 2002). This inflated—to say the least—confidence is unrelenting for the entirety of the *Brain Gym®: Teacher's Edition* book (Dennison & Dennison, 2010) which displays some of the most egregious instances of the ferrous fallacy we have found to date.

[1] A revealing interview on BBC's *Newsnight* with Paul Dennison, the founder of Brain Gym®, can be found at the following URL: http://www.youtube.com/watch?v=YjRhYP5faTU

Box 15.2 Left- and right-brained learning

Everyone would love to be more creative. Could one gain inspiration just by performing Brain Gym® exercises, or simply by breathing through the left nostril to direct the oxygen to the right brain hemisphere, as 'life coach' Angela M. Zakon has apparently claimed (http://bit.ly/hlvueP)? Or perhaps—as it would be remiss of us, given the theme of this chapter, to fail to note—by using magnetic pills to 'balance the hemispheres of the brain' (http://www.magneticpill.com)?

Although the most recent edition of the Brain Gym® manual (Dennison & Dennison, 2010) has thankfully had some of the more absurd statements of previous editions on this matter removed, the misconception about left- versus right-brained learning has been spread widely in the past, and remains one of the most common myths about the brain (Corballis, 1999, 2007 and Chapter 13, this volume; Lilienfeld, Lynn, Ruscio & Beyerstein, 2010). While the left hemisphere of the brain is somewhat more efficient at particular tasks than the right, and vice-versa, there is no reason to think—as many seem to—that the left hemisphere epitomizes the military-industrial complex of the West while the right hemisphere has all the glamour and mystery of the East ('bureaucracy', declared Marshall McLuhan in 1977, 'is all left hemisphere. The day when bureaucracy becomes right hemisphere will be utopia' (quoted in Whitaker, 1982)). There is also no evidence for the belief that some more imaginative people are more accomplished at 'accessing' their right hemisphere, the hidden font of creativity and insight.

The irrational desire to classify items and concepts into circumscribed groups, common in popular conceptions of the brain, has been called 'dichotomania' (Whitaker, 1982) and 'the Tyranny of the Dichotomous Mind' (Dawkins, 2004). We are, as Bozzi (1990) explains, 'biologically Aristotelian beings, prone to categorize our world into black and white, matter and form, action and thought, health and disease, natural motions and violent motions, Athenians and Barbarians' (p. 286).

Tellingly, a recent comprehensive review of experiments investigating creativity at the brain level (Dietrich & Kanso, 2010) concluded that if one wished to suggest the right hemisphere was the wellspring of creativity, one could find studies supporting the idea, but one could also find plenty of studies which point to the *left* hemisphere having that function. As well as the hemispheric confusion, myriad studies associate a variety of brain areas with creativity. In other words, the literature on creativity is both confused and confusing, and there is as much evidence against any particular creativity theory as there is for it. Dietrich & Kanso (2010) suggest that researchers may have to revise their notions of creativity away from it being one single 'monolithic' process (p. 846). Therefore, any claim that the evidence on creativity points in any particular direction is, at this stage, false. Incidentally, it should be noted that even if the simplistic right-left hemisphere dichotomy were correct, there is no reason to think that any physical exercises, Brain Gym® or otherwise, could 'stimulate' the different hemispheres and enhance the creative process.

It is tempting to suggest that before accepting these beliefs, we stop trying to think with our left or right brain, and start thinking with what's left of our brain. . .!

Twenty-six separate movements and activities are described by Dennison and Dennison (2010), and all are claimed to have impressive effects on a variety of an individual's physical, cognitive and social abilities. Examples include the 'Thinking Cap', where the ears are pulled and massaged (this activity apparently enhances listening comprehension and public speaking), 'Brain Buttons', where children rub areas under their collarbones (to improve 'ease of eye movements' and hand-eye coordination (p. 57)), 'Think of an X', wherein children imagine an 'x'-shape linking their hips and shoulders (to improve 'spatial awareness and visual discrimination' (p. 51)), and 'Sipping Water', of which more below.

A host of nonsensical—and sometimes outright bizarre—assertions accompany the descriptions of the movements. To pick just a few examples, Dennison and Dennison (2010) claim that children need to be taught to 'listen with both ears' (p. 38), while 'excessive exposure to electronic sounds... will "switch off" the ears' (p. 66); that riding in a car or bus can 'adversely affect depth perception and binocular vision' (p. 96); that liquids other than water 'are said to be processed in the body as food' (p. 54); that yawning can aid creative writing skills (p. 64); and that Brain Gym® can be used effectively for children with learning disabilities, individuals with depression, survivors of natural disasters and attendees of drug rehabilitation clinics (pp. 97–98). They cite no scientific evidence for any of these claims; indeed, the evidence appears to be non-existent.

Clearly some logical leaps are in effect, here. While it is plausible (and as we will see later, evidenced) that incorporating periodic gaps in lessons for fun activities will reduce boredom in children, potentially enhancing learning, this should not be taken as proof that the very specific activities described by Brain Gym® proponents are effective. In addition, while the brain is split into two hemispheres, there is no evidence that there is a 'midline' between them which children must be taught to utilize, as Dennison and Dennison (2010, p. 10) claim (see also Box 15.2). This patina of pseudoscientific claims on top of sensible ideas and facts is a paradigm example of the ferrous fallacy.

The lack of a theoretical basis for Brain Gym® would not necessarily matter if there were well-conducted peer-reviewed studies showing the programme's effectiveness at improving learning. There are, however, no such studies. Indeed, Dennison and Dennison (2010) advise readers that the effectiveness of the activities 'can be personally validated by anyone who takes a few minutes to do [them]. . . Brain Gym® International doesn't conduct research on its own methodologies' (p. xiv). However, we—and we suspect, many others—believe that children deserve better than mere assertions and anecdotal evidence. A small number of studies have been performed on Brain Gym®, and are reviewed by Hyatt (2007). Unfortunately, all the studies have been of extremely poor quality, gaining few marks on our evidence checklist. For example, Wolfsont's (2002) study had only four participants, one of whom was the lead experimenter, while Sifft and Khalsa (1991) did not measure the pre-experiment baseline performance of their participants, leaving them unable to show an improvement over time. Hyatt (2007) also notes that some statistical techniques are used that make the findings difficult to interpret; as noted in Box 15.1, gain scores are a controversial way of reporting results (Thorndike & Thorndike-Christ, 2010, p. 140). The review concludes that there is no sound research evidence for the positive effects of Brain Gym®.

There is, however, a small set of higher-quality studies investigating the effects of exercise breaks—not specific Brain Gym® activities—on children's learning. Aside from the obvious, though very important, effects that regular exercise has on children's health (Strong et al., 2005), classroom behaviour can be improved by incorporating exercise into the school day. A very large epidemiological study by Barros, Silver and Stein (2009) found that, in a cohort of over 10,000 children, a break of 15 minutes or more per day was significantly associated with more positive behavioural ratings by teachers. This enhancement translates to behaviours that are conducive to educational attainment—studies by Pellegrini and Davis (1993) and Jarrett et al. (1998) showed that most children who have a daily 'recess' period display significantly less fidgeting and listlessness when they return to class (see also the reviews of exercise and cognition by Hillman, Erikson & Kramer, 2008 and Howard-Jones, 2010).

There is less research on in-class exercise techniques. Stewart, Dennison, Kohl and Doyle (2004) showed that interspersing the school day with several 10-minute periods of physical exercise during class is a simple way for children to achieve a substantial amount of healthy physical activity, but did not assess the children's behaviour. More recently, Mahar et al. (2006) performed a similar study on a classroom-based exercise programme, and showed that several short periods of exercise per day not only provide children with meaningful physical activity, but also increase 'on-task' behaviour in the classroom. Neither of these studies tested whether the exercise techniques had measurable effects on academic achievement, however. A correlational study in a sample of over 200 children (Coe, Pivarnik, Womack, Reeves & Malina, 2006) has shown that regular exercise is positively correlated with academic performance, though more intervention studies are required before one can conclude anything more specific about the effects of exercise—and especially in-class exercise sessions—on children's learning.

As we can see from this brief review of exercise and school performance, the results so far look fairly encouraging. However, this is far from support for the pseudoscience peddled by the proponents of Brain Gym®, and much less an endorsement of their programme. Coupling fun breaks—which may reduce boredom and improve concentration—with demonstrably false claims about 'Brain Buttons', 'midlines', and such is irresponsible, unethical and humiliating from teachers who should know better. The fact this is encouraged for an *educational* environment is deeply ironic.

Given the number of re-writings and backtrackings in recent years—as noted earlier, the Brain Gym® manual has been revised and updated to remove some of the more conspicuous errors (see Dennison & Dennison, 1994; Sense About Science, 2008)—one imagines that the proponents of Brain Gym®, as well as the policymakers and educationalists who uncritically support them, are somewhat embarrassed by their previous claims. They should be, but this embarrassment should also carry over to the unscientific claims they are currently promulgating. Until all such claims are removed, Brain Gym® remains an exercise in pseudoscience.

15.4 **Water on the brain**

The claim that regularly drinking water can aid children's classroom performance can be found in Brain Gym® literature, but is also a fairly widespread belief amongst the public (Grandjean, Reimers, Haven & Curtis, 2003). While it is true that the popular idea that eight glasses of water are required per day is a myth (Valtin, 2002), and studies of the effects of dehydration on cognition in adults are inconsistent (Cian, Barraud, Melin & Raphel, 2001; Lieberman, 2007), it is not necessarily unreasonable to test the idea in children.

Observational studies indicate that many schoolchildren are underhydrated (e.g. Kaushik, Mullee, Bryant & Hill, 2007), and dehydration can certainly result in a variety of negative behaviours such as irritability and restlessness in children (D'Anci, Constant & Rosenberg, 2006). Addressing hydration needs is obviously important for nutritional reasons, but a more interesting question is whether providing children with water in the classroom will immediately improve cognitive function.

This has been directly addressed by only two studies to date. The first (Benton & Burgess, 2009) was a repeated-measures experiment where 40 children were tested on memory and attention tests on two consecutive days, on one day having drunk 300 millilitres of water just prior to testing, and on the other having had no extra water. Scores on memory—but not attention—tests were significantly higher on the day where water was drunk. The second study (Edmonds & Jeffes, 2009) used a between-subjects design, splitting 23 children into two groups who were tested twice in one day on memory, visual attention, visual search and visuomotor tasks. One group drank a glass of water in between the testing sessions, and the authors reported a significant effect of group on an increase in scores on visual attention and visual search, but not memory or visuomotor tasks.

Thus, the two small-scale studies that have been carried out so far are to some extent contradictory in their results, and certainly do not represent a generalizable literature, not least because the studies were not double-blinded (although, short of hooking children up to an intravenous drip, as suggested for adult participants by Lieberman (2007), it is difficult to imagine how such blinding could be achieved!). Additionally, in neither study was each child's dietary or hydration status at the time of testing taken into account. In any case, as Edmonds and Jeffes (2009) note, even if it is conclusively shown that drinking water facilitates classroom learning, a host of questions remain: how much water is necessary, and is it dependent on body mass? How long before starting schoolwork should a child drink? Lastly, and importantly, by what mechanism does drinking water improve cognitive function? A great deal more research should be done before these findings can be utilized in regular classroom situations.

15.5 Fish oils: piscine in the wind?

Two separate, though related, instances of the ferrous fallacy can be found in the use of fish oils, specifically the two 'omega-3' polyunsaturated fatty acids, EPA (eicosapentanoic acid) and DHA (docosahexaenoic acid) to improve educational outcomes. These are 'essential' fatty acids—they cannot be synthesized by the body, and must be obtained through diet—which are involved in various metabolic processes, including many of those in the brain (e.g. Das, 2003), and appear to be particularly important during early brain development in utero (e.g. Dunstan, Simmer, Dixon & Prescott, 2008).

A marked reduction in consumption of oily fish in modern, as compared to prehistoric, diets has seen the quantity of omega-3 fatty acids decrease, while omega-6 fatty acids—found in poultry, eggs and vegetable oils—have become far more common. Omega-3 and omega-6 fatty acids compete for the same enzyme in metabolic processes, so an excess of one type inhibits the body's ability to use the other (Kirby, Woodward & Jackson, 2009). The underlying hypothesis here states that many individuals—especially those not eating a balanced diet—are deficient in omega-3s, and therefore supplementation could have measurable benefits; since omega-3s are important in brain processes, these benefits could be cognitive or neurological.

The first example of the ferrous fallacy came before any conclusive results from tests of the omega-3 hypothesis had been published, when media reports publicized fish oils as a potential treatment for dyslexia, dyspraxia, attention deficit hyperactivity disorder (ADHD) and autism (e.g. Cassidy, 2002). The leap thus was made from the essential nature of the nutrients, and a few anecdotal accounts, to claiming their effects on a wide variety of serious conditions, even in a newspaper article informing readers that a study of those effects had just begun. It should be obvious to science journalists that the results of experiments should never simply be assumed a priori, and certainly not extrapolated to a variety of conditions not even examined by the study in question.

The study in question—the 'Oxford–Durham Study'—was described by Richardson & Montgomery (2005). A total of 117 children with a diagnosis of developmental coordination disorder (dyspraxia) were split into two groups, one receiving capsules containing omega-3 and omega-6 fatty acids, and one receiving an olive oil placebo for 3 months. Pre- and post-study testing failed to show any change in the main outcome measure—coordination improvements in the treatment group were no different from those in the placebo group—but the treatment group's reading and spelling abilities were significantly better.

A number of other studies on groups of children with various developmental disorders such as autism, dyslexia and ADHD have since been performed. Studies on fish oil supplementation and autism spectrum disorders are reviewed by Bent, Bertoglio and Hendren (2009), while Kirby et al. (2009) provide an excellent review of the totality of the evidence. Ultimately, due to the wide variation in supplement compositions, sample characteristics, experimental methodologies, time periods and outcome measures—not to mention the inconsistent results reported—reviewers find it impossible to come to any firm conclusions about the evidence for the use of the supplements. There is simply no adequate body of evidence upon which to base the claim that fish oil supplementation can improve any cognitive or behavioural outcome for children with special educational needs.

A great many claims have nonetheless been made on the basis of this inconsistent evidence. Most notably, it has become widely assumed that even typically-developing children could benefit from fish oil supplements (see Figure 15.2)—a second example of the ferrous fallacy as, until recently, no proper fish oil trials had been undertaken with 'neurotypical' children as participants.

Media coverage reached fever pitch when Durham Education Authority in England announced they would be administering regular fish oil supplements to a sample of up to 5000 children who were sitting their GCSE exams in 2007/8, in an effort to improve their performance. This 'study' had no control group, was not double blind, and was not ultimately published in a scientific journal—clearly getting a dreadful mark on our evidence checklist. Indeed, it was so poor that Goldacre (2008) uses it as an example to instruct a popular audience how *not* to conduct research. Nevertheless, the impression members of the public would have taken from the media coverage—published before the study had even begun—was that fish oils are an intervention almost guaranteed to improve children's school performance. This was not helped by eager media appearances by the educational psychologists carrying out the 'research' (Goldacre, 2008).

Luckily, the Durham GCSE trial is not the only evaluation of fish oils' ability to enhance the learning of typically-developing children. Osendarp et al. (2007) and Kirby, Woodward, Jackson, Wang and Crawford (2010) have investigated this alleged effect in large samples with high-quality study designs, and both reported discouraging results. In the Osendarp et al. (2007) study, while supplementing children with a variety of micronutrients (iron, zinc and several vitamins) did improve learning and memory performance, simply using omega-3 fatty acids did not. This study used two samples, one from a developed country with good nutrition (Australia) and one from a developing country with a poor nutrition (Indonesia). Perhaps counter-intuitively, the Indonesian children, despite their inferior diet, did not benefit substantially more from the micronutrient supplements than the Australian children. Of a range of over 35 cognitive, physical and emotional outcomes measured, Kirby et al. (2010) found three significant differences between their omega-3 supplemented group and a placebo group after 16 weeks of supplementation. Interestingly, one of the differences was that the behaviour of the placebo group, according to one teacher-rated measure, had become significantly *better* than that of the supplemented group. Small and inconsistent outcomes like these indicate that the results may have been simply due to chance. It is worth noting that as one measures more and more outcomes, the chances of finding 'false positive' results—in any direction—are multiplied; with 35 measures, it is not surprising that a few reached statistical significance.

Results from the Osendarp et al. (2007) study, along with others (e.g. Gesch, Hammond, Hampson, Eves & Crowder, 2002) point towards the fact that omega-3 fatty acids alone may not be enough to produce significant cognitive or behavioural benefits. These studies only showed a difference in ability and behaviour if nutrients *other* than omega-3 fatty acids were provided. Kirby et al. (2009) note that human nutritional biochemistry is deeply complex, and involves an enormous variety of contingent, interacting reactions. Several other micronutrients are required for the body to properly utilize fatty acids, and if a child is also deficient in these, it is unsurprising if omega-3-only studies are disappointing.

In a perfect world, all parents would have the knowledge and the means to provide their children with a balanced diet, including oily fish. In reality, supplementation may turn out to be an important

Fig.15.2 Is fish oil the secret to academic success? We can only find out using well-designed experiments. © Dario Battisti, 2011.

method of providing children with essential nutrients like omega-3 fatty acids. However, at this time, it is impossible for anyone to come to firm conclusions about whether supplementation might reliably improve children's academic performance, or especially about which supplements might have which effects. While the best research currently points towards fish oils having no such benefits, especially in typically-developing children, a great deal more high-quality research is required before anyone can make conclusive statements about the efficacy of the technique. Certainly, at present, the supplements cannot be recommended for school use without committing the ferrous fallacy.

15.6 **Brain training: that's entertainment (and nothing more?)**

Given the huge—and growing—popularity of video games amongst children (e.g. Lenhart et al., 2008), it is not surprising that educators have begun to incorporate similar technologies in their classrooms. Numerous recent studies and reviews have looked at the effects of video games on children, in particular focusing on the potentially negative consequences of playing violent games, an area that has become an ideological and scientific battleground (Ferguson, 2010). However, as well as being entertaining, could video games be used to aid learning in school? If so, which types of game are best? Recently, it has been suggested that 'brain training' games could be of particular educational use, potentially improving children's learning abilities and self-esteem (e.g. Miller & Robertson, 2010). Is this really the case?

Box 15.3 Origins of *Brain Training*

Kawashima's original study (Kawashima et al., 2005) included two groups of 16 people suffering from Alzheimer's disease (a severe degenerative condition that progressively damages the brain), in a nursing home in Japan. One group—the controls—continued with their usual activities, while the other group engaged in a 'learning therapy' method consisting in reading fairy tales aloud and doing some simple mental calculations—a precursor to the *Brain Training* tasks. Follow-up testing after 6 months showed that, while the control group continued to deteriorate, the 'learning therapy' group improved on a standard test of cognitive function in dementia, and some even showed improvements in communication and independence.

However, from the results of this study, it is impossible to disentangle whether this advantage was due to the extra mental stimulation from the therapy, or to the extra attention and social interaction that the treated group received. Furthermore, the communication and independence improvements were only reported for two of the participants.

Of course, claiming that some mental stimulation which allegedly slows the havoc caused by dementia should also improve the performance of healthy people—or indeed schoolchildren—requires quite a ferrous-fallacious leap.

Nintendo's *Brain Training* (or, in the US, *Brain Age*) games are a phenomenally popular series for the hand-held Nintendo DS console. They are designed to be played for a short time each day, as gamers train themselves to complete various tasks such as basic arithmetical problems, calculations of the difference in time between two clocks and counting the numbers of syllables in words. Then, the player can progress to the 'brain age check', which involves more arithmetic, Stroop tests (the console has an in-built microphone to record the player's voice), and memory tests. Poor performance will result—somewhat embarrassingly, no doubt—in the player being assigned a 'brain age' far higher than their chronological age. With enough practice, this brain age can be reduced.

Nintendo is quick to point out that the concept of brain age is not a scientific one (it appears to be merely a marketing strategy; Lorant-Royer, Munch, Mesclé & Lieury, 2010). However, the company does use Ryuta Kawashima, a Japanese neuroscientist, as the public face of the *Brain Training* games, which base their content on a book he authored (Kawashima, 2005). The brain age concept has obvious connotations—Kawashima has published gerontological research (see Box 15.3)—and various other studies have assessed the efficacy of *Brain Training* for slowing the effects of dementia (e.g. Smith et al., 2009; Willis et al., 2006). Butcher (2008) notes that concerns have been raised about Nintendo's marketing of *Brain Training*, especially in advertisements in which actors are seen failing to remember names, then using *Brain Training*, after which their memory swiftly improves.

Exaggerated claims like these are incongruous with the experimental data into the efficacy of brain training, which are equivocal (Butcher, 2008; Nacke, Nacke & Lindley, 2009). A recent large study in adults (Owen et al., 2010) recruited 11,430 participants online, and after assessing them on a battery of cognitive tests, organized them into three groups: those who were to receive training in reasoning and problem-solving, those who were to receive *Brain Training*-like exercises in arithmetic, memory and attention, and those who were to have no training and instead were simply to answer trivia questions on a range of subjects. The groups performed these tasks for 6 weeks, at which point they completed the test battery once more. The test scores of all three groups improved to a broadly similar degree—that is, not only did the type of training make no difference, but even those undergoing no specific training performed just as well.

The Owen et al. (2010) study illustrated a very important point about brain training. While gamers do show improvements on the game itself—in the study, the brain training group showed significantly enhanced performance on the brain training tasks—this does not imply that those benefits will be apparent outside of the circumscribed gaming context. Anyone claiming—on the basis of their gaming score—that the benefits can translate into daily life, or that the benefits are peculiar to brain training, is committing the ferrous fallacy; they are not providing any direct evidence for these suggestions, basing their ideas on mere analogy. Indeed, Fuyuno (2007) quotes neuroscientist Michael Marsiske: 'users may get the illusion of huge gains when starting with *Brain Age*, but these have more to do with learning the device than actual mental improvements.' (p. 20).

A recent set of experiments (e.g. Jaeggi, Buschkuehl, Jonides & Perrig, 2008; Jaeggi, Buschkuehl, Jonides & Shah, 2011) have shown that training children on specific 'working memory' tasks known as '*n*-backs' appears to improve their fluid intelligence, a component of general intelligence (*g*) which involves reasoning and problem-solving abilities (see Morrison & Chein, 2011, for a review, while Moody, 2009 and Conway and Getz, 2010 offer substantial criticisms of this approach). The studies received a great deal of media attention, much of it making the leap from 'a specific task has been shown to work for a specific measure', to in effect stating 'Nintendo's *Brain Training* has been validated by scientists' (e.g. Derbyshire, 2011). However, the studies did not in fact test Nintendo's *Brain Training*, so are of limited relevance to our discussion in this chapter.

In any case, the ferrous fallacy is committed once again if it is claimed that brain training can boost learning of specific classroom subjects. The best way to combat the fallacy is, of course, to perform research in the appropriate contexts. To our knowledge, three studies have been performed on the effects of Nintendo's *Brain Training* in the classroom: two in France (Lorant-Royer, Spiess, Goncalves & Lieury, 2008; Lorant-Royer et al., 2010) and one in Scotland (Miller & Robertson, 2010). They provide conflicting data, but are illustrative of the issues one might encounter in carrying out and reporting such classroom studies.

Lorant-Royer et al. (2008) divided 49 10-year-old children, who had been tested on two school subjects (science and geography), self-esteem questionnaires, and on measures from an intelligence test, into four groups: a group playing *Dr Kawashima's Brain Training*, a group playing a similar Nintendo game, *Brain Academy*, a group doing paper-and-pencil puzzles, and a control group who had normal school lessons. When tested after a period of 11 weeks, there were some increases in performance in the *Brain Training* groups, but also in the paper-and-pencil and control groups. No between-group differences were found on any of the measures. *Brain Training*, then, was found to be no better than paper-and-pencil games, or indeed normal school activities, at improving academic achievement, or self-esteem. The authors concluded that *Brain Training* is a game, and nothing more.

Miller and Robertson (2010) reported a study with a somewhat similar design, which tested 71 10–11-year-old children from three separate classes at three separate schools. In the first class, 21 children used *Dr Kawashima's Brain Training* game for 20 minutes each school day for 10 weeks. In the second, 31 children performed a set of Brain Gym® exercises each day for the same time period. In the third, which was the control group, 19 children had normal school lessons. Their mathematics performance, as well as their self-esteem, was assessed at both the start and the end of the 10-week period.

The maths scores of all groups increased; the *Brain Training* group by 10%, the Brain Gym® group by 2%, and the control group by 5% (for self-esteem, no significant differences were found). Surprisingly, as noted by Logie & Della Sala (2010) in a critique of the study, the authors did not perform a between-groups test on the results. That is, they did not test whether the increase of 10% in the *Brain Training* group was statistically significantly greater than the increase of 5% in the control group. Subsequent analyses showed that the difference was not, in fact, significant, meaning that the researchers had essentially reported null results. Clearly, this study has failed to check the 'appropriate statistical analysis' box on our evidence checklist.

If one uses Miller and Robertson's (2010) logic of simply comparing the raw size of the gain in score (and it should be pointed out again that the use of gain scores is contentious; Thorndike & Thorndike-Christ, 2010), it would appear that Brain Gym® has a *detrimental* effect on maths performance, as the no-treatment control group achieved a greater score gain than the Brain Gym® group. Puzzlingly, Miller and Robertson (2010) suggest—albeit cautiously—that Brain Gym® proponents may 'find encouragement' in the results (p. 252). However, a simple increase in a score, which may appear impressive on its own, should not be considered outside of the context of other results; in this case, the surrounding context makes the Brain Gym® group's result look far from remarkable.

This was not the only flaw in Miller and Robertson's (2010) results. Even if the results of their study hold, they do not show the effectiveness of *Brain Training* per se—a control group who played a different game on the same console should have been included (Logie & Della Sala, 2010). We could then see whether the results could be caused by *any* computer game—perhaps an effect of gaming in general, or the non-specific placebo and Hawthorne effects of being given a desirable console and being part of a study (for an example of this latter effect in education, see Rosas et al., 2003). More recently, Lorant-Royer et al. (2010) performed a study with controls of this type, and found no particularly noteworthy cognitive test score differences between no-treatment controls and groups playing *Dr Kawashima's Brain Training*, another Nintendo game (*Super Mario Bros.*), or pencil-and-paper games.

Miller and Robertson (2010) mention that a randomized controlled trial of brain training in a much larger sample of 800 children is now underway. We await their results with interest, but question whether the results found in, and the quality of, their preliminary work really justifies the effort and funding that will be required to carry out such a large study.

The evidence for the effectiveness of brain training in the classroom is at best questionable. There is, however, considerable interest in other forms of video games as learning aids; for teachers to neglect investigating the educational potential of such a popular form of entertainment would be foolish. Annetta (2010) provides an interesting theoretical discussion of the aspects of video gaming that are likely to enhance learning. It could be argued that brain training does not quite fit the bill; due to the fairly trivial nature of the puzzles, important factors such as 'immersion' and 'instruction' are likely to be low (see also Howard-Jones, 2010, for more discussion of educational games). Spence and Feng (2010), in a review of the effects of video games on spatial cognition, note that several studies show that 'first-person shooter' (FPS) action gamers have improved visuospatial abilities which are both persistent through time and generalize to outside of the game's particular context. Of brain training, they note that the simplicity of the tasks—especially when compared to complex modern action games—would not likely lead to any general cognitive benefits: 'games modelled on the FPS genre have much more promise for remedial training than the kinds of simple puzzle games promoted by most brain-training enterprises' (p. 102).

We are some way yet from designing the perfect educational video game, either for typically-developing children or for children with developmental disorders (Durkin, 2010). But once again, an 'alternative' educational technique provides us with a peculiar irony: it may be that brain training is, despite its promises, the type of game *least* likely to translate to more general increases in cognitive abilities. Naturally, if individuals enjoy the puzzles and challenges presented by brain training, we would not attempt to discourage them. However, as we have seen before, enjoyment does not necessarily translate into enhanced learning.

15.7 **Chewing gum: rumination for rumination?**

Teachers are likely to run a mile upon hearing this, but it has been suggested that chewing gum—that notorious classroom irritation—can aid important learning skills like memory and attention, and should be encouraged in schools (e.g. James, 2007; Jensen, 2007). In December 2010, *The Guardian* ran a story under the headline 'German pupils told to keep chewing as scientists extol virtues of gum' (Connolly, 2010), which described a Bavarian study of chewing gum's effects on concentration in the classroom. We have seen when discussing fish oils that media reporting of studies can often be misleading; this headline is a good example, since the study it alludes to had not yet been completed when it was written. Despite the enthusiasm of the German scientists and teachers quoted in the news report, we should—as always—adopt a healthy scepticism towards such claims. Is it the case that chewing gum can aid learning, and if so, how does it have this effect?

The claims are made on the basis of laboratory studies such as that of Wilkinson, Scholey and Wesnes (2002), who showed that adult volunteers who chewed sugar-free gum did better on memory tests—but not attention tests—than others who did not chew. It should be noted that another group, participants who mimicked chewing movements but without any gum, performed only slightly worse than those with the gum. The authors suggested that the jaw activity involved in chewing might increase cortical blood flow, thereby enhancing cognition. It is certainly true that the action of chewing is accompanied by neurophysiological changes (e.g. Onozuka et al., 2003; Sakamoto, Nakata, Hondam & Kakigi, 2009). However, the research into the effects themselves is controversial: a study published in the same journal, by Tucha, Mecklinger, Maier, Hammerl and Lange (2004), failed to replicate Wilkinson et al. (2002)'s results.

More recent results are similarly inconclusive. In two studies by Smith (2009, 2010), while gum-chewing did appear to improve to a modest degree alertness, reaction time and intelligence scores, it had no effect on memory tasks compared to a 'no chewing' control condition. Allen, Norman and Katz (2008) also found no effect of chewing gum on memory. Studies of context-dependent learning—those which look at whether chewing whilst learning facts enhances recall of those facts whilst chewing later—have generally been negative (e.g. Johnson & Miles, 2007; Overman, Sun, Golding & Prevost, 2009).

Perhaps chewing gum does not directly influence learning, but instead reduces stress and anxiety—rather like using a 'stress ball' toy—and allows learning to go ahead unimpeded? Celebrities such as Alex Ferguson, manager of Manchester United Football Club and noted gum-chewer during tense matches, would certainly attest to gum's stress-diminishing properties. Zibell and Madansky (2009) found that many people do report using chewing gum to lower stress levels. The two studies that investigate the effects of chewing gum on stress during difficult tasks (Scholey et al., 2009; Torney, Johnson & Miles, 2009) are directly contradictory; different assessment tasks were used, however, which may explain the inconsistency. Unfortunately, no studies have directly investigated the effects that reduced stress due to chewing might have on cognitive ability or, for that matter, classroom performance.

The studies discussed here, then, in no way represent a conclusive evidence base for the beneficial effects of chewing gum on cognition. Even if they did, however, anyone recommending the use of chewing gum in the classroom is committing the ferrous fallacy, taking the results out of the laboratory context. No published studies, to our knowledge, have yet directly assessed the effects of chewing gum on learning either in a sample of children or in a school classroom setting—indeed, the one study that mimicked a classroom setting (though with adults; Allen et al., 2008) reported no effect. To make matters even more complicated, the brain activation caused by chewing in young adults is significantly different from that in children (Onozuka et al., 2003)—one cannot, then, expect data from studies using adult participants to generalize to children.

The practice of using chewing gum in the classroom to indirectly improve learning is not utterly implausible, but it is currently unproven. However, any benefits future studies may discover will in all likelihood be rather small. We would ask how these would weigh against the practical aspects of encouraging children to chew more gum. As Primo Levi notes in the essay 'Signs on Stones' (Levi, 1989), discarded chewing gum is 'practically indestructible' (p. 180), not to mention highly unattractive. Proponents of gum in the classroom may have to chew over the potential for some unintended and unpleasant consequences if their technique ever becomes popular.

15.8 **Conclusion**

We have no doubt that, often, proponents of the 'alternative' educational techniques discussed in this chapter are well-meaning, and genuinely have the best interests of children at heart. Sadly, their enthusiasm too frequently gives the impression that their unproven techniques are effective, and sales pitches are often highly irresponsible, no matter their intention. Unregulated websites and reckless reports in the media—frequently written before any proper trials have been conducted—only add to consumers' misconceptions.

Unfortunately, the 'alternative' techniques discussed here, and those like them, seem particularly convincing and easy to sell. This seems to be due to a variety of factors. Firstly, a 'quick-fix' mentality is extremely attractive. We would all like to be able to rapidly improve the memory and intelligence of our children by getting them to perform a set of simple exercises, giving them fish oil pills, chewing gum, or glasses of water or purchasing the latest computer game. The sad reality is that genuine learning takes long hours of hard work, and 'quick-fixes' are very rare indeed. Tempering our fervent desires with scientific evidence is not always easy.

Secondly, there is a common misunderstanding of the scientific method, and where the burden of proof lies. One of us (author SDS) recently encountered this line of argument upon being forwarded a letter, sent by Glasgow City Council Education and Social Work Department, to parents concerned about the use of Brain Gym® in Glasgow classrooms. Brain Gym® has 'working theories that stand to be disproved', claimed the Council employee. 'This is how science works and evolves. So in stating there is no evidence for using specific types of movement, this does not mean that the movement does not help learning [sic]'.[2] This particular line of argument—that the claims are not *dis*proved, so there is no problem in recommending their use—is made regularly to support controversial educational treatments (Hyatt et al., 2009). However, following this reasoning, one should be prepared to uncritically accept for classroom use literally *any* technique, whether pseudoscientific or otherwise, as long as there is no evidence either way of its efficacy (see also chapters 12, 13 and 14, this volume). Once again, we think the public, and especially children, deserve better than this illogical reasoning.

Thirdly, there is something specific about the beguiling mystery of the brain which seems to encourage credulity. Studies have shown how receptive people are to arguments which include 'scientific'-seeming information about the brain (Weisberg, Keil, Goodstein, Rawson & Gray, 2008), or indeed pictures of the brain (McCabe & Castel, 2008), compared to unaccompanied arguments—even if this information is unnecessary and provides no new facts. This may be due

---

[2] Of course, no scientist would recognize Brain Gym®'s rationale as a 'working theory'. Theories which are worth investigating are based on evidence, not uninformed speculation.

to the intuitive appeal of 'reductionist' terminology that explains something apparently ethereal—like a thought process—at a hard, physical level (Weisberg et al., 2008). An apparently 'neuroscientific' rationale for an educational technique—as all of the methods discussed in this chapter have to a greater or lesser extent, from Brain Gym®'s 'midlines' to *Brain Training*'s 'brain age'—can seem very impressive. However, another sad reality is that science is much messier and more ambiguous than proponents of 'alternative' techniques imply (for a discussion of interviews with teachers on their understanding of neuroscience, see Howard-Jones, 2010).

Fourthly, and as we have seen throughout this chapter, advocates of alternative treatments are prone to using the ferrous fallacy, claiming that techniques work based not on hard evidence—as defined by the evidence checklist in Box 15.1—but on inappropriate generalization from other contexts.

It is certain that more of these techniques will appear in future. Thus, critical responsibilities for researchers are both to perform rigorous studies of the techniques,[3] and then communicate the evidence, along with the skills to properly evaluate new claims, to the public. We echo the suggestion of Hyatt et al. (2009), that a well-designed, easily-accessible online resource—similar to the Cochrane Collaboration website, which provides reviews of medical research—should be set up to collect high-quality reviews of the evidence for controversial educational techniques. Such a website would be of great assistance to parents, teachers and other educators who are inundated by the claims of various proponents, and simply do not have the time to research the evidence for each individual technique.

Evidence being available is one thing, but teacher-training courses should be responsible for instructing prospective teachers how to evaluate evidence when it comes to controversial educational techniques. In the context of special education, Mostert and Crockett (2000) provide a strong argument for educators being informed of the history of their discipline, and the many methods that had been adopted before being discredited by subsequent scientific investigation. Regrettably, a great many illustrative examples of the use of disproven techniques can be found in the annals of special education—one particularly tragic example being the use of 'facilitated communication'.

Offit (2008) describes how, in the early 1990s, the hopes of thousands of parents of autistic children worldwide were raised by the appearance of this practice, where a 'facilitator' helps an autistic child type words on a keyboard. It appeared as if the deeply distressing communication problems in autism had at last been solved. Sadly, controlled experiments showed that the facilitator was—sometimes subconsciously—doing the typing; the child was not, in fact, communicating in any way (see also Jacobson, Foxx & Mulick, 2005). Facilitated communication techniques have also been used for coma patients[4], with similarly disappointing results. Extreme cases like this show the seductive power of 'alternative' techniques, but also illustrate the heartbreaking disappointment—not to mention the wasted time, effort and money—that is all too often associated with them.

While it is the case that children and families with special educational needs are particularly vulnerable to overblown and irresponsible claims about interventions, we would argue that Mostert and Crockett's (2000) argument should be extended to all educators. The profound contribution

---

[3] Though it is useful to note Wolpert's (1992) argument that some claims which are absurd in the light of current scientific knowledge are not necessarily worth investigating—Wolpert illustrates the point with the hypothetical argument that 'eating hamburgers makes you a good poet'.

[4] One particularly upsetting case, that of Rom Houben in Belgium, is shown in the following YouTube video: http://www.youtube.com/watch?v=SWKd7tynIYE and described by Boudry, Termote and Betz (2010).

made by education to the lives of children must not be marred by the use of techniques that may raise hopes, but are at best unproven or at worst disproved. In the words of Mostert and Crockett (2000):

> We [educators] cannot inspire confidence in our university colleagues, preservice teachers, or students and parents by hawking the baubles of every latest fad that washes through the popular media, latest in-service workshop, or educational journal. We must be able to discern the effective wheat from the useless chaff – discernment enhanced by attending to the history of our practices over time, what we have achieved, and perhaps most importantly, by closely attending to the historical mistakes we have made (p. 138).

Acknowledgements

We are very grateful to Dario Battisti for the illustrations in this chapter. Gary Lewis provided helpful comments on an earlier draft.

References

Allen, K.L., Norman, R.G., & Katz, R.V. (2008). The effect of chewing gum on learning as measured by test performance. *British Nutritional Foundation Nutrition Bulletin*, 33, 102–107.

Annetta, L. A. (2010) The 'I's' have it: A framework for serious educational game design. *Review of General Psychology*, 14(2), 105–112.

Barros, R. M., Silver, E. J., & Stein, R. E. K. (2009). School recess and group classroom behaviour. *Pediatrics*, 123, 431–36.

Bent, S., Bertoglio, K., & Hendren, R. L. (2009). Omega-3 fatty acids for autistic spectrum disorder: A systematic review. *Journal of Autism and Developmental Disorders*, 39, 1145–54.

Benton, D., & Burgess, N. (2009). The effect of the consumption of water on the memory and attention of children. *Appetite*, 53, 143–46.

Bozzi, P. (1990). *Fisica Ingenua*. Milan: Garzanti.

Boudry, M., Termote, R., & Betz, W. (2010). Fabricating communication. *Skeptical Inquirer*, *34* (4). Available at http://www.csicop.org/si/show/fabricating_communication.

Butcher, J. (2008) Mind games: Do they work? *British Medical Journal*, 336, 246.

Cassidy, S. (2002). Fish oil pills may aid pupils with learning difficulties. *The Independent*, 14 May Available at http://www.independent.co.uk/news/education/education-news/fish-oil-pills-may-aid-pupils-with-learning-difficulties-651130.html (accessed on 4 January 2011).

Cian, C., Barraud, P. A., Melin, B., & Raphel, C. (2001). Effects of fluid ingestion on cognitive function after heat stress or exercise-induced dehydration. *International Journal of Psychophysiology*, 42, 243–51.

Coe, D. P., Pivarnik, J. M., Womack, C. J., Reeves, M. J., & Malina, R. M. (2006). Effect of physical education and activity levels on academic achievement in children. *Medicine and Science in Sports and Exercise*, 38(8), 1515–1519.

Connolly, K. (2010). German pupils told to keep chewing as scientists extol virtues of gum. *The Guardian*, 2 December. Available at http://www.guardian.co.uk/world/2010/dec/02/germany-pupils-chewing-gum-bavaria (accessed 4 January 2011).

Conway, A. R. A., & Getz, A. J. (2010). Cognitive ability: Does working memory training enhance intelligence? *Current Biology*, 20(8), R362–64.

Corballis, M. C. (1999). Are we in our right minds? In S. Della Sala (Ed.), *Mind myths: Exploring popular assumptions about the mind and brain* (pp. 25–42). Chichester: Wiley.

Corballis, M. C. (2007). The dual-brain myth. In S. Della Sala (Ed.), *Tall tales about the mind and brain: Separating fact from fiction* (pp. 291–314). Oxford: Oxford University Press.

D'Anci, K. E., Constant, F., & Rosenberg, I. H. (2006). Hydration and cognitive function in children. *Nutrition Reviews*, 64(10), 457–64.

Das, U. N. (2003). Long-chain polyunsaturated fatty acids in the growth and development of the brain and memory. *Nutrition*, 19(1), 62–65.

Dawkins, C. R. (2004). *The Ancestor's Tale: A Pilgrimage to the Dawn of Life*. London: Phoenix.

Dennison, P. E., & Dennison, G. E. (1994). *Brain Gym® Teacher's Edition- Revised*. Ventura, CA: Edu-Kinesthetics.

Dennison, P. E., & Dennison, G. E. (2010). *Brain Gym®: Teacher's Edition*. Ventura, CA: Hearts at Play, Inc.

Derbyshire, D. (2011). Brain training computer games 'can improve your child's grades within weeks'. *Daily Mail*, 14 June. Available at http://www.dailymail.co.uk/sciencetech/article-2003155/Brain-training-games-improve-child-s-grades-weeks.html (accessed 17 June 2011).

Dietrich, A., & Kanso, R. (2010). A Review of EEG, ERP, and neuroimaging studies of creativity and insight. *Psychological Bulletin*, 136(5), 822–48.

Dunstan, J. A., Simmer, K., Dixon, G., & Prescott, S. L. (2008). Cognitive assessment of children at age 2½ years after maternal fish oil supplementation in pregnancy: A randomised controlled trial. *Archives of Disease in Childhood—Fetal and Neonatal Edition*, 93, 45–50.

Durkin, K. (2010). Videogames and young people with developmental disorders. *Review of General Psychology*, 14(2), 122–40.

Edmonds, C. J., & Jeffes, B. (2009). Does having a drink help you think? 6–7-year-old children show improvements in cognitive performance from baseline to test after having a drink of water. *Appetite*, 53, 469–72.

Ferguson, C. J. (2010). Blazing Angels or Resident Evil? Can violent video games be a force for good? *Review of General Psychology*, 14 (2), 68–81.

Fuyuno, I. (2007). Brain craze. *Nature*, 447(7140), 18–20.

Gesch, C. B., Hammond, S. M., Hampson, S. E., Eves, A., & Crowder, M. J. (2002). Influence of supplementary vitamins, minerals and essential fatty acids on the antisocial behaviour of young adult prisoners—randomised, placebo-controlled trial. *British Journal of Psychiatry*, 181, 22–28.

Goldacre, B. (2008). *Bad Science*. London: HarperCollins.

Grandjean, A. C., Reimers, K. J., Haven, M. C., & Curtis, G. L. (2003). The effect of hydration on two diets, one with and one without plain water. *Journal of the American College of Nutrition*, 22(2), 165–73.

Hillman, C. H., Erickson, K. I., & Kramer, A. F. (2008). Be smart, exercise your heart: Exercise effects on brain and cognition. *Nature Reviews Neuroscience*, 9(1), 58–65.

Howard-Jones, P. (2010). *Introducing Neuroeducational Research*. Oxford: Routledge.

Hyatt, K. J. (2007) Brain Gym®: Building stronger brains or wishful thinking? *Remedial and Special Education*, 28(2), 117–24.

Hyatt, K. J., Stephenson, J., & Carter, M. (2009). A review of three controversial educational practices: Perceptual motor programs, sensory integration, and tinted lenses. *Education and Treatment of Children*, 32(2), 313–42.

Jacobson, J. W., Foxx, R. M., & Mulick, J. A. (2005). Facilitated communication: The ultimate fad treatment. In J.W. Jacobson, R. M. Foxx & J. A. Mulick (Eds.), *Controversial Therapies for Developmental Disabilities: Fad, Fashion, and Science in Professional Practice*. London: Routledge.

Jaeggi, S. M., Buschkuehl, M., Jonides, J., & Perrig, W. (2008). Improving fluid intelligence with training on working memory. *Proceedings of the National Academy of Sciences of the United States of America*, 105(19), 6829–33.

Jaeggi, S. M., Buschkuehl, M., Jonides, J., & Shah, P. (2011). Short- and long-term benefits of cognitive training. *Proceedings of the National Academy of Sciences of the United States of America*, 108(25), 10081–10086.

James, A. N. (2007). *Teaching the male brain: How boys think, feel, and learn in school*. Thousand Oaks, CA: Corwin Press.

Jarrett, O. S., Maxwell, D. M., Dickerson, C., Hoge, P., Davies, G., & Yetley, A. (1998) Impact of recess on classroom behaviour: group effects and individual differences. *The Journal of Educational Research*, 92(2), 121–26.

Jensen, E. (2005). *Teaching with the Brain in Mind* (2nd ed.). Alexandria, VA: Association for Supervision and Curriculum Development.

Johnson, A.J., & Miles, C. (2007). Chewing gum and context-dependent memory: The independent roles of chewing gum and mint flavour. *British Journal of Psychology*, 99(2), 293–306.

Kaptchuk, T. J. (1998). Intentional ignorance: A history of blind assessment and placebo controls in medicine. *Bulletin of the History of Medicine,* 72, 389–433.

Kawashima, R. (2005) *Train Your Brain: Better Brainpower, Better Memory, Better Creativity.* Teaneck, NJ: Kumon Publishing.

Kawashima, R., Okita, K., Yamakazi, R., Tajima, N., Yoshida, H., Taira, M., Sugimoto, K., et al. (2005). Reading aloud and arithmetic calculation improve frontal function of people with dementia. *Journal of Gerontology: Medical Sciences*, 60A(3), 380–84.

Kaushik, A., Mullee, M. A., Bryant, T. N., & Hill, C. N. (2007). A study of the association between children's access to drinking water in primary schools and their fluid intake: can water be 'cool' in school? *Child: Care, Health, and Development,* 33(4), 409–415.

Kirby, A., Woodward, A., & Jackson, S. (2009). Benefits of omega-3 supplementation for schoolchildren: Review of the current evidence. *British Educational Research Journal*, 36(5), 699–732.

Kirby, A., Woodward, A., Jackson, S., Wang, Y., & Crawford, M. A. (2010). A double-blind, placebo-controlled study investigating the effects of omega-3 supplementation in children aged 8–10 years from a mainstream school population. *Research in Developmental Disabilities*, 31, 718–30.

Koester, C. (2002) Have you heard of Brain Gym? Available at http://www.iamthechild.com/articleresearch/heard.html (accessed 29 December 2010).

Lenhart, A., Kahne, J., Middaugh, E., Macgill, A. R., Evans, C., & Vitak, J. (2008). Teens, video games and civics: Teens' gaming experiences are diverse and include significant social interaction and civic engagement. Available at: http://www.pewinternet.org/PPF/r/263/report_display.asp (accessed 4 January, 2011).

Levi, P. (1989). *Other People's Trades.* London: Abacus.

Lieberman, H. R. (2007). Hydration and cognition: A critical review and recommendations for future research. *Journal of the American College of Nutrition*, 26(5), 555–61.

Lilienfeld, S. O., Lynn, S. J., Ruscio, & Beyerstein, B. L. (2010). *50 Great Myths of Popular Psychology: Shattering Widespread Misconceptions about Human Behaviour.* Chichester: Wiley-Blackwell.

Logie, R. H., & Della Sala, S. (2010). Brain training in schools, where is the evidence? *British Journal of Educational Technology*, 41(6), 27–28.

Lorant-Royer, S., Munch, C., Mesclé, H., & Lieury, A. (2010). Kawashima vs. 'Super Mario'! Should a game be serious in order to stimulate cognitive aptitudes? *Revue Européene de Psychologie Appliquée*, 60, 221–32.

Lorant-Royer, S., Spiess, V., Goncalves, J., & Lieury, A. (2008). Programmes d'entrainment cérébral et performances cognitive: efficacité, motivation… ou marketing? De la Gym-Cervau au programme du Dr. Kawashima… *Bulletin de Psychologie*, 61(6), 531–49.

Mahar, M. T., Murphy, S. K., Rowe, D. A., Golden, J., Shields, A. T., & Raedeke, T. D. (2006). Effects of a classroom-based program on physical activity and on-task behaviour. *Medicine and Science in Sports and Exercise*, 38(12), 2086–94.

McCabe, D. P., & Castel, A. D. (2008). Seeing is believing: The effect of brain images on judgements of scientific reasoning. *Cognition*, 107(1), 343–52.

Miller, D. J., & Robertson, D. P. (2010). Using a games console in the primary classroom: Effects of 'Brain Training' programme on computation and self-esteem. *British Journal of Educational Technology*, 41(2), 242–55.

Moody, D. E. (2009). Can intelligence be increased by training on a task of working memory? *Intelligence*, 37, 327–28.

Morrison, A. B., & Chein, J. M. (2011). Does working memory training work? The promise and challenges of enhancing cognition by training working memory. *Psychonomic Bulletin and Review*, 18, 46–60.

Mostert, M. P., & Crockett, J. B. (2000). Reclaiming the history of special education for more effective practice. *Exceptionality*, 8(2), 133–43.

Nacke, L. E., Nacke, A., & Lindley, C. A. (2009). Brain training for silver gamers: Effects of age and game form on effectiveness, efficiency, self-assessment, and gameplay experience. *Cyberpsychology and Behaviour*, 12(5), 493–99.

Offit, P. A. (2008) *Autism's False Prophets: Bad Science, Risky Medicine, and the Search for a Cure*. New York: Columbia University Press.

Onozuka, M., Fujita, M., Watanabe, K., Hirano, Y., Niwa, M., Nishiyama, K., & Saito, S. (2003). Mapping brain region activity during chewing: A functional magnetic resonance imaging study. *Journal of Dental Research*, 81, 743–46.

Osendarp, S. J., Baghurst, K. I., Bryan, J., Calvaresi, E., Hughes, D., Hussaini, M., Wilson, C., et al. (2007). Effect of a 12-mo micronutrient intervention on learning and memory in well-nourished and partially-nourished school-aged children: 2 parallel, randomised, placebo-controlled studies in Australia and Indonesia. *The American Journal of Clinical Nutrition*, 86, 1082–93.

Overman, A.A., Sun, J., Golding, A.C., & Prevost, D. (2009). Chewing gum does not induce context-dependent memory when flavour is held constant. *Appetite*, 53, 253–55.

Owen, A. M., Hampshire, A., Grahn, J. A., Stenton, R., Dajani, S., Burns, A. S., Ballard, C. G., et al. (2010). Putting brain training to the test. *Nature*, 465(7299), 775–78.

Pellegrini, A. D., & Davis, P. D. (1993). Relations between children's playground and classroom behaviour. *British Journal of Educational Psychology*, 63(1), 88–95.

Pittler, M. H., Brown, E. M., & Ernst, E. (2007). Static magnets for reducing pain: Systematic review and meta-analysis of randomised trials. *Canadian Medical Association Journal*, 177(7), 736–42.

Richardson, A. J., & Montgomery, P. (2005). The Oxford-Durham study: A randomised, controlled trial of dietary supplementation with fatty acids in children with developmental coordination disorder. *Pediatrics*, 115, 1360–66.

Rosas, R., Nussbaum, M., Cumsille, P., Marianov, V., Correa, M., Flores, P., Salinas, M., et al. (2003). Beyond Nintendo: Design and assessment of educational video games for first and second grade students. *Computers and Education*, 40, 71–94.

Sakamoto, K., Nakata, H., Hondam Y., & Kakigi, R. (2009). The effect of mastication on human motor preparation processing: A study with CNV and MRCP. *Neuroscience Research*, 64(3), 259–66.

Scholey, A., Haskell, C., Robertson, B., Kennedy, D., Milne, A., & Wetherell, M. (2009). Chewing gum alleviates negative mood and reduces cortisol during acute laboratory physiological stress. *Physiology and Behaviour*, 97, 304–312.

Sense About Science (2008). Brain Gym. Available at http://www.senseaboutscience.org.uk/pdf/braingym.pdf (accessed 4 January 2011).

Sifft, J. M., & Khalsa, G. C. K. (1991). Effect of educational kinesiology upon simple response times and choice response times. *Perceptual and Motor Skills*, 73, 1011–1015.

Smith, A. (2009). Effects of chewing gum on mood, learning, memory and performance of an intelligence test. *Nutritional Neuroscience*, 12, 81–88.

Smith, A. (2010). Effects of chewing gum on cognitive function, mood and physiology in stressed and non-stressed volunteers. *Nutritional Neuroscience*, 13, 7–16.

Smith, G. E., Housen, P., Yaffe, K., Ruff, R., Kennison, R. F., Mahncke, H. W., & Zelinski, E. M. (2009). A cognitive training program based on principles of brain plasticity: Results from the Improvement of Memory with Plasticity-based Adaptive Cognitive Training (IMPACT) study. *Journal of the American Geriatrics Society*, 57(4), 594–603.

Spence, I., & Feng, J. (2010). Video games and spatial cognition. *Review of General Psychology*, 14(2), 92–104.

Stewart, J. A., Dennison, D. A., Kohl, H. W., & Doyle, J. A. (2004) Exercise level and energy expenditure in the Take 10! in-class physical activity program. *Journal of School Health,* 74(10), 397–400.

Strong, W.B., Malina, R. M., Blimkie, C. J. R., Daniels, S. R., Dishman, R. K., Gutin, B., Trudeau, F., et al. (2005). Evidence based physical activity for school-age youth. *The Journal of Pediatrics*, 146(6), 732–37.

Thorndike, R. M., & Thorndike-Christ, T. M. (2010). *Measurement and Evaluation in Education and Psychology* (8th ed.). Boston, MA: Allyn & Bacon.

Torney, L.K., Johnson, A.J., & Miles, C. (2009). Chewing gum and impasse-induced self-reported stress. *Appetite*, 53, 414–417.

Tucha, O., Mecklinger, L., Maier, K., Hammerl, M., & Lange, K.W. (2004). Chewing gum differentially affects aspects of attention in healthy subjects. *Appetite*, 42, 327–29.

Valtin, H. (2002). 'Drink at least eight glasses of water a day'. Really? Is there scientific evidence for '8 x 8'? *American Journal of Physiology—Regulatory, Integrative and Comparative Physiology*, 283, 993–1004.

Weisberg, D. S., Keil, F. C., Goodstein, J., Rawson, E., & Gray, J. R. (2008). The seductive allure of neuroscience explanations. *Journal of Cognitive Neuroscience*, 20(3), 470–77.

Whitaker, H. A. (1982). Dichotomania: An essay on our left and right brains. *Journal of Visual Verbal Languaging*, 2(1), 7–13.

Wilkinson, L., Scholey, A., & Wesnes, K. (2002). Chewing gum selectively improves aspects of memory in health volunteers. *Appetite*, 38, 235–36.

Willis, S. L., Tennstedt, S. L., Marsiske, M., Ball, K., Elias, J., Koepke, K. M., … Wright, E. (2006). Long-term effects of cognitive training on everyday functional outcomes in older adults. *Journal of the American Medical Association*, 296(23), 2805–2814.

Wolfsont, C. (2002). Increasing behavioural skills and level of understanding in adults: A brief method integrating Dennison's brain gym balance with Piaget's reflective processes. *Journal of Adult Development*, 9, 187–202.

Wolpert, L. (1992). *The Unnatural Nature of Science*. London: Faber & Faber.

Zibell, S., & Madansky, E. (2009). Impact of gum chewing on stress levels: online self-perception research study. *Current Medical Research and Opinion*, 25, 1491–1500.

Ziliak, S. T., & McCloskey, D. N. (2008). *The Cult of Statistical Significance*. Ann Arbor, MI: University of Michigan Press.

Section 6

Current conjectures from educational neuroscience

Chapter 16

Bridging between brain science and educational practice with design patterns

Michael W. Connell, Zachary Stein and Howard Gardner

Overview

The current 'neuroscience and education' dialogue seems to centre largely on the question of how (or whether) neuroscience research can inform mainstream educational practice. Building on Dewey's (1929) analysis of educational science in *The Sources of a Science of Education*, we reframe the question to ask: 'How can research in the special sciences (such as neuroscience) and insights from educational practice both inform a science of education?' We point to *explanatory mental models* as the point of overlap between teacher perception, informal expertise, scientific theory and teacher action, and argue that these mental models in the heads of educators are both the site of educational science proper and a leverage point for driving more desirable educational outcomes in a scalable manner. Through our analysis we identify six 'gaps' that must be bridged to catalyse a sustainable science of education. Three of these gaps represent obstacles to collaboration between scientists and educators, and the other three gaps inhibit educators' widespread adoption, application and validation of scientific theories. We propose that design patterns, thoughtfully crafted, can help bridge all six gaps. A design pattern is a description of a recurring problem (such as how to assess specific competencies reliably) plus a description of a general solution that can be applied flexibly to many instances of the problem across diverse contexts. Design patterns have had a tremendous impact on applied domains such as architecture and software engineering. We believe they can play a similarly important role in developing a sustainable interdisciplinary science of education.

16.1 Bridging between brain science and educational practice with design patterns

Psychologists, politicians and educationists have pursued the prospect of a true science of education for over a century. Neuroscientists have joined the conversation in recent decades, giving rise to a movement rooted in efforts to address educational problems using models and methods from the brain sciences. A sizeable and growing literature associated with that movement focuses specifically on the relationship (or lack thereof) between neuroscience and education (Ansari & Coch, 2006; Bruer, 1997, 2002; Byrnes & Fox, 1998; Connell, 2005; Gabrieli, 2009; Geake & Cooper, 2003; Goswami, 2004; Hall, 2005; Katzir & Pare-Blagoev, 2006; Mayer, 1998; Schunk, 1998; Stanovich, 1998; Willingham, 2008). It appears that in the post-industrial West, questions about the prospects of a science of education are slowly being transformed into questions about how to arrange ways for cognitive science, brain science and genetics to inform—or even to determine—what happens in schools. In the context of these shifting questions and debates about the role of scientific research in educational practice, we offer an analysis and a set of recommendations. We first look back to Dewey to situate the issue of 'neuroscience in education' within the larger context of 'neuroscience, educational practice and educational science', and then we look forward toward a science of education built around a library of design patterns that integrate systematic, interdisciplinary scientific research into flexible and effective educational solutions that educators can readily apply in practice.

In the first section to follow we summarize and elaborate on some of Dewey's central arguments concerning the nature of educational science. Dewey (1929) argues that a science of education must draw on *general descriptions*, such as general scientific models of learning and motivation, to inform *particular prescriptions*, such as specific techniques for helping Johnny manage his math anxiety and ADHD well enough in an overcrowded and noisy classroom that he can come to understand the Pythagorean theorem and its importance. This should be accomplished, suggests Dewey, in a manner that supports the efficient development and effective application of the educator's expert judgement instead of seeking to override that judgement. We should not, in particular, expect neuroscience or any other type of scientific research to provide either a 'stamp of approval' for specific educational practices or detailed 'recipes' for achieving particular educational objectives.

Dewey's (1929) incisive and perspicacious analysis of what educational science should be provides a fresh and relevant perspective on the current conversation about neuroscience and education. In particular, he distinguishes between educational science on the one hand and the *sources* of educational science on the other hand. He classifies the specific sciences (e.g. neuroscience, psychology) as well as educational practice as sources of educational science. This distinction may perhaps seem subtle, but it shifts the terms of the dialogue quite radically. Whereas much of the conversation to date has focused on questions along the lines of 'what is the role of neuroscience *research* in educational *practice*?', Dewey suggests we should instead be asking 'what are the roles of *neuroscience research* and *educational practice* in *educational science*?'. Dewey's educational science resides in the considered judgement of the educator, who draws on the results of relevant sources of scientific research in conjunction with the collective experience of reflective educational practitioners. His analysis thus illuminates a novel way to understand how the dialogue between neuroscientists and educational practitioners can be mediated to move both forward productively.

Dewey's treatment leaves off, however, at quite a conceptual, theoretical and strategic level. He does not, in particular, discuss how his insights can be made useable to practising scientists and educators, either individually or in collaboration. In sections 16.2 through 16.4, therefore, we seek to pick up where Dewey left off, using his analysis as the basis for motivating and describing a very practical framework that we believe can facilitate a robust two-way dialogue between research scientists (including neuroscientists) and educational practitioners. Specifically, we suggest that *design patterns* can catalyse the kind of interdisciplinary dialogue envisioned by Dewey, providing a practical framework supporting the synthesis of insights from basic neuroscience research and educational practice, and fostering the kind of systematic accumulation of valid, useable, public knowledge that is associated with mutually supportive scientific and technological progress in other domains such as medicine, agriculture and engineering.

A design pattern describes a recurring problem in a domain and the core of a solution to that problem in a way that allows the solution to be applied flexibly to a wide range of situations in which the problem occurs. In education, for example, design patterns might address recurring problems such as how to design educational materials that are accessible to all learners, how to build formative assessments of student understanding, or how to engage and motivate students in certain key learning processes. In fields where design patterns are used, the stakeholders agree upon a template, which codifies the basic form of the useable knowledge they collaborate to produce. This is a unique way of representing both what is known scientifically and what has been done in practice; it allows for the cumulative and collaborative construction of useable knowledge at the interface of specific sciences and context sensitive problem-focused domains of application. We introduce a design pattern template that we think could help bring Dewey's ideas about educational science into current practice.

Finally, we illustrate how design patterns can be used to bridge between neuroscience and educational practice. We use one element of the *Universal Design for Learning* framework (Rose & Meyer, 2002) to construct an example of a neuroscience-informed design pattern for addressing the ubiquitous educational problem of accommodating individual learning differences and disabilities. This example is offered as a way of making clear just what design patterns are and how they can be useful in furthering the science of education.

16.2 **The virtuous cycle of educational science**

Dewey (1929) defines educational science in terms of two central elements: *explanatory models*[1] that educators use to guide their practice and *systematic methods of inquiry* that they use to improve those models. One can glean from Dewey's writings a vision of educational science as a kind of progressive, self-correcting system constructed around these two central elements that provides educators with immediately useable knowledge while also driving a virtuous cycle in which the cumulative store of educational expertise is systematically expanded and progressively refined over time.

Building on Dewey's philosophical analysis in an effort to make his ideas more practically accessible, we find it useful to identify explanatory models as the central organizing structures in educational science. These structures can be seen as the point of overlap between two distinct processes or loops (see Figure 16.1). The *application loop* corresponds to educational practice, in which educators apply explanatory models to make sense of their observations about students and to make informed decisions about what educative actions to take next. The *adaptation loop* corresponds to scientific inquiry, wherein the stock of explanatory models is adapted (that is, expanded and refined) through an ongoing systematic process of problem identification, solution generation and solution validation. Note that the adaptation loop involves a dialogue between educators and scientists in the specific sciences, such as neuroscience. In this dialogue, educators are responsible for identifying worthwhile problems and testing the validity of proposed solutions. Neuroscientists and other scientists outside of education, in contrast, are responsible for generating explanatory models of phenomena associated with the patterns and problems identified by educators.

---

[1] Dewey actually calls them 'explanatory laws', but we prefer the term 'explanatory models' because it seems to have less of a normative connotation, especially given the association with fundamentally normative civil and criminal laws.

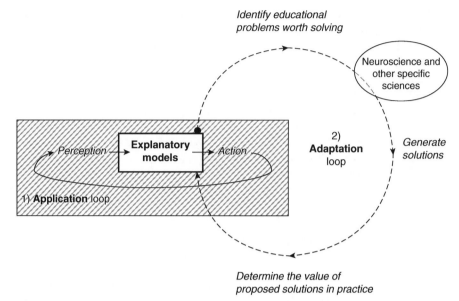

Fig. 16.1 Educational science as a system of two feedback loops (application and adaptation) organized around explanatory models. The *application loop* (1) corresponds to the process of applying the explanatory models in educational practice, and the *adaptation loop* (2) corresponds to the process of expanding and refining the stock of practically useful explanatory models through systematic scientific inquiry. Reproduced with permission of Michael W. Connell. © 2010. All rights reserved.

The system of educational science depicted in Figure 16.1 draws heavily on Dewey's (1929) philosophical analysis. The motivation for this proposal derives, in part, from analogies to other domains such as engineering and medicine, where this kind of progressive system linking theory and application has to date been realized more fully and successfully than in education. Our proposal, simply put, is that the kind of progressive, self-correcting science of education envisioned by Dewey is feasible, and that closing the application and adaptation loops depicted in Figure 16.1 is certainly necessary—and may be sufficient—to establish it.

Closing the adaptation loop means facilitating the dialogue between educators, neuroscientists and others that could generate relevant scientific explanatory models and integrate them into effective educational solutions. Closing the application loop means supporting teachers in integrating these solutions (and the explanatory models embedded within them) into their regular practice. But the work of educational practitioners and research scientists differs in a number of fundamental ways. These differences can be thought of as 'gaps' that make it challenging to close the two loops. In the following sections, we identify key gaps that need to be bridged to close the loops, and then we argue that design patterns—when organized specifically to address these gaps—can be used to bridge virtually all of them.

16.3 **Obstacles to establishing a sustainable science of education**

The adaptation and application loops depicted in Figure 16.1 represent the two fundamental processes involving explanatory models—that is, systematically changing (generating and refining) the models over time, and systematically applying them to guide educational practice, respectively. In this section, we elaborate the idealized system of Figure 16.1 to identify some of the practical obstacles that must be overcome to close these loops.

Adaptation loop: systematically generating and refining explanatory models

The adaptation loop involves a complex collaborative dialogue between educators and researchers in the specific sciences, such as neuroscience, cognitive psychology and economics. These researchers operate outside of education, but nonetheless have insights that can help educators solve practical problems and achieve educational objectives.

An educational practitioner's role in closing the adaptation loop differs from that of neuroscientists and other special scientists (see Chapter 3, this volume). Specifically, educators, in the course of their classroom practice, should simultaneously be carrying out two functions of systematic inquiry, namely:

1. Problem identification: systematically identifying recurring practical problems worthy of solution.

2. Solution validation: systematically testing the value of proposed solutions in improving practice.

Researchers in the special sciences, in contrast, are responsible for systematically generating *explanatory models* of phenomena relevant to the educational problems identified by educational practitioners.

For example, imagine a language arts teacher who notices that over the years a few students seem to have persistent and profound difficulties with reading compared to their age mates. This individual teacher might learn to recognize the signs of this language difficulty and develop ad hoc strategies for responding to it—by coordinating one-on-one tutoring services for such students, or arranging for them to participate in a less advanced reading group, perhaps in a lower grade. Having taken such 'common sense' actions in response to these particular students' needs, the teacher might consider her responsibility fulfilled to the best of her ability and available resources. This is an example of educational practice that does *not* meet the criteria of educational science, which specifically entails applying explanatory models in practice and employing systematic methods of inquiry.

The teaching scenario described falls short of educational science in the first respect because the educator lacks any explanatory model she can use to reason about the observed pattern of struggling readers—that is, her response does not derive from an understanding of *why* the children might be having extra difficulties. Often, such explanatory models can be found outside of education proper. For example, neuroscientists and cognitive psychologists have developed models of memory (Anderson, 1983; Atkinson & Shiffrin, 1968; Baddeley, 1976; Eichenbaum, 1997; Eichenbaum, Otto & Cohen, 1994; Miller, 1956), attention (Pashler, 1997; Posner & Peterson, 1990) and visual processing (Frost & Katz, 1992; Seidenberg, 1995; Seidenberg & McClelland, 1989) that might be relevant to the pattern the teacher observes. Poor reading has been associated with disorders in a wide range of cognitive and neural systems, including impaired working memory that inhibits a student's ability to hold the words online long enough to extract their meaning, attentional problems, and problems recognizing or decoding the written language symbols (Grigorenko, 2001; Paloyelis, Rijsdijk, Wood, Asherson & Kuntsi, 2010; Rose & Meyer, 2002). Each explanatory model would lead the educator to seek additional data further afield than she might otherwise consider relevant—such as in the student's performance in math or science classes, or related to the student's comprehension of stories during 'circle time' where the teacher does the reading compared to comprehension during independent reading time. Applying such explanatory models would thus lead the educator to *perceive* the situation in a more systematic, comprehensive and generally intelligent way, and would guide her to *respond* or *act* in a more nuanced, individualized and generally more effective manner than the general 'common sense' response that might otherwise be applied across the board. This example illustrates how explanatory models provide 'a light to the eyes and a lamp to the feet' (Dewey, 1929), simultaneously engaging and supporting processes of careful observation, systematic reasoning and thoughtful judgement on the part of the educator as opposed to triggering rigid educational scripts or 'recipes'.

The teacher in this hypothetical scenario also fails to fulfil the second criterion of educational science, which requires her to employ systematic methods of inquiry. Having identified a recurring problem in the classroom—in this case a small subset of students who exhibit unusual difficulties with reading, as an educational scientist she would endeavour to isolate or even formalize this as a problem requiring solution, and then seek solutions, and then test the validity of those solutions in practice. Solutions to such problems may already exist, or explanatory models may exist elsewhere that can be integrated into useable educational solutions, or basic research may need to be initiated in other domains to generate explanatory models of the phenomenon of interest. Regardless of the current state of scientific understanding, the practicing educator is in the best position to identify important problems of practice and to test the value of available or proposed solutions (Dewey, 1929).

This example surfaces some of the practical obstacles to closing the adaptation loop of educational science, especially given that educators need explanatory models, most of which will come from scientific domains outside of educational practice (see Figure 16.2). In particular, in addition to the physical separation between groups of people (especially educators and scientists), there are also fundamental differences in the nature of each group's work. We highlight three

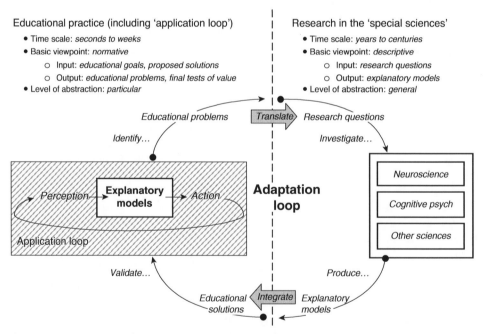

Fig. 16.2 Closing the 'adaptation loop'. With respect to the adaptation loop of educational science, classroom practice is a source of educational *problems* worth solving, and of the final *test of value* of proposed solutions. The special sciences are sources of *explanatory models* that can inform practice when integrated into comprehensive solutions, strategies and techniques. To close the adaptation loop of educational science, it is necessary to bridge across the distinct *time scales*, *basic viewpoints* and *levels of abstraction* that distinguish educational practice from research in the special sciences. Reproduced with permission of Michael W. Connell. © 2010. All rights reserved.

dimensions along which domain differences create practical obstacles to productive dialogue and collaboration:

◆ Time scales: scientific research follows its own course, typically over *years*, *decades* or even *centuries*, whereas educators need immediate solutions to guide their practice from *moment-to-moment*, *day-to-day* and *week-to-week*.

◆ Levels of abstraction: scientific explanatory models are by definition general in scope (tying together and explaining many particular data points) and therefore tend to be relatively *abstract* and *context-free*, whereas classroom practice typically requires educators to respond to very *particular* and *context-specific* situations.

◆ Basic viewpoints: scientific research is a fundamentally *descriptive* enterprise, whereas education is fundamentally *normative*—that is, goal-oriented, value-laden and ethically and morally charged (see Stein, Connell & Gardner, 2008 for a discussion of basic viewpoints in the context of interdisciplinary educational research and practice).

We must bridge the gaps along these three dimensions (time scales, levels of abstraction and basic viewpoints) in order to close the adaptation loop and catalyse a self-sustaining science of education in which important educational problems are identified, systematic scientific research produces robust explanatory models that can be incorporated into useable educational strategies and solutions, and the value of those strategies and solutions is evaluated through application in actual practice so that the solutions (and the explanatory models embedded within them) can be further refined. Before discussing a constructive proposal for bridging the gaps we have identified, we first discuss another set of challenges related to educators' application of explanatory models— namely, how explanatory models become integrated into educators' repertoire of practical strategies and tactics to influence actual practice.

Application loop: how explanatory models influence educational practice

On the surface, the application loop of educational science corresponds to educational practice as people generally conceive it (that is, teachers educating students). In terms of educational science, we define the application loop more specifically as an iterative process in which educators observe students (gauging student motivation and understanding from one moment to the next, for example) and act responsively in light of their own understanding (or 'mental model') of the situation as they perceive it (Figure 16.3). In general, we assume that teachers act in ways that they believe will help students achieve specific educational goals. Teacher actions change the classroom situation, which leads to new observations, which in turn drive decisions about new actions, and so on. As already suggested, this iterative process—and in particular the educator's decision-making about what actions to take—occurs against a normative backdrop that includes educational goals, moral and ethical concerns, cultural values, social norms and the like.

Although educational practice always involves some kind of mental model linking perception to action, not all such mental models are *explanatory* models. The application loop of educational science (as defined in this essay and following Dewey, 1929) depends specifically upon *explanatory* mental models.

What are mental models?

Mental models are representations in people's minds of some part of the world 'out there'. More specifically, mental models function as simplified simulations of some small aspect of reality, thereby supporting understanding, reasoning and decision-making (Craik, 1943; Johnson-Laird, 1983).

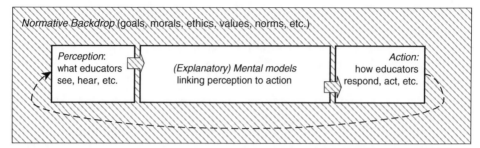

Fig. 16.3 The 'application loop' of educational science is an iterative process in which educators respond to what they *perceive* in the classroom, in light of the *mental models* they use to make sense of their perceptions and decide which *actions* will be most effective given the *normative backdrop* of educational goals, moral and ethical concerns, cultural values, social norms, etc. Although educational practice always involves some kind of mental model linking perception to action, not all such mental models are *explanatory models*. The application loop of educational science depends specifically upon *explanatory* mental models. Reproduced with permission of Michael W. Connell. © 2010. All rights reserved.

Educators, in particular, use these mental models to make sense of their observations about student behaviour and performance, to predict what will happen in response to possible actions they might take, and therefore to decide what educative actions are appropriate in a particular situation in light of their immediate and long-term educational objectives.

For example, imagine a language arts teacher has just assigned an essay for a group of students to read. He then hands out a comprehension test. One student answers all of the questions correctly, another answers half correctly, a third produces all incorrect answers and a fourth doodles on the worksheet without writing any answers at all.

Different teachers placed in this situation would behave differently. One teacher might infer that the student who produced all correct answers has mastered the strategies for reading comprehension covered in class and that the other three students have not. Based on that interpretation of the situation, one teacher might decide to have the first student work independently while he repeats the prior instruction more slowly and in greater detail with the other three. Another teacher making the same interpretation might decide on a different strategy, having the first student explain the strategies to the other three. A third teacher might draw the same conclusion about the first student, but infer that the other three students have distinct challenges with reading comprehension that need to be diagnosed and remediated individually, and proceed accordingly. A fourth teacher might interpret the doodling behaviour not as a problem primarily of understanding but as a problem with lack of engagement or motivation, and respond very differently to address that student's need. And so on.

The point is that even though the objective classroom situation is identical in all the cases just described, there are myriad ways teachers might *interpret* this situation, and for each interpretation, there are myriad ways teachers might *respond* to it. These manifest differences in teacher behaviour can be explained by differences in their underlying mental models, which determine how they make sense of the students' behaviour and decide what actions to take next. As Dewey (1929) observes '. . . the final reality of educational science is not found in books, nor in experimental laboratories, nor in the classrooms where it is taught, but in the minds of those engaged in directing educational activities' (p. 32).

We argue that the form educational science takes in the minds of educators can be productively conceived in terms of mental models—and more specifically, *explanatory* mental models. Moreover, we submit that such explanatory mental models are a natural point of convergence where teacher perception, teacher action, informal teacher expertise and formal scientific theory come together, which makes them a potentially powerful leverage point for driving improved educational outcomes in a scalable manner—if we can figure out how to close the application loop of educational science by bridging between explanatory models as they are represented in the textbooks, literature, laboratories, etc. of the special sciences and the useable explanatory mental models of educators. Closing the application loop requires closing at least three gaps:

- Educators must *internalize* explanatory scientific models as useable mental models.
- Educators must develop *conditional expertise* in selecting an appropriate mental model and adapting it to the particulars of a situation encountered in practice.
- Any new mental models must *displace* less effective ones that educators are currently using to inform their practice.

16.4 **Proposal for supporting a sustainable science of education**

Through the preceding analysis we have identified several tactical requirements that must be implemented to close the two loops of educational science and make explanatory models both useable by and progressively more useful to educators in practice. These requirements can be organized into two groups: process and infrastructure.

Process requirements

To support a sustainable science of education, educators need to be supported in:

- Identifying recurring educational problems requiring solutions, and, if necessary, communicating these to the people who can conduct relevant scientific investigations.
- Accessing and learning how to use available educational solutions embodying explanatory models.
- Testing the utility of proposed educational solutions and providing feedback in some systematic, cumulative form.

In order to bridge between the different basic viewpoints of educational practice and the special sciences (normative and descriptive, respectively), some group of people[2] needs to be supported in:

- Translating normative educational problems into descriptive scientific research questions.
- Integrating descriptive/explanatory scientific models into normative educational solutions that educators find accessible, learnable, useable and useful.

Scientists in the special sciences need to be supported in:

- Becoming aware of the set of research questions derived from educational problems so they can initiate scientific investigations based on those questions.
- Making relevant explanatory models accessible to the people who can integrate them into educational solutions.

---

[2] This role is analogous to the M.D.-Ph.D. in the medical domain, who is trained in both theory and practice and helps facilitate bi-directional transfer between them. It seems like an excellent role for graduates of interdisciplinary graduate programs in education emerging around the globe, such as the Master's programme in Mind, Brain and Education at the Harvard Graduate School of Education (Blake & Gardner, 2007). Note also that some teams of people conducting design experiments in education also seem to be performing this kind of role (Brown, 1992; Cobb, Confrey, diSessa, Lehrer & Schauble, 2003).

Infrastructure requirements

Given the highly interdisciplinary and distributed nature of the sources of educational science that need to be coordinated in the minds of educators at the point of application, a supportive infrastructure is absolutely critical to bridging many of the gaps and overcoming the obstacles described in previous sections (recall Figure 16.2, in particular). Many of these challenges arise because the different groups of contributors work in fundamentally distinct ways (on different time scales, with different goals, etc.) and make qualitatively different kinds of contributions to educational science. Conferences and other socially-based venues are useful for connecting members of the different groups together to promote dialogue and collaboration, but such events are based on a fundamentally 'synchronous' model of interaction where all parties must bring their contributions to the table simultaneously. The probability that just the right people will come together at just the right time around just the right problem and everyone will be willing and able to follow through immediately with a collaborative project is quite low if everything depends on this kind of synchronous collaboration model.

Progress would be greatly facilitated if different parties could make their contributions asynchronously, largely independently of one another—for example, if educators could generate a running 'wish list' of recurring problems they would like solved, the appropriate people could, at their convenience, translate these into a running list of scientific research questions that are linked to the original problems, scientists could access the list of research questions and investigate them as they have interest and resources and later contribute their explanatory models as they develop and validate them, the appropriate people could integrate those explanatory models into educational solutions, educators could access the relevant solutions when they encounter a specific problem in practice and come back later to annotate the solution with application examples and data on their experience of the solution's usability and effectiveness, etc.—all independently of each other.

The basic infrastructure that would be required to support asynchronous collaboration and communication between disparate contributors includes the following two components:

♦ A database or library of problem specifications and associated solutions (or solution fragments) asynchronously accessible to and independently updateable by all parties.

♦ A standard format or template for entries in the library.

In the next section, we describe *design patterns*, which have been used to achieve similar ends in other applied domains such as architecture and software engineering. We argue that design patterns could be used to bridge many of the gaps and overcome many of the obstacles to establishing a sustainable educational science as depicted in Figures 16.1–16.3.

16.5 Design patterns as a medium for coordinating the diverse sources of educational science and making them useable by practitioners

The system of relationships between educational science and educational practice laid out in Figure 16.2 has analogues in other practical domains, such as engineering, architecture and software design. In all these professions, there are explanatory laws from various scientific domains that inform better practice without over-specifying tactics for the practitioner in the form of 'recipes' or scripts. In some domains, such as chemical and electrical engineering, there is a fairly tight coupling between one or more scientific disciplines (chemistry and physics, respectively) and the practical applications that are typically developed. In other domains, the relationship between explanatory science and practical application is looser. In some of these latter domains, most notably architecture and software engineering, people have introduced the idea of 'design patterns' to facilitate the process of integrating the kinds of theoretical and practical elements shown in Figure 16.2 and making them useable by practitioners to support systematic perception and better decision-making. In what follows, we explain what design patterns are and illustrate through a detailed example how they might be used to support the development, acquisition, and application of educational science by teachers.

Overview of design patterns

Alexander and colleagues (Alexander, Ishikawa & Silverstein, 1977) are credited with originating the idea of design patterns in the domain of architecture. They describe the basic idea thus:

> Each pattern describes a problem which occurs over and over again in our environment, and then describes the core of the solution to that problem, in such a way that you can use this solution a million times over, without ever doing it the same way twice (p. x).

An architectural example of a recurring problem is the outdoor porch. Porches serve a variety of purposes—for example, some porches are small and meant to shade entryways from the rain and snow, while others are large covered areas where people can sit outside shaded from the sun, and still others connect the interior of the building to a specific exterior space such as a courtyard. In addition, every porch design is unique. In terms of design patterns, what porches have in common is that they provide a transitional space that is neither inside nor outside, and these transitional spaces are important both practically (for example, to shelter people from the elements while waiting at the door) and psychologically (for instance, the transition from inside to outside or vice versa is less jarring if it is mediated by a space that has elements of interior spaces—like a roof— and exterior spaces—like open walls). Viewed in this way, the *Porch* design pattern provides much more useful information to support an architect than would a series of examples alone, because it specifies the criteria of a good porch design without constraining the specific implementation details unnecessarily. The design pattern also formalizes and subjects to public scrutiny and nego- tiation a set of well-defined and revisable criteria for distinguishing between better and worse designs, which would otherwise only be implicitly defined in the heads of experts. Finally, by creating meaningful categories applicable to diverse exemplars to which simple names can be attached, design patterns support the development of a common vocabulary to facilitate commu- nication among members of the field.

   Design patterns have had an even more dramatic impact in computer science than in architecture where they originated, facilitated greatly by the publication of a now classic book compiling many common and useful software design patterns in one comprehensive reference, all organized around a standard pattern template (Gamma, Helm, Johnson & Vlissides, 1994). The demon- strated utility of design patterns in domains such as architecture and software engineering provides a proof-of-concept that they could also add significant value in education.

Design pattern specifications

Other people have started working to apply the idea of design patterns to education (Anthony, 1996; Bergin et al., n.d.; Mislevy et al., 2003). In general, design patterns are specified using a standardized template. Various standards have been proposed by different camps within and across domains. There is variation in these proposed standards, but there is also significant overlap. Standard elements include, for example, a short descriptive name, to facilitate learning of the patterns and efficient communication among practitioners; a description of the recurring problem that the pattern helps to address; a description of the general solution; information on when and how to apply the pattern; and examples of applications.

Our analysis has generated additional constraints and design goals for a design pattern framework, above and beyond the elements common to many existing frameworks. Specifically, we propose that design patterns in educational science should:

♦ Help close the adaptation loop of educational science by supporting coordination across the characteristic time scales, basic viewpoints and levels of abstraction that distinguish the world of educational practice from the world of scientific research—by allowing, for example, educators and scientists to make their contributions independently and asynchronously; and

♦ Facilitate the integration of explanatory models into educator practice by providing representations that educators find accessible, learnable, useable and useful.

The tactical role of design patterns in closing the two loops of educational science is illustrated in Figure 16.4.

Design pattern template

Combining the key common elements of other design pattern templates with the design considerations summarized at the end of the previous section leads us to propose the following design pattern template for educational science:

♦ Name: a short, descriptive name for the pattern.

♦ Intent: a succinct description of what applying the pattern is intended to accomplish.

♦ Motivation: a description of the educational problem or opportunity the pattern is meant to address, illustrated with a representative example scenario.

♦ Explanatory model(s): brief description of the scientific explanatory models that support good reasoning and decision making with respect to the problem or opportunity specified in the Motivation section, plus references to relevant scientific literature describing and substantiating the models.

♦ Applicability: conditions under which this pattern might be applicable.

♦ Validation: criteria for testing the value of the pattern in practice, plus cumulative data on its value in practice:

  • How to test the value of the pattern.

  • Informal feedback on its utility and/or effectiveness.

  • Formal research on educational outcomes resulting from application of the pattern.

♦ Research questions: a list of open research questions that follow from the educational problem statement or that have been generated in the course of applying it.

♦ Additional resources: examples, supporting materials, techniques, tactics, technologies, standards, guidelines, etc. on applying the pattern in general or specific cases.

♦ Related patterns: other design patterns that are complementary to or need to be differentiated from this one.

Example of a design pattern in educational science

In this section, we illustrate what a design pattern in educational science looks like using one of the three foundational principles of Universal Design for Learning (UDL)—the principle of representing information using multiple formats and media to make it accessible to all learners (Rose & Meyer, 2002). We selected this principle because UDL is a research-based framework grounded

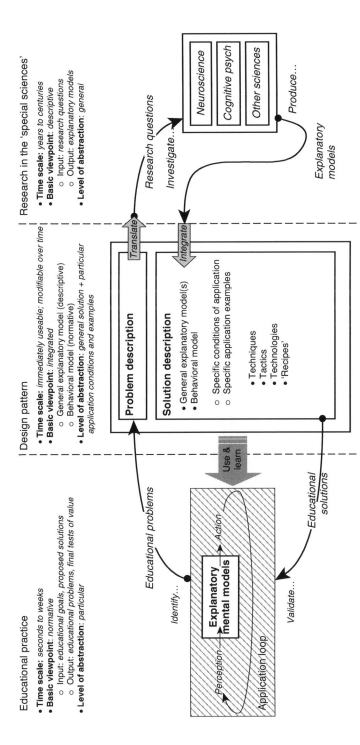

Fig. 16.4 Closing the 'adaptation loop' and 'application loop' of educational science with design patterns. Design patterns describe recurring problems plus the core of a general solution that can be applied repeatedly across a wide range of contexts and situations. They provide a persistent medium that can bridge across the time scales, basic viewpoints and levels of abstraction that distinguish educational practice from research in the special sciences. They also provide a standard format for representing and accessing immediately useable solutions to recurring educational problems that facilitates educator learning of the embedded explanatory models over time. Reproduced with permission of Michael W. Connell. © 2010. All rights reserved.

in multiple scientific sources—including neuroscience—that supports the development of practical tools for use by educators. We call this design pattern 'Perceptual Accessibility' for short.

Name: Perceptual Accessibility

Intent: separate the storage medium of educational content from its delivery mode (including perceptual modality, communication medium and content format) so that the content and its delivery mode can be changed independently of each other.

Motivation: historically, much educational content has been bound inflexibly to a particular delivery mode (including perceptual modality, medium and format) at the time it is first generated. For example, the content of a classic printed textbook is inextricably bound to the visual modality, in the medium of fixed text supplemented with static images, and in a particular format (such as 12-point Times New Roman font).

Binding the storage medium in this way to the specific delivery mode selected by the producer limits access by some groups of learners (those with specific sensory impairments, for example) and to all learners under some learning conditions (where lighting is poor, movement is constricted, etc.). This binding of storage medium to delivery mode makes it difficult to accommodate the full range of learner needs, preferences, and study environments and limits the ability of educators to re-use content and presentation components independently of each other.

Educational content designers should be able to create and store educational content without committing to a concrete set of fixed decisions about delivery modality, media and format. Only the delivery configuration in a particular learning *instance* should depend on these specific decisions. Therefore, educational content specifications should define content without mentioning particular delivery characteristics.

The Perceptual Accessibility pattern addresses these concerns by separating the specification of the content storage medium from the specification of delivery parameters. A textbook might be *stored* on a computer server in one format (as digitized text, for example), and *delivered* (or *accessed*) in a variety of perceptual *modalities*, including visual (for example, printed text), auditory (for example, using text-to-speech), or tactile (translated into Braille, for instance). Within a single modality, such as visual, the stored content might be rendered in a variety of *media*, such as animations, static images, or text. Finally, within any medium, the *format* of the content can be varied (rate of speech slowed, size of text increased, contrast of images enhanced, colour palette customized, etc.).

Explanatory model(s): learning depends upon the coordination of parallel streams or 'channels' of perceptual information coming in from the senses, such as sight and hearing. Each channel comprises a sequence of processing stages. Cognitive psychologists, neuroscientists and researchers in other domains have characterized some of the processes and systems involved in the perceptual processing and maintenance of incoming information that is necessary for effective learning, particularly in the visual and auditory channels (and to a lesser extent the tactile systems involved in reading Braille, for example). Key systems include:

- Physical sensory systems (eyes, ears etc.).
- Perceptual buffers (large-capacity, very short-duration memory systems not accessible to conscious control).
- Attention.
- Working memory (limited-capacity, short-duration memory systems that can be consciously manipulated and maintained).
- Executive function (necessary, for example, for allocating limited attention efficiently and effectively and for actively managing working memory).
- Long-term memory (large capacity, durable memory systems, of which there are several subtypes).

Although gross sensory impairments such as profound blindness or deafness are generally self-evident, problems with any other processing subsystem in this list can be much less obvious, but can nonetheless inhibit learning within a specific modality or modalities. By enabling access to content through a variety of modalities, media, and formats, the Perceptual Accessibility pattern enables educators and learners to switch the mode of instruction flexibly between different perceptual modalities or 'channels' to support cognitively diverse learners, and can also support informal diagnosis on the educator's part so she can take appropriate next actions (such as making referrals to appropriate specialists).

Applicability: use the Perceptual Accessibility pattern when. . .

- The learning infrastructure supports it—for example, when a digital delivery medium is reliably available in all instances where the content will be delivered.

- There are individual students who have profound limitations in one or more perceptual modalities, or in any modality-specific supporting systems including attention, working memory, executive function or long-term memory; these might be identifiable by formal assessments and clinical diagnoses, and in some cases may be initially identified through informal observations by educators.

- The learner population is or may be diverse with respect to perceptual strengths, limitations and/or preferences.

- The same content will be delivered across a variety of delivery platforms with different learning affordances and limitations, such as PCs, e-books, smartphones, iPods and on-demand hard-copy printouts.

- Students appear disengaged or frustrated, or are having trouble comprehending the educational content, and there is reason to believe that the delivery mode, medium or format is contributing to the problem (this is an especially good time to give the learners some control over how the content is delivered so they can match the delivery to their preferences).

Validation:

- See Rose and Meyer (2002) for a review of some of the formal research demonstrating that applying the Perceptual Accessibility pattern can produce significant educational benefits compared to control conditions where modality, medium and/or format of the educational materials are inflexible.

Research questions:

- *Computer science*: are there more effective and universal formats we could use to store the abstract educational content than digital text, images, etc.? For example, instead of converting an image to text via a pre-stored caption for auditory delivery, would it be possible to provide an abstract conceptual description in place of a string of text or a particular image, and then based on the delivery parameters set for a given learning session, to search the web at the time of delivery for the best possible instance of the specified content in the desired delivery mode (image, explanation, model, etc.) that is available at that time?

- *Cognitive psychology, neuroscience, cognitive science*: are certain kinds of content more easily or effectively learned via certain cognitive/neural pathways? What kinds of delivery configurations most effectively support certain learner profiles?

- *Computational neuroscience*: given that the different perceptual modalities have very different 'bandwidth' or information capacities, are there general models to understand tradeoffs when switching between them and strategies for doing so most effectively?

Additional resources: see CAST's web site at http://www.cast.org for a list of relevant research and resources.

Related patterns: if we had a database of design patterns in educational science, we would refer in this section, for example, to more specific patterns in the database related to each processing component mentioned earlier (attention, working memory etc.). Such references would provide more specific diagnostic criteria educators could use to isolate specific problems and/or refer learners to appropriate specialists for follow-up diagnosis and support.

16.6 Discussion

How design patterns close the two loops of educational science

Design patterns help close the *adaptation loop* of educational science by bridging gaps in time scales, basic viewpoints and levels of abstraction that distinguish educational practice from research in the special sciences.

- ◆ Time scales: design patterns are immediately useable, supported by guidelines on when to apply them and what to look for to assess effectiveness, application examples etc. Specific sub-components of design patterns provide connections for scientific researchers in other domains to draw from (e.g. research questions) in their research and to update with relevant information as it becomes available (e.g. explanatory models).

- ◆ Basic viewpoints: the translation from educational problem (described in the 'intent' and 'motivation' sections of the pattern) to scientific research question (listed in the 'research ques- tions' section of the pattern) is made explicit to facilitate the transition from the normative basic viewpoint of education to the descriptive basic viewpoint of scientific research. Going the other way, the explanatory models can be contributed by research scientists and the integration into educational solutions is at least in part accomplished by supplementing the descriptive explana- tory models with information on when to apply the model, examples of application, suggestions and data on evaluating the utility of the model, etc.

- ◆ Levels of abstraction: research questions and explanatory models are general and relatively context-free, while the other components of the pattern provide more particular context (when to apply, particular examples of application, etc.), making it easier for educators to know when and how to apply, and how to know if the pattern is adding value in practice.

Design patterns help close the *application loop* of educational science by making scientific explanatory models accessible, learnable, useable and useful to educators.

A design pattern can be thought of as an externalized mental model, or a 'tool to think with'. The intent is that through the process of applying a design pattern, educators will over time internalize the explanatory models at the heart of the pattern and be able to apply them flexibly. The tech- niques, tactics, examples and other supporting materials provide initial scaffolds (Fischer & Bidell, 1998; Vygotsky, 1978) that can be dispensed with as the explanatory models become inte- grated into educators' repertoires and conditionalized on appropriate application conditions.

16.7 Conclusion

We began by describing a vision of educational science as a system organized around scientifically rigorous *explanatory models* that are *applied* by educators in their moment-to-moment practice and progressively *adapted* (that is, broadened in scope and refined in terms of usefulness) through a systematic process of scientific inquiry involving a collaboration between educators and scien- tists in other domains, such as neuroscience. We argued that supporting such a system of educa- tional science could catalyse a virtuous cycle of progress in educational practice grounded in a cumulative store of transparent, high-quality domain knowledge of the sort available to experts in other complex applied domains, such as medicine and engineering.

Following Dewey (1929), our analysis distinguishes *educational science* proper from its *sources*, which include both educational practice and the special sciences such as neuroscience and cognitive psychology. In addition, we argue that educational science is located in the brains and minds of educators, whereas its sources can be found in various forms, represented in various media, and distributed across a wide range of domains. This key distinction between educational science and its sources helps clarify a number of important but otherwise potentially confusing issues. For example:

♦ What is educational science? Educational science is practical educational expertise grounded in explanatory mental models derived from systematic scientific research, plus systematic methods of inquiry used by educators to identify educational problems worth solving and to test the practical value of proposed solutions.

♦ What is the role of neuroscience in education? In our analysis, neuroscience is one of many scientific sources of educational science which can make quite specific contributions to it, in the form of explanatory models that can be integrated into solutions to recurring educational problems. Neither neuroscience nor any other scientific source of educational science should be confused with educational science proper, which exists in the brains and minds of educators.

♦ What is the role of 'recipes' in educational science? Detailed educational scripts or 'recipes' for action that are meant to be followed by educators without judgement or reflection—even if generated through scientific means—do not constitute educational science. They can, however, serve as useful examples (as in the Perceptual Accessibility design pattern specification) that help educators know how to apply explanatory models in particular cases, thereby helping them to internalize and become fluent with applying the explanatory models in more general and nuanced ways with experience over time.

♦ Why does Dewey (1929) state that science cannot provide 'stamps of approval' for particular educational practices? As illustrated in Figure 16.2, educational practice and scientific research embody different basic viewpoints—normative and descriptive, respectively. Stated another way, scientific research can help us understand how the world is and why it is that way, but we cannot discover through research how the world should be—including what educational ends to pursue, or even in the final analysis how we should behave—that is, what means we should pursue to achieve specific educational objectives. We do of course recognize that scientific research can answer questions such as whether flash cards or constructivist activities are more efficient at teaching children their maths facts according to some strict operational criterion. Our point is that this kind of scientific evidence is always an insufficient basis upon which to choose an educational intervention, because—just as an example—we are also choosing to subject the student to a certain kind of experience. And that choice necessarily has moral implications—whether we are talking about subjecting them to flash cards or pharmaceuticals.

Note that even as we package up some of the explanatory models into technologies such as automated assessments or computer tutoring systems, we do not decrease the need for or the status of teachers, any more than we decrease the need for or status of engineers or architects as we understand more about physics. On the contrary, the net effect of such technological advances in other scientific domains tends to be to expand the available toolkit, the leverage of the individual practitioner, and the range of goals, challenges and opportunities that are within their professional reach (see Chapter 19, this volume). Given the inherent challenges of education compared even to other very complex domains such as medicine and engineering, there is every

reason to believe that moving toward a bona fide science of education would have an expansive—
not a diminishing—effect on the professional status and capabilities of educators.

References

Alexander, C., Ishikawa, S., & Silverstein, M. (1977). *A Pattern Language: Towns, Buildings, Construction.*
New York: Oxford University Press.

Anderson, J. R. (1983). *The architecture of cognition.* Cambridge, MA: Harvard University Press.

Ansari, D., & Coch, D. (2006). Bridges over troubled waters: education and cognitive neuroscience. *Trends
in Cognitive Sciences,* 10(4), 146–51.

Anthony, D. L. G. (1996). Patterns for classroom education. In J. M. Vlissides, J. O. Coplien, & N. L. Kerth
(Eds.), *Pattern languages of program design 2* (pp. 391–406). Reading, MA: Addison-Wesley.

Atkinson, R. C., & Shiffrin, R. M. (1968). Human memory: A proposed system and its control processes.
In K. W. Spence & J. T. Spence (Eds.), *The psychology of learning and motivation, Vol. 2* (pp. 89–115).
New York: Academic Press.

Baddeley, A. D. (1976). *The psychology of memory.* New York: Basic Books.

Bergin, J., Eckstein, J., Manns, M., Sharp, H., Voelter, M., Wallingford, E., et al. (n.d.). *The pedagogical
patterns project.* Available at: http://www.pedagogicalpatterns.org/.

Blake, P., & Gardner, H. (2007) A first course in Mind, Brain, and Education. *Mind, Brain, and Education,* 1,
61–65.

Brown, A. L. (1992). Design experiments: Theoretical and methodological challenges in creating complex
interventions in classroom settings. *Journal of the Learning Sciences,* 2, 141–78.

Bruer, J. T. (1997). Education and the brain: A bridge too far. *Educational Researcher,* 26(8), 4–16.

Bruer, J. T. (2002). Avoiding the pediatrician's error: How neuroscientists can help educators (and
themselves). *Nature Neuroscience Supplement,* 5(November 2002), 1031–33.

Byrnes, J.P., & Fox, N.A. (1998). The education relevance of research in cognitive neuroscience. *Educational
Psychology Review,* 10(3), 297–342.

Cobb, P., Confrey, J., diSessa, A., Lehrer, R., & Schauble, L. (2003). *Educational Researcher,* 32(1), 9–13.

Connell, M. W. (2005). Foundations of Educational Neuroscience: Integrating Theory, Experiment, and
Design. Unpublished doctoral thesis. Harvard University Graduate School of Education: Cambridge,
MA. Available online through ProQuest at http://gradworks.umi.com/32/07/3207712.html.

Craik, K. (1943). *The nature of explanation.* Cambridge: Cambridge University Press.

Dewey, J. (1929). *The sources of a science of education.* New York: Horace Liveright.

Eichenbaum, H. (1997). Declarative memory: Insights from cognitive neurobiology. *Annual Review of
Psychology,* 48, 547–72.

Eichenbaum, H., Otto, T., & Cohen, N. J. (1994). Two functional components of the hippocampal memory
system. *Behavioral and Brain Sciences,* 17, 449–517.

Fischer, K. W., & Bidell, T. R. (1998). Dynamic development of psychological structures in action and
thought. In R. M. Lerner (Ed.) & W. Damon (Series Ed.), *Handbook of child psychology: Vol. 1:
Theoretical models of human development* (5th ed.), 467–561. New York: Wiley.

Frost, R., & Katz, L., (Eds.) (1992). *Orthography, phonology, morphology, and meaning.* Amsterdam:
North-Holland.

Gabrieli, J. D. E. (2009). Dyslexia: A new synergy between education and cognitive neuroscience. *Science,*
325(5938), 280–83.

Gamma, E., Helm, R., Johnson, R., & Vlissides, J. M. (1994). *Design patterns: Elements of reusable object-
oriented software.* Reading, MA: Addison-Wesley.

Geake, J., & Cooper, P. (2003). Cognitive neuroscience: Implications for education? *Westminster Studies in
Education,* 26(1), 7–20.

Goswami, U. (2004). Neuroscience and education. *British Journal of Educational Psychology,* 74, 1–14.

Grigorenko, E. L. (2001). Developmental dyslexia: An update on genes, brains, and environments. *Journal of Child Psychology and Psychiatry and Allied Disciplines,*42(1), 91–125.

Hall, J. (2005). Neuroscience and education. *Education Journal,* 84, 2729.

Johnson-Laird, P. N. (1983). *Mental models: Towards a cognitive science of language, inference, and consciousness.* Cambridge, MA: Harvard University Press. Cambridge: Cambridge University Press.

Katzir, T., & Pare-Blagoev, J. (2006). Applying cognitive neuroscience research to education: The case of literacy. *Educational Psychologist,* 41(1), 5374.

Mayer, R.E. (1998). Does the brain have a place in educational psychology? *Educational Psychology Review,* 10(4), 389–96.

Miller, G. A. (1956). The magical number seven plus or minus two: Some limits on your capacity for processing information. *Psychological Review,* 63, 81–96.

Mislevy, R., Hamel, L., Fried, R., G., Gaffney, T., Haertel, G., Hafter, A., . . .Wenk, A. (2003). *Design patterns for assessing science inquiry* (PADI Technical Report 1). Menlo Park, CA: SRI International. Available from Principled Assessment Designs for Inquiry (PADI) website: http://padi.sri.com/downloads/TR1_Design_Patterns.pdf.

Paloyelis, Y., Rijsdijk, F., Wood, A., Asherson, P., & Kuntsi, J. (2010). The genetic association between ADHD symptoms and reading difficulties: The role of inattentiveness and IQ. *Journal of Abnormal Child Psychology,* 38(8), 1083–95.

Pashler, H. (1997). *The psychology of attention.* Cambridge, MA: MIT Press.

Posner, M. I., & Petersen, S. E. (1990). The attention system of the human brain. *Annual Review of Neuroscience,* 13, 25–42.

Rose, D. H., & Meyer, A. (2002). *Teaching Every Student in the Digital Age: Universal Design for Learning.* Alexandria, VA: ASCD.

Schunk, D.H. (1998). An educational psychologist's perspective on cognitive neuroscience. *Educational Psychology Review,* 10(4), 411–417.

Seidenberg, M. S. (1995). Visual word recognition: An overview. In J. L. Miller & P. D. Eimas (Eds.), *Speech, language, and communication.* San Diego, CA: Academic Press.

Seidenberg, M. S., & McClelland, J. L. (1989). A distributed developmental model of visual word recognition and naming. *Psychological Review,* 96, 523–68.

Stanovich, K. E. (1998). Cognitive neuroscience and educational psychology: What season is it? *Educational Psychology Review,* 10(4), 419–26.

Stein, Z., Connell, M., & Gardner, H. (2008). Exercising quality control in interdisciplinary education: Toward an epistemologically responsible approach. *Journal of Philosophy of Education,* 42(3–4), 401–14.

Vygotsky, L. S. (1978). *Mind in Society: The development of higher psychological processes.* Cambridge, MA: Harvard University Press.

Willingham, D. (2008). When and how neuroscience applies to education, *Phi Delta Kappan,* 89(06), 421–23.

Chapter 17

Assuring successful lifelong learning: can neuroscience provide the key?

Christiane Spiel, Barbara Schober, Petra Wagner and Monika Finsterwald

Overview

Using the term 'neuro' in scientific contexts becomes more and more popular. Many politicians and practitioners, especially in the field of education, seem to believe that scientific work gets a specific aura if it is neuroscientifically based. The present chapter questions whether, and if so, *how* neuroscience can contribute to explaining or promoting complex learning processes. To illustrate complex learning processes and the requirements associated with their enhancement, we focus on lifelong learning (LLL). The needs of the 'knowledge society' have placed LLL at the centre of an intensive ongoing political debate. Essential constituents of LLL are a persistent motivation to learn and the skills to realize this motivation. Researchers in the field of education agree that schools provide the basis for LLL and therefore teachers play a decisive role for imparting relevant competences. Consequently, LLL trainings should also target teachers. As a concrete example a training programme (TALK) for teachers to enhance LLL is presented. Based on this example, we discuss how bridges can be built between neuroscience and education to cooperate in explaining and promoting complex learning processes.

17.1 What is necessary for successful lifelong learning?

Lifelong learning (LLL) is a central sociopolitical concern and has been an explicit focus of European educational policies since 2000. Accordingly, the promotion of LLL has been designated a Europe-wide strategy (Commission of the European Communities, 2000). While most of the programmes in this field have concerned topics of further education, there is now more and more demand to also focus on school, as this is the place, where cornerstones for LLL are laid: 'It is where they gain the basic knowledge, skills and competences that they need throughout their lives, and the place where many of their fundamental attitudes and values develop' (European Commission, 2010; see also Prenzel, 1994). But what is it in detail that should be promoted for successful LLL and how can schools contribute?

Existing definitions of LLL stem primarily from task force groups charged with the development of educational policies for the European Union and can be summarized as follows: LLL refers to 'all learning activities undertaken throughout life, with the aim of improving knowledge, skills, and competence within a personal, civic, social and/or employment-related perspective' (Commission of the European Communities, 2000, p. 9). Additionally one also finds slightly differing enumerations of key competences (like languages, information technology-competences, mathematical skills, skills for self-organization, interpersonal skills, cultural awareness, etc., see, e.g. European Commission, 2010). From a psychological perspective, LLL cannot be considered a new construct, but rather a concept which is relevant in the context of psychological research and intervention—in particular concerning educational psychology, developmental psychology and organizational psychology. Currently, there is no theory that explicitly concerns basic psychological parameters determining successful LLL for individuals. But what can consistently be found in the wide-ranging and diverse body of literature on the topic of LLL (e.g. Artelt, Baumert, Julius-McElvany & Peschar, 2003; Weinstein & Hume, 1998) are two central components that seem to be pivotal for LLL, independent from age or learning contexts: (1) the persisting *motivation* for and an appreciation of learning and education (= will to learn); and (2) those capacities which enable individuals to successfully translate the motivation to learn into concrete activities which enhance existing knowledge and talent levels (= skill to learn). Accordingly, individuals will be capable of successfully meeting the summons to LLL when they consider learning and the acquisition of knowledge to be attractive and valuable, and when they can draw on the abilities needed for *self-regulated learning* as well as the related capacity to utilize effective knowledge management.

If school (as mentioned earlier) is considered to be responsible for providing the basis for successful LLL in the sense of positive learning motivation and competence to self-regulate learning, the question is, how well does school presently succeed? A large number of studies examined a wide range of motivational beliefs as well as self-regulated learning behaviour among school-aged boys and girls (see overview by, e.g. Schunk, Pintrich & Meece, 2008). Although results differ in detail, one finds strong evidence that the motivation of boys and girls decreases the longer they stay in school, especially after the transition to secondary school (Fischer & Rustemeyer, 2007; Schunk et al., 2008; Wigfield, Eccles, Schiefele, Roeser & Davis-Kean, 2006). In some studies also a decline in pupils' self-regulated learning behaviour has been reported (e.g. Peetsma, Hascher, Van der Veen & Roede, 2005). As teachers are the key players in everyday school life (e.g. Schober et al., 2007), who have the opportunity to influence the development of their pupils' LLL-competences, a clear need appears to foster and impart teachers' competences in this field. But what is it they should be trained in besides their basic education in didactics? It is increasingly discussed that findings from neuropsychological research on learning offer a promising solution. So what could neuroscience provide to make teachers more successful in teaching learning motivation and self-regulation as decisive parameters of LLL?

17.2 How neurosciences contribute to learning and education

In the last decade, tremendous progress in the field of neuroscience in general and numerous publications on the connections of neuroscience to education in specific have raised the hope for '*evidence-based*' *education* (Ansari & Coch, 2006; Goswami, 2004; Stern, 2005). However, expectations of policymakers, teachers and the public are often unrealistic and oversimplified, and the claims for 'brain-based learning' allow for '*neuromyths*' to flourish (Ansari & Coch, 2006; Geake, 2008; Goswami, 2004, 2006). The report of the Organisation for Economic Cooperation and Development (OECD, 2002) pays most attention to three myths: the belief in 'left brain' versus 'right brain' learning, the notion that there are 'critical periods' for learning certain matters, and the idea that the most effective education interventions need to be timed with periods of synaptogenesis (see also Goswami, 2004). As a consequence, specific methods as e.g. Brain Gym® are widely and unreflectively used in education (Geake, 2008).

The high and often unrealistic expectations about what neuroscience can contribute to education and the dominance of neuromyths are related to two risks: (1) that the important strides being made by neuroscience in many areas relevant to education are obscured (Goswami, 2006 & see Chapter 4, this volume); (2) that the importance of research on learning and instruction and the knowledge this research has provided will be ignored (Stern, 2005).

The tools of neuroscience offer various possibilities to education, in particular concerning the early diagnosis of special educational needs as well as the analyses of the effects of different kinds of educational input on learning in specific academic domains (Goswami, 2004). Therefore, neuroscientific investigations particularly contribute to the diagnosis and explanation of developmental cognitive and learning deficits (Schumacher, 2007). Despite many neuroscientific studies focusing on specific domains, e.g. language, reading and mathematics (for summaries, see e.g. Goswami, 2004, 2006; Schumacher, 2007), there are also studies that provide findings useful for all teachers (Ansari & Coch, 2006): Links between emotion and cognition (Ansari & Coch, 2006; Goswami, 2004) are of specific importance for learning. Findings indicate that different aspects of memory are activated in different emotional contexts (Erk et al., 2003).

Applications of these findings in school settings have been proven to be difficult and unsatisfying (e.g. Bruer, 1997). One simple reason is that large-scale studies in natural settings are not possible so far because of technical limitations as, for example, reading a book in school or at home is hardly the same as reading words in a laboratory study of reaction time (Fischer et al., 2007). Even more important is the fact that neuroscientific investigations in fact provide information about *when* treatments are needed but do not provide specific information *what* exactly has to be done (Schumacher, 2007). Here, the framework of psychological and pedagogical theories is needed as individuals' learning is a complex process in which, for example, motivational and social aspects are decisive, too. Another reason might be that neuroscience primarily focuses on learning and not on successful teaching, which is the natural counterpart of successful learning (Goswami, 2004). However, for fostering LLL competences in school successfully teachers are the key persons (Schober et al., 2007).

Altogether, little attention has been paid to developing an overarching framework bridging neuroscience to education (Ansari & Coch, 2006 & see Chapter 3, this volume). Neuroscience alone cannot provide the specific knowledge required to design powerful learning environments (Stern, 2005). Up to now the studies mostly focused on single aspects of a learning process. But multiple methodologies and levels of analysis in multiple contexts combining the knowledge from different disciplines are needed (Ansari & Coch, 2006). While research on learning and instruction has provided precise and applicable knowledge on how to design powerful learning environments in many content areas, neuroscience can help to explain why some learning environments work while others fail (Stern, 2005).

When addressing the design of powerful learning environments in school and other natural learning settings it has to be taken into account that in this context the human brain is only part of a larger system (Schumacher, 2007; Stern, 2004). According to Schumacher (2007) there is a distinction between *biologically privileged learning* (speech, motor skills etc.) and *cultural learning* (all contents and abilities taught at school such as reading, writing and maths). Preconditions for cultural learning are primarily knowledge preconditions (Schumacher, 2007). To acquire, for example, new knowledge on the physical concept of density, knowledge on other physical concepts like weight and volume already must have been acquired and the respective knowledge base must be well-organized. A further precondition is that the related knowledge base is well-organized: Experts and novices differ especially in the organization of their knowledge. Furthermore, experts have more efficient strategies for integrating new information into their already existing knowledge base.

Good teachers have an idea about their students' previous knowledge in the academic content they teach and consider it in their teaching approaches. They also have ideas on the misconceptions underlying certain failures of students and react on them with specific tasks (Stern, 2004). Here, findings from research on developmental psychology and on self-regulated learning, one of the key components of LLL are very helpful. There is generally accepted knowledge in the research field of learning and instruction on how a 'good teacher' should act and how learning environments have to be designed to promote insightful learning (Baumert et al., 2004).

To sum up, even if present neuroscience research can provide new insights into learning, the question about how to design learning environments to enhance determinants of learning, especially LLL in the sense of motivation and self-regulation remains rather open from this perspective. Here the knowledge of pedagogy and psychology is needed. In these disciplines well-grounded knowledge is available on the parameters of a framework for insightful learning and the promotion of LLL competences, but there is a lack of research on the implementation of this knowledge into school (Stern, Grabner & Schumacher, 2005). Our training programme TALK[1] is an example of how it can be done. In the following sections we shortly describe the programme with the main focus on the didactic principles used to create an optimal learning environment for the participating teachers.

17.3 The TALK-programme as an example for systematic transfer of scientific knowledge into school

TALK is oriented towards concrete needs of teachers and the specific framework of school (for more details of the programme see Schober et al., 2007). It is systematically founded in current educational research and evaluated in a sophisticated field study. Therewith TALK is geared to basic principles of 'evidence-based' research, which becomes more and more important in the field of interventions in educational areas. The programme is characterized by a consequent integration of relevant research approaches and a focus on ensuring sustainability and implementation in teachers' daily work life. Furthermore, TALK does not only focus on the individual teacher but also on school development.

Target variables and theory

TALK pursues the central target to systematically promote teacher competences to arrange their teaching affairs and their schools according to the main goal of fostering their pupils' LLL-competences. The main *target variables of the intervention* result from the pertinent literature on common *core factors for LLL*. As already mentioned at the beginning of this chapter, two core determinants for successful LLL are essential: (1) enduring motivation for learning and (2) skills for self-regulated learning. In addition, social and cognitive skills are important accompanying factors for fostering motivation and self-regulated learning (for more details see Lüftenegger et al., 2010; Schober et al., 2007). TALK focuses these four core factors.

As the *basic theoretical framework for teacher competences*, a well-known model was used that was conceptualized in the so-called COACTIV study (Cognitive activation in the classroom; see Kunter et al., 2007). In order to assess and develop teacher competences: (a) teacher knowledge,

---

[1] Training programme to foster teacher competences to encourage lifelong learning—translation of the German title: Trainingsprogramm zum Aufbau von LehrerInnenKompetenzen zur Förderung von Lebenslangem Lernen. TALK was financially supported by a grant from the Austrian Federal Ministry for Education, Arts, and Culture.

(b) teacher beliefs and (c) their psychological functioning, in particular their motivation, are addressed (Baumert et al., 2006; Kunter et al., 2007). According to the model, and in line with the target variables of TALK, teachers who are characterized as being competent in fostering LLL should: (a) possess knowledge about enhancing LLL (its core and accompanying factors, how to design appropriate instructional processes, etc.), (b) be convinced that LLL is a matter for schools and (c) be motivated and feel themselves capable of fostering their pupils' LLL-competences.

Furthermore, and in terms of ensuring the effects of the training programme, TALK also focuses on increasing cooperation among teachers. To ensure that challenging changes will be mastered successfully in schools, providing an opportunity for cooperation among teachers is important: if problems surface during the implementation of new skills, teachers can support one another in finding solutions (Meirink, Meijer & Verloop, 2007). In addition, discussions with colleagues are perceived as relevant for teachers and valuable for the improvement of their own teaching (Dunn & Shriner, 1999; Gräsel & Parchmann, 2004; Meirink, Meijer & Verloop, 2007). Not surprisingly, positive relationships between cooperation and teachers' motivation were observed (Chambers Cantrell & Hughes, 2008; Henson, 2001). Consequently, a fruitful and effective LLL training programme considers that a group of teachers cooperates over a long period of time (see also Gräsel & Parchmann, 2004; Lipowsky, 2006). Thus, TALK requires that each school does not only send one single teacher to attend the programme, but rather a critical mass of teachers from each school is involved.

TALK pursues *goals on different levels*. On the proximal level, modifications in teacher competences are the primary goals. On the distal level, TALK aims to trigger changes among the pupils. This means modifications in teacher competences should lead to perceivable changes in their teaching behaviour/lessons and this should also lead to corresponding changes in their pupils' LLL competences. Figure 17.1 summarizes this 'working model'.

Implementation of TALK

Curriculum

TALK is conceptualized for teachers instructing in secondary schools. It was conducted within the framework of a three-term course of studies in a university setting and embraced a total of 130 hours which were blocked into one- or two-day workshops. These took place every 3 weeks in the first term, and every four to five weeks during the second and third terms. In accordance with recommendations by Mandl, Prenzel and Gräsel (1991) in the context of a 'cognitive apprenticeship', the support offered to the teachers was increasingly reduced ('fading').

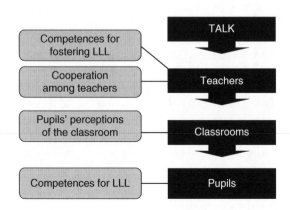

Fig. 17.1 The working model of the TALK programme.

TALK consists of an 'intensive phase' (terms 1 and 2), which aimed to optimize classroom instruction by developing teachers competences for fostering LLL, and a 'supervision phase' (term 3) where the focus is on ensuring the sustainability of effects on teacher competences. The topics of the workshops during the first two terms covered the areas of motivation, self-regulated learning, social and cognitive skills (for more details see Table 17.1). The topics of most sessions were determined in advance, however for two of these sessions the participants had the opportunity to discuss previously addressed material in elective seminars (in the sense of setting individual focuses). In the third term, the teachers initiated projects at their schools which were directly tied into

Table 17.1 Overview of the topics addressed in the TALK sessions

Session	Topics
1st term (Intensive phase)	
1	Introduction (overview) and definition of lifelong learning
2	Learning and motivation (interests, goals)
3	Motivation (self-esteem, confidence in own abilities)
4	Motivation (reference norms, attributions, IPT, feedback)
5	Social skills (cooperative learning)
6	Social skills (conflict management, making decisions)
7	Elective 1
8	Self-regulated learning 1 (learning skills: learning strategies, meta-cognition)
9	Self-regulated learning 2 (planning a SRL learning project)
10	Self-regulated learning 3 (presenting the SRL learning projects) and intermediate reflection on the training
2nd term (Intensive phase)	
11	Cognitive skills (critical and creative thinking/problem solving)
12	Social skills (diversity)
13	Elective 2
14	Planning a minor project for classroom implementation
15	Schools as organizations and a review of the training
16	Presenting the results of the minor classroom projects, Development of school projects for the 3rd term and intermediate reflection on the training
3rd Semester (Supervision phase)	
17	Implementation of a school project: project planning
18	Project execution and direction; coaching
19	Project execution and direction; coaching
20	Project execution and direction; coaching
21	Final meeting (presentation of the results of the school projects, reflections on the projects and the training)

*Note: on meetings scheduled for electives, the participating teachers were able to work in discussion groups organized around topics of general or specific interest. IPT = implicit theory of intelligence; SRL = self-regulated learning.

TALK, and passed the content on to colleagues who did not attend the programme. In doing so the TALK teachers could and should choose special focuses to address points relevant to their individual schools. They were supervised across all phases of project development, from seeking out a topic, to planning and execution, up through the evaluation of their projects. At the end of the programme, there was a final meeting where the individual school initiatives were presented and the entire group reflected on and discussed how the individual projects were executed.

Finally and by request of the teachers a 'refreshing unit' was realized half a year after the end of the supervision phase. The goal of this refreshing unit was to discuss the experiences with realizing TALK-topics in daily life and to optimize and deepen the knowledge again. Within this unit concrete examples for lessons according to TALK-principles have also been developed and written down with a view to the multiplication of knowledge in school.

Didactic principles in designing the TALK workshops

In TALK, four didactic principles were systematically applied: (1) the encouragement of various types of knowledge; (2) reliance on instructional psychology in structuring the TALK sessions; (3) the utilization of transfer technologies; and (4) the explicit and implicit impartment of learning contents.

(1) By working through learning material which takes into consideration the existence of *various types of knowledge*, a concrete acquisition of declarative, procedural, contextual and meta-cognitive knowledge was actively supported (see Steiner, 2006): conceptualizations and contents were not merely theoretically imparted. Teachers were stimulated to reflect on their own learning processes and to come up with concrete applications for the material being covered. Furthermore, the relevance of the learning material for a concrete implementation in their own classrooms was emphasized. The main concern was to support the teachers in preparing their pupils for a successful realization of LLL, and to consolidate the belief among these teachers that LLL is a topic relevant for scholastic education.

(2) To ensure a successful increase in knowledge and skills, care was taken that the TALK sessions were based on solid *principles* developed in the field of *instructional psychology*. An integration of approaches developed by Klauer (1985) and Gagné (1985) resulted in seven teaching functions, which served as the basis for the general structure of every TALK session (see Table 17.2).

(3) In order to secure a transfer of the learning material to the classroom, conventional *transfer technologies* were utilized (Hager & Hasselhorn, 2000). Thus, in accordance with the 'anchored instruction' approach (see, e.g. Bransford et al., 1990) the work done addressed authentic learning situations to the highest degree possible (e.g. by working on current problems being dealt with in the classroom/school). Furthermore, during the TALK sessions, the fundamental principles of the specific topics were developed and opportunities to implement them were discussed (for instance, encouraging interest), which should be tested in the subsequent school lessons. Following each TALK session, the teachers were to set individual goals (e.g. observation of own feedback behaviour; maintaining learning diaries for seminar projects). In the following TALK session, time was reserved for the participants to exchange their experiences in pursuing these objectives. With the aim of encouraging motivation, emphasis was placed on the successes the teachers achieved in their classroom according to the training contents; for 'failures' collective searches for suggestions to optimize these efforts were conducted. Additionally, reflections on what elements of TALK one could pass on to colleagues should inspire close cooperation within the school teams. This was indispensable for the transfer of the training into the schools.

Table 17.2 Overview of the general structure of the TALK workshops based upon the steps of learning and teaching developed by Klauer (1985) and Gagné (1985)

General structure	Teaching/learning step
Short review of the process and the subordinate goals	
Review of the previous block and a retrospective exchange of experiences	(Secure transfer of the previous session)
Short introduction of the next topic	Attract attention
Formulate specific goals	Explicitly inform the participants of the learning goals
Present the unit	Activate prior knowledge
	Present learning material
	Illustrate the significance of the material
Differentiation and exercises	Offer guidance while learning
	Adaptation/implementation
Elaborate on consequences for instruction and the school	Secure transfer
Summarize topic and set goals for concrete implementation over the coming weeks	Secure retention and transfer
Close the session	

(4) The *learning contents were imparted* both *explicitly and implicitly*. For instance, in a TALK unit on encouraging motivation, the high degree of relevance for giving pupils opportunities to make choices in their learning activities was stressed. In parallel, the participating teachers were often supplied with the freedom of choice in conjunction with their own learning activities (e.g. topics, tasks, learning forms). Also the topic of learning progress was not only addressed explicitly but targeted implicitly through the process of maintaining a learning diary during the training as well as via the reflection rounds at the beginning and end of each TALK session. This simultaneously served to encourage motivation. Also the central elements of self-regulated learning were imparted both explicitly and implicitly. The teachers selected a topic in the first term which they were to work through over the course of 2 months (either individually or in small groups). Various types of material to support the work on this topic were introduced (e.g. worksheets for setting goals and making plans, learning diaries, reflection worksheets), which were then opened to discussion on their appropriate use in the classroom after the project ended. Subsequently, the viability of the adapted materials for use in the classroom could be tested out.

Evaluation of TALK

TALK was subjected to both a summative and a formative evaluation, realizing a multi-informant (teachers, pupils, colleagues) and multimethod approach (quantitative: closed questions in surveys; qualitative: open-ended questions in surveys, focus groups, interviews and portfolios). In order to evaluate the summative effectiveness of TALK, surveys were conducted within the framework of a training group versus control group comparison (pre-test post-test design) as well as a follow-up for the training group. The control group consisted of a sample of teachers and pupils who were parallelized according to pivotal variables using a matching procedure (Spiel et al., 2008) and were not participating in the intervention.

A total of 40 teachers from 14 schools participated in the TALK programme (in three training groups), each with one class. The summative evaluation of the intensive phase of TALK (terms 1 and 2) showed positive *effects on the proximal level* according to the training goals, namely the teachers. At the beginning of TALK the teachers showed a clear need for more knowledge regarding the question on how to enhance LLL in school. After the intensive phase the teachers not only reported that they had learned a lot, there was also clear objective evidence for a gain of knowledge on how to design and arrange lessons and instructional processes in the sense of reaching more learning motivation and self-regulation (Schober, Finsterwald, Wagner & Spiel, 2009). Besides the knowledge increase, teacher belief consolidated that fostering LLL competences among pupils is in fact a task for the school and their own motivation for their job and their self-efficacy was enhanced. Teacher cooperation was positively affected by the training as well. In particular, TALK teachers spent much more time with their colleagues at school discussing the contents of lessons and tests, teaching and evaluation methods. Last but not least, the observed training effects of the intensive phase remained or even improved through the subsequent supervision phase of TALK (term 3).

On a *distal level*, the changes in teacher competences were reflected, e.g. in a higher perceived autonomy in the lessons (= essential precondition for a motivating learning environment) assessed by pupils in TALK classes in comparison to pupils in control classes. With regard to self-regulated learning, pupils in TALK classes reported that they were encouraged more to set their own goals, to learn and apply adaptive learning strategies and to monitor and reflect about their own learning process (= essential precondition for a learning environment that foster self-regulated learning). Furthermore, sleeper effects occurred for several variables measuring pupils' motivation. With a view to training effects on the schools, results of the school projects (conducted within the third term, documented in portfolios and analysed via expert ratings) and the surveys of the colleagues and headmasters suggested changes in the intended direction.

17.4 **How to build bridges between neuroscience and education**

Many (educational) researchers fear that the high and often unrealistic expectations about the contributions of neuroscience to education may result in an ignorance of the importance of research on learning and instruction and the knowledge this research has provided (Stern, 2005). The description of TALK exemplifies that promoting competences for LLL in schools is a complex and demanding task. One cannot only focus on a singular learning aspect, the learning process as a whole has to be considered. Teachers should not only know methods about how to enhance their pupils' knowledge, but also about how to provide a learning environment in which pupils are motivated to learn and able to acquire skills for self-regulated learning. Just focusing on the brain or the mind would be too simple if the aim is the enhancement of competences for (lifelong) learning. Furthermore, TALK is a programme that was conducted and evaluated within real school settings. In contrast, research in neuroscience is limited to laboratory studies; the focus mostly is on learning deficits. But there is clearly a potential for neuroscience to contribute more extensively to educational research (e.g. the discussion about phonics versus whole language in reading instruction). However, it will need time to extend the neuroscience methodology in such a way that it can be used in classrooms, e.g. to identify whether students are in a motivating learning environment or not (Stern, 2004). Concretely, educational programmes like TALK could benefit from neuroscience and vice versa if methods were available that could be used reasonably in real-life settings to measure motivational parameters which are not (yet) conscious, e.g. early stages of helplessness or anxiety. This would be helpful for the evaluation and optimization of such programmes as well as for teacher education. Another starting point for bridging both domains with such measures would be a systematic integration of field-experimental results into training programmes and teacher education showing, for example, how direct feedback or different kinds of instruction impact various parameters of students' motivation. In particular, for explanation of gender differences in motivation and cognitive achievement it might be very useful to combine motivational research and results of studies in neuroscience showing that the amount of these differences is clearly linked to the presence of gender stereotypes (e.g. Hausmann, 2011).

Recently, a growing interest in and a general debate about the relation between cognitive neuroscience and education respectively educational psychology occurred; opinions are especially voiced across researchers in the field of neuroscience to advance studies in the field of mind, brain *and* education (e.g. Ansari & Coch, 2006; Bruer, 1997; Fischer et al., 2007; Goswami, 2004; Schumacher, 2007). They call for building bridges between the different disciplines and actors including teachers and policymakers as the latter are especially influential but also prone to neuromyths. So far, bridges have been established between education (including educational psychology) and cognitive psychology for more than 50 years, and between cognitive psychology and neuroscience for about 20 years. Therefore, according to Bruer (1997), cognitive psychology is the firm ground for anchoring these bridges. By all means, to connect neuroscience and education respectively educational psychology, research must move beyond the ivory tower into real school settings, and educational practice and institutions must become available for scientific scrutiny (Shonkoff & Philipps, 2000). Interweaving the perspectives of research and practice is required. Neuroscience and cognitive psychology have to learn from educational practice and vice versa. Ansari and Coch (2006) argued for new teacher-training programmes in which courses are specifically designed to allow the investigation and discussion of how to link research and education: in the same way that medical professionals are trained in molecular biology and organic chemistry, training in neuroscience is expected to influence teachers' thinking about their practice in ways that are indirect and unpredictable a priori—because of differences in individual cognitive preconditions and domain-specific knowledge—but eventually measurable. TALK clearly illustrates that a theoretically based training programme (including the promotion of basic knowledge about how to foster competences for lifelong learning) provides such measurable outcomes. However, as teachers are keen for simple solutions it is very important to avoid presenting simple guidelines but to move them from spontaneous, unreflective teaching and education to scientific based reflective teaching.

For building maintainable and successful bridges between educationalists and neuroscientists respectively cognitive psychologists, communication needs to be fully bidirectional (Ansari & Coch, 2006). Therefore, scientists have to be trained in understanding educational process and practice with all its real-world constraints. As practical ways of starting a dialogue, Ansari and Coch (2006) recommend focus groups and workshops on specific, mutual interests, classroom and school visits by neuroscientists and laboratory visits by educators. The TALK project has successfully realized these kinds of actions including teachers and educational psychologists. Ansari and Coch (2006) additionally recommend that educational researchers participate in neuroscientist training programmes. Such an approach might also blur the traditional boundaries between 'basic' and 'applied' research. In 1997, Stokes proposed a model that rejects the one-dimensional perspective (a fundamental field versus an implementation field) as being too simple and postulates two dimensions involving understanding and usefulness (see also Spiel, 2009a).

Last but not least, building bridges has to include the society as a whole (Spiel, 2009b; Spiel, Lösel & Wittmann, 2009). Society is mostly represented by policymakers and journalists who have high and often unrealistic expectations about what neuroscience can bring to education

(Fischer et al., 2007). However, scientific research alone cannot answer important questions in education. Decisions about how to educate require not only scientific information about what is effective but also decisions about what is valuable including what should be taught and how schools and institutions of teacher educations should be organized (Fischer et al., 2007, see Chapter 19, this volume). Therefore, instead of using neuroscience as a basis of speculations about principles of 'brain-based' education, it should rather be used to contribute to interdisciplinary collaborations on learning and instruction by revealing characteristics of the learning brain that are not observable at the level of behaviour (Schumacher, 2007).

References

Ansari, D., & Coch, D. (2006). Bridges over troubled waters: Education and cognitive neuroscience. *Trends in Cognitive Sciences*, 10, 146–51.

Artelt, C., Baumert, J., Julius-McElvany, N., & Peschar, J. (2003). *Learners for life: Student approaches to learning. Results from PISA 2000*. Paris: OECD.

Baumert, J., Kunter, M., Brunner, M., Krauss, S., Blum, W., & Neubrand, M. (2004). Mathematikunterricht aus Sicht der PISA-Schülerinnen und -Schüler und ihrer Lehrkräfte [Students' and teachers' view on maths-education in the context of PISA]. In M. Prenzel, J. Baumert, W. Blum, R. Lehmann, & D. Leutner (Eds.), *PISA 2003: Der Bildungsstand der Jugendlichen in Deutschland—Ergebnisse des zweiten internationalen Vergleichs* [PISA 2003: The educational level of German adolescents—Results of the 2nd international comparison.] (pp. 314–54). Münster. Waxmann.

Baumert, J., Kunter, M., Brunner, M., Krauss, S., Furtak, E., Dubberke, T., et al. (2006). Research area IV: Cognitive activation in the classroom. In Center for Educational Research (Ed.), *Annual Report 2006* (pp. 100–110). Available at: http://www.mpib berlin.mpg.de/en/forschung/eub/pdfs/EUB_annual_report_2006.pdf (accessed 15 September 2010).

Bransford, J. D., Sherwood, R. D., Hasselbring, T. S., Kinzer, Ch. K., & Williams, S.M. (1990). Anchored Instructions: Why we need it and how technology can help. In D. Nix & R. Spiro (Eds.). *Cognition, Education and Multimedia: Exploring ideas in high technology* (pp. 163–205). Hillsdale, NJ: Erlbaum.

Bruer, J.T. (1997). Education and the brain: A bridge too far. *Educational Researcher*, 26, 4–16.

Chambers Cantrell, S., & Hughes, H.K. (2008). Teacher efficacy and content literacy implementation: An exploration of the effects of extended professional development with coaching. *Journal of Literacy Research*, 40(1), 95–127.

Commission of the European Communities (2000). *Memorandum on lifelong learning*. Brussels: Commission of the European Communities.

Dunn, T. G., & Shriner, C. (1999) Deliberate practice in teaching: what teachers do for self improvement. *Teaching and Teacher Education*, 15, 631–51.

European Commission (2010). *School education: equipping a new generation* (Doc 64). Available at: http://ec.europa.eu/education/lifelong-learning-policy/doc64_en.htm (accessed 15 September 2010).

Erk, S., Kiefer, M., Grothe, J., Wunderlich, A.P., Spitzer, M., & Walter, H. (2003). Emotional context modulates subsequent memory effect. *Neuroimage* 18, 439–47.

Fischer, N., & Rustemeyer, R. (2007). Motivationsentwicklung und schülerperzipiertes Lehrkraftverhalten im Mathematikunterricht [The impact of perceived teacher behavior on motivational development in mathematics]. *Zeitschrift für Pädagogische Psychologie*, 21, 135–44.

Fischer, K. W., Daniel, D. B., Immordino-Yang, M. H., Stern, E., Battro, A., & Koizumi, H. (2007). Why mind, brain, and education? Why now? *Mind, Brain, and Education*, 1, 1–2.

Gagné, R. M. (1985). *The conditions of learning and theory of instruction*. New York: Holt, Rinehart & Winston.

Geake, J. (2008). Neuromythologies in education. *Educational Research*, 50(2), 123–33.

Goswami, U. (2004) Neuroscience and education. *British Journal of Educational Psychology*, 74, 1–14.

Goswami, U. (2006). Neuroscience and education: From research to practice? *Nature Reviews Neuroscience,* 7, 406–411.

Gräsel, C., & Parchmann, I. (2004). Implementationsforschung—oder: Der steinige Weg, Unterricht zu verändern, [Implementation research—or: The rocky path to changing classroom instruction]. *Unterrichtswissenschaft, 32,* 196–214.

Hager, W., & Hasselhorn, M. (2000). Psychologische Interventionsmaßnahmen: Was sollen sie bewirken können? [Psychological intervention measures: What should be their effects?]. In W. Hager, J.-L. Patry & H. Brezing (Eds.), *Evaluation psychologischer Interventionsmaßnahmen* [Psychological intervention measures] (pp. 41–85). Bern: Verlag Hans Huber.

Hausmann, M. (2011). Sex oder Gender? Neurobiologie kognitiver Geschlechtsunterschiede [Sex or gender? Neurobiology of cognitive gender differences]. In G. Magerl, R. Neck, & C. Spiel (Eds.), *Wissenschaft und Gender* [Science and gender] (pp. 55–79). Wien: Böhlau.

Henson, R. (2001). The effects of participation in teacher research on teacher efficacy. *Teaching and Teacher Education, 17,* 819–36.

Klauer, K. J. (1985). Framework for a theory of teaching. *Teaching and Teacher Education, 1,* 5–17.

Kunter, M., Klusmann, U., Dubberke, T., Baumert, J., Blum, W., Brunner, M., et al. (2007). Linking aspects of teacher competence to their instruction: Results from the COACTIV project. In M. Prenzel (Ed.), *Studies on the educational quality of schools. The final report on the DFG Priority Programme* (pp. 39–56). Münster: Waxmann.

Lipowsky, F. (2006). Auf den Lehrer kommt es an. Empirische Evidenzen für Zusammenhänge zwischen Lehrerkompetenzen, Lehrerhandeln und dem Lernen der Schüler [It depends on the teacher. Empirical evidence for the relations between teacher competence, teacher behavior, and student learning]. *Zeitschrift für Pädagogik, 51,* 47–65.

Lüftenegger, M., Wagner, P., Finsterwald, M., Schober, B., & Spiel, C. (2010). TALK—Ein Trainingsprogramm für Lehrkräfte zur Förderung von Lebenslangem Lernen in der Schule [TALK—A training program for teachers to encourage lifelong learning in school]. In F. H. Müller, A. Eichenberger, M. Lüders, & J. Mayr (Eds.), *Lehrerinnen und Lehrer lernen. Konzepte und Befunde zur Lehrerfortbildung* [Teachers' Learning. Concepts and findings concerning teacher trainings] (pp. 327–343). Münster: Waxmann.

Mandl, H., Prenzel, M. & Gräsel, C. (1991). *Das Problem des Lerntransfers in der betrieblichen Weiterbildung (Forschungsbericht Nr. 1)* [The problem of transfer of learning in further education]. München: Ludwig-Maximilians-Universität, Lehrstuhl für Empirische Pädagogik und Pädagogische Psychologie.

Meirink, J.A., Meijer, P.C., & Verloop, N. (2007). A closer look at teachers' individual learning in collaborative settings. *Teachers and Teaching, 13*(2), 145–64.

OECD (2002). *Understanding the brain: Towards a new learning science.* Paris: OECD Publishing.

Peetsma, T., Hascher, T., van der Veen, I., & Roede, E. (2005). Relations between adolescents' self-evaluations, time perspectives, motivation for school and their achievement in different countries and at different ages. *European Journal of Psychology of Education, 20,* 209–225.

Prenzel, M. (1994). Mit Interesse in das 3. Jahrtausend! Pädagogische Überlegungen [With interest into the 3rd millennium! Pedagogic deliberations]. In N. Seibert, & H. J. Serve (Eds.), *Erziehung und Bildung an der Schwelle zum 3. Jahrtausend* [Education and training on the cusp to the 3rd millennium] (pp. 1314–1339). München: PimS-Verlag.

Schober, B., Finsterwald, M., Wagner, P., Lüftenegger, M., Aysner, M., & Spiel, C. (2007). TALK—A training program to encourage lifelong learning in school. *Journal of Psychology, 215* (3), 183–93.

Schober, B., Finsterwald, M., Wagner, P., & Spiel, C. (2009). Lebenslanges Lernen als Herausforderung der Wissensgesellschaft: Die Schule als Ort der Förderung von Bildungsmotivation und selbstreguliertem Lernen, [Lifelong learning as a challenge for the knowledge society: The school as a place for educational motivation and self-regulated learning]. In W. Specht (Ed.), *Nationaler Bildungsbericht Österreich 2009, Band 2, 121–39.* Graz:Leykam.

Schumacher, R. (2007). The brain is not enough: potential and limits in integrating neuroscience and pedagogy. *Analyse & Kritik,* 29, 38–46.

Schunk, D. H., Pintrich, P. R., & Meece, J. L. (2008). *Motivation in Education: Theory, Research and Application* (3rd ed., pp. 168–207). Englewood Cliffs, NJ: Prentice Hall.

Shonkoff, J. P., & Phillips, D. A. (Eds.) (2000). *From neurons to neighborhoods: The science of early childhood development.* Washington, DC: National Academy Press.

Spiel, C. (2009a). Evidence-based practice: A challenge for European developmental psychology. *European Journal of Developmental Psychology,* 6, 11–33.

Spiel, C. (2009b). Evidenzbasierte Bildungspolitik und Bildungspraxis—eine Fiktion? Problemaufriss, Thesen, Anregungen [Evidence based politics and educational practice—is this fiction? Problems, theses, suggestions]. *Psychologische Rundschau,* 60, 255–56.

Spiel, C., Lapka, D., Gradinger, P., Zodlhofer, E. M., Reimann, R., Schober, B., Wagner, P. & von Eye, A. (2008). A Euclidean distance-based matching procedure for non-randomized comparison studies. *European Psychologist,* 13(3), 180–87.

Spiel, C., Lösel, F., & Wittmann, W.W. (2009). Transfer psychologischer Erkenntnisse in Gesellschaft und Politik [Transfer of psychological knowledge in society and politics]. *Psychologische Rundschau,* 60, 241–42.

Steiner, G. (2006). Lernen und Wissenserwerb [Learning and knowledge acquisition]. In A. Krapp & B. Weidenmann (Eds.), *Pädagogische Psychologie* [Educational psychology. A textbook] (5th ed., pp. 137–202). Weinheim: Beltz.

Stern, E. (2004). Wie viel Hirn braucht die Schule? Chancen und Grenzen einer neuropsychologischen Lehr-Lern-Forschung [How much brain is needed in school? Prospects and limits of neuropsychological research in teaching and learning.]. *Zeitschrift für Pädagogik,* 50, 531–38.

Stern, E. (2005). Pedagogy meets neuroscience. *Science,* 310, 745.

Stern, E., Grabner, R., & Schumacher, R. (2005). *Bildungsreform Band 13: Lehr-Lern- Forschung und Neurowissenschaften—Erwartungen, Befunde und Forschungsperspektiven* [Educational reform 13: Research on teaching, learning and neurosciences—Expectations, findings and perspectives]. Bonn: Bundesministerium für Bildung und Forschung.

Weinstein, C. E., & Hume, L. M. (1998). *Study strategies for lifelong learning.* Washington, DC: American Psychological Association.

Wigfield, A., Eccles, J. S., Schiefele, U., Roeser, R. W. & Davis-Kean, P. (2006). Development of achievement motivation. In N. Eisenberg (Ed.), *Handbook of child psychology. Vol. 3: Social, emotional, and personality development* (pp. 933–1002). Hoboken, NJ: Wiley.

Chapter 18

Educational cognitive neuroscience: designing autism-friendly methods to teach emotion recognition

Simon Baron-Cohen, Ofer Golan and Emma Ashwin

Overview

Children and adults with autism spectrum conditions (ASC) have difficulties with empathy, but have intact or even above average development of 'systemizing', that is, the drive to analyse or build a system. In this chapter we review the evidence for both the empathy deficits and the systemizing strengths, and present two examples of educational software designed to help people with ASC improve in empathy by playing to their strengths in systemizing. These are the *Mind Reading* and *The Transporters* DVDs, respectively. Treatment trials of both are summarized which confirm that even after relatively short-term intervention with these programmes, improvements in empathy (specifically the cognitive element of emotion recognition) are seen. We conclude that if interventions use the knowledge from cognitive neuroscience to design autism-friendly formats for teaching, then even complex aspects of social cognition can be improved.

18.1 Introduction

It is regrettable that all too often education and cognitive neuroscience barely communicate, since educationalists are entrusted with the most valuable of tasks, to enable a child to fulfil his or her potential, whilst cognitive neuroscientists frequently study children who are not neurologically 'wired' in the typical way, and who therefore process information differently. In this chapter we argue that cognitive neuroscience can assist educators to design teaching materials that are a good 'fit' with the autistic mind; specifically we illustrate how children with ASC may need tailored teaching methods to help them learn aspects of 'empathy', since part of their disability is that they find it difficult to read other people's minds, and make sense of other people's behaviour. We do not wish to suggest that communication between cognitive neuroscience and education is a one-way street however, since in the best-case scenario, educational materials and methods are co-designed by the scientists and practitioners each contributing their expertise (see Chapters 3 and 16, this volume). Before looking at specific methods of teaching empathy to children with ASC, we turn to the question of the nature of empathy.

18.2 What is empathy?

We define empathy as the ability to attribute mental states to others, and to respond with an appropriate emotion to the other person's mental states (Baron-Cohen and Wheelwright, 2004).

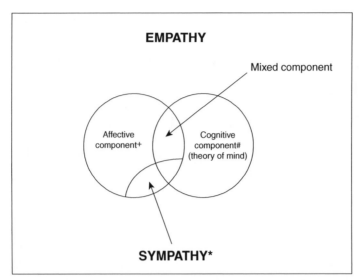

Fig. 18.1 Fractions of empathy. + Feeling an appropriate emotion triggered by seeing/learning of another's emotion. # Understanding and/or predicting what someone else might think, feel, or do. * Feeling an emotion triggered by seeing/learning of someone else's distress which moves you to want to alleviate their suffering. With kind permission from Springer Science+Business Media: *Journal of Autism and Developmental Disorders*, The Empathy Quotient: An investigation of adults with Asperger syndrome or high functioning autism, and normal sex differences, 34 (2), 2004, pp. 163–75, Simon Baron-Cohen.

This definition of empathy suggests the two main 'fractions' of empathy are a 'cognitive' component (the recognition of another person's mental state) Figure 18.1 illustrates this distinction and an 'affective' component (the emotional reaction to another person's mental state). The cognitive component is sometimes also called a 'theory of mind' (Dennett, 1987). Mental states include thoughts and emotions, thoughts being traditionally fractionated into beliefs, desires, intentions, goals and perceptions (Dennett, 1987; Baron-Cohen, 1995). Emotions are traditionally fractionated into six 'basic' emotions (happy, sad, angry, afraid, disgusted and surprised) (Ekman, 1999), and numerous 'complex' emotions that are acquired at different points in childhood (Baron-Cohen, Golan, Wheelwright, Granader & Hill, 2010). Complex emotions involve attributing a cognitive state as well as an emotion, and are more context and culture dependent (Griffiths, 1997). The basic emotions are held to be so because they are universally recognized and expressed in the same way. It may be that more emotions are universally recognized and expressed than these six but have been overlooked because of how expensive, time-consuming and difficult cross-cultural research is (Baron-Cohen et al., 2006). Indeed, research into complex emotions (usually towards developing taxonomies) has been mostly language and culture specific (Ortony, Clore & Foss, 1987; Storm & Storm, 1987). Our own work described the development of the emotional lexicon in the English language (Baron-Cohen et al., 2010), suggesting there are at least 412 distinct emotions and related mental states (each with their own descriptor that is not just a synonym for another emotion) that are recognizable by independent judges within the UK (Baron-Cohen et al., 2010).

18.3 **Empathy in autism spectrum conditions**

Some individuals in the population may be delayed in the development of empathy, for different reasons. These include people with ASC who for genetic—and ultimately neurological—reasons have difficulties in putting themselves into someone else's shoes and knowing how to respond to another's feelings, in real time. Since such deficits may have a significant impact on their social functioning, this raises the challenge of whether aspects of empathy can be facilitated or taught to individuals with ASC. We summarize some evidence that the first component of empathy—cognitive empathy—can indeed be taught. This task is made easier through the design of educational resources (including computer-based methods) that tap into systematic areas of interest, characteristic of ASC, that are therefore intrinsically motivating. Whilst we do not rule out that the second component of empathy—affective empathy—can be taught, it remains the case that all efforts have so far been focused on the cognitive fraction, so that it is unknown if the second could also be taught.

People with ASC have social-communication difficulties alongside circumscribed interests ('obsessions') and a strong preference for sameness and repetition (APA, 1994). Underlying these characteristics are difficulties understanding the emotional and mental states of others (Baron-Cohen, 1995). Individuals with ASC have difficulties recognizing emotions from facial expressions, vocal intonation, body language, separately (Baron-Cohen et al., 2001; Hobson, 1986a; Hobson, 1986b; Yirmiya et al., 1992), and in context (Golan, Baron-Cohen & Golan, 2008; Klin et al., 2002). Although some individuals with ASC recognize basic emotional expressions (Baron-Cohen, Spitz & Cross, 1993; Grossman et al., 2000), difficulties in identifying more complex emotions persists into adulthood (Baron-Cohen et al., 2001; Baron-Cohen, Wheelwright & Jolliffe, 1997; Golan, Baron-Cohen & Hill, 2006).

The emotion recognition difficulties are in part the result of altered face processing (Langdell, 1978; Dawson et al., 2004; Klin et al., 2002), which in itself may be due to a failure to interpret the mentalistic information conveyed by the eyes (Baron-Cohen, 1995). Others' facial expressions may also be less intrinsically rewarding. Children with ASC show reduced attention to faces and to eyes in particular (Swettenham et al., 1998). The result of this reduced experience with faces is that children with ASC thus do not become 'face experts' (Dawson, Webb & McPartland, 2005). For example, whilst the typically developing brain shows an electrophysiological response to upright faces called the N170 wave form, the autistic brain shows a reduced N170 (Grice et al., 2005).

18.4 Systemizing in autism spectrum conditions

In contrast to their difficulties in emotion recognition, individuals with ASC have intact or even enhanced abilities in 'systemizing' (Baron-Cohen, 2002, 2006). Systemizing is the drive to analyse or build systems, allowing one to predict the behaviour of the system and control it. Systems may be mechanical (e.g. vehicles), abstract (e.g. number patterns), natural (e.g. the tide) or collectible (e.g. a library classification index). The 'obsessions' or narrow interests of children with ASC cluster in the domain of systems (Baron-Cohen & Wheelwright, 1999). These include vehicles, spinning objects and computers, all of which are attractive to individuals with ASC. At the heart of systemizing is the ability to detect patterns or rules of the form 'if a, then b'. The systemizing theory of autism relates this affinity to their systematic and predictable nature. In the study summarized in the next section, we illustrate how these special interests can be harnessed when teaching children with ASC, using computer-based or multimedia formats, to keep them intrinsically motivated.

The systemizing theory of ASC has been supported by different studies: children with ASC have been found to outperform matched controls on tests of 'intuitive physics' (Baron-Cohen et al., 2001), and adults with ASC were at least intact on such tests (Lawson, Baron-Cohen & Wheelwright, 2004), as well as on other tests that involve excellent attention to detail (Happe & Frith, 2006; Mottron et al., 2006), a prerequisite for good systemizing (Baron-Cohen, 2008; Jolliffe &

Baron-Cohen, 1997; O'Riordan et al., 2001; Shah & Frith, 1983). In addition, individuals with ASC score above average on the Systemizing Quotient (SQ), a self-report (or parent-report) measure of how strong one's interests are in systems (Baron-Cohen et al., 2003; Wheelwright et al., 2006; Wakabayashi et al., 2007; Auyeung et al., 2009).

18.5 From cognitive neuroscience to intervention

Lego® Therapy

If children with ASC possess intact or enhanced systemizing skills, it may be possible for them to use such skills to facilitate their empathy, particularly in the cognitive component of emotion recognition. Lego® Therapy (Owens et al., 2008) is an example that encourages young children with ASC, to build Lego® models in groups of 3, thereby gaining opportunities for social interaction. Children participating in Lego® Therapy are intrinsically motivated by Lego® because it involves constructional systems that can be assembled in predictable and repeating sequences.

Mind Reading DVD

A method harnessing systematic skills to teach empathy to individuals with ASC is the *Mind Reading* DVD. This comprises educational software that was designed to be an interactive, systematic guide to emotions (Baron-Cohen et al., 2004) (http://www.jkp.com/mindreading). It was developed to help people with ASC learn to recognize both basic and complex emotions and mental states from video clips of facial expressions and audio recordings of vocal expressions. It covers 412 distinct emotions and mental states, which are organized developmentally and classified taxonomically to be attractive to a mind that learns through systemizing. The principle behind this was that individuals with ASC may not learn to recognize emotional expressions in real time during live social situations because emotions are fleeting and do not repeat in an exact fashion, which may reduce the number of opportunities to systematically learn from repetition. Putting emotions into a computer-based learning environment enables emotions to be played and replayed over and over again in an identical fashion, such that the learner can have control over their speed and the number of exposures they need in order to analyse and memorize the features of each emotion.

Furthermore, since emotions vary depending on who is expressing them, in the real world it can be difficult to see what defines each specific emotion. *Mindreading* helps its users overcome this problem by having each of the 412 emotions portrayed by six different actors (male and female, old and young, different ethnicities), to facilitate learning to recognize emotions independently of the identity of the person expressing that emotion. In addition, in the real world emotions can appear unlawful (some people smile when they are happy, other people smile when they are pretending to be happy, and yet others are happy when they are not smiling at all) so *Mind Reading* imposes some lawfulness onto emotions by assigning a clear label to each emotional expression, including masked, or insincere emotional expressions (e.g. emotions in the 'sneaky' category). Finally, emotions in the real world can be hard to classify so *Mind Reading* offers the user a pre-designed classification system, to assist in finding patterns among inherently unpatterned emotional information.

Using *Mind Reading* over a 10-week intervention (2 hours usage per week), individuals with ASC improved in their ability to recognize a range of complex emotions and mental states (Golan & Baron-Cohen, 2006). In a follow-up conducted 1 year after the completion of the intervention period, individuals with ASC who used *Mind Reading* reported an improved ability to form friendships and relationships, and increased awareness of the importance of emotions and emotional expressions in everyday life, improving their understanding of emotions and their corresponding expressions, and affecting their ability to function socially (Golan & Baron-Cohen, 2007). These findings have been replicated with children in the United States, supporting the cross-cultural validity of this intervention (LaCava, Golan, Baron-Cohen & Myles, 2007). These are encouraging results because they suggest that at least one cognitive component of empathy can be taught, and that it may have a long-term effect and affect social functioning. It is not known if such improvement would be seen if the intervention was shorter in duration, or if the users had not just ASC but additional learning difficulties (below average intelligence quotient (IQ)). Finally, it could reasonably be objected that learning to recognize emotions in the simplified context of a computer screen, devoid of the 'noise' of a real social situation like a school playground or a birthday party or an argument, is likely to be both simplified and therefore easier to achieve. This objection is important since it raises the question of whether such learning from 'artificial' contexts generalizes to more natural settings. Guarding against the risk of artificiality, *Mind Reading* used real faces rather than cartoon or schematic faces. However, future work using the *Mind Reading* DVD could assess the benefits of a longer intervention than just 10 weeks. The DVD could also be used with more interactive teaching methods such as social skills groups, or as part of dramatic role-play.

The Transporters DVD

Difficulties with generalization from taught material to everyday life have been found both in computer-based intervention programmes (Bölte et al., 2002; Silver & Oaks, 2001) and in social skills training courses (Barry et al., 2003; Bauminger, 2002). The limited effectiveness of these interventions could be related to a lack of intrinsic motivation, since they utilize explicit rather than implicit teaching methods. The study reviewed next and reported in detail elsewhere (Golan et al., 2010) evaluates the effectiveness of another DVD, this time an animation series created to motivate young children with ASC to learn about emotions and facial expressions by embedding them in a world of mechanical vehicles. Again, this is based on the premise that the reason children with ASC love to watch films about vehicles (according to parental report) may be because they are strong 'systemizers' (Baron-Cohen, 2006, 2008). That is, they are drawn to predictable, rule-based systems, whether these are repeating mathematical patterns, or repeating electrical patterns (e.g. light switches), or repeating patterns in films. Kanner's first descriptions of ASC drew attention to their 'need for sameness' and their 'resistance to change' (Kanner, 1943). At the core of ASC may be an *ability* to deal effortlessly with systems because they do not change and produce the same outcome every time; and by the same token, a disabling *difficulty* to deal with the social world because it is always changing unpredictably and because the outcome is different every time.

According to the hyper-systemizing theory (Baron-Cohen 2006), vehicles whose motion is determined only by physical rules (such as vehicles that can only go back and forth along linear tracks) would be much preferred by children with ASC over vehicles like planes or cars whose motion could be highly variable, moving at the whim of the human driver operating them. In vision neuroscience this relates to the distinction between physical-causal/mechanical motion (Michotte, 1963) vs. animate/biological motion (Castelli et al., 2000; Premack, 1990). The former requires intuitive physics (Saxe, Carey & Kanwisher, 2004; Wellman & Inagaki, 1997) whilst the latter requires intuitive psychology, in particular the ability to detect others' goals, desires and intentions (Baron-Cohen, 1995).

We therefore created a children's animation series, *The Transporters* (http://www.thetransporters. com), based around eight characters who are all vehicles that move according to rule-based motion. Onto these vehicles we grafted real-life faces of actors showing emotions. We tested whether creating an autism-friendly context of predictable mechanical motion could render facial expressions of emotion more learnable and increase the motivation to learn them. The different toy vehicles (two trams, two cable cars, a chain ferry, a coach, a funicular railway and a tractor) had motion that was constrained in a linear manner (all the vehicles moved on tracks or cables).

The Transporters is a high-quality three-dimensional children's animation series and consists of 15 5-minute episodes, each of which focuses on a key emotion or mental state. The 15 key emotions depicted on the vehicles are *happy, sad, angry, afraid, disgusted, surprised, excited, tired, unfriendly, kind, sorry, proud, jealous, joking* and *ashamed*. The emotions selected include the six 'basic' emotions (Ekman, 1999), emotions that are more 'complex' but still developmentally appropriate (e.g. *jealous, proud, ashamed*), and emotions and mental states that are important for everyday social functioning (e.g. *kind, unfriendly, tired, joking*). These emotions were chosen because typically developing children recognize and understand these between 2–7 years of age (Bretherton & Beeghly, 1982; Ridgeway, Waters & Kuczaj, 1985).

In the study by Golan et al. (2010), three groups were assessed twice: at Time 1 and then 4 weeks after at Time 2. In each assessment participants were tested at different levels of generalization, one testing participants' emotional vocabulary, and the other three testing their ability to match a socioemotional situation to the appropriate facial expression. Level 1: *emotional vocabulary*—participants were asked to define 16 emotion words and give examples of situations that evoked them. These were the 15 key emotions from the series (listed earlier), in addition to *worried*. Level 2: *familiar close generalization*—participants had to match familiar situations taken from the intervention series to facial expressions of familiar characters from the series. Level 3: *unfamiliar close generalization*—participants had to match novel situations with novel expressions from *The Transporters* characters. These expressions were *not* shown by these characters in the intervention series. Level 4: *distant generalization*—to test generalization to facial expressions that are not attached to vehicles, participants had to match novel situations with novel expressions using a selection of human non-*Transporters* faces taken from the *Mind Reading* software (Baron-Cohen et al., 2004). Examples of items from Levels 1 and 3 are shown in Figure 18.2.

Three groups took part in the study: an ASC intervention group, an ASC control group and a typically developing control group. Participants in the two clinical groups were randomly assigned and took part according to the following test conditions: (1) *ASC intervention group*: the parents of 20 participants were given the intervention series and DVD guide to use with their child at home. Children were asked to watch at least 3 episodes per day over a period of 4 weeks. (2) *ASC control group:* 19 participants did not participate in any intervention during the 4-week interval, except for their standard school curriculum. One participant dropped out of the study after the first session. (3) *Typical control group:* 18 participants were recruited for this group. The three groups were matched for sex, age and verbal ability (using the British Picture Vocabulary Scale (BPVS); Dunn, Whetton & Burley, 1997).

At Time 1, there were significant differences between groups on the emotional vocabulary task and on the three situation-expression matching tasks. These differences were due to the significantly higher scores of the typical controls on all tasks compared to the two clinical groups, which did not differ from each other. Analysis of results after Time 2 testing revealed significant time by group interactions, with the ASC intervention group showing significant improvement across all task levels between Time 1 and Time 2. Furthermore, this improvement was comparable to levels of performance found in the typical control group. In contrast, the ASC and typical control groups showed no significant improvement on any of the tasks between test sessions. These effects are illustrated in Figure 18.3.

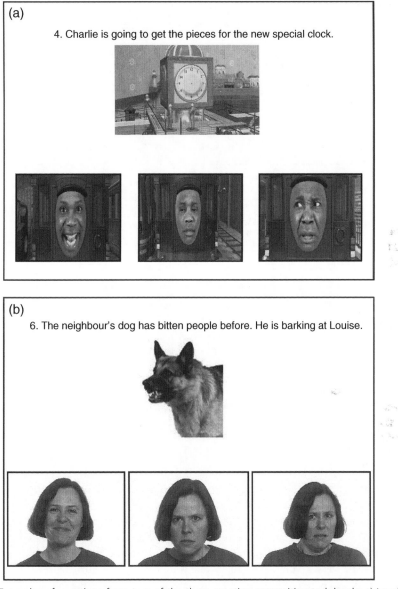

Fig. 18.2 Examples of questions from two of the three emotion recognition task levels. a) Level 1 task: match familiar scenes from the series with familiar faces. b) Level 3 task: match novel scenes and faces using real human faces.

The study we have reviewed (reported in detail in Golan et al., 2010) investigated the effectiveness of individual use of *The Transporters* animated series (with parental support) over a 4-week period. The results show that use of the DVD led children with ASC to improve significantly in their emotion comprehension and recognition skills on tasks including the emotions presented by *The Transporters*: from the same level of ability seen with the ASC control group at Time 1, to a level that was indistinguishable from the typically developing group at Time 2.

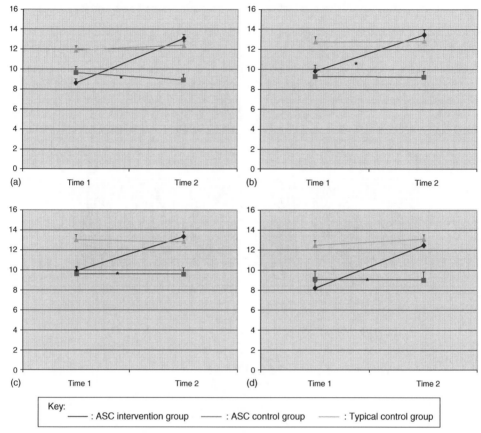

Fig. 18.3 Graphs to show mean scores (with standard error bars) for each group on the four tasks. a) Situation-expression matching task—Level 1. b) Situation-expression matching task—Level 2. c) Situation-expression matching task—Level 3. d) Emotional vocabulary task. * $p<0.001$.

The improvement of the intervention group was not limited to tasks that required close generalization; these participants were also able to generalize their knowledge to perform at the level of typical controls on the distant generalization task, which required emotion recognition from naturalistic clips of human characters that were not attached to vehicles. *The Transporters* may have facilitated generalization because the series was designed using intrinsically motivating media, such that the children enjoyed watching the vehicles whilst learning about emotions from real faces grafted onto them (incidental rather than explicit learning). *The Transporters* used characters and an environment that appealed to a preference for order, systems, and predictability that is characteristic of ASC. Anecdotal evidence from the parents of the intervention group suggests that their children became more willing to discuss emotions, and became more interested in facial expressions. Parents also noticed a change in their children's behaviour, and in their ability to interact with others. Such anecdotal changes need formal evaluation.

We expect the integration of *The Transporters* with other educational or therapeutic methods for children with ASC may improve its effect even further. We conclude that the use of systemizing as an intrinsically motivating method for learning about empathy allows affective information, which would otherwise be confusing, to become more intelligible and appealing to the autistic mind.

18.6 **Future directions**

If *The Transporters* is having such a positive effect on the learnability of emotional expressions by children with ASC, might there be other ways to harness the same preference for systemizing in the teaching of emotions to these children? Clearly vehicles are not the only kind of systems that children with ASC enjoy, and others might include robots (Dautenhahn & Werry, 2004), or rules (Hadwin et al., 1996). We see these sorts of interventions as part of an adaptation of the mainstream environment to be more suited to people with ASC, and such environmental adaptations need not be restricted to the teaching of emotions. An example outside the emotional domain might involve sensory perception, where people with ASC may experience hypersensitivity (Baron-Cohen et al., 2009). In this case, as with the use of specifically oriented media for the teaching of emotions, classrooms may need to be specially designed to ensure information, which may otherwise be easily processed by the neurotypical brain, is not over stimulating and therefore aversive to the autistic brain. We cannot expect learning to proceed smoothly or even to occur at all if the information is in a form that causes distress or is even painful. Equally, we need to recognize the limitations of what interventions like *The Transporters* can achieve, since whilst methods like this might improve emotion-recognition we do not yet know whether it would have far-reaching effects into other aspects of social cognition (such as understanding pragmatics of language, or turn-taking in dialogue) or have any impact in non-social areas of the condition, such as the narrow interests and repetitive behaviour. We suspect not, and it is important to have realistic expectations. Nevertheless, having an impact on emotion recognition might still be of considerable value to a child who otherwise might be anxious or confused in the face of other people's changing facial expressions. We conclude that bringing cognitive neuroscience and education together, in partnerships around the design of teaching materials and methods, can facilitate the development of skills that are traditionally viewed as 'impaired', such as empathy in children with ASC.

Acknowledgements

All the authors were affiliated with the Autism Research Centre at University of Cambridge during the study. This work was conducted in association with the NIHR CLAHRC for Cambridgeshire and Peterborough NHS Foundation Trust. Parts of this paper appeared in the Journal of Autism and Developmental Disorders (Golan et al., 2010) and in the Proceedings of the Royal Society of London (Baron-Cohen et al., 2010). We are grateful to Yael Granader, Suzy McClintock, Kate Day and Victoria Leggett for help with data collection, and to Gina Owen, Kimberly Peabody and Ben Weiner for useful discussions. Ofer Golan now holds a position at the Department of Psychology, Faculty of Social Sciences, Bar-Ilan University in Israel. We are grateful to Culture Online and the Department for Culture, Media and Sport (DCMS) for funding, and to Catalyst Ltd and Culture Online (particularly Claire Harcup, Paul Bason, Khairoun Abji and Professor Jon Drori) for their production of *The Transporters*. We are also indebted to the families who participated in the intervention. *The Transporters* series was nominated for a BAFTA in the Children's Awards Category, November 2007. Forty thousand copies of *The Transporters* DVD were distributed for free in 2007 in the UK to families with a child on the autistic spectrum. It is now available for sale via Changing Media Development Ltd, of which Simon Baron-Cohen, Jon Drori and Claire Harcup are Directors. Profits from sales go to autism research and development of new autism educational products.

References

APA. (1994). *DSM-IV Diagnostic and Statistical Manual of Mental Disorders, 4th Edition*. Washington DC: American Psychiatric Association.

Auyeung, B., Wheelwright, S., Allison, C., Atkinson, M., Samaranwickrema, N., & Baron-Cohen, S. (2009). The Children's Empathy Quotient (EQ-C) and Systemizing Quotient (SQ-C): Sex differences in typical development and in autism spectrum conditions. *Journal of Autism and Developmental Disorders*, 39,1509–21

Baron-Cohen, S. (1991) Do people with autism understand what causes emotion? *Child Development* 62, 385–95.

Baron-Cohen, S. (1995). *Mindblindness: an essay on autism and theory of mind*. Boston, MA: MIT Press/ Bradford Books.

Baron-Cohen, S. (2002). The extreme male brain theory of autism. *Trends in Cognitive Science* 6(6), 248–54.

Baron-Cohen, S. (2006). The hyper-systemizing, assortative mating theory of autism. *Progress in Neuro-Psychopharmacology & Biological Psychiatry* 30(5), 865–72.

Baron-Cohen, S. (2008). Autism, hypersystemizing, and truth. *The Quarterly Journal of Experimental Psychology* 61(1), 64–75.

Baron-Cohen, S., Golan, O., Wheelwright, S., Granader, Y., & Hill, J. (2010) Emotion word comprehension from 4 to 16 years old: a developmental survey. *Frontiers in Evolutionary Neuroscience*, 2, 109.

Baron-Cohen, S., Golan, O., Wheelwright, S., & Hill, J. J. (2004). *Mind Reading: the interactive guide to emotions*. London: Jessica Kingsley Limited (http://www.jkp.com/mindreading).

Baron-Cohen, S., Richler, J., Bisarya, D., Gurunathan, N., & Wheelwright, S. (2003). The Systemising Quotient (SQ): An investigation of adults with Asperger syndrome or high functioning autism and normal sex differences. *Philosophical Transactions of the Royal Society, Series B, Special issue on 'Autism: Mind and Brain'*, 358, 361–74.

Baron-Cohen, S., Riviere, A., Cross, P., Fukushima, M., Bryant, C., Sotillo, M., Hadwin, J., & French, D. (1996) Reading the mind in the face: A cross-cultural and developmental study. *Visual Cognition*, 3, 39–59.

Baron-Cohen, S., Spitz, A., & Cross, P. (1993). Can children with autism recognize surprise? *Cognition and Emotion* 7, 507–16.

Baron-Cohen, S., Tavassoli, T., Ashwin, E., Ashwin, C., & Chakrabarti, B. (2009) Talent in autism: hyper-systemizing, hyper-attention to detail, and sensory hypersensitivity. *Proceedings of the Royal Society, Philosophical Transactions, Series B*, 364(1522), 1377–83.

Baron-Cohen, S., & Wheelwright, S. (1999). Obsessions in children with autism or Asperger syndrome: a content analysis in terms of core domains of cognition. *British Journal of Psychiatry* 175, 484–90.

Baron-Cohen, S., & Wheelwright, S. (2004) The Empathy Quotient (EQ). An investigation of adults with Asperger syndrome or high functioning autism, and normal sex differences. *Journal of Autism and Developmental Disorders*, 34, 163–75.

Baron-Cohen, S., Wheelwright, S., Hill, J., Raste, Y., & Plumb, I. (2001). The 'Reading the Mind in the Eyes' Test revised version: a study with normal adults, and adults with Asperger syndrome or high-functioning autism. *Journal of Child Psychology and Psychiatry* 42(2), 241–51.

Baron-Cohen, S., Wheelwright, S., & Jolliffe, T. (1997). Is there a 'language of the eyes'? Evidence from normal adults and adults with autism or Asperger syndrome. *Visual Cognition* 4, 311–31.

Baron-Cohen, S., Wheelwright, S., Spong, A., Scahill, V. L., & Lawson, J. (2001). Are intuitive physics and intuitive psychology independent? A test with children with Asperger syndrome. *Journal of Developmental and Learning Disorders*, 5, 47–78.

Barry, T. D., Klinger, L. G., Lee, J. M., Palardy, N., Gilmore, T., & Bodin, S. D. (2003). Examining the effectiveness of an outpatient clinic-based social skills group for high-functioning children with autism. *Journal of Autism and Developmental Disorders* 33(6), 685–701.

Bauminger, N. (2002). The facilitation of social-emotional understanding and social interaction in high-functioning children with autism: Intervention outcomes. *Journal of Autism and Developmental Disorders* 32(4), 283–98.

Bölte, S., Feineis-Matthews, S., Leber, S., Dierks, T., Hubl, D., & Poustka, F. (2002). The development and evaluation of a computer-based program to test and to teach the recognition of facial affect. *International Journal of Circumpolar Health 61 Suppl* 2, 61–8.

Bretherton, I., & Beeghly, M. (1982). Talking about internal states: The acquisition of an explicit theory of mind. *Developmental Psychology*, 18(6), 906–21.

Castelli, F., Happe, F., Frith, U., & Frith, C. (2000). Movement and mind: a functional imaging study of perception and interpretation of complex intentional movement patterns. *Neuroimage* 12(3), 314–25.

Dautenhahn, K., & Werry, I. (2004) Towards interactive robots in autism therapy. *Pragmatics & Cognition*, 12(1), 1–35.

Dawson, G., Toth, K., Abbott, R., Osterling, J., Munson, J., Estes, A., & Liaw, J. (2004). Early social attention impairments in autism: Social orienting, joint attention, and attention to distress. *Developmental Psychology* 40(2), 271–83.

Dawson, G., Webb, S. J., & McPartland, J. (2005). Understanding the nature of face processing impairment in autism: insights from behavioral and electrophysiological studies. *Developmental Neuropsychology* 27(3), 403–24.

Dennett, D. (1987). *The Intentional Stance*. Cambridge, Mass: MIT Press/Bradford Books.

Dunn, L. M., Whetton, C., & Burley, J. (1997). *The British Picture Vocabulary Scale, Second Edition*. Windsor: NFER-Nelson.

Ekman, P. (1999). Basic emotions. In T. Dalgleish & M. Power (Eds.), *Handbook of Cognition and Emotion* (pp. 45–60). Sussex: John Wiley & Sons Ltd.

Golan, O., Ashwin, E., Granader, Y., McClintock, S., Day, K., Leggett, V., & Baron-Cohen, S. (2010) Enhancing emotion recognition in children with autism spectrum conditions: an intervention using animated vehicles with real emotional faces. *Journal of Autism and Developmental Disorders, 40,* 269–79

Golan, O., & Baron-Cohen, S. (2007). Teaching adults with autism spectrum conditions to recognize emotions: Systematic training for empathizing difficulties. In E. McGregor, N. Nunez, K. Cebula, & J. C. Gomez (Eds.), *Autism: An Integrated View* (pp. 236–59). Oxford: Blackwell Publishing.

Golan, O., Baron-Cohen, S., & Hill, J. J. (2006). The Cambridge Mindreading (CAM) Face-Voice Battery: testing complex emotion recognition in adults with and without Asperger syndrome. *Journal of Autism and Developmental Disorders*, 36(2), 169–83.

Golan, O., & Baron-Cohen, S. (2006). Systemizing empathy: teaching adults with Asperger syndrome or high-functioning autism to recognize complex emotions using interactive multimedia. *Development and Psychopathology* 18(2), 591–617.

Golan, O., Baron-Cohen, S., & Golan, Y. (2008). The 'Reading the Mind in Films' Task [Child Version]: Complex emotion and mental state recognition in children with and without autism spectrum conditions. *Journal of Autism and Developmental Disorders*.

Grice, S. J., Halit, H., Farroni, T., Baron-Cohen, S., Bolton, P., & Johnson, M. H. (2005). Neural correlates of eye-gaze detection in young children with autism. *Cortex* 41(3), 342–53.

Griffiths, P. (1997). *What emotions really are: the problem of psychological categories*. Chicago, IL: University of Chicago Press.

Grossman, J. B., Klin, A., Carter, A. S., & Volkmar, F. R. (2000). Verbal bias in recognition of facial emotions in children with Asperger syndrome. *Journal of Child Psychology and Psychiatry and Allied Disciplines* 41, 369–79.

Hadwin, J., Baron-Cohen, S., Howlin, P., & Hill, K. (1996) Can we teach children with autism to understand emotions, belief, or pretence? *Development and Psychopathology*, 8, 345–65.

Happe, F., & Frith, U. (2006). The weak coherence account: detail-focused cognitive style in autism spectrum disorders. *Journal of Autism and Developmental Disorders* 36(1), 5–25.

Hobson, R. P. (1986a). The autistic child's appraisal of expressions of emotion. *Journal of Child Psychology and Psychiatry* 27, 321–42.

Hobson, R. P. (1986b). The autistic child's appraisal of expressions of emotion: a further study. *Journal of Child Psychology and Psychiatry* 27, 671–80.

Holm, S. (1979). A simple sequentially rejective multiple test procedure. *Scandinavian Journal of Statistics*, 6, 65–70.

Jolliffe, T., & Baron-Cohen, S. (1997). Are people with autism or Asperger's syndrome faster than normal on the Embedded Figures Task? *Journal of Child Psychology and Psychiatry* 38, 527–34.

Kanner, L. (1943). Autistic disturbances of affective contact. *Nervous Child*, 2, 217–50.

Klin, A., Jones, W., Schultz, R., Volkmar, F., & Cohen, D. (2002). Visual fixation patterns during viewing of naturalistic social situations as predictors of social competence in individuals with autism. *Archives of General Psychiatry* 59(9), 809–16.

LaCava, P. G., Golan, O., Baron-Cohen, S., & Myles, B. S. (2007). Using assistive technology to teach emotion recognition to students with Asperger syndrome: A pilot study. *Remedial and Special Education*, 28(3), 174–81.

Langdell, T. (1978). Recognition of faces: an approach to the study of autism. *Journal of Child Psychology and Psychiatry* 19, 225–38.

Lawson, J., Baron-Cohen, S., & Wheelwright, S. (2004). Empathising and systemising in adults with and without Asperger syndrome. *Journal of Autism and Developmental Disorders* 34(3), 301–10.

Lord, C., Rutter, M., & Le Couteur, A. (1994). Autism Diagnostic Interview-Revised: a revised version of a diagnostic interview for caregivers of individuals with possible pervasive developmental disorders. *Journal of Autism and Developmental Disorders,* 24(5), 659–85.

Michotte, A. (1963). *The perception of causality*. Andover: Methuen.

Mottron, L., Dawson, M., Soulieres, I., Hubert, B., & Burack, J. (2006). Enhanced perceptual functioning in autisM: An update, and eight principles of autistic perception. *Journal of Autism and Developmental Disorders*, 36(1), 27–43.

O'Riordan, M. A., Plaisted, K. C., Driver, J., & Baron-Cohen, S. (2001). Superior visual search in autism. *Journal of Experimental Psychology, Human Perception and Performance* 27(3), 719–30.

Ortony, A., Clore, G., & Foss, M. (1987). The referential structure of the affective lexicon. *Cognitive Science*, 11, 341–64.

Owens, G., Granader, Y., Humphrey, A., & Baron-Cohen, S. (2008). LEGO therapy and the social use of language programme: an evaluation of two social skills interventions for children with high functioning autism and Asperger Syndrome. *Journal of Autism and Developmental Disorders* 38(10), 1944–57.

Premack, D. (1990). The infants theory of self-propelled objects. *Cognition*, 36(1), 1–16.

Ridgeway, D., Waters, E., & Kuczaj, S. A. (1985). Acquisition of emotion descriptive language: Receptive and productive vocabulary norms for ages 18 months to 6 years. *Developmental Psychology*, 21, 901–8.

Saxe, R., Carey, S., & Kanwisher, N. (2004). Understanding other minds: linking developmental psychology and functional neuroimaging. *Annual Review of Psychology,* 55, 87–124.

Scott, F. J., Baron-Cohen, S., Bolton, P., & Brayne, C. (2002). The CAST (Childhood Asperger Syndrome Test): preliminary development of a UK screen for mainstream primary-school-age children. *Autism,* 6(1), 9–31.

Shah, A., & Frith, U. (1983). An islet of ability in autistic children: a research note. *Journal of Child Psychology and Psychiatry,* 24(4), 613–20.

Silver, M., & Oakes, P. (2001). Evaluation of a new computer intervention to teach people with autism or Asperger syndrome to recognize and predict emotions in others. *Autism,* 5, 299–316.

Storm, C., & Storm, T. (1987). A taxonomic study of the vocabulary of emotions. *Journal of Personality and Social Psychology*, 53(4), 805–816.

Swettenham, J., Baron-Cohen, S., Charman, T., Cox, A., Baird, G., Drew, A., Rees, L., & Wheelwright, S. (1998). The frequency and distribution of spontaneous attention shifts between social and nonsocial stimuli in autistic, typically developing, and nonautistic developmentally delayed infants. *Journal of Child Psychology and Psychiatry and Allied Disciplines,* 39(5), 747–53.

Wakabayashi, A., Baron-Cohen, S., Uchiyama, T., Yoshida, Y., Kuroda, M., & Wheelwright, S. (2007). Empathizing and systemizing in adults with and without autism spectrum conditions: cross-cultural stability. *Journal of Autism and Developmental Disorders,* 37(10), 1823–32.

Wheelwright, S., Baron-Cohen, S., Goldenfeld, N., Delaney, J., Fine, D., Smith, R., Weil, L., & Wakabayashi, A. (2006). Predicting Autism Spectrum Quotient (AQ) from the Systemizing Quotient-Revised (SQ-R) and Empathy Quotient (EQ). *Brain Research,* 1079(1), 47–56.

Wellman, H., & Inagaki, K. (1997). *The emergence of core domains of thought: Children's reasoning about physical, psychological, and biological phenomena.* San Francisco, CA: Jossey-Bass Inc.

Yirmiya, N., Sigman, M., Kasari, C., & Mundy, P. (1992). Empathy and cognition in high functioning children with autism. *Child Development,* 63, 150–60.

Chapter 19

Schools and the new ecology of the human mind

Domenico Parisi

Overview

The human mind does not function in a void but it functions in a specific ecology and the ecology of the human mind does not remain always the same but it changes historically. In the last decades digital information technologies have created a new ecology for the human mind and today's schools have problems in achieving their goals because what they offer to students is the old ecology of the mind: books, lessons and a 'vertical' social structure with the teacher at the top. Learning in schools happens almost exclusively through language but learning through language, for many students, is not sufficiently accessible and motivating. The new digital technologies (visualizations, animations, computer simulations, virtual reality, computer games, Internet) make it possible to learn by participating in a 'horizontal' social structure, to supplement language with other communication channels, and to learn by doing rather than by being told. Scientists should help to identify advantages and limitations of the new digital technologies for learning and education, to indicate how to overcome the many resistances which are encountered when one tries to question and to radically change an age-old educational system and, since the new informational technologies have tremendous potential for learning, to participate in the design, implementation and testing of new digital technologies that realize this potential.

19.1 The ecology of mind

It is a general rule of biology that ecology shapes living organisms. The ecology of an animal is the physical environment in which the animal lives and the specific adaptive niche that the animal occupies in that environment and that is determined by the animal's body, sensory/motor systems and past adaptive history. What an animal is and what it does depends on the interactions of the animal with its ecology. The rule also applies to humans and in particular to their minds. The human mind has its own ecology and it is shaped by this ecology. Humans are special among animals because they substantially modify their environment and, since they learn most of what they do, instead of inheriting it encoded in their genotype, humans learn in an ecology which may change greatly from one generation to the next. While the genetic basis of the human mind is the result of changes which have happened over hundreds of thousands of years, learning occurs in a much more rapidly changing environment, and this is particularly true today when science and technology cause changes that may occur in years or even months. What we call the ecology of the mind includes how information is accessed and retrieved, what different types of information exist, how information can be manipulated by the individual and how the individual interacts with informational artefacts. And the ecology of the mind also includes the social environment and what society expects from the individual—how an individual interacts with other individuals and, since much of human learning is learning from others, who the other individuals are from whom the individual learns and is expected to learn. All of this has changed greatly in the last decades and it continues to change today at an accelerating pace.

19.2 **School systems and the new ecology of mind**

These simple observations explain the current crisis of educational systems in the West and, given the increasing Westernization of the world, in the entire world. Today's school systems were conceived when the ecology of the mind was very different from the present ecology and they were adapted to that ecology (see Chapter 20, this volume). How can we expect that educational systems which presuppose a particular ecology of the mind will function appropriately in a society where the ecology of the mind has so radically changed? There are in fact two ecologies of the mind today. One is the ecology of the mind which exists in the society outside the school and the other one is the ecology of the mind that the students find when they enter their school. These two ecologies are very different, and their differences inevitably create great difficulties for students, for their teachers and for the entire school system.

Let us consider some key differences between the old and new ecologies of the mind. A crucial difference concerns the role of verbal language. In the old ecology of the mind verbal language was the main or exclusive means for absorbing and communicating information, for thinking and for interacting with others. Educational systems and educational transmission not only were based on verbal language, the written language of books and the heard language of teacher's lessons, but they required a particular use of language, a 'sustained', intellectually challenging use, made up of analyses, arguments, distinctions, abstractions and cultural references. In the new ecology of the mind informational technologies have greatly increased the importance of visual language with respect to verbal language. In the new ecology of the mind the role of verbal language is not only more restricted but verbal language is used in a much more fragmented and intellectually unchallenging way.

This explains an important aspect of the current educational crisis. In the past, school education was aimed at a restricted minority of the entire population of girls and boys, those belonging to the families of the cultural and power elite. The extension of school education to the entire population—mass education—has been an important achievement of modernity but it has created a problem which is at the root of the current crisis of educational systems. For genetic or social reasons only a subset of all girls and boys can function adequately in a school system which is based on verbal language and, in fact, on a very sustained and intellectually challenging use of verbal language. If there are genetic reasons behind that, the problem is impossible to solve. If the reasons are social, it can be solved but it remains very difficult to solve and, in fact, is not solved except on rare occasions. The existence of a new ecology of the mind outside the school which is not based on verbal language makes the problem even more intractable because it indicates that language is not the only way in which the mind can function and communicate and convinces students that they can do without it.

The new (digital) informational technologies not only reduce the importance of verbal language but they create new artefacts which imply a much more active role on the part of their users, which is unlike the passive role required by reading books, listening to lessons and even seeing films and watching TV. This is true for almost all the more recently introduced digital artefacts such as the Internet, mobile phones and especially computer games and simulations. This is another important difference between the old and the new ecology of the mind which creates problems for traditional educational systems. Traditional educational systems presuppose a passive role of the learner while the new ecology of the mind allows or requires a two-way interaction between the learner and the source of his/her learning.

But the new ecology of the mind is not only new from a cognitive point of view, that is, in terms of the type of language which it uses and of the type of interactions with informational artefacts which it makes possible, but also from a social point of view. Schools are based on a *star-like* pattern of social interactions, where the teacher is at the centre of the star and he or she interacts with his/her students, while the interactions among students are discouraged. This schema of interactions has two justifications. One is organization and discipline in the classroom. If every student were to interact (talk, discuss, work with) with every other student in the classroom, this would be incompatible with the traditional way of working in the classroom and with the maintenance of discipline. The other justification is more important and it is the idea that teachers are the only source of knowledge and understanding for students, and therefore it is entirely appropriate that students interact with the teacher but not with the other students.

The new ecology of mind abandons this star-like pattern of social interactions and privileges a *network-like* pattern where everyone interacts with everyone else. This is part of a more general change which is taking place in today's societies and which is due to economic and political reasons: in present-day societies all individuals are equal because everyone is a consumer and a voter. The change can be described as the shift from a vertical to a horizontal society. In traditional, vertical, societies some individuals—parents, teachers, old people—were recognized as being at a 'higher level' so that it was appropriate to learn from them and to be told by them how to behave. In today's horizontal societies this hierarchical organization tends to be cancelled and everyone is entitled to learn from, and to be told how to behave by, everyone else. The new informational technologies are at the same time the cause and effect of these changes. They make possible a network of interactions 'among equals' where people chat, create social networks, are a 'crowd', and which is not compatible with the vertical organization of knowledge and understanding which characterizes school systems. This is especially clear in how the Internet is used today. The Internet and its search engines have become an important source of information for (almost) everyone, and especially for students. And on the Internet information is horizontally and not vertically created, for example, in Wikipedia. How can this new, horizontal, ecology of the mind be compatible with the vertical ecology of the mind still existing in schools and in school culture?

And, finally, the new informational technologies require changes in the physical and organizational aspects of educational systems. Schools are buildings, students must be together in classes, lessons have specific contents, and they take place at specific times. Recent technological innovations, such as portable laptop and tablet computers and smartphones, make it possible to learn by ignoring all these constraints. In fact, they make it possible for everyone to learn everything, in all places, at all times.

19.3 The role of the sciences of mind

We have said that the ecology of mind of students outside the school is very different from the ecology of mind the students find when they enter the school building, and this contrast creates many problems which are difficult to solve. One often hears today of a crisis in education (low test scores, high dropout rates, deficits in such basic skills as reading, mathematics and history, young people who do not read books and who are unable to articulate their thoughts), a crisis which creates specific problems for a society which is said to be increasingly based on knowledge

(Stewart, 1998). There may be many causes for the current crisis in education but, if one accepts the idea of an ecology of mind, the distance between the ecology of mind outside schools and the ecology of mind inside schools clearly is an important factor. What should we do about it?

One thing is clear. If we want to eliminate the discrepancy between the school's ecology of mind and the society's ecology of mind, we cannot change the society's ecology of mind. It is impossible to think that what has created the new ecology of the mind can simply disappear. The new informational technologies and the underlying economic and political reasons that create a horizontal society are here to stay and, if they were to change, certainly they would not change by returning to the informational technologies and the vertical society of the past. Therefore it is the school system which has to change.

And here we are confronted with two problems. The changes are radical and they involve all aspects of the school system: how students learn, the role of adults in their learning, where they learn, when they learn, what they learn. The other problem is that the school system is one of the institutions of society which is more resistant to change. It is very ancient, it has served Western societies well for millennia, it is an important part of the Western cultural tradition, there are many 'vested interests' which are opposed to change: teachers, parents, constructors of school buildings, publishers of school books, pedagogists, politicians. These problems are so difficult to tackle that today we do almost nothing to address the real, and very radical, need for change of our educational systems.

This is very bad because if we were to see clearly that the real problem of school systems is the clash between the school's ecology of mind and the society's ecology of mind, some steps towards solving the problem could be made. In a very tentative way we will end this chapter by indicating some of these steps.

The first step is to try to better understand the psychological, neural and social consequences of the new ecology of mind, raising the following questions, among others:

- Digital technologies make it possible to not have to store information in one's head because one can always and easily find information on the Internet when one needs it. But if mental creative work relies on the often non-conscious interaction of different pieces of information in one's head so that something new can be generated as a consequence of these interactions, how can creative work be performed when information in not in one's head but on online servers (see Chapter 11, this volume)?

- Does the capacity to find information on the Internet become an ability that replaces other types of mental abilities and creates new types of differences among individuals?

- What are the consequences of the new digital technologies for such basic cognitive capacities as attention, memory, reasoning and planning (Bauerlein, 2009; Jackson & McKibben, 2009; Small & Vorgan, 2008; Tapscott, 2008)?

- Do the new digital technologies reduce the impact of reading and writing or modify how people read and write? If the answers are affirmative, what are the consequences? If, as we are told, reading and writing have modified the manner of functioning of the mind since they were first invented, how will the new ways of expressing, communicating and using information modify the mind?

- Digital technologies greatly expand the power of non-verbal information and communication, of learning not by listening or reading but by seeing and doing. What are the consequences?

- The Internet tends to create a 'horizontal society' where there are no experts and, more generally, 'vertical' authorities but the interactions among users determine classifications, orderings, evaluations, choices (Surowiecki, 2005). What are the consequences?

Trying to answer these questions should be an important task for scientific research but today these questions are in many cases only the object of popular essays, not of serious and systematic scientific study (Prensky, 2010; Tapscott, 1999, 2008; the essays on 'neuroeducation' published in Battro, Fischer & Léna, 2007, do not mention information technologies.). The reason why psychologists and neuroscientists do not dedicate more of their time to the scientific study of the new ecology of mind (and, as we will see in a moment, of its extraordinary potential for learning) is that, generally, the ecological approach to the study of the human mind, and of behaviour more generally, plays a very marginal role in these disciplines and, in fact, behavioural ecology is a discipline which is mainly concerned with non-human animals. This is due to a variety of reasons but the critical one appears to be the central role that the use of laboratory experiments plays in psychology and in the neurosciences. Laboratory experiments are of critical importance for these disciplines but they are intrinsically non-ecological. The conditions in which the experimenter collects his/her data in the laboratory have the advantage of being controlled and measured but the disadvantage of being artificial, i.e. non-ecological. More specifically, the conditions to which participants are exposed in laboratory experiments are extremely simple and they cannot even attempt to reflect what makes the new ecology of the mind different from the old one. Psychologists and neuroscientists have the idea that in their experiments they capture the essential structure of the human mind, and that this essential structure is the same whatever the ecology in which the human mind lives. It is not clear if this is really true, but in any case we would like to know what the consequences are for the human mind to be exposed to different ecologies, and this is particularly true today when the new digital informational technologies create an entirely new ecology for the human mind.

But what we should ask scientists of the mind to do is not simply to find out what are the consequences of the new ecology for the human mind. If we knew these consequences, we would be in a better position to judge if they are good or bad. It is very possible that the new ecology of the mind does not have all good consequences, and therefore what psychologists and neuroscientists should do is suggest how to modify the new ecology of the human mind so that we can avoid its negative consequences and exploit its potential for learning.

But if one has the goal to try to influence the future development of the new informational technologies, one has to keep two considerations in mind. The first consideration is that the new informational technologies, like all present-day technologies, are like 'tigers' that need to be 'tamed' (to use an expression of Paul A. Samuelson, a Nobel Prize-winning economist). Today society is characterized by technological innovation but technological innovation is something which we (most of us) have little control over. Technologies are developed with mainly economic goals in mind and little attention is dedicated to their consequences beyond their specific area of application. The other consideration is that new digital informational technologies have not been developed as tools for learning and, in particular, as educational tools to be used in schools. Hence, if we want the new ecology of the mind to penetrate the school system, we should appropriately modify the new informational technologies so that they can be used as *the* educational tool of a new school system. So a crucial task for psychologists and neuroscientists would be to participate, in one way or another, in the development of the new informational technologies so as to help people have some control on their development and to realize their potential for learning.

19.4 The learning potential of the new ecology of mind: computer simulations

In fact, opening schools to the new ecology of the mind would not only eliminate a crucial contrast between school and society which is the main cause of the present educational crisis but would make it possible to explore the great potential for learning of the new digital informational technologies. Let us give only one example.

Most school learning is learning through language. Students learn by reading books and other written material and by listening to the teacher's lessons. The new informational technologies make it possible to learn not only by using language but by seeing and doing. The student sees something visualized on the screen of the computer and he or she responds by doing something and observing the consequences. This is a very basic way of learning in everyday life and it is also the basic way of learning of science because in the laboratory the experimenter manipulates the conditions and observes the consequences of his/her manipulations. As we have already said, one of the problems of learning through language is that, especially in today's mass education, not all students have the motivation and the capacity to learn by manipulating words which are completely detached from what they refer to. The new digital technologies can help us solve this problem. There are many different ways of having students learn by seeing and doing with the new digital technologies (visualizations, animations, virtual reality, computer games, Internet; for an account of how computer games can help children learn, see Shaffer, 2008) but one of the most promising ones is to construct computer simulations of all sorts of phenomena, to design an appropriate interface for these simulations and to have students learn by interacting with the simulations. Computer simulations are a powerful new tool for performing scientific research which is being increasingly used in almost all scientific disciplines. Theories and models are not expressed verbally or by using mathematical symbols but they are used to construct computer simulations which run on the computer and produce results that, like the predictions derived from a theory or model, must match the empirical data. But computer simulations can also be an important new educational tool. Everything can be simulated, from the phenomena of nature studied by physics, chemistry and biology, to the phenomena of human behaviour and human societies which are studied by the cognitive and social sciences, including history. Students can learn about all these different phenomena by interacting with simulations that reproduce the phenomena on a computer (Aldrich, 2005; Thorson, 1979). And simulations can be virtual experimental laboratories where the student does experiments by manipulating the conditions in which the simulated phenomena take place and observing the results of his/her manipulations. This has many advantages with respect to real experiments in the real school laboratory: (1) while only a very restricted subset of phenomena can be studied in the real laboratory, all phenomena can be studied in the virtual laboratory of a computer simulation, including those studied by the social and historical sciences; (2) virtual experiments are much less expensive and organizationally costly than real experiments; (3) only real phenomena can be studied in the real laboratory while both real and counterfactual phenomena can be studied in the virtual laboratory so as to explore answers to such questions as 'How would physical reality look like if there were no gravitation?' or 'What if Napoleon never existed?'. Using simulations to learn can increase the motivation to learn, which is one of the most serious problems in today's education system. Learning from language is a passive form of learning, especially if the student is not able or not willing to elaborate in his/her mind what he or she has read. Using simulations to learn is learning 'in the active mode': the student decides what he or she wants to see and to do and, what is crucially important, he or she will know what the effects are of what he or does. As psychologists tell us, learning by doing increases our motivation to learn. This is especially true if the simulations are used to create games which challenge the abilities of the student and can be competitive games with the help of Internet (Moreno-Ger, Burgos & Torrente, 2009; for a classical view of the relationship between motivation and learning in 'pure' psychology, see Miller, 2007).

Computer simulations are an important new tool for learning but there are many other important and useful ways in which the new informational technologies can change educational systems which are unable to function appropriately in a changed society. The new digital technologies make it possible to learn by participating in a 'horizontal' social network including other students, teachers and experts, compared with the 'star-like' network with the teacher at the centre of the star, and they can be used to implement radical changes in the educational curriculum such as those proposed by Roger Shank from traditional school subjects to the learning of skills (Shank, 2005). (For a recent account of the impact, or lack of impact, of the new digital technologies on the school system, see Collins and Halverson, 2009.)

References

Aldrich, C. (2005). *Learning by Doing: A Comprehensive Guide to Simulations, Computer Games, and Pedagogy in e-Learning and Other Educational Experiences*. New York: Wiley.

Bauerlein, M. (2009). *The Dumbest Generation: How the Digital Age stupefies Young Americans and Jeopardizes Our Future*. London: Tarcher.

Battro, M., Fischer, K.W., & Léna, P.J. (Eds) (2007). *The Educated Brain: Essays in Neuroeducation*. Cambridge: Cambridge University Press.

Collins, A., & Halverson, R. (2009). *Rethinking Education in the Age of Technology: The Digital Revolution and Schooling in America*. New York: Teachers College Press.

Jackson, M., & McKibben, B. (2009). *Distracted: The Erosion of Attention and the Coming Dark Age*. Amherst, NY: Prometheus Books.

Miller, N. (2007). *Learning, Motivation, and their physiological mechanisms*. Piscataway, NJ: Aldine Transaction.

Moreno-Ger, P., Burgos, D., Torrente, J. (2009). Digital games in elearning environments: Current uses and emerging trends. *Simulation Gaming*, 40, 669–87.

Prensky, M.R. (2010). *Teaching Digital Natives: Partnering for Real Learning*. London: Corwin Press.

Schank, R. (2005). *Lessons in Learning, e-Learning, and Training: Perspectives and Guidance for the Enlightened Trainer*. New York: Pfeiffer.

Shaffer, D.W. (2008). *How computer games help children learn*. Basingstoke: Palgrave Macmillan.

Small, G., & Vorgan, G. (2008). *iBrain: Surviving the Technological Alteration of the Modern Mind*. New York: Collins Living.

Stewart, T.A. (1998). *Intellectual Capital: The New Wealth of Organizations*. New York: Random House.

Surowiecki, J. (2005). *The Wisdom of the Crowds*. Garden City, NY: Anchor.

Tapscott, D. (1999). *Growing Up Digital: The Rise of the Net Generation*. New York: McGraw Hill.

Tapscott, D. (2008). *Grown Up Digital: How the Net Generation is Changing Our World*. New York: McGraw Hill.

Thorson, E. (Ed.) (1979). *Simulation in Higher Education*. Miami, FL: Exposition Press of Florida.

Chapter 20

Brain-science and education in Japan

Hideaki Koizumi

Overview

From the viewpoint of neuroscience, learning and education can be defined as the processes of making neuronal circuits in response to external stimuli, and of controlling or preparing appropriate stimuli, respectively. Learning and education can thus be studied as a new field with brain science that includes disciplines related to the brain, e.g. neuropsychology, developmental cognitive neuroscience, behavioural sciences and child neurology as well as neuroscience. We named this new discipline 'brain-science and education'. National programmes run by the Japan Science and Technology Agency (JST) in 'brain-science and education' began in December 2000 and 19 projects were completed by March 2011. In this Japan initiative, not only were conventional non-invasive brain function imaging techniques such as functional magnetic resonance imaging (fMRI), magnetoencephalography (MEG) and electroencephalogram (EEG) used, but also a new methodology, near infrared spectroscopic optical topography (NIR-OT) was employed, resulting in the elucidation of various learning processes in the brains of infants and children. Various longitudinal studies were performed, including studies in collaboration with local areas in Japan, with a large number of monozygotic and dizygotic twins, and with school children on second language acquisition. The neuronal basis of learning and education was also studied with neonates, infants and children to elucidate early localization in the brain, early learning process for language and early development of working memory. Some results started to be applied to educational politics. In this chapter, the author would like to share information on the special situation in Japanese education and some of the 'brain-science and education' initiative's results that may be important in future neuroscience in education.

20.1 Introduction

In the Edo era of Japan (1603–1868), children in samurai families studied academic subjects and received lessons to acquire specific skills from early childhood at schools called *hankō*, which were run by local governments called *han*, whereas many children in the general public learned reading and mathematics and acquired practical knowledge and skills at private educational facilities called *terakoya*. It is said that approximately 15,000 *terakoya* schools existed nationwide at the end of the Edo era. At the time of the Meiji Restoration (1868), the national literacy rate for the Japanese language was estimated to be 43% for boys and 10% for girls, which at the time was the world's highest level for native language literacy (Dore, 1965). This is why the ongoing educational improvement campaign conducted by the United Nations Educational, Scientific and Cultural organization (UNESCO) is called the 'UNESCO World Terakoya Movement'. After the Meiji Restoration, following the long-term national seclusion from the outside of the world, Japan adopted an administrative policy to prioritize education to quickly absorb Western cultures and scientific technologies. Such historical background is considered to be part of the reason for the Japanese government's early promotion of brain-science and education research, which I discuss in this chapter.

In the late 1960s, Jushichiro Naito (1906–2007), a paediatrician and director of Aiiku Hospital, invited neuroscientist Toshihiko Tokizane (1909–1973), a professor at the Faculty of Medicine at the University of Tokyo, to begin a workshop on childcare and neuroscience, which was the first step toward the application of neuroscience for childcare and education. His disciples compiled the fruits of such efforts and published *Noh to Hoiku* ('Brain and Childcare') in 1974 after Tokizane passed away. Since then, the application of neuroscience in studies on childcare and education studies has long been a dream of Japanese neuroscientists.

In 1992, fMRI for humans (Ogawa et al., 1992) was practically applied (Yamamoto et al., 1992), and was recognized as a cue to develop neuroscience for the humanities and social sciences. Until then, it was only a neuroscientist's dream to apply neuroscience to education as it was impractical to do so based solely on animal experiments. This dream came into fruition with the discovery of non-invasive higher-order brain-function imaging methods beginning with fMRI.

As mentioned earlier, 'brain-science and education' started with the relationship between childcare and neuroscience, as information on changes taking place in the human brain in the early stages of development is extremely important for childcare. With fMRI, however, higher-order brain-function imaging can only be conducted by confining the subject within a restricted measurement space and therefore, its application to babies and young children was difficult. As a solution, optical topography (NIRS-OT) which applied near-infrared spectroscopy (NIRS) was developed as a new higher-order brain-function imaging method (Koizumi et al., 1999; Maki et al., 1995).

Moreover, to clarify the relationship between the brain and learning/education, a longitudinal study—the tracking and observation of groups of individuals over time—is important, and various cohort studies were conducted as part of a national project. Through this initiative, based on brain-science and education, many different fields such as paediatric neurology, psychology, behavioural development studies and linguistics were bridged and fused in a transdisciplinary manner. In this way, a new academic field, 'brain-science and education', which incorporates the perspective of actual application to education sites, was established.

20.2 Concept of brain-science and education

To promote this research, the concepts of *learning* and *education* were newly defined from a biological viewpoint. *Learning* was defined as the 'process of forming neural circuits' based on the environmental stimuli received (everything surrounding the self) and *education* as the 'process of controlling and supplementing necessary stimuli to configure neural circuits' (Koizumi, 2000).

Learning, that is, the biological process to form neural circuits, consists of morphological and functional connectivities of the synapses. Morphological connectivity is noticeably formed during early childhood, where new interneural connectivity is formed and further modified, resulting in more environmentally adjusted neural circuits. Functional connectivity is noticeable in adults. Conductivity between existing synapses is strengthened or weakened, resulting in the formation of new neural circuits.

Moreover, the concept of *learning* contains both passive and active learning. An example of passive learning, which starts right after birth, is to configure perceptual neural circuits, e.g. vision, while adjusting them to the environment receiving environmental stimuli. In contrast, active learning is conscious learning out of motivation. The direct influence of passions and aspirations on learning efficiency and the importance of such influence have been clarified by neuroscience.

Education is inherently a human-specific activity. Even *Pan paniscus* (bonobos) and *Pan troglodytes* (chimpanzees), which are the closest living relatives to *Homo sapiens*, have not shown any active educational activity. A rare exception is that of songbirds which teach songs to their chicks. Other than that, educational activities are conducted only by humans. Even Immanuel Kant (1724–1804) noted it in *On Pedagogy* (*Über Pädagogik*). In an experiment that involved nurturing by a substitute parent, Kant recognized a prototype of education in songbirds after discovering that sparrows nurtured by a canary sing.

The concept of brain-science and education is to pursue more desirable learning, education and child care for humans, as well as a nurturing environment and new curricula that society should prepare by fully utilizing developmental cognitive neuroscientific knowledge acquired by the use of the non-invasive, high-level brain-function measurement methods that have been rapidly developed in recent years and behavioural development-related accumulated research results. Brain-science and education is a new attempt to bridge and fuse education, which traditionally occurred in the humanities and social sciences, with the natural sciences (Ansari and Coch, 2006; Goswami, 2006; Koizumi, 2000, 2004, 2010).

A significant portion of *The Republic* by Plato (428 BC–348 BC), is devoted to theories on education. Much historical data have shown that the survival of a nation or an organization depends on education. Children's sound mental and physical growth is important because the future of the nation and the world will eventually be entrusted to the next generation.

The environment surrounding children, however, has changed significantly in recent years. Developed nations including Japan have seen remarkable trends in the volume of information available, pursuit of higher efficiency, increasing individualism and the aging of society. The reality is that children are being nurtured in an environment that humans have never before experienced.

Educational problems and child-related social problems are increasingly serious, which is partially a result of the radical changes in the environment surrounding children and their parents. Evidence-based support and initiatives from the perspective of brain-science and proper direction are needed.

Child care and education have been discussed based on various experiences to date. Educational principles and philosophy have been mainly based on the humanities such as the study of education, philosophy and psychology. In the rapidly changing environment surrounding children, however, discussions on the direction of practical education tend to result in circular argument and educational policies fluctuate like a pendulum. Given such a dire situation, evidence-based educational and practical theories that involve the natural sciences are necessary (Koizumi, 2008).

20.3 **The dawn of research on neuroscience in education**

The Organization of Economic Cooperation and Development (OECD), consisting of 30 member countries, has shown a keen interest in issues on education. It has been more than 40 years since the establishment of the Center of Educational Research and Innovation (CERI) in 1968, after which, the Directorate of Education was launched in 2003 and CERI was positioned as a major division in this directorate. Three international conferences related to learning science and brain research have been held in 2000 and 2001 under CERI's research programme 'Learning Sciences and Brain Research'. In April 2002, Phase II of the research programme began and its central roles have been followed through by Bruno della Chiesa, based on the idea conceived by Jarl Bengston, who was a councillor and the head of CERI. Phase II consists of three research networks: (1) literacy (learning of reading and writing), (2) numeracy (learning of mathematics) and (3) lifelong-learning (learning that continues throughout one's life), and an advanced international team of approximately 30 countries were involved in aggressive research activities. As a result, numerous theses were produced and the OECD compiled these results in an easily accessible book-form to facilitate the distribution and application of the results in society and the general public. The English and French editions were initially published in 2007, and translations into other languages are being planned.

In November 2003, an international conference, 'Mind, Brain and Education', was held by the Pontifical Academy of Sciences, commemorating the 400th anniversary of the world's oldest science academy, which was established in the era of Galileo Galilei. Nearly half of the approximately 80 members of the academy are recipients of the Nobel Prize. 'Mind, Brain and Education' was chosen as the most important theme for current science (Battro, Fischer & Lena, 2008). As a result of this conference, another international conference was held in October 2004 at Harvard University School of Education. Concurrently the International Mind, Brain and Education Society (IMBES) was launched, and Kurt Fischer and Antonio Battro have led this group since its inception.

In 2007, the *Journal of Mind, Brain, and Education* was first published by Blackwell (Fischer et al., 2007). This academic periodical is suitable for new fields that comprehensively bridge and fuse psychology, neuroscience, the study of education and the practical applications of the results of the study of education. Expectations are high for this periodical against the backdrop of the fact there are not many periodicals on transdisciplinary bridging and fusion of different fields, and it was awarded The Best New Journal Award by the American Association of Publishers (AAP).

China has also recently begun to focus on brain-science and education, and these efforts have been led by Dr Wei Yu, vice-minister of education of China. She attended the LSBR programme Phase II launch symposium as an observer (The Royal Institution of Great Britain, 2002). Ministry of Education-level and national-level centres have been established at universities and similar research and educational institutions. Education and national defence are the pillars of national security.

In Japan, based on the concept 'Importance of interaction between environment and the brain' which was presented at the 1996 JST transdisciplinary forum on global environmental, the 2000 JST transdisciplinary forum entitled 'Developing the Brain: The Science of Learning & Education' was held in 2000 under the chairmanship of the author. The conference was deliberately set in the final year of the 20th century to develop a new research field for the 21st century. During this 4-day forum, attended by the former Japanese Minister of Education, the formation of new academic fields such as 'Brain Science and Education' and 'Developing the Brain' was clarified. An innovative research programme, 'Brain-Science and Education', which invited public participation in proposing ideas and undertaking research, was prepared by JST, and this programme was launched officially in 2001. From among the many public proposals, 12 innovative and intriguing research themes were adopted. This trailblazing programme was completed in 2011 by adding seven themes. The aforementioned activity is summarized in Figures 20.1–20.3 and Tables 20.1 and 20.2.

Applied Research

JST*/RISTEX**
(Contract-type research)

From 2001 ~
**"Brain-Science &
Education Type I"**
(12 projects)

JST*/RISTEX**

"Healthy development of the mind and brain,
including physical and linguistic contexts"

From 2004 ~(6 projects+ JCS)
"Brain-Science & Education Type II" (6 projects)
"Japan Children's Study: JCS" (Birth Cohort Study)

July 2003 Council final report

OECD/CERI
"Learning Sciences
& Brain Research"
Phase I from 1999
Phase II from 2002
(3 networks, ~30 projects)

"Life-long Learning"
(Coordinated by Japan)

JST/CREST
"Elucidation of a learning mechanism based
on the knowledge of the development of
brain functions "   From 2003 ~(15 projects)

RIKEN Brain Science Institute
"Nurturing the brain"
                    From 2003 (~ 6 projects)

July 2002 Council intermediate report

Basic Research

*JST: Japan Science & Technology Agency
**RISTEX: Research Institute of Science and Technology for Society
***MEXT's Council for Brain-Science and Education

Fig. 20.1 'Brain-Science & Education' initiative in Japan.

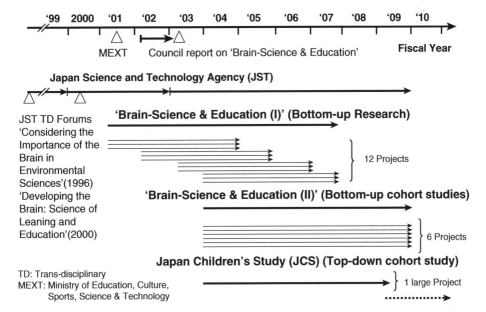

Fig. 20.2 Brain-Science & Education projects in 'Brain-Science & Society' division at Institute of Science and Technology for Society in Japan Science and Technology Agency.

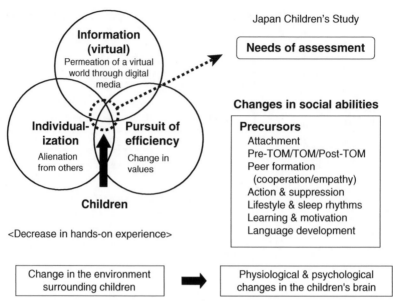

Fig. 20.3 Objective of birth cohort studies in three local areas (Japan Children's Study: JCS).

Table 20.1 Brain-Science & Education Type (I) Programme in JST selected by public announcement in collaboration with neuroscientists, educators and practitioners.

Themes of 12 projects:	
1. Development of prefrontal cortex functions (working memory and origins of basic learning)	2001–2005
2. Development of communication (language and brain cross modalities)	
3. Plasticity of the brain (sensitive periods of human brain functions)	
4. Development of memories (bases of various kinds of learning)	2002–2006
5. Molecular bases of plasticity (biological mechanisms of critical periods)	
6. Practical problems in schools (non-attendance at school, etc.)	
7. Effects of media on the brain (TV watching & video games)	2003–2007
8. Brain mechanisms on learning disabilities (interventions, etc.)	
9. Environmental influence to gene expression (genetic and epigenetic)	
10. Mechanism of facial recognition and its learning and development	2004–2008
11. Phonological language acquisition (mechanisms and applications)	
12. Analysis of non-verbal communications between mother and baby	

Applied brain-function-imaging methods
fMRI: 1, 2, 3, 4, 6, 7, 8, 9, 10, 11, 12 MEG: 2, 10, 11 NIR-OT: 1, 2, 3, 6, 7, 9, 11, 12

20.4 Research methods and neuroethics

Non-invasive brain function imaging

The advantage of fMRI is relatively high spatial resolution (3–5 mm), and its limitations are a need to restrain subjects, low temporal resolution and the use of a strong magnetic field (Ogawa et al.,

Table 20.2 Brain-Science & Education Type (II) Programme in JST selected by public announcement in collaboration with neuroscientists, educators and practitioners. (Cohort studies with brain-function imaging.)

Themes of six projects:	
1. Tokyo Twin Cohort Project (ToTCoP) —longitudinal study of twins in infancy & childhood –	2004–2010
2. Cohort studies on language acquisition, cerebral specialization and language education	
3. A cohort study of autism spectrum disorders: a multidisciplinary approach to the exploration of social origin in atypical and typical development	
4. Cohort studies of higher brain function of normal elders and children with learning disabilities	
5. Development of a new biomedical tool for student mental health	
6. Cohort study with functional neuroimaging on motivation of learning and learning efficiency	

Applied brain-function-imaging methods:
  fMRI: 2, 3, 4, 5, 6 MEG: 2 NIR-OT: 1, 2, 3, 4, 5

1992; Yamamoto et al., 1992). We also developed near infrared spectroscopic optical topography (NIRS-OT) (Maki et al., 1995; Koizumi et al., 1999), and applied it to the field of brain-science and education. NIRS-OT is a non-invasive brain function imaging technology which observes the change in rCBV (the regional cerebral blood volume) which is linked to neuronal activities. A near infrared light irradiated from the scalp through the skull is reflected on the cerebral cortex, and then detected on the scalp again between hair roots. A NIRS-OT head cap uses fine optical fibres which are light and completely non-invasive.

The advantages of NIRS-OT are an ability to measure in natural environments, safety (Ito, Kennan, Watanabe & Koizumi, 2000; Kiguchi et al., 2007), the freedom for subjects to move (Atsumori et al., 2010). The limitation of this non-invasive imaging method is its relatively low spatial resolution (15–25 mm) due to a large light scattering on the bio-tissue, which limits observation to the surface of the cerebral cortex. NIRS-OT is needed for research on development during early childhood and in natural environments.

In the field of brain-science and education, NIRS-OT is frequently used as there are no other non-invasive brain-function imaging methods that can be utilized. Although one of the most basic questions on *tabula rasa* posed by John Locke (1632–1704)—at what stage does the region-specific properties of the brain start?—was a mystery, we are gradually obtaining new knowledge on the appearance of brain localization in infant brains (Taga, Asakawa, Maki, Konishi & Koizumi, 2003), and the early development of the prefrontal cortex (Nakano, Watanabe, Homae & Taga, 2009). Further, NIRS-OT was used to measure brain functions related to newborn babies' special response to their mother tongue, resulting in a finding that babies show a noticeable response to their mother tongue only a few days after birth (Pena et al., 2003). Recently, it was discovered that babies can recognize human frontal faces at about 5 months after birth, whereas it takes about 8 months from birth for babies to be able to recognize side faces (Nakato et al., 2009). More recently, optical measurements of neural connectivity between functional areas have become possible and expectations are high for future development (Homae, 2010). Figure 20.4 shows measurement examples of newborn babies' brain activities.

At a research level, new wearable methods that enable the simultaneous measurement of multiple brain activities (i.e. wearable optical topography: WOT) have been developed (Atsumori et al., 2009).

Although the observation of brain functions has been limited to the brain of an individual, research on the interaction of multiple brains is about to begin as a new concept in brain-science research. The day will soon come when a bilateral relationship (e.g. a mother-and-child relationship) or a trilateral relationship (e.g. a mother-and-child relationship involving a third party or an object as an intermediary) will be discussed based on the data from brain-function imaging.

As various non-invasive brain function measurement technologies progress, research on brain-science and education is developing rapidly. All non-invasive brain function measurement technologies, however, still have limitations. It is therefore necessary to optimize the strengths of each method and combine multiple methods. In particular, with regard to fMRI and NIRS-OT, the mechanism of local blood volume and the blood flow volume of the brain relative to neural activity has yet to be fully unravelled. It is an urgent issue to unravel this mechanism in line with the neurophysiology that has been achieved in animal experiments. As to the capillary blood flow adjustment mechanism to carry energy (oxygen and glucose combined with intrablood carrier haemoglobin) for neural activity, we must take into account neurotransmission routes including neuroglial cells and blood flow adjustments based on the physical pressure by swelling.

fMRI is superior in terms of the comprehensive understanding it provides on brain functions using its relatively high spatial resolution and the three-dimensional imaging of the entire brain. The tractography of neural fibres and the unravelling of connectivity between functional cerebral areas through the use of anisotropic diffusion MRI, is an important theme of brain-science and education in the process of understanding learning and education. Diffusion MRI is also important for understanding the development of functional cerebral areas and comprehensive neural circuits because it enables observation of myelination, or the membrane-coating of unmyelinated nerves.

Cohort study

The JST research on brain-science and education has evolved into a wider framework of Brain-Science and Society. In 2004, a longitudinal study to observe developments and learning over time was launched. Figure 20.3 shows the concept and objective of the new Birth Cohort Study (Japan Children's Study: JCS). Six projects included in this research combine the cohort study (the tracking and observation of groups of individuals over time) and high-level brain-function imaging shown in Table 20.2. Furthermore, the development cohort study JCS to track social abilities including precursor elements from birth, was conducted in the Brain-Science and Society research area. The objective of this study was to elucidate the development of social abilities including precursors. This new cohort study is unique in that professional researchers such as medical doctors (paediatricians, child-neurologists), nurses, and development psychologists continue to observe children develop, thus differing from conventional cohort study methods which use survey sheets. Specialized research groups collaborate to deepen the methodologies for observation and analyses involving real-life situations. A deductive verification was made on the relationship between exposure to the environment (everything except self) and its outcome that occurs on children using the appropriate measures, so called instruments, based on working hypotheses. Unknown development mechanisms were inductively searched (Yamagata, Sadato, Anme & Maeda, 2010) from the results of multifaceted observations that closely follow children's growth.

In the social sciences and the humanities, for which there are many related parameters, it is often difficult to clarify cause-and-effect relationships even though various relationships can be identified. It becomes important here that the time arrow, which is a basic principle of physics, flows only from the past to the future because we can clearly state that cause always precede result. Cohort studies that are conducted for a certain extensive period of time bring forth knowledge to clarify cause-and-effect relationships using time factors. Although few neuroscientists are interested in cohort studies at this time, I believe that will change in the future. Most data in current neuroscience research derive from experiments using rodents such as mice and rats. There are significant gaps, however, between the evolution of rodents and that of primates. For instance, from the anatomical point of view, rodents have a small cortex that corresponds to primates' prefrontal cortex. At the evolutionary stage of *Homo sapiens*, the cerebral areas responsible for mental functions such as languages and ethics, which relates to human dignity, are important. Ultimately, human-related research can only be conducted using human subjects, in which lies the essential importance of cohort studies.

Brain-science and ethics

In cohort studies, personal information is managed strictly and ethical considerations are prioritized. From the beginning of the cohort studies in 2004, we established a dedicated research group on neuroethics, independent from the Ethics Committee governed by neutral external specialists on ELSI (ethical, legal and social issues). This neuroethics research group consisted of a group leader (Osamu Sakura, Professor of the University of Tokyo) with a fulltime researcher, and more than 10 consulting board members. The group worked to solve actual new ELSI problems in the Brain-Science & Education cohort studies, and also conducted several international or domestic symposiums on neuroethics. As a result, much effort has been focused on the creation of a new research field called Brain-Science and Ethics (Koizumi, 2007). Brain-Science and Ethics consists of two difference concepts, such as 'Ethics of Brain-Science' (neuroethics) (Illes, 2005) and 'Brain-Science of Ethics' (Aoki, Funane & Koizumi, 2010; Della-Chiesa, 2010). Due to the careful deliberations and activities with the neuroethics research group as well as the personal data handling group, no serious ELSI problems were experienced in our numerous cohort studies which involved the direct cooperation of more than 10,000 children and babies, and their care givers.

20.5 Examples of research on brain-science and education

In the new field of brain-science and education, many attempts (including those overseas) have been made in the last decade to bring forth clearer comprehensive concepts and enable the development of various specific themes. Brain-function maps were originally based on the study of the brain area that was responsible for illness. Starting with special cases, once knowledge is accumulated, concepts are generalized and general applications become possible. Likewise, brain-science and education can be launched first in relation to special education involving the brain-damaged illness and developmental disorders, and then extended to general elementary, middle-school and higher education and life-long learning and education.

As shown in Tables 20.1 and 20.2, we worked on various themes deemed important and urgent. We investigated time and budget for nominating these themes by organizing working group, committees as well as our own investigation and hearings. The results of the 19 projects over 10 years were reported on the homepage of JST (JST, 2011). These research programmes were evaluated by external official committees on peer review and accountability. Since most of our reports are written in Japanese, I would like to select and share some examples which might be important for the future of neuroscience in education.

Unravelling of the intrabrain mechanism to form a critical period

The mechanism of a critical period has also been unravelled by several of our projects and projects in brain-science and education initiative in Japan. For example, animal experiments have clarified that the neurotransmitter GABA (gamma-amino-butyric acid) contributes to the beginning and closing of a critical period. Benzodiazepine—the GABA agonist that promotes receptor connectivity—accelerates the start of the critical period, whereas the antagonist accelerates the closing of the critical period (Hensch & Stryker, 2004). Whether it is acceptable for humans to control such a natural phenomenon as the critical period is an issue for brain-science and ethics. Moreover, research under the CREST programme by the JST revealed that as neural myelination progresses, morphological neural plasticity (the creation of neural circuits by forming new dendrites and synapses) ceases and functional neuroplasticity (the creation of neural circuits by changing the information transmission efficiency of neural synapses) becomes a major part of learning.

Morphological neural plasticity during early childhood and its critical period, which is the basis for learning during the period, are an essential part of learning and education. Although the research on the critical period appears to be a basic study, it can provide various hints to actual learning activities (Kato, Watabe & Manabe, 2009).

With regard to high-level brain functions such as the acquisition of languages other than one's mother tongue, uniform discussion on the critical period is not necessarily appropriate. The critical period for intuitive distinction and the categorical cognition of consonants and vowels is approximately 1 year from birth. However, many functions need to be integrated to enable the acquisition of foreign languages and a simple concept of the critical period does not work for this development. Our experiments have revealed that the motivation or will to learn can substantially change the efficiency of learning, so rather than a '*sensitive period*' the term '*optimal period*' is considered more suitable.

Dynamic interaction of heredity and environment

A large-scale cohort study on twins was conducted in the brain-science and education programme. Monozygotic twins and dizygotic twins were observed and investigated using a cohort study. We asked the parents of twins born in Tokyo and surrounding prefectures for cooperation. The care givers of 1760 pairs of twins responded to the recruiting from a database of 47,000 pairs of twins, and the twins and the care givers were observed at 6–7, 9, 12, 15, 18, 24–30, 36, 42 and 48 months old during the 4-year research period. A significant volume of basic knowledge was obtained. The educational effects on development, learning and child care involve both environmental and genetic factors. Given that there is no genetic difference between monozygotic twins, developments that show a significant difference between monozygotic twins can be understood to have been strongly affected by unshared environmental factors. On the other hand, if no difference is recognized, we can understand that even environmental factors did not have any effect. Given that there is a difference in the genetic factors of dizygotic twins, when the observation results on one development item differ significantly for dizygotic twins, we can understand that both genetic and environmental factors influenced them to a certain extent, the ratio of which is clarified by comparing the results of observation from monozygotic twins. If no difference is found, there are two possibilities—neither genetic nor environmental factors have influenced them or those factors may have offset each other, which can be clarified by comparison with the observation results from monozygotic twins. With regard to the development items that were observed, a dynamic mechanism through the interaction of genetic and environmental factors became apparent. These results are considered to be significant in the origin of education and child care. A huge amount of data is still under investigation (Ozaki, Toyoda, Iwama, Kubo & Ando, 2011).

Mechanism and cure of developmental disorders

New knowledge in brain-science has enabled a new focus on development disorders. Indeed, the number of patients with autism spectrum disorders (ASD), which include Kanner-type autism,

Asperger's syndrome and others, is increasing worldwide. One reason for this is the increasing detection of patients due to progress in diagnostic methods and clinical conditions. It is necessary to unravel the mechanism behind ASD, determine early diagnostic markers and develop effective scientific data-based therapies and nurturing methods. Noticeable symptoms of disorders during development are seen especially in the formation of sleep rhythm, physical abilities such as crawling, linguistic development and the theory of mind (e.g. mind-reading and empathy). Over the last decade, research on the responsible genes has also been aggressively promoted, leading to results which suggest that such disorders are caused not by particular genes but rather by deep-rooted problems such as, for instance, an interneural connectivity (synapse)-related genetic problem (Fujita et al., 2008).

Brain activity processes in the acquisition of a second language

A longitudinal observation was made on the intrabrain language processing methodology for the acquisition of a second language with more than 500 subjects using optical topography and a brain EEG (multiple-point electroencephalograph) at public schools that provide early second-language (English) education and special schools that conduct immersion education. Our first approach was to observe the N400-level signals that surface when we hear words without going deep into complicated grammar and then we analysed those signals. As a result, it was discovered that the pattern of activating the entire brain changes in the process of acquiring a second language and that language acquisition proficiency indicators can be obtained. A subsequent study has continued in different programmes to yield progress in this research area (Ojima, et al., 2011) Recently, NIRS-OT succeeded in resolving more precise language areas in second language acquisition resulting in more meaningful data than has been obtained previously by EEG (Sugiura, 2011).

From special education to general education

To apply brain-science to learning and education, the difficulties of extending the application from special cases to general cases has become apparent.

In the field of brain-science and education, it has been shown that moving from research on noticeable specific symptoms to that on normal cases is suited for the exploration of new research fields related to humans. It is not easy, however, to conduct research on healthy people, including babies and infants due to ethical considerations which are a priority. Direct experiments with humans are therefore minimized. It has become apparent that cohort studies—the observation of groups of children over time in line with their growth—can be an important avenue of research.

Nearly a decade of research has revealed the significance of first establishing research fields for developmental and/or learning disorders caused by damage to brain functions such as special support education for babies and young children and special education before presenting new learning methods that directly contribute to general compulsory education. On the other hand, knowledge on brain-science contributes directly to various initiatives for sound quality-of-life aging and measures to alleviate or delay the symptoms of cognitive disorders. As disorders make it easy to clearly identify brain functions, it is relatively easy to directly apply knowledge of brain-science. New brain-function research fields require a process of discovering new knowledge from cases of particular disorders, followed by the careful application of such knowledge to general cases.

Intrabrain mechanism for intrinsic learning and extrinsic learning

Actual learning often results in a pleasurable sensation via a reward system. Learning driven by a reward system has existed since the beginning of animal evolution. With pleasurable and unpleasurable sensations as a compass for choosing a higher probability of survival, positive learning was promoted in animal evolution. In human evolution, the motivation for learning further developed to include a mental function to predict rewards. In the study of education, a classification of intrinsic learning and extrinsic learning is sometimes made. Intrinsic learning is self-motivated learning by feeling pleasure in learning itself, whereas extrinsic learning is learning motivated by the expectation of a reward as a result of learning. In our 'Brain-Science and Education' programme at RISTEX/JST, a non-invasive brain-function imaging has shown that the striatum (the putamen and caudate nucleus), which is regarded as the centre of the reward system, is activated not only by a physical reward but also by a mental reward (Izuma, Saito & Sadato, 2008; Mizuno et al., 2008). In the study of education, the fundamental difference between a physical reward and a mental reward has been discussed. In the future, the difference between intrinsic learning and extrinsic learning may have to be reviewed based on new knowledge in brain-science.

20.6 **Beyond neuromythology**

Science has not been perfect in any era. If we insist that science represents facts without flaws, it becomes a religion rather than science. Scientists are by nature modest. To approve something as a fact, evidence and documentation must always be attached to show that 'this is a fact within our current knowledge'. Antoine Laurent Lavoisier (1743–1794), who was called the father of modern chemistry, proposed, when he defined *elements,* that something that cannot be further decomposed should be called elements. Even with such attention to detail, what was regarded as an element at that time turned out later to be a compound. Even today, many basic particles called elementary particles have been discovered, which shows the fundamental development of science. In this sense, brain-science has many unknown and unravelled issues.

The same can be said of brain-science. Although it is increasingly clear that the applications of brain-science are diverse, there is a tendency to promote research not based on neuroscientific evidence or to exaggerate certain research findings. Such misleading or inaccurate theories are referred to as '*neuromythology*'. They are believed as if they were proven in brain-science, which is a fundamental problem. In a report, 'Understanding the Brain: The Birth of a Learning Science', by the OECD (English and French translations published in 2007), the wide applicability of brain-science to learning and education is mentioned, whereas one chapter is dedicated to 'Dispelling "Neuromyths"'. Confusion caused by neuromythology is serious in Japan, too.

The first myth involves 3-year-old children and says that the most basic cerebral neural system is formed until the age of 3 and nurturers (usually the mothers) must be devoted to child care during this period. This myth was debunked by a US development cohort study (the tracking and observation of groups of individuals over time) (Friedman & Haywood, 1994), which proved that there is no problem in raising a child outside the child's own family if the child is nurtured properly. A phrase similar to 'As the boy, so the man' is found in almost all countries. From my experience, child care until the age of 3 is extremely important and various kinds of knowledge exist in developmental cognitive neuroscience. However, caution is required when that discussion is applied to social problems.

Another myth is that 'a critical period exists in learning and learning becomes impossible forever if we miss that period'. This myth is used, for example, when a salesperson of learning kits wants to sell products. This myth was influenced by a discovery of Konrad Z. Lorenz (1903–1989) of birds' 'imprinting' (i.e. birds follow what they spot first after hatching), but there is a significant difference between birds and humans. The 'imprinting' of Greylag geese has a critical period of 24 hours after hatching. In the case of humans, although the 'sensitive period' (which is not as significant as the 'critical period') for learning has been recognized with regard to senses such as vision and hearing, the critical period for learning has not been discovered concerning various kinds of high-level functions. With regards to language acquisition and other sophisticated learning activities, the concept of an 'optimal period' for learning is being studied. Concerning language learning, we see foreign sumo wrestlers visiting Japan after graduating from high school speaking Japanese fluently in a relatively short period.

These are my explanations on specific examples of neuromythology. Sometimes information has not been accurately dispatched even within the field of neuroscience. For instance, the frontal cortex evolved significantly at the stage of primates and particularly when we look at the frontal pole of human beings only, the ratio of the frontal pole to the entire brain is twice as high as that of chimpanzees. The anatomical structure of the prefrontal area of rodents, which were phylogenetically diverted much earlier, is not similar to that of primates. We must always take caution in directly applying the results of experiments with rodents such as mice and rats to humans, who are phylogenetically distant from rodents.

A problem in research on brain-science and education that we have studied for more than 10 years is how to separate research that is based on facts from neuromythology that either lacks scientific evidence as explained earlier or exaggerates certain research findings. At present, 'brain talk' is booming worldwide. Actual research on brain-science and education faces an enormous number of books and discussions on TV programmes on neuromyths. Understanding that brain research is a powerful means through which to address our daily lives and social problems is a concept that should be promoted. In reality, however, neuromythology is a problematic issue in terms of widely disseminating research on brain-science and education to society. We foresee the most difficulty on this point.

20.7 **Conclusion**

The results of the 'brain-science and education' initiative have started to be used for educational politics and the improvement of school education. As one of early effects in Japan, teachers/nurses in infant/nursery schools started to learn the basics of development cognitive neuroscience and try to follow a scientific mind in education and childcare.

Our strong belief is 'if we promote learning and education, capable human resources will be nurtured and the nation will flourish'. This belief is regarded to directly connect with Japan's visions. Japan is characterized by a lack of natural resources, a small amount of land, a dense population and a language that differs significantly from the Indo-European languages. When we look at the history of Japan, we first learned foreign languages, then absorbed the world's knowledge, used such knowledge to add value to imported raw materials, and produced and exported value-added products. With many people working hard, Japan became one of the world's economic leaders. Today, the country is facing serious economic liabilities and is in a critical situation with a seriously aging society. It is projected that by 2030, about 20 years from now, two young people will have to support one senior. The author believes that brain-science and education should help provide an objective solution to finding an appropriate way for the type of education required by future nations of the world.

History has shown that there were times when powerful countries expanded their territories seeking natural resources. Even today, conflicts in the pursuit of energy resources such as petroleum are ongoing. Natural resources, however, are limited and will be depleted in time. To vie for them does not lead to a solution. It requires human wisdom to create new forms of energy, use resources rationally, and adjust lifestyles to actual circumstances. In so-called advanced countries, precious food is left uneaten as leftovers and people complain because of their incessant desires.

Education is necessary to remind and return humankind to an original state of 'appreciating what we have'. Results of research on the human reward system gave us an important hint to understand the mechanisms of strong emotion, satisfaction and happiness.

To create a better world, the optimal use of all human resources is required. To that end, I would like to reemphasize the importance of brain-science and education and conclude this chapter.

Acknowledgements

The author thanks to Japanese Ministry of Education, Culture, Sports, Science (MEXT) and Japan Science and Technology Agency (JST) for supporting the brain-science and education initiative. He also expresses his sincere gratitude to advisors and his colleagues appearing in MEXT and JST homepages through this Initiative.

References

Ansari, D., & Coch, D. (2006). Bridges over troubled waters: education and cognitive neuroscience. *Trends in Cognitive Sciences*, 10, 146–51.

Aoki, R., Funane, T., & Koizumi, H. (2010). Brain science of ethics: Present status and the future. *Journal of Mind, Brain, and Education*, 4, 188–195.

Atsumori, H., Kiguch, M., Katura, T., Funane, T., Obata, A., Sato, H., et al. (2010). Noninvasive imaging of prefrontal activation during attention-demanding tasks performed while walking using a wearable optical topography system. *Journal of Biomedical Optics*, 15, 046002 1–7.

Atsumori, H., Kiguchi, M., Obata, A., Sato, H., Katura, T., Funane, T., & Maki, A. (2009). Development of wearable optical topography system for mapping the prefrontal cortex activation. *Review of Scientific Instruments*, 80, 043704, 1–6.

Battro, A., Fischer, K., & Lena, P. (Eds.). (2008). *The Educated Brain*. Cambridge: Cambridge University Press.

Della Chiesa, B. (2010). 'Facilis descensus Averni' Mind, Brain, Education, and Ethics: Highway to hell, stairway to heaven, or passing dead end? *Journal of Mind, Brain, and Education*, 4, 45–8.

Dore, R. P. (1965). *Education in Tokugawa Japan, in International Library of Sociology and Social Reconstruction*. Berkeley, CA: University of California Press.

Fischer, K. W., Daniel, D. B., Immordino-Yang, M. H., Stern, E., Battro, A., & Koizumi, H. (2007). Why mind, brain, and education? Why now? *Mind, Brain and Education*, 1, 1–3.

Friedman, S., & Haywood, H.C. (Eds.). (1994). *Developmental follow-up: Concepts, domains, methods*. New York: Academic Press.

Fujita, E., Tanabe, Y., Shiota, A., Ueda, M., Suwa, K., Momoi, M.Y., & Momoi, T. (2008). Ultrasonic vocalization impairment of Foxp2 (R552H) knockin mice related to speech-language disorder and abnormality of Purkinje cells. *Proceedings of the National Academy of Sciences of the United States of America*, 105, 3117–22.

Goswami, U. (2006). Neuroscience and education: from research to practice? *Nature Reviews Neuroscience*, 7. 406–13.

Hensch, T.K., & Stryker, M.P. (2004). Columnar architecture sculpted by GABA circuits in developing cat visual cortex. *Science*, 303, 1678–81.

Homae, F., Watanabe, H., Otobe, T., Nakano, T., Go, T., Konishi, Y., & Taga, G. (2010). Development of global cortical network in early infancy. *Journal of Neuroscience*, 30, 4877–82.

Illes, J. (2005). *Neuroethics: Defining the issues in theory, practice and policy*. Oxford: Oxford University Press.

Ito, Y., Kennan, R., Watanabe, E., & Koizumi, H. (2000). Assessment of heating effects in skin during continuous wave near-infrared spectroscopy. *Journal of Biomedical Optics*, 5, 383–90.

Izuma, K., Saito, D.N., & Sadato, N. (2008). Processing of social and monetary rewards in the human striatum, *Neuron*, 58, 164–5.

JST Homepage on Brain-Science & Education Programmes in Research Institute of Science and Technology for Society (RISTEX). (2011). http://www.ristex.jp/result/brain/index.html, Results of the Research: http://www.ristex.jp/result/brain/program/index.html. Evaluation Report: http://www.ristex.jp/result/brain/program/pdf/ind01.pdf. Symposiums: http://www.ristex.jp/eventinfo/pasrelative/index.html. Interview for the Director 'Brain-Science Reforms Education':http://scienceportal.jp/HotTopics/interview/interview52/index.html, http://scienceportal.jp/HotTopics/interview/interview52/06.html

Kato, H.K., Watabe, A.M., & Manabe, T. (2009). Non-Hebbian synaptic plasticity induced by repetitive postsynaptic action potentials. *Journal of Neuroscience*, 29, 11153–60.

Kiguchi, M., Ichikawa, N., Atsumori, H., Kawaguchi, F., Sato, H., Maki, A., & Koizumi, H. (2007). Comparison of light intensity on the brain surface due to laser exposure during optical topography and solar irradiation. *Journal of Biomedical Optics*, 12, 062108.

Koizumi, H. (2000). Developing the brain: The science of learning and education. *Kagaku* (Science), 70, 878–84.

Koizumi, H. (2004). The concept of 'Developing the Brain': a new natural science for learning and education. *Brain & Development*, 26, 434–41.

Koizumi, H. (2007). The concept of 'Brain-Science & Ethics'. *Journal of Seizon and Life Sciences*. 17, 13–32.

Koizumi, H. (2008). Developing the brain: A functional imaging approach to learning and educational sciences. In A. Battro, K. Fischer, & P. Lena (Eds.), *The Educated Brain* (pp. 166–80). Cambridge: Cambridge University Press.

Koizumi, H. (2010). Toward a new educational philosophy. In M. M. Suarez-Orozco & C. Sattin-Bajaj (Eds.), *Educating the Whole Child for the Whole World* (pp. 81–96). New York: New York University Press.

Koizumi, H., Yamashita, Y., Maki, A., Yamamoto, T., Ito, Y., Itagaki, H., & Kennan, R. (1999). Higher order brain function analysis by trans-cranial dynamic near-infrared spectroscopy imaging. *Journal of Biomedical Optics*, 4, front cover and 403–13.

Maki, A., Yamashita, Y., Ito, Y., Watanabe, E., Mayanagi, Y., & Koizumi, H. (1995). Spatial and temporal analysis of human motor activity using noninvasive NIR topography. *Medical Physics*, 22, 1997–2005.

Mizuno, K., Tanak, M., Ishii, A., Tanabe, H.C., Onoe, H., Sadato, N., & Watanabe, Y. (2008). The neural basis of academic achievement motivation. *Neuroimage*, 42, 369–78.

Nakato, E., Otsuka, Y., Kanazawa, S., Yamaguchi, M.K., Watanabe, S., & Kakigi, R. (2009). When do infants differentiate profile face from frontal face? A near-infrared spectroscopic study. *Human Brain Mapping*, 30, 462–72.

Nakano, T., Watanabe, H., Homae, F., & Taga, G. (2009). Prefrontal cortical involvement in young infants' analysis of novelty. *Cerebral Cortex*, 19, 455–63.

Ogawa, S., Tank, D.W., Menon, R., Ellermann, J.M., Kim, S.G., Merkle, H., & Ugurbil, K. (1992). Intrinsic signal changes accompanying sensory stimulation: functional brain mapping with magnetic resonance imaging. *Proceeding of the National Academy of Sciences the United States of America*, 89, 5951–5.

Ojima, S., Nakamura, N., Matsuba-Kurita, H., Hoshino, T., & Hagiwara, H. (2011). Neural correlates of foreign-language learning in childhood: a 3-year longitudinal ERP study. *Journal of Cognitive Neuroscience*, 23, 183–99.

Ozaki, K., Toyoda, H., Iwama, N., Kubo, S., & Ando, J. (2011), Using non-normal SEM to resolve the ACDE model in the classical twin design. *Behavioral Genetics*, 41, 329–39.

Pena, M., Maki, A., Kovacic, D., Dehaene-Lambertz, G., Koizumi, H., Bouquet, F., & Mehler, J. (2003). Sounds and silence: an optical topography study of language recognition at birth. *Proceedings of the National Academy of Sciences of the United State of America*, 100, 11702–5.

Sugiura, L., Ojima, S., Matsuba-Kurita, H., Dan, I., Tsuzuki, D., Katura, T., & Hagiwara, H. (2011). Sound to language: Different cortical processing for first and second languages in elementary school children as revealed by a large-scale study using fNIRS. *Cerebral Cortex*, First published online 24 February.

Taga, G., Asakawa, K., Maki, A., Konishi, Y., & Koizumi, H. (2003). Brain imaging in awake infants by near-infrared optical topography. *Proceedings of the National Academy of Sciences of the United States of America*, 100, 10722–7.

Yamagata, Z., Sadato, N., Anme, T., & Maeda, T. (Eds.). (2010). Japan Children's Study 2004–2009, a developmental cohort study of early childhood. *Journal of Epidemiology*, 20, Supplement 2.

Yamamoto, E., Takahashi, T., Takiguchi, K., Onodera, Y., Itagaki, H., & Koizumi, H. (1992). Non-invasive brain function imaging by ultra fast magnetic resonance imaging. *Image Technology and Information Display*, 24, front cover & 1466–7.

Section 7

Final remarks

The good, the bad and the ugly in neuroscience and education: an educator's perspective

Paul Howard-Jones

21.1 A caveat

This chapter is intended to provide an educational perspective on some of the issues covered in this volume, but a caveat should be provided from the outset. I am a researcher whose work involves some neuroscience as well as education, so I cannot claim to be representative of currently practising teachers and educators. However, I have tried to use this opportunity to focus more on the educational than the scientific issues involved with the neuroscience and education enterprise. This educational perspective is, therefore, a personal one, although one that is formed by past experience as a teacher, a trainer of teachers and an inspector of schools. It is also, inevitably, informed by my own research in areas intersecting neuroscience and education. Some of my arguments views are supported by the results of surveys and consultations with teachers, but ultimately it is a personal view, and I am grateful to the editors for allowing me to voice it.

21.2 The good and the bad

The title 'The good, the bad and the ugly' summons up images in my head of Clint Eastwood having a shoot out in a desolate ghost town. At first, I must confess, I didn't see the link. But, in the new territories opening up between neuroscience and education, law and order is still in short supply and you can meet a variety of characters here in search of fame and fortune. These include entrepreneurs roaming with a merciless approach to exploiting the innocent. Yes, there sure are a lot of cowboys around here.

Leaving Clint behind, the title 'The Good, the bad and the ugly' also emphasizes that the theme is about judgement. Most of the contributors to the present text are scientists. Anyone who regularly experiences peer review in science will know that science is not free of subjectivity, but it does strive *towards* a 'value-free' objective identification of what is fact, based on a reasoned set of criteria. There is almost a formal rule book (often expressed with the help of statistics) about what is, and is not, acceptable to claim in science. That rule book helps scientists converge, usually comfortably, on what is 'good' science and what is 'bad' science. Scientific reviews abound that convincingly demolish pseudoscience that has flouted these rules, but this pseudoscience can continue to flourish (in economic terms at least) on, or beyond, the fringes of science and away from the harsh glare of its scrutiny. We are fortunate, then, to have writers like McIntosh and Ritchie (Chapter 14, this volume) who are prepared to venture into the badlands, round up suspect outlaws and put them on trial.

But this book is not just about what comprises good science. It is about the application of science, and in particular neuroscience, in education. And what is good and bad about the application of science is a more complex issue than examining the rigour of the science itself. We can scrutinize scientific statements about nuclear fission and we should be able to reach a clear consensus about what can reasonably claimed, i.e. what is good and bad science. If we wish to understand what is good and bad about the applications of nuclear fission, it requires interfacing the science, and its potential to generate cheap energy, with a range of other diverse fields: politics, history, economics, risk and the perception of risk etc. Ultimately, the extent to which it is good or bad to build a nuclear reactor on a particular site requires establishing some level of *informed* consensus about *values*. That requires dialogue between specialists in many different areas and often non-specialists who may be involved with using the outcomes and may be affected by them.

I will argue in this chapter that such a dialogue needs to begin early in the process of knowledge construction and that the best way to generate and exploit scientific research which is useful to education is to involve educators from start to finish. The knowledge that must be constructed must be informed by both neuroscience and educational understanding, and this knowledge will be more than the sum of these individual parts. As identified by Beauchamp and Beauchamp (Chapter 2, this volume), such a transdisciplinary approach has become a strong emerging theme in discussions seeking solutions to the many challenges presented by the neuroscience and education venture. Before that, however, it seems worth examining an idea that arises from some contributors to this volume: that psychology alone may be sufficient in informing education. Then, the potential role of dialogue and collaboration will be examined at each stage in the knowledge construction process: to clear out the myth, to identify tractable and relevant questions, to develop theory, to collect evidence and to apply it. Finally, a few thoughts will be offered about the types of statement that educators can find 'ugly' and the importance of moral aesthetics in this area.

21.3 Does education need neuroscience?

The extent to which cognition is represented in this volume attests to the vital role of cognitive models in making the link between brain and behaviour, including learning behaviour (Morton & Frith, 1995). Indeed, whatever contribution can be made by neuroscience, Seron (Chapter 6, this volume) emphasizes the vital role of developing more appropriate cognitive models in furthering understanding of our mathematical learning processes. The emphasis on cognition (see also Spiel et al., Chapter 17, this volume) raises the question of whether educational thinking might benefit from pulling back from its growing interest in neuroscience and focusing instead on the mind. Indeed, in their chapter, Roediger et al. (Chapter 8, this volume) make it clear from the outset that they will not review much neuroscience, but discuss cognition instead. Their review highlights the richness and quantity of the valuable insights from cognitive psychology about memory and how different strategies can enhance memory performance. Reading this chapter, one could indeed be almost convinced that educators should put the neuroscience books back on the shelf and return to reading more about psychology. Roediger et al. suggests moving from cognition to instruction would mean an important 'first bridge' has been built. However, I would argue that the 'cognition to neuroscience' bridge needs building at the same time and that, by keeping company with the relevant neuroscience, cognitive science has a better chance of achieving appropriate impact. The reasons range from pragmatism to principle. Firstly, authentic neuroscience may help engage the interest of educators with concepts that can appear abstract when presented purely as mental models of behaviour. Authentic neuroscience is also vital in dispelling the neuromyth that has been generated in its absence and, most importantly, it can contribute insights into underlying learning processes that can help inform practice.

Take, for example, 'spaced learning'. Roediger et al. remind us that we learn more when we break up our learning into chunks separated in time compared with learning all the material within a single period. Twenty years ago, Dempster (1988) described educational uptake of this idea as 'a failure to apply the results of psychological research'. Dempster might now be very comforted by the efforts of Paul Kelley, a headteacher in the UK, who has incorporated spaced learning as a key concept in lesson organization at his school. Kelley's efforts have drawn the attention of the national and international press to spaced learning as never before. How this started, however, was because Kelley read about 'discoveries in the chemical and genetic processes of creating memory' by 'one of the world's leading neuroscientists' in *Scientific American* in 2005 (Kelley, 2009). So, despite being reported by a psychologist Ebinghaus in 1885 and researched in detail by other psychologists ever since, it took an article (Fields, 2005) appearing to link this effect to an achievement of neuroscience before it became the focus of educational and media attention (Braid, 2007; Kellaway, 2010; Kelley, 2009). Neuroscience, it seems, really does have an allure (Weisberg, Keil, Goodstein, Rawson & Gray, 2008) that can sometimes leave psychology, somewhat unfairly, in the shade. Understandably, this might make cognitive psychologists feel their trade is undervalued. But, perhaps more importantly, education loses out by ignoring the psychology.

Kelley's school is doing an admirable job of exploring new approaches in education, but attention to psychological concepts could help here. When we look more closely at the practice of spaced learning espoused by Kelley, we find that 10 minutes is recommended as the optimum duration of the space between learning sessions. The 10-minute figure is drawn from the work by Fields. However, the validity of this link to the work of Fields appears debatable, since it refers to an interval arbitrarily chosen by experimenters to stimulate a rat's neuron in a Petri dish. If we look at the considerable psychological research on the effect of different durations, the findings do not converge clearly on this, or any other, figure. For example, Dempster (Dempster, 1988) reports that 30 minutes was better than 5 minutes, Bahrick and Phelps (1987) found 30 days was better than 1 day, while others (Challis, 1993) have pointed out that the optimal interval depends on the level of processing (and probably, I would suspect, many other unidentified factors). In the absence of information derived from scrutinized interdisciplinary research, unestablished links to neuroscience are made and neuromyths are born. And yet, there is recent work from neuroscience that can add to understanding in this area. An imaging paper by Callan and Schweighofer (2008) shows the spacing effect in verbal learning is due to enhanced maintenance rehearsal in spaced relative to massed presentations, suggesting a much more reasonable explanation that leaves open matters of optimal timing and undermines the notion of 10 minutes as a magic number. Neuroscience should be seen as a potential ally for those who wish that cognitive psychological concepts could make more impact in education. Apart from engaging audiences and helping to communicate concepts, it can add to authentic scientific understanding of the concepts themselves and also help pre-empt the generation of more neuromyth.

Spiel et al. provide another chapter where neuroscience is conspicuous by its absence. They review their own project demonstrating how teachers' competences were fostered through collaboration and cooperation on issues derived from their own classrooms and students, resulting in improved practice in the classroom and consequently better experiences for students. Appropriately, the project was called TALK and it emphasized social interaction between its organizers and the teachers, between teachers and between teachers and their students. The project avoided transmitting simple guidelines to the teachers but, instead, successfully moved teachers from spontaneous unreflective teaching to a scientifically-based reflective practice of their own design, providing an excellent example of how educational psychologists can inform and improve the practice of teachers in the classroom. Although the programme was successful in using concepts from educational psychology, it would be interesting to know whether including perspectives from cognitive neuroscience would have generated further interest and helped transfer to other schools. It seems apparent that, despite early hopes (Thorndike, 1926) and many decades, the present impact of psychological concepts on teacher's practice has been disappointing. Indeed, taking the UK as an example, psychology was dropped from teacher training in the Education Reform Act of 1988 and, despite calls for its return, postgraduate trainee teachers are unlikely to encounter it. On the other hand, there are now powerful voices arguing that all teachers should receive some basic neuroscience in their training (Royal Society, 2011). Spiel et al. ask 'what could neuroscience provide to make teachers more successful in teaching learning motivation. . .?'. My answer would be that neuroscience can contribute additional insight and impact to cognitive theories, such as those used by Spiel et al. Concepts from the sciences of mind and brain may, together, be more effective in developing, promoting and applying ideas about memory, motivation and self-regulation in educational thinking. Indeed, in a programme aimed at modifying teenagers' theories about their own mental development, a study by Blackwell et al. (2007) has shown that teaching adolescents about the plasticity of their own brain can improve self-concept and academic achievement.

Although Spiel et al. did not attempt to integrate neuroscience into their TALK intervention, their description of TALK illustrates how developing understanding and practice amongst teachers requires investment in extended collaborative projects based on interdisciplinary dialogue. But developing practice might be considered the final stage in a wider set of knowledge construction processes required for integrating neuroscience into education. Next, I will argue that, taken together, the chapters in this book demonstrate the need for collaboration and dialogue between neuroscience and education at every stage in this neuroeducation venture, from clearing out the neuromyth that education is already suffering, to identifying researching questions, developing theoretical frameworks, collecting evidence and developing practice.

21.4 Clearing out the myth

Neuromyths may abound more freely in education but the original term can be accredited to the neurosurgeon Alan Crockard, who became frustrated in the 1980s by how easily some unscientific ideas about the brain were embedding themselves in medical culture. He noticed, for example, that Lhermitte's sign[1] had become almost synonymous with a diagnosis of multiple sclerosis, despite also being generated in a number of other conditions. Crockard used the term 'neuromyth' in his lectures, and later in written articles, to describe a misleading type of 'received wisdom' within medical circles about clinical symptoms and causes (Crockard, 1996). In 2002, the Brain and Learning project of the Organisation for Economic Co-operation and Development (OECD) drew attention to the many misconceptions around the mind and brain arising outside of the medical and scientific communities. They re-defined the term 'neuromyth' as:

> a misconception generated by a misunderstanding, a misreading or a misquoting of facts scientifically established (by brain research) to make a case for use of brain research, in education and other contexts (OECD, 2002, p. 111).

Neuromyths have had a major influence on shaping the perceptions and views of educators about neuroscience and its potential role in education. This volume contains contributions by Corballis (Chapter 13) and Ritchie, Chudler and Della Sala (Chapter 15) reviewing many strange and wonderful ideas that have proliferated in the public domain and in schools. These include ideas about how exercise and rubbing parts of the body integrate hemispheric function, and the efficacy

---

[1] A reported electrical sensation in the spine and limbs produced by bending the neck forward.

of categorizing students in terms of learning preferences based on hemispheric dominance or on their auditory/visual/kinaesthetic processing. In education, the term 'neuromyth' might also refer to a story about the brain formed and reformed by many retellings, rather than some fundamentally misconceived notion. Indeed, neuromyths frequently originate from, and to a lesser extent retain, some genuine scientific understanding. This residual element of real science remains an important part of their enduring power. It affords them credibility long after they have divorced themselves from the scientific community and become effectively protected from its scrutiny. Since some original science remains in the myth, the task here becomes not a matter of determining if an idea is right or wrong, but what part of the story can be supported by valid evidence and what cannot. That makes the scrutiny of neuromyth a job that requires the judicious interrelation of neuroscientific and educational understanding, as demonstrated here by Corballis and Ritchie et al. This interrelation helps provide the authority required to dismiss what is untrue, but also ensures that the subtleties of the scientific facts, including those with genuine educational relevance, are well communicated.

In this volume, Koizumi (Chapter 20) also makes a significant contribution in this regard, by clearly communicating a number of scientific insights from Japan that may enlighten and improve educational approaches, but also by helping to separate the fact from fiction. One example of scientific progress derives from the CREST (Core Research for Evolutional Science and Technology) programme run by the Japan Science and Technology Agency (JST), which is helping us understand more about sensitive periods. The concept of sensitive periods has stimulated considerable interest amongst educational researchers, practitioners and policymakers in the UK (Howard-Jones et al., 2011). It has been used to justify a range of educational and parenting approaches, and even economic models of economic investment that are becoming influential (Heckman, 2007, 2008). Worryingly, as Koizumi points out, it often prompts myths and misunderstanding so the clear explanations provided in his chapter are very welcome. In terms of the neuroscience and education venture, Koizumi emphasizes there are considerable difficulties to be faced in finding ways to disseminate neuroscience to avoid such misunderstandings. However, using sensitive periods as an example, he also provides part of the solution: scrutinizing and, where necessary, creating appropriate language. The increasing difficulty faced by learners of a second language after about the age of about seven has often been explained as deriving from a sensitive period for language learning (Thomas & Knowland, 2009). If sensitive periods play a part, there are likely to be many overlaid on top of each other, but the neural evidence for such periods has still to be identified. A factor contributing to this educational misunderstanding may be that 'sensitive period' can acquire a less precise meaning within education, leading to confusion. Koizumi suggests the use of the phrase 'optimal period' (referring to periods when educational curriculum is most easily learnt) might be helpful in distinguishing between the meanings usually intended in these two contexts. The successful integration of neuroscientific insights into educational thinking may benefit greatly from such reflection and creativity regarding language.

McIntosh and Ritchie (Chapter 14, this volume) devote a whole chapter to scrutiny, focusing on the dramatic claims for using colour filters to treat reading difficulties, despite the small and inconsistent effects that are reported in the experimental literature. A lack of high-quality evidence prompted the authors to mount their own double-blind study. The intervention failed to provide evidence of positive benefit. It also suggested the likelihood of a placebo effect associated with tinted overlays, further undermining the credibility of previous results derived from less judiciously constructed research designs. McIntosh and Ritchie point to the 'magic bullet' simplicity of the idea of using colour filters to explain its popularity, combined with a mass of anecdotal evidence that may be linked to the placebo effect they identify. They warn that dyslexia is a complex disorder and that, while effective treatments are being developed and neuroscience can contribute to these, there are no easy short cuts and no magic bullets. I would go further and suggest that, since human development itself is complex and outcomes (including those classified as normal) are so diverse, this warning against quick fixes might be extended to apply to interventions across the ability range.

21.5 **What's the question?**

Although scepticism is good, scepticism alone will not ensure progress in moving neuroscience towards achieving something useful in education (Howard-Jones, 2009). Some scientists feel safer within the bounds of scepticism, and avoid reaching out to understand what *may* be possible. Similarly, I would argue that many educators and researchers within the social sciences (e.g. Davis, 2004; Schumacher, 2007) can prefer to deconstruct the scientific concepts, rather than help construct interrelations between these concepts and ideas from other perspectives on human behaviour. But this type of positive interdisciplinary dialogue is vital even when making the first step along the road of knowledge construction in this area, i.e. when identifying a question that is educationally relevant and scientifically tractable.

For example, in this volume, Cowan (Chapter 7) draws attention to working memory (WM) ability being more limited than previously thought, and emphasizes the vulnerability of a learner's WM to distracting influence when it is being to used to form new long-term knowledge. WM allows a learner who is solving a problem to temporarily hold relevant information in their attention while using it work out a correct answer. Cowan highlights many other ways in which it influences our educational abilities, including the processing of complex thoughts and the maintenance of goals. The concept of WM helps provide insight into the importance for efficient encoding of developing good categories for information being presented, the additional difficulty that can be encountered when problems are abstract, why we often fall back on unsound heuristics, and why mnemonic memory strategies develop with age. Understanding of WM appears to be providing concepts that can potentially be interrelated with pedagogic thinking. In this respect, one significant advance is a move away from considering a limited set of different types of WM (visuospatial, phonological) towards the notion of myriad instances of temporarily activated portions of long-term memory. In this model, the focus of attention holds only a small number of items in an integrated form, whereas many features in long-term memory outside of attention can be temporarily activated at the same time, but in a less-integrated form. This move away from modularity to a more distributed form of WM makes it easier for educators to conceptualize how any idea can be held in one's attention, without considering how it must be coded in terms of terms of a limited number of WM faculties. Cowan does an excellent job of highlighting the potential relevance of WM to learning performance.

However, given this potential relevance, a question of key interest to educators would be whether WM can be trained. In the final section, Cowan touches on this issue, referencing two studies that show it can be trained and two suggesting that improvement is 'narrow'. Cowan suggests positive results of WM training may simply reflect improvement in task performance but, from an educational point of view, it seems surprising that this issue is not given more considered treatment. We understand that, as reviewed briefly in this chapter, there is evidence that WM develops during childhood, and it does not appear unreasonable to speculate that this development can be positively influenced by intervention. It is important for education to be advised of how seriously it should take studies such as Jaeggi et al. (2008, but not reviewed by Cowan) who claim rapid improvements in WM (and fluid intelligence) from computer-based training. There are issues related to this question that involve the underlying neural processes. For example, if we conceptualize WM as temporary activation of long-term memory distributed across the cortex, what are the neuroscientific arguments for, and against, the efficiency and/or capacity of such processes being susceptible to environmental influence? Having presented very persuasive evidence for the important role of WM in educational learning processes and outcomes, Cowan devotes less space to the educational issue of training it, preferring to implicate self-belief and attitude as the factors that US educational institutions should focus upon. One leaves the chapter having learnt much about working memory, but also with the impression that our most talented scientific researchers may not always wish to focus on the questions of greatest concern to education. And this, in many respects, does not seem an unreasonable position. Good science is about the rigorous determination of new fundamental knowledge. For scientists, perhaps the immediate value of their work in terms of real world application should be a secondary concern. But, if neuroscience and education is to be about science and application, a dialogue that addresses both of these aims at the outset may be needed. It is a dialogue in which educators express what they need to know, and scientists express what is known, that is most likely to identify those questions of educational relevance that can be addressed with methods and concepts that possess scientific validity.

This issue of bringing together the professional aims of different disciplines is helpfully explored by Parisi (Chapter 19, this volume), who reviews some of the issues that increasingly preoccupy educators (and the wider public) about the impact of technology on how and what we learn. Technology, with neuroscience approaching in the distance, are two rapidly developing streams of knowledge whose impact on education can be expected to grow. So perhaps it is strange that, as highlighted by Parisi, there have been few studies from cognitive neuroscience that address learning with technology. Part of the reason, Parisi argues, is that laboratory experiments are intrinsically non-ecological and artificial (echoing some of my concerns about randomized-controlled trials as a 'gold-standard' method of assessing educational approach—see later in this chapter). Parisi points out that learning with technology often occurs as part of a horizontal network and involves the individual modifying their own environment. In some ways, this is at odds with the controlled types of experiment beloved of scientists. However, in other ways, learning with technology may be more amenable to being studied by techniques such as brain imaging than other types of learning, such as that derived from face-to-face classroom teaching. Indeed, functional magnetic resonance imaging (fMRI) studies often use learning tasks that are computer based, with communication with participants occurring via technology, and some studies now use online communication as a feasible means of studying social interaction in a scanner (Redcay et al., 2010). Education, as Parisi notes, is likely to follow the rest of society and use technology to move away from a pure language approach towards learning via more seeing and doing. These types of activity, via a computer interface such as a keyboard, may also be more straightforward to study using methods such as fMRI than contexts involving teacher–pupil face-to-face interaction. Historically, however, the preoccupations of neuroscience have not focused on normal learning processes, but more on abnormal learning and development. Parisi's highlighting of a range of very new and important issues for education again raises this question of how neuroscience might choose issues of educational interest to focus upon. Parisi's analysis queries whether the educational questions we would like neuroscience to help us address now might be out of date by the time the answers are known, with more pressing and important questions having arisen in the interim. There appears to be a clear and immediate implication of this analysis: Not only must neuroscience engage with educators' current practice to determine relevant questions for research, but it may also be necessary to scan horizons *with* educators to identify those questions we are likely to face in the future.

21.6 **What's the theory?**

If we have an educational question about learning that neuroscience appears poised to help address, we might devise collaborative research to address it (such programmes have been recently funded in Germany, the Netherlands and elsewhere). But a first barrier to progress is that neuroscientists and educators inhabit different worlds and speak different languages, even if they use many of the same words. Even the term 'learning', while associating itself in the scientist's mind with changes in an individual's neural connectivity, means something entirely different to an educator. Educators do not usually consider the biological processes involve with learning. Instead, there is an emphasis on social construction, on learning within groups and communities, and on the importance of context. Additionally, educators concern themselves with issues that are often problematic in neuroscience research, such as personal meaning, the will to learn, values and the distributed nature of these and other aspects of learning beyond the level of the individual. So, assuming that educators will not wish to leave such ideas behind, it means that concepts from neuroscience must somehow find some theoretical basis for coexisting with them.

In this volume, Goswami (Chapter 4) confronts the thorny question of how neuroscience should position itself amongst the different perspectives that currently inform educational thinking. She rejects the idea that neuroscience is based upon a particular world view, and that its approaches are merely one 'discourse' amongst many, and that these are based on professional identity and power relationships rather than science and knowledge. There are many that argue that the production of scientific knowledge is not (and/or should never be) a value-free enterprise and that it is susceptible to issues such as identity and power that effect knowledge production in other domains. To me, it seems unwise to consider that neuroscience is wholly immune to such influence, but its aspiration to seek objective knowledge is surely one its strengths. Biological perspectives can claim to be generating a set of determined facts whose number and quality is increasing, even if the relation of these facts to real world learning often appears complex and even, sometimes, remote.

Goswami begins her chapter with a convincing case that neuroscience should be in principle of interest to education, drawing particular attention to the value of counter-intuitive and surprising findings. It may be these neuroscience findings that help challenge established educational thinking, demonstrating its value as a 'touchstone' of fresh ideas. Goswami indicates two categories of neural finding that can provide this value: structure and mechanism. In the terms of the first category, she draws attention to the activation of neural structures for spoken language when novice readers perform tasks with print, challenging logographic theories of reading acquisition in which children learn to read by going directly from the visual word form to meaning, without recoding the print into sound first. Moreover, it appears the brain tunes itself to print such that, soon after reading instruction begins, electroencephalography (EEG) can detect differences in children at genetic risk from dyslexia before any diagnosis has taken place. Goswami highlights the potential usefulness of such measurements to identify children at risk before symptoms are manifest in behaviour, and that such neural markers may be helpful in counteracting arguments that a child is lazy, stupid or not trying. Further unexpected findings arise from research on mathematical development. Here, connectionist models of neural mechanisms have provided insights into how the brain can form generic (or prototypical) representations of experiences in its networks of cell assemblies, but Goswami also points out that we have the ability to influence these ourselves through thinking and 'inner speech', without external simulation. This is an important point. Connectionist computational modelling of neural networks is one of the most technical fields of neuroscience, often representing human cognition on software with inputs (e.g. sensory) and outputs (e.g. motor). As discussed previously, educational models of the learner place emphasis on autonomy, which is not easily represented in such computational models—since we presently have little understanding how consciousness and free will can be incorporated. Nevertheless, this is not evidence of fundamental conflict between neuroscientific and educational models of learning, but it certainly reflects a difference in emphasis with the respect to how learning is conceptualized. Drawing attention to how our neural networks must, in some respects, be influenced by our own will contributes to the conceptual bridge building that is needed here. And, incorporating connectionist models in our educational thinking can also prompt us to seek out behaviours we might not have otherwise looked for that are, as Goswami suggests, highly relevant to educational theorizing. These include higher-order spelling consistency and syntactic rules that are acquired implicitly from environmental stimulus without directed teaching strategy.

Goswami predicts neuroscience will enable a componential understanding of the cognitive skills taught by education. Having suggested that neuroscience is more than 'one discourse amongst many', she makes clear its contribution will be at one level of enquiry only, and will not replace social, emotional and cultural analyses of learning. This still leaves open, of course, the question of how these different levels might be interrelated. This is a challenging issue, especially if one perspective claims a value-free approach and others, such as humanitarian perspectives on learning, might aspire to be value-centred. In the present volume, this question is taken up by Coch and Ansari (Chapter 3), who appear to agree with Goswami that neuroscience is not here to replace other levels of explanation. Coch and Ansari outline how a Mind, Brain and Education (MBE) approach might integrate across levels of analysis, constraining interpretation at each level with evidence from other levels. They provide an example of how speed in a mathematical task was similar for adults and children but regional variation of brain activity differed, providing clues about development where no behavioural evidence could be observed. But Coch and Ansari also emphasize how limited such studies are in terms of their interrelation with many 'macro' factors such as culture and socioeconomic status. A challenge for research at the interface of neuroscience and education, they suggest, is tackling the issues of 'grain size' between neuroscience and these other levels of explanation. They suggest that research in this area must systematically compare and contrast brain, behavioural and sociocultural level changes in a wide range of domains of interest to education, with a focus on how interactions across levels can help better understand and improve learning processes. In this statement, Coch and Ansari identify the fundamental challenge that should be occupying researchers who claim an interest in enriching educational understanding with neuroscience. We can enjoy tearing apart neuromyths, and tease ourselves with reviews of the potential relevance of neuroscience for education, but it seems that there is another type of 'core' intellectual work to be done: we must tackle the practical and theoretical issues faced when developing explanations built on interrelating changes at these different levels. It will not be a simple matter of reduction. In terms of our present understanding, we cannot expect biology to be wholly explained by, or explain, cognition. Sociocultural aspects of development will not be wholly explained by, or explain, biology.

The alternative to reducing educational explanations to a single level is to accept that educational, biological, cognitive, emotional, contextual and sociocultural aspects of development can be considered as interrelated and interdependent. But, it seems to me that a practical set of research challenges flows naturally from this theoretical position, suggesting a rich variety of research designs and methodologies are needed, combining different research approaches. Elsewhere it has been suggested that methods drawn from different perspectives can be adapted to serve the interrelation of findings at one level with data derived from other levels of analysis, without detracting from validity as defined by the 'home' perspective (Howard-Jones, 2008).

Coch and Ansari identify how these interdisciplinary journeys must include gaining an understanding of the theoretical frameworks, methods, bodies of knowledge, vocabulary and 'grain size' of the different perspectives involved, suggesting interdisciplinary training could help here. They suggest that balancing and integrating theories from neuroscience and education, however difficult, has to be at the core of this work. This type of interdisciplinary challenge can help a novel conceptual field emerge that goes beyond its original constituents. In this sense, it may more appropriately be described as transdisciplinary (see Beauchamp and Beauchamp, Chapter 2 in this volume for a more detailed review of this term). My personal experience would support this: it can be as unfruitful when neuroscientists, without educational input, discuss the potential educational applications of their work as when educators, without neuroscientific expertise, attempt to draw on neuroscience to support their own ideas. On the other hand, when workers from different disciplines sit down to try and relate their different levels of explanation, then something of educational relevance can emerge that is not scientifically erroneous. And yet, a quick scan through the pages of the *Journal of Mind, Brain and Education* would tend to suggest this type of dialogue is still in its infancy. Although growing in number, instances of experts from education working with scientists to develop these connections are still few.

Coch and Ansari suggest that progress in neuroscience and education will essentially 'blur' the boundaries of the disciplines concerned. As they point out, unprincipled attempts to do this already feature in several programmes claiming a brain-basis. The word 'blur' may need some unpacking—how can you blur in a principled way? Clearly Coch and Ansari do not mean creating a fog of pseudoscience. I interpret the word 'blur' here to mean a loss of interest (self-serving and/ or professional) in maintaining sharply delineated boundaries of enterprise. However, I wonder whether this process of blurring might begin, ironically perhaps, with a discussion that delineates more clearly the boundaries between levels of current explanation and analysis, clarifying their limitations and their potential to constrain each other, in order to allow a principled journey across them to make links? Perhaps, only after several decades of developing links between levels, a type of conceptual blurring might begin. The number of connections in Coch and Ansari's metaphorical interchange might then be so many that distinguishing between levels of explanation will become unhelpful and/or meaningless. The word 'blur' might require some caution in its interpretation, but Coch and Ansari's essential point is well made: multifaceted understanding from multilevel analyses is needed if we want neuroscience to have real world significance in education—and that will mean workers from different perspectives leaving their comfort zones.

21.7 **Where's the evidence?**

Evidence in neuroscience must usually test a hypothesis. In education, evidence often has to do other things as well. It must persuade different groups (researchers, teachers, schools, governments and the general public) and help communicate messages to this broad audience. Findings from the laboratory can help determine a principle, but most teachers draw their evidence from what they and their colleagues observe around them, and from reflections of the learners themselves upon their own experience. Laboratory findings are inevitably too removed from the classroom to be sufficiently meaningful for educators to apply them directly. Quasi-experimental studies can be used to help bridge this gap between laboratory and classroom. For example, one such study was prompted by evidence from brain imaging experiments that a link may exist between numbers and fingers, suggesting a special cognitive role for fingers in the learning of mathematics. Gracia-Bafalluy and Noel (2008) carried out a cognitive behavioural study in a school to test this idea, focusing on finger gnosis (awareness of one's fingers). Following the experimental method, they divided 47 school children into three groups: children with poor finger gnosis, high finger gnosis and a control group (Gracia-Bafalluy & Noel, 2008). After an 8-week intervention consisting of two weekly half-hour sessions of finger training, the children with poor finger gnosis improved on mathematical tests, in contrast to the control group who did not receive the intervention, or the group who were already very competent with their fingers. These trends, confirmed as statistically significant, support the hypothesis for a functional link between finger gnosis and numerical skill development. It also brings the theoretical model closer to being useful in the classroom, by demonstrating the relevance of this brain–mind–behaviour relationship in a school environment. But, as a controlled experiment, this is not likely to get educators as excited as the successful evaluation of an educational programme based on such ideas, developed and delivered by educators in a variety of different schools as part of the curriculum. The experiment lacks such authenticity. We know, for example, very little about the interaction and discourse between teacher and students that was used, and even less about what teaching style would be optimum to deliver this approach. The study is an example of how scientists are beginning to research more 'real-world' contexts, as a means to test and modify concepts around brain–mind–behaviour relationships. But, for many teachers, it is likely to fall short of providing an example of best practice that can convince, inspire and support them.

I understand, with the half of me that is a scientist, why this can be frustrating for the experimenters. Neuroscience is a progressive body of generalizable knowledge. It has much to commend it. It doesn't go around in circles according to fads and fashions like, it might be argued, some educational policy making. It aspires to (and mostly succeeds) in maintaining the best standards of rigor in its methodology, and it commands a respect which is often the envy of many other fields of research. So, if we are considering importing some of its concepts to educational thinking, why not go a little further while we're at it—and also accept these type of quasi experimental studies as the best way of evaluating practice in the classroom? Experimental methods can certainly be informative for educational research so perhaps, as Coltheart and McArthur (Chapter 12, this volume) seem to suggest, we might grab a long-awaited opportunity to radically revise the philosophy that characterizes much of educational research. Perhaps we should adopt the random and double-base lined controlled trials as the gold standards of methodology in education, as they have been in medicine and many other areas?

As Coltheart and McArthur point out, the unscientific and educationally unsound programmes that claim a brain basis appear to have been increasing in number. The testimonials support them but, as Coltheart and McArthur emphasize, these should not be trusted. Instead, they suggest these programmes just need controlled evaluation on the basis of cognitive testing, and there is no need for neuroscience to get involved. In some ways, this appeal to science is alluring, much like neuroscience itself. Such a proposal has a no-nonsense ring of authority about it. Indeed, controlled trials have a very important role to play in education, since they can provide a particular type of evidence that, in the ways outlined by Coltheart and McArthur, appears convincing. Like all single types of educational evidence, however, such data are limited in their ability to persuade. A clue about one obvious limitation, of course, is in the title: 'controlled'. Schools are not laboratories, so any control is partial at best and attempts to impose it will usually detract proportionally from ecological validity. How are teacher–pupil relationships, attitudes at home controlled? More of a problem, however, is what the control condition is. There has to be a comparison against something—whether it is alternative type of intervention or the status quo. In the latter case, the result is simply telling you that the intervention is better than the usual approach, which may have included some aspects of poor practice otherwise unrelated to the issue under investigation. Unlike in medicine, there is no clearly defined baseline condition of health or normality.

Whatever their limitations, controlled trials can give rise to valuable and interesting findings, but this type of evidence should probably be always considered alongside other types of evidence, rather than assume its trustworthiness renders other evidence redundant. For example, Coltheart and McArthur refer to a study by Bull (2007) that illustrates subjective measures of improvement can diverge from those derived using cognitive tests. In this study, a rise in self-esteem appears at odds with cognitive tests that show no improvement. However, supposing another type of intervention had produced the opposite result (rise in cognitive test result, drop in self-esteem). In that scenario, the subjective measures are suggesting something potentially important: we may have an intervention that improves cognitive function but leaves the learners feeling a bit crushed. It is not difficult to imagine such a thing, and clearly the experiential perspective would severely tarnish the usefulness and appropriateness of the intervention. Moreover, suppose in the study by Bull (2007) that educational outcomes and self-esteem had, despite a drop in cognitive measures, improved after the intervention. This would possibly suggest that the rise in self-esteem had improved motivation, irrespective of the lack of improved cognitive function. It could be argued that, in educational terms, this is a positive result. In dialogue between education and neuroscience, it has been suggested that even the meaning of the word 'placebo' needs some reconsideration when crossing from medical to educational models of evaluation (Howard-Jones, 2010, p. 71).

Controlled trials are an important part of building evidence to make judgements, but methods derived from other perspectives are also needed. Neuroscience can shed light on processes, and an understanding of the processes can help inform judgement (as, for example, in Chapter 5 by Ashby and Rayner in this volume). In the example of the Miracle Belt™ described by Coltheart and McArthur, the link to neuroscience may be a key reason for its success in terms of sales, and this alleged scientific basis might be the best place to start when it comes to examining the evidence. Teachers care about the 'why' of what they do (Pickering & Howard-Jones, 2007). They generally implement ideas which they believe have a reasonable chance of improving learning, and if the reasoned basis of an intervention is exposed as false and misleading, this should help dissuade them from using it. So, providing well-communicated and accessible information about the brain and whether different interventions are based on genuine scientific understanding would be a helpful and cost-effective way to begin supporting their decisions with neuroscience. Controlled trials can form an important part of this information, but it is considerably easier for researchers to wrangle about them and educators to dismiss them than Coltheart and McArthur appear to suggest. Controlled trials are often worthwhile, but the idea that the natural sciences can provide them as a gold standard of methodology for education is, sadly, just another type of myth. Education policy, and a large part of educational research, will never be a natural science. It will always be about making informed judgements in very situated contexts, according to a diverse range of different types of evidence.

21.8 **What do we do with the evidence?**

Having identified questions, constructed theories and collected the evidence to explore and test them, the question remains of how education might incorporate the findings into policy and practice. Ashby and Rayner (Chapter 5, this volume) provide an example of how this might occur, reviewing the neuroscience evidence in one important area of education (learning to read). They relate this evidence to an ongoing debate in need of illumination: whether pedagogy should be emphasizing immersion in literacy-rich environment that avoids any focus on systematic phonics, or whether learning to read benefits more from dividing time between phonics instruction and exposure to children's literature. Their chapter provides a scrupulous examination of the pedagogical question and, in so doing, makes good use of concepts and principles at neural, cognitive and behavioural levels. One important point to note is that the behavioural evidence, often using the type of well-designed methodological approaches suggested by Coltheart and McArthur only provide part of the answer to this apparently simple pedagogical question. The analysis of Ashby and Rayner demonstrate how the evidence from neuroscience, and from computer models drawing on hypothesized neurocognitive processes, can clearly help answer important pedagogical questions. Rather than seeking a conclusive answer from a single perspective, Ashby and Rayner build their arguments by looking for convergence of evidence from different perspectives.

This strategy results in a convincing and educationally-helpful response to the question, using instances of divergence as a useful prompt for reconsideration of the emerging model.

There are examples of neuroscientists already translating their work into educational application by developing multimedia or software (e.g. Wilson, Dehaene, Dubois & Fayol, 2009; Wilson et al., 2006). Baron-Cohen, Golan and Ashwin (Chapter 18, this volume) review how they developed *The Transporters,* an animated series for children with autistic spectrum disorders. This approach to translation allows greater control over the implementation of the science than when teachers use the scientific principles to develop their own strategies and resources. However, despite the success of the controlled appraisal of *The Transporters,* Baron-Cohen et al. make the point that it might be even more effective when integrated as part of other educational methods and so how the teacher should apply their own understanding of the evidence remains an issue. Interdisciplinary work by experts in interrelating findings across research in neuroscience, psychology and education may help generate useful resources but is not likely to result in prescriptions for developing lessons plans. There will still be work for the teacher in interpreting expert advice within the contexts of their own classroom and in terms of individual students they are currently teaching. Connell et al. (Chapter 16, this volume) present an idea on how this could be achieved, and explore whether the teacher's role could extend beyond being merely an end-user of the knowledge. They suggest that, through their idea of 'design patterns', a new 'educational science' can involve teachers in catalyzing interdisciplinary dialogue through engaging with and contributing to its own systematic accumulation of valid, useable, public knowledge, similar to that associated with other scientific domains. There are, then, at least two new ideas here. One is the idea of teachers using a design pattern approach to applying knowledge in the classroom, the other is to get neuroscience and education talking to each other by making education part of the science.

Certainly, many educators would benefit from being more systematic in their approach to using research evidence. Teachers' mental models, which include ideas about the brain, could be improved by better training for the adaptation loop described by Connell et al., i.e. by learning to better identify the recurring educational problems they face, to access and to develop their ability to use solutions that include explanatory models and to determine the value of these solutions by testing them in practice and providing feedback. Design patterns may support this learning and help 'close' the adaptation loop discussed by Connell et al. in a number of different ways. These ways include encouraging educators to become more systematic and to contribute to a more accessible, more easily applied body of educational understanding that can also more readily accumulate. In terms of developing the scientific contribution to these patterns, Connell et al. suggest a partial solution may be to employ people to 'translate' the educational problem into science, and then integrate the science into an educational solution (in the form of a design pattern). I suspect that the translational process referred to here will still require interdisciplinary and collaborative research, but generating a design pattern might be a suitable objective for such research to pursue.

In terms of helping to develop an educational science, however, I feel more cautious. This type of educational science would surely need to be a moral science about what should be done, based on personal collective experience enriched by relevant sources of expert information that might, as suggested by Connell et al. include neuroscience. But, of course, this is very different from a natural science such as neuroscience, which moves forward by developing, through hypothesis testing in controlled conditions, a collective and progressive body of general concepts. Although it would be helpful if educators contributed to, and used, research findings more systematically, it seems debatable whether this is the major barrier for integrating neuroscience into classroom practice. The gap is not simply a lack of translation and reflection-based integration on behalf of teachers. It may derive more from a lack of research-based knowledge that draws expertly and credibly on both perspectives.

21.9 **The ugly**

I have tried to argue that, while good and bad science can be discriminated by scientific scrutiny alone, the judgement that its application in education is good or bad should be reached through genuine interdisciplinary dialogue. Additionally, this judgement is more likely to be favourable if the application itself is a product of such dialogue. So, who are the ugly? From an educational perspective, there is a certain category of statement produced by scientists that teachers might find unattractive. These statements can be scientifically justified and, in the sense intended by the scientist at least, supported by evidence. Yet statements in my ugly category can still produce an aversive response in educators. The aversion of their gaze may not simply derive from a lack of comprehension or a Luddite unwillingness to accept progress. It may, instead, be that educators find unpalatable any perspective that fails to reflect the values they work hard to promote and, perhaps more importantly, fails to acknowledge how their students' achievements depend upon these values. Just as teachers can be naïve in their conceptions of the brain, some neuroscientists can lack awareness about educational issues and, it would seem, the political and moral sensitivities within the field. For example, consider the article in *Science Mind* written by the renowned neuroscientist Michael Gazzaniga, promoting that idea that we should unconditionally welcome cognitive enhancing drugs (Wilson, 2005). Gazzaniga justified his position by pointing to unmedicated high-achievers: 'We accept the fact that they must have some chemical system that is superior to ours or some neural circuitry that is more efficient. So why should we be upset if the same thing can be achieved with a pill?'. At the level of the individual this seems to make perfect sense. However, educational assessment has the task of characterizing individuals on the basis of their academic achievement, such that they are inevitably compared with each other. Attitudes to educational assessment are intertwined with the values of our society, with many stakeholders in assessment beyond the individual being assessed. Academic assessment provides information to the individual about the outcomes of their efforts and contributes to self-perception, but also to their educational institution about the impact of its teaching and support for learning, and to potential employers regarding each individual's relative suitability for a particular professional role. It seems likely that considering artificial enhancement in the context of educational assessment will raise some complex issues involving all these stakeholders, each with their own different set of priorities. Additionally, the fact that socio-economic status is likely to mediate access to cognitive enhancers will make the issues even more inflammatory. In terms of our educational system, how should we be promoting achievements derived using drugs rather than determination and hard work? (In our survey of 100 educators, 76% felt that grades achieved with the help of cognitive enhancers should not be valued as highly as grades achieved without them (Howard-Jones & Fenton, 2011)). I suggest educators might find ideas such as Gazzaniga's ugly, not because cognitive enhancers are necessarily good or bad for education—that remains to be determined—but because the presentation of his ideas fails to consider, or even acknowledge, the values that underpin our educational system.

Genetics is an area where particular educational sensitivities exist, and with good reason. Teachers strive daily to encourage children to work their hardest and achieve their best, often despite the learner's sense of their own limitations. Perhaps, on the basis of the scientific facts, a genuine understanding of the role of genetics in educational achievement should not impact on their adherence to this maxim—but it can. When we surveyed 158 postgraduate trainee teachers about to start their professional careers, we found a correlation between their strength of belief in genetics as a factor influencing outcome, and a belief in their own students being biologically limited in what they could achieve (Howard-Jones, Franey, Mashmoushi & Liao, 2009). It is, therefore, essential for the future role of genetics in education that messages are sensitively and carefully constructed. Some of the messages provided by Bates (Chapter 11, this volume) are problematic in this sense. Bates emphasizes that it is important that education understands the genetic limits to skill, stating 'While behaviour can change, learning is unlikely to be able to roll-back development and re-engineer the billions of cells and their complex migration and connections involved in neocortex'. To illustrate this, Bates points to evidence suggesting that the biological basis of dyslexia lies in failure of neuronal migration. However, we already know that some interventions for dyslexia have succeeded in partially remediating both behavioural and neural function, although probably more in terms of connectivity than cell migration (e.g. Shaywitz et al., 2004; Temple et al., 2003). Clearly, if we accept that no 'roll back' is possible, children will never benefit from such interventions. More generally, given that all human learning can be assumed to have biological substrates, some serious 're-engineering' must take place when we are educated, even if it does not include the processes that Bates refers to as limiting us. To some extent, Bates's initial position appears to oversimplify issues that he attends to in more detail later in his article, but the same biological message is also explicit in his model, with no arrow leading from the psychological to the biological. Yet external influences (e.g. school) have an effect on our psychology and, if we believe memory is encoded in the brain, we should assume that this influences our biological processes and capacities. From the well-publicized research on London taxi drivers, we know their posterior hippocampus enlarges as they gather years of experience in finding their way around (Maguire et al., 2000). Structural changes achieved over shorter time scales have been found in young adults learning to juggle (Draganski et al., 2004). Evidence for this direction of change is something of a footnote in cognitive neuroscience, whose brain–mind–behaviour model traditionally pursues change operating in the opposite direction, with unidirectional arrows linking brain→mind→behaviour in similar fashion to Bates (Morton & Frith, 1995). But these studies of plasticity are of particular interest in education, because they help counter misconceptions that science preaches a deterministic doctrine of 'biology as destiny', whereas the notion of genetically limited learning appears to support this doctrine.

From an educational perspective, some of the evidence referred to by Bates to support his notion of biological limitation appears to contradict his position. The fact that original differences are magnified by similar amounts of intensive and organized training does not feel like evidence for a well-defined biological limit. It suggests, instead, that everyone can benefit from education, albeit it at different rates. Being slower to learn does not mean we can predict a biological-defined limit to a student's ultimate learning achievement. Although Bates admits that connectivity is almost continuously remodelled throughout the brain, he points out that some molecular processes cannot be reversed and that this implies that there is a genetic limit to skill. He suggests his conclusion will protect educators from being blamed for 'leaving children behind' because, presumably, it is their genetics that is leaving them behind. But I would warn that, rather than relieve teachers, the message promoted by Bates might actually exacerbate the plight of the learners further, by reducing both teachers' and learners' sense of agency within the learning process.

I would judge that this unattractive quality of Bates' message has arisen due to a misunderstanding about the discourse and language within education, rather than a flaw in the arguments that Bates follows. This misunderstanding also comes to the fore when he attacks beliefs he considers common within education, some of which I have difficulty recognizing. For example, who are the educators who believe their job is 'equalizing children'? Rather, the teachers I have met and worked with tend to believe in equal opportunity, and that equal value should be placed on efforts to help different children achieve their potential. I feel it is unlikely that educators would argue that children are identical in their learning potential, especially since teachers aim to promote the achievement of their pupils, all uniquely different in their capacities, in so many diverse areas.

The educational meaning of the statement 'all children face the limits and potentials of their brain mechanisms for learning' is problematic because, intended or otherwise, it carries a message of predetermination. In educational and psychological testing, both the terms 'limit' and 'potential' tend to be avoided. In educational discourse, 'potential' is used as a driver of teacher and learner motivation, rather than to describe a preordained barrier to progress—or 'limit' in the educational sense. If these words are to be used in some special scientific sense, this new sense needs to be very carefully defined. Even in the case of learning disorders such as dyslexia, dyscalculia and attention deficit hyperactivity disorder (ADHD) etc., we will probably never be able to define a fixed brain-based boundary of achievement up to which children suffering from these disorders can achieve and no further. So the limit that Bates refers to is hypothetical, ill-defined and, if it is ever quantifiably predicted for an individual child, is likely to be expressed in probabilistic terms. And what is the meaning of the word 'learning' here? Does it refer to fundamental types of processing ability (e.g. visual processing or working memory capacity), or complex higher-order skills such as producing language and meaning? There is considerable evidence supporting the ability of the cortex to drastically reengineer itself to support such higher order skills (Immordino-Yang, 2007). So, although 'all children face the limits and potentials of their brain mechanisms for learning' sounds authoritative and unassailable, as a scientific statement it appears in need of considerable qualification. As a message to education, it is unhelpful since it may dissuade teachers and their students from believing that every individual can reach beyond their apparent limitations. One might say, as reasonably, that 'all children face the limits and potentials created by the attitudes of those around them' and that Bates' message is not likely to contribute positively to these attitudes.

Kovas and Plomin (Chapter 9, this volume) also suggest education is shying away from genetics with a simplistic and deterministic view of it. Clearly, educators should not shy away from considering the potential influence of genetics because, as pointed out by all contributors on genetics in this volume, its present rate of progress suggests significant opportunities for education in the future. However, it would be disingenuous to claim anxiety amongst educators about genetics is due solely to miscommunication and ignorance. There are at least two well-justified reasons why many in education might prefer to see enthusiasm around genetics and education tempered with a greater sense of caution.

Firstly, as acknowledged by Kovas and Plomin, a knowledge gap between what we know about a child's future and what sorts of interventions should be provided is likely to exist for some time. Bates states that a fraction of the annual educationally-relevant information flowing from biological and genetic approaches could fill an encyclopaedia, but it might also be said that the advice genetics can presently offer to educators could be written on a postcard. And it would probably read 'watch this space'. The sciences of mind, brain and genetics will undoubtedly contribute to filling this vacuum and it seems likely that Kovas and Plomin, and Bates, are correct that genetics may help avoid assigning blame for poor performance solely to environmental failures of school and family. However, for 'blame for poor performance', one might also substitute 'praise for good performance' or even just 'responsibility'. And, apart from conflicting with education's aim higher philosophy, responsibility is potentially being shifted to a place where no-one may be offering solutions. The analysis by Grigorenko (Chapter 10, this volume) emphasizes how far we have to travel before such genetically informed solutions are available. She takes the example of public health genomics (PHG) and uses its four stages of translating discovery into application to identify the challenges that lie ahead for education. While, as in PHG, the majority of work is still at the first 'discovery' stage, there are a very few examples of stage 2 ('clinical validity') work in education, although a degree of existing association between candidate genes and reading disability is being validated in variety of samples and contexts. However, this work has not extended to the point of developing specific guidelines for detecting or predicting genetic vulnerability to reading disability or, as crucially for education, for applying pharmacological and/or individualized education in response to this knowledge. Grigorenko points out that, for education, the behemoth tasks of stages 3 (clinical utility) and 4 (accessing outcomes in practice) are still awaiting us. There will, therefore, be a gap between being able to detect genetic vulnerability and being able to apply a genetically-informed intervention to do something about it.

Secondly, although the earliest intervention for medical conditions and for disabilities might be justified more easily, early interventions to promote mainstream 'optimized education' on the basis of genetic information are more questionable. It is not clear how far up the 'several continuous dimensions' Kovas and Plomin consider would be appropriate for genetic information to be used for educational intervention. Whether or not Kovas and Plomin are directly advocating interventions for students without developmental disorders, they make clear that the knowledge being derived from genetics research will help facilitate such approaches. And what—some may ask—is wrong with that? Is this no different from sending one's child to the best school possible, checking their work and rewarding achievement? The key difference here, as brought out clearly by another contributor to this book in a previous article, is between designing a child and rearing a child. In designing a child, there is no need to make use of relationships built on communication, the child does not participate in shaping his/her life, but is acted on from the outside by processes beyond his/her control with results he/she cannot feel responsible for. Stein (2010) points out that these scientific advances have created an unprecedented state of affairs: we can change the behaviour of a child without establishing shared goals or a situation of mutual understanding, overriding the child's emerging autonomy. Even when we issue a punishment, this is usually with a communicative intent, yet very early interventions based on highly sophisticated interpretations of specialist information entirely go *around* the child's judgement and choice. The potential implications of this novel path will need to be considered carefully.

However genetics comes to be applied in education, the increasing need for greater discussion between genetic experts and education, and the broader public, appears indisputable. The three chapters by Bates, Kovas and Plomin, and Grigorenko all contribute valuably to that discussion, by presenting expert opinions on the processes involved and what is known about the considerable role of genetic determinants in educational outcome. They also alert us to a future in which children are born into a society where much more can be predicted about their developmental and educational trajectory than is presently possible, where they may be subject to very early educational interventions to prevent as yet unobservable behaviours, and where there may be a significant shifting of responsibility away from home, school and the individual, towards causes whose remediation is the subject of nascent research. As Grigorenko points out, educating the public and providers (i.e. medical and educational professionals), as well as stimulating discourse at the intersection of PHG and education, are all tasks that might need to begin now.

21.10 Conclusions

From clearing out neuromyths, to identifying research questions and applying the results, the territory between neuroscience and education is best traversed with interdisciplinary dialogue and collaborative work between experts and practitioners in both fields. Neuroeducational dialogue and research is emergent, and it is presently the stuff of pioneers. That makes it exciting but it also means that the territory will be populated by the good, the bad and the ugly for a few years to come, as we struggle to reach consensus about how to tell these apart. Ultimately, in the film, Clint Eastwood's character Blondie dons a poncho and, breaking a tense Mexican stand off in Sad Hill Cemetery, sorts out the good from the bad with guns blazing (saving the 'ugly' despite their misdeeds). Establishing and policing a field of research and enterprise involving neuroscience and education may be a more extended process than this. Also, hopefully, it will be negotiation and discussion that provides the solutions. No single field has demonstrated that it has a monopoly on truth, so we may have to take turns in wearing Clint's poncho. Each of us, whether we arrive from neuroscience, genetics, psychology or education, has much to learn from the other fields that will be contributing to education in the future.

References

Bahrick, H. P., & Phelps, E. (1987). Retention of Spanish vocabulary over 8 years. *Journal of Experimental Psychology – Learning Memory and Cognition,* 13(2), 344–49.

Blackwell, L. S., Trzesniewski, K. H., & Dweck, C. S. (2007). Implicit theories of intelligence predict achievment across an adolescent transition: A longitudinal study and an intervention. *Child Development,* 78(1), 246–63.

Braid, M. (2007). A break is as good as a test. *The Sunday Times,* 15 July. Available at: http://www.timesonline.co.uk/tol/news/uk/education/article2075200.ece

Bull, L. (2007). Sunflower therapy for children with specific learning difficulties (dyslexia): a randomised, controlled trial. *Complementary Therapies in Clinical Practice,* 13, 15–24.

Callan, D. E., & Schweighofer, N. (2008). Positive and negative modulation of word learning by reward anticipation. *Human Brain Mapping,* 29(2), 237–49.

Challis, B. H. (1993). Spacing effects on cued-memory tests depend on level of processing. *Journal of Experimental Psychology-Learning Memory and Cognition,* 19(2), 389–96.

Crockard, A. (1996). Review: Confessions of a brain surgeon. *New Scientist,* 21 December.

Davis, A. J. (2004). The credentials of brain-based learning. *Journal of Philosophy of Education,* 38(1), 21–36.

Dempster, F. N. (1988). The Spacing effect—a case-study in the failure to apply the results of psychological research. *American Psychologist,* 43(8), 627–34.

Draganski, B., Gaser, C., Busch, V., Schuierer, G., Bogdahn, U., & May, A. (2004). Changes in grey matter induced by training. *Nature,* 427, 311–312.

Fields, R. D. (2005). Making memories stick. *Scientific American,* 292, 58–65.

Gracia-Bafalluy, M., & Noel, M.-P. (2008). Does finger training increase young children's numerical performance? *Cortex,* 44, 368–75.

Heckman, J. J. (2007). The economics, technology, and neuroscience of human capability formation. *Proceedings of the National Academy of Sciences of the United States of America,* 104(33), 13250–55.

Heckman, J. J. (2008). Schools, skills, and synapses. *Economic Inquiry,* 46(3), 289–324.

Howard-Jones, P. A. (2008). Philosophical challenges for researchers at the interface between neuroscience and education. *Journal of Philosophy of Education,* 42(3–4), 361–80.

Howard-Jones, P. A. (2009). Scepticism is not enough. *Cortex,* 45(4), 550–51.

Howard-Jones, P. A. (2010). *Introducing Neuroeducational Research.* Abingdon: Routledge.

Howard-Jones, P. A., & Fenton, K. D. (2011). The need for interdisciplinary dialogue in developing ethical approaches to neuroeducational research. *Neuroethics,* 1–16.

Howard-Jones, P. A., Franey, L., Mashmoushi, R., & Liao, Y.-C. (2009). The neuroscience literacy of trainee teachers. Paper presented at the British Educational Research Association Annual Conference. Available at: http://www.leeds.ac.uk/educol/documents/185140.pdf

Howard-Jones, P. A., Washbrook, L. E., & Meadows, S., (in press). The timing of educational investment: A neuroscientific perspective, Developmental Cognitive Neuroscience, available on-line, doi:10.1016/j.dcn.2011.11.002.

Immordino-Yang, M. H. (2007). A tale of two cases: Lessons for education from the study of two boys living with half their brains. *Mind, Brain and Education, 1*(2), 66–83.

Jaeggi, S. M., Buschkuehl, M., Jonides, J., & Perrig, W. J. (2008). Improving fluid intelligence with training on working memory. *Proceedings of the National Academy of Sciences of the United States of America, 105*(19), 6829–33.

Kellaway, K. (2010). Paul Kelley: Give kids the space to learn. *The Observer,* 30 May. Available at: http://www.guardian.co.uk/technology/2010/may/30/paul-kelley-monkseaton-space-learning

Kelley, P. (2009). One hour: time it took Year 9 to crack GCSE science. *Times Educational Supplement,* 30 January.

Maguire, E. A., Gadian, D. S., Johnsrude, I. S., Good, C. D., Ashburner, J., Frackowiak, R. S., et al. (2000). Navigation related structural change in the hippocampi of taxi drivers. *Proceedings of the National Academy of Sciences of the United States of America, 97*(8), 4398–4403.

Morton, J., & Frith, U. (1995). Causal modelling: A structural approach to developmental psychopathology. In D. Cicchetti & D. J. Cohen (Eds.), *Manual of Developmental Psychopathology* (Vol. 1, pp. 357–390). New York: Wiley.

OECD. (2002). *Understanding the Brain:Towards a New Learning Science*. Paris: OECD Publications.

Pickering, S. J., & Howard-Jones, P. A. (2007). Educators' views on the role of neuroscience in education: Findings from a study of UK and international perspectives. *Mind, Brain and Education, 1*(3), 109–113.

Redcay, E., Dodell-Feder, D., Pearrow, M. J., Mavros, P. L., Kleiner, M., Gabrieli, J. D. E., et al. (2010). Live face-to-face interaction during fMRI: A new tool for social cognitive neuroscience. *Neuroimage, 50*(4), 1639–47.

Royal Society. (2011). *Brain Waves Module 2:Neuroscience:implications for education and lifelong learning*. London: Royal Society.

Schumacher, R. (2007). The brain is not enough: Potential and limits in integrating neuroscience and pedagogy. *Analyse & Kritik, 29*(1).

Shaywitz, B. A., Shaywitz, S. E., Blachman, B. A., Pugh, K. R., Fullbright, R. K., Skudlarski, P., et al. (2004). Development of left occipitotemporal systems for skilled reading in children after a phonologically-based intervention. *Biological Psychiatry, 55*(9), 926–33.

Stein, Z. (2010). On the difference between designing children and raising them: Ethics and the use of educationally oriented biotechnology. *Mind, Brain and Education, 4*(2), 53–67.

Temple, E., Deutsch, G., Poldrack, R. A., Miller, S. L., Tallal, P., & Merzenich, M. M. (2003). Neural deficits in children with dyslexia ameliorated by behavioral remediation: Evidence from functional fMRI. *Proceedings of the National Academy of Sciences of the United States of America, 100,* 2860–65.

Thomas, M. S. C., & Knowland, V. C. P. (2009). Sensitive periods in brain development—Implications for education policy. *European Psychiatric Review, 2*(1), 17–20.

Thorndike, E. L. (1926). *Educational Psychology. Volume 1: The original nature of man*. New York: Teachers College.

Weisberg, D. S., Keil, F. C., Goodstein, J., Rawson, E., & Gray, J. (2008). The Seductive Lure of Neuroscience Explanations. *Journal of Cognitive Neuroscience, 20*(3), 470–77.

Wilson, A. J., Dehaene, S., Dubois, O., & Fayol, M. (2009). Effects of an adaptive game intervention on accessing number sense in low-socioeconomic-status kindergarten children. *Mind, Brain, and Education, 3*(4), 224–34.

Wilson, A. J., Dehaene, S., Pinel, P., Revkin, S. K., Cohen, L., & Cohen, D. (2006). Principles underlying the design of 'The Number Race', an adaptive computer game for remediation of dyscalculia. *Behavioral and Brain Functions, 2*(19).

Wilson, R. S. (2005). Mental challenge in the workplace and risk of dementia in old age: is there a connection? *Occupational and Environmental Medicine, 62,* 72–73.

Chapter 22

Of all the conferences in all the towns in all the world, what in heaven's name brought us to *neuroeducation*?

Mike Anderson and Mary Oliver

During our respective careers that have spanned the fields of education and the scientific study of cognitive development we have witnessed at least three different major waves of influence on how teachers think about educating children. Each of these waves has been said to be inspired by science (however contentious they have been within the specialist discipline of cognitive development). The first and probably by far the most influential has been *Piagetian theory*. We think it fair to say that Piaget's theory of how the structure of children's knowledge changes through development influenced several generations of teachers and indeed educational systems. While Piagetian theory is no longer the dominant force it once was in the study of cognitive development it is also fair to say that no single theory of child development has emerged to replace it—and certainly not one that has the same potential impact on education. So to the extent that education has any base in the science of child development that base is Piagetian. The second wave was Gardner's theory of *multiple intelligences*. This theory fed on an educator's natural bias against the dominance of an all-encompassing, heritable, immutable general intelligence or intelligence quotient (IQ) (more of this later). Again the fact that multiple intelligences theory was the subject of serious scientific dispute from the get-go did little to dampen the enthusiasm of teachers. And one thing we would like to emphasise is the *enthusiasm* that teachers have for new ideas. We see this as a vital marker of the committed teacher and educator. However, it is also something that can be exploited for ideological or financial reasons (as documented in a number of places in this volume). And unfettered enthusiasm for new ideas can lead to a worrying suspension of critical faculties. Although it is early days, we see the third wave as potentially the most influential yet—*the brain*. Despite the 'bad' and the 'ugly' documented in this volume we think that *neuroeducation* is here to stay.

Our chapter will provide a whistle-stop tour of the major issues and concerns in this meeting of neuroscience and education. However, our main goal is to draw out what it is about neuroscience that is so beguiling for educators. It may be something as simple as this little syllogism:

- Education is about children's learning.
- The brain is responsible for learning.
- Neuroscience (the scientific study of the brain) is fundamental to education.

Yet the way learning is instantiated in brain processes, and even more pertinently how the brain develops, and how either relates to the way children are best educated, is more akin to speculation than scientific hypothesis—never mind fact. Be that as it may, it is clear to us that neuroscience will only grow in influence. You will not find a detailed argument or a scholarly presentation of data relevant to a particular topic in this chapter. Rather, in a modern pastiche of Plato's dialogues, we ask you to imagine the bar at the end of a long day's conference presentation on 'The brain and education'. We have two delegates, one a developmental cognitive neuroscientist and the other an educator, sitting in a booth by the bar. One is drinking a beer, the other a pina colada. One is nibbling on peanuts, the other pistachios. Despite their differences they are both motivated to come to this conference and inevitably they begin to talk . . .

So what brought you along to this conference, then?

We, as teachers are bombarded with ideas about 'brain-based learning' in schools now and somehow it all seems very persuasive, but I am interested in how we can distinguish the wheat and the chaff. I started looking into these programmes, and reading about interesting things like the effect of 'learning the knowledge' on the brains of London taxi drivers. You must know this study—the one where they found that cabbies have a larger hippocampus and that the longer you have been a cabbie the larger it is. The clear implication was that the need for highly developed spatial memory skills produced structural changes in relevant areas of the brain. Anyway, I started wondering about what effect teachers could be having on the brains and minds of children: what do studies show about different school systems, the timing of the school day for teenagers, bilingualism, dyslexia, dyscalculia, attention deficit hyperactivity disorder, etc.? How much is interesting and/ or relevant for teachers? Teachers are in a unique position to 'mess with minds' of children so what sort of relevant or interesting research ought teachers have access to? For example, is the hyperconnectivity reported in individuals with absolute perfect pitch and autism a spurious finding or relevant for educators?[1] Regions of the brain associated with counting are close to parts involved in using fingers may help be relevant to early years' teachers.[2] So those kinds of things brought me here. Trouble is that it is difficult to know what is interesting and relevant and what is not.

I take your point entirely. It is not clear a priori what piece of neuroscience research is likely to be relevant in the classroom, or indeed how involved individual teachers should be with developments in neuroscience. I have come across two schools of thought on this. One is that teachers themselves must not only become scientifically literate in domains relevant to their educational goals but must be themselves part of the research process. Personally, I think this is utopian thinking. Teachers are mostly too busy for either of these, and way underskilled for the latter. The other view is that the stuff that neuroscientists do that is likely to actually be effective in improving learning is in the long term going to be discoveries of how learning is implemented in neural structures in the brain and what chemical and genetic processes influence this. On this view the impact of neuroscience is unlikely to be in the classroom but rather in the pharmacy or stem-cell laboratory—more utopian thinking in my view. I think that it is more realistic that we might expect a domain of psychology and neuroscience to open up where researchers develop expertise in testing hypotheses about classroom interventions that are informed by a more general cognitive neuroscience research—and also by at least some appreciation of how classrooms actually work.

I don't think I can agree with your first point or indeed its tone. Teachers are busy, yes, but reflective practice is what we try and engender in pre-service teachers. Do you really think that it is beyond teachers' wit to engage with the scientific process of teaching and reflection? It is time to get scientific about teaching. If there are good methods, these should be evaluated, validated and celebrated. It's up to leaders in education at school to policy level, to talk about effect size, value added and promote best practice in education.

---

[1] Loui, P., Li, H. C., Hohmann, A. & Schlaug, G. (2011). Enhanced cortical connectivity in absolute pitch musicians: A model for local hyperconnectivity. *Journal of Cognitive Neuroscience, 23*(4), 1015–26.

[2] Butterworth, B. (1999). A head for figures. *Science, 284*(5416), 928–9.

And in this kind of context I believe that teachers are perfectly capable of engaging with the relevant neuro research.

I certainly didn't mean any offence, but I have to say that some of my colleagues in psychology find the educationalists' new-found obsession with neuroscience a bit rich. From their point of view education has pretty much been an evidence-free zone. Apparently a high proportion of trainee teachers believe that children should be taught in their 'preferred learning style'—a proposition for which there is absolutely no evidence. And what about the local council or health boards who are only too willing to shell out on coloured lenses to cure dyslexia? Contrast this with the wilful disregard of the 'science of IQ'. We have known for the better part of a century that children differ generally in their ability to learn and moreover that an important part of this difference may be genetically based. Yet educationalists in the last 40 years have refused to take this seriously.

Don't get me started on the genetics issue. Teachers work with so many children every day in their classrooms, many of whom come to school with specific or non-specific learning needs. The IQ stuff, and maybe what you would call the old genetics, seem to teachers to be an abrogation of responsibility for education. What is the point of educating if it is all IQ and genetics anyway? Modern neuroscience doesn't feel like that to educators. The growth in educational neuroscience has certainly brought educators and neuroscientists into a new sort of research dialogue to both affirm and suggest ways of intervening.

Yes but what is it about neuroscience that is so different to any previous attempt to introduce a scientific evidence base? Why wouldn't teachers be interested in things that can and cannot change IQ scores? Would that not be more important than producing measurable effects on the brain?

I do agree that we are still at the stage in educational neuroscience where any educational interventional programme needs to have appropriate and measurable learning outcomes (such as behavioural changes, improved and sustained changes in learning—call it IQ if you must) and neuroscience can confirm whether there are any changes at the level of the brain, either structurally, or in terms of efficiency or levels of activity. It is certainly true that not all interventions or educational programmes seek a neurological answer to evaluate their effectiveness: we can rely on other measures; we don't need to brain scan to justify the development or implementation of a particular programme. However, evidence provided for a scientific basis is always rather compelling: I think it is reasonable to cost educational improvements and so triangulating the data to support claims made using the tools of neuroscience may be reasonable.

I think there is an implicit danger here and that is in your use of the term 'scientific'. Unless I am mistaken this is predicated on finding changes not only in behaviour or performance but changes at the level of the brain. So you seem to be saying that teachers classify evidence about the brain as 'scientific' whereas evidence from the behavioural sciences might be regarded as ideas or hypotheses at best. If I am right you have put your finger on one of the reasons that neuroscience may be attracting such a lot of attention from educationalists—it is unequivocally scientific and teachers as a profession do value science. So what is the problem with that? It implies that evidence from the brain is real, lasting and significant but evidence from behaviour is ephemeral, woolly and provisional. And I don't think this is just a taste thing. What if you were to find that some intervention programme has spectacular results in terms of improved learning but a parallel set of neuroscientific observations led to conclusions that there were no changes to underlying brain structure? On the other hand another intervention that has minimal impact of learning outcomes nevertheless produced measurable changes in the neuroscientific measures. Which is to be preferred? I hope it is a 'no-brainer' [laughs uproariously spitting nuts rather disgustingly over the drinks] that the successful change in the behaviour (and not the brain) is the point of any educational intervention.

Well, I do take that point but it can be difficult to recognize the real science from the phoney. The distinction seems clearer for neuroscience research, although we have seen in the conference that this can often be exploited to mask nonsense. But consider for the moment other forces that are acting on the evidence base for teachers. What about the acolytes of postmodernism that eschew the scientific method altogether? You know what I mean—all knowledge is a social-construction/science has no privileged status/school is about reflecting power structures not children's learning—that kind of thing. The problem with much educational evidence in the literature is that it is heavily derived from qualitative analysis, situated in some sort of sociocultural theoretical framework and not actually very accessible to classroom teachers.

Well then surely you should be inclined to look again at some of the stuff on intelligence, genetics and development of the brain. I am not saying that there has been no history of abuses of intelligence testing and indeed fallacious conclusions drawn from the unassailable fact that there are large genetic contributions to variation in intelligence. But I think we are moving on from all that. There is a recognition that there is no inherent conflict between the idea of significant individual differences in the capacity to learn and developmental change itself. And anyway, the modern approach to behaviour genetics is not just about saying A or B is X% or Y% heritable, it is elucidating where in the educational process there may be genetic AND not genetic contributions to learning. And what about the research that links developmental change in the thickness of the cortex to IQ scores?[3] Surely this research is both significant and exciting for educators?

And I think educators are prepared to look again at hard evidence that is useful for promoting change and development. Teachers know that students are not a 'tabula rasa', but come to school with a range of dispositions, attitudes and curiosity that impact on their readiness to learn. To learn that the socioeconomic status exerts powerful long-lasting effects on children's IQ is confronting for educators, as is knowing that maternal nutritional status during pregnancy and maternal educational levels, may be determinants of a child's educational success. The Programme for International Student Assessment (PISA) comparatively samples 15-year-old students across Organisation for Economic Co-operation and Development (OECD) countries every 3 years. Data from these tests inform educational practice and reveal patterns such as the effects of social inequity, selective schooling and pre-school education.[4] These data are important given that you rather pointedly describe us as working in an evidence-free zone. Widening gaps in performance between rich and poor, are consequences of social policies; pre-school intervention programmes continue to show positive impacts in adolescence, and ending school selection at 11 years in Poland has improved the performance of their students in literacy which has economic consequences for the country's future. The Hattie report published in 2009 comprised a meta-analysis of more 800 studies to determine the effects of different educational practices.[5] Diet, out-of-school curricula experiences and perceptual motor programmes have really small effects; on the other hand, self-reported grades, Piagetian programmes and formative assessment have demonstrated efficacy and impact. Clearly, the evidence for how to teach better and how to help students to learn better is available in the academic literature and the gap between the literature and the preparation of teachers needs to be closed.

None of that it seems to me is particularly inspired by neuroscience.

---

[3] Shaw, P., Greenstein, D., Lerch, J., Clasen, L., Lenroot, R., Gogtay, N. *et al.* (2006). Intellectual ability and cortical development in children and adolescents. *Nature*, 440, 676–9.

[4] http://www.pisa.oecd.org.

[5] Hattie, J. (2009). *Visible Learning; A synthesis of over 800 meta-analyses relating to achievement*. London: Routledge.

Yes, but let's get serious about how science might impact on education. Bluntly it would be to provide some sort of hypotheses about how to intervene to promote children's learning and then to provide some evidence that the intervention actually works. So if we think that cognitive development can be accelerated or improved over and above normal classroom practice, then we should certainly examine this critically. And think how much more precisely we could test these claims if we think the reason for any success is that neural processing is faster, or more efficient or that the prefrontal cortex is greyer. I don't think it would be unreasonable to want to see evidence of this. You were scathing earlier about the value of neuroscientific outcomes—do we want measurable changes in behaviour, attitude or attainment or changes to neural physiology detected with an imaging technique? As a teacher, I am happy if my students are better behaved, more committed, self-regulating their learning and are optimistic about their future. So you are right, the neuroscience doesn't really need to impact me if I can see that the 'what', 'how' and 'way' I teach has a positive impact on students. But if you can show me that there are neural correlates with student change, then that is most interesting. A lot of work into the neural basis of dyslexia has given rise to some positive remedial intervention methods. The fact that brain scans can identify poor neural connections well before signs of dyslexia are manifest enable early remediation to take place—surely that has to be a positive outcome of neuroscience? Is that more than interesting? I would suggest it is helpful. Likewise, I think it is helpful to learn and reflect on the neural effects of social ostracism. These are powerful findings that could lead to modified practice and behaviours.

While I can see how some of the neuro work on learning disorders such as dyslexia might be of interest, how relevant is it really for mainstream education of typically developing kids?

Maybe only time will tell. But be careful that you are not hoist by your own petard here. It is also hard in the middle of a conference like this to realize how such a small part neuroscience, or cognitive psychology for that matter, plays in the bigger picture of education and indeed individual children's learning. Let's think about classrooms: what happens there is the second most private social activity we engage in. I say that because it is very difficult to describe all the interactions, effects and outcomes that take place there. But I am optimistic that good ideas or theories of learning might translate into better classroom practice. To give just one recent example—a Piagetian programme, cognitive acceleration through science education (CASE), has shown that students' improvement in science is long term and that transfer to other curriculum areas with concomitant improvements in achievement.[6] Data from interventions such as cognitive accelerations and the Philosophy for Children (P4C) programmes where students achievement is improved across the curriculum and the impact continues long after the intervention, suggest that general intelligence has been improved. That such improvement takes place at all levels of ability in many different school settings is compelling and interesting—what exactly has happened to bring about improvements that are pervasive and transferable? Surely if we can specify what we mean by general intelligence and it is as you imply a property of the brain, then would we not expect to see concomitant changes at the level of the brain? I see the growth of educational neuroscience having great possibilities if some of the cross disciplinary boundaries are truly breached. We could see a paradigm shift which would be very helpful in elucidating classroom practice, educational interventions to better understand best practice.

Well I think we do agree more than we disagree. That is my idea of serious triangulation: a theory of the nature of an important cognitive function (e.g. general intelligence) that claims some basis in brain structure or function or whatever; an educational intervention inspired by an attempt to change/develop that particular cognitive mechanism (e.g. some aspect of executive functioning); and

---

[6] Oliver, M., Venville, G. & Adey, P. (2010). Thinking Science Australia: Improving teaching and learning through science activities and reasoning. Paper presented at the Australasian Science Education Research Association Annual Conference, Shoal Bay, NSW.

an outcome test that is both behavioural (did the intervention lead to changes in the operating characteristic of that mechanism?) and has predicted neural correlates. Then we really would be getting somewhere. Talking of getting somewhere—would you like another beer?[7] Yes? Play it again Sam.[8]

---

[7] Who spotted the psychologist's trick of exploiting gender stereotypes for dramatic effect?

[8] And yes we know this is a misquote but did you spot the connection with the title?

Author Index

Note: 'n.' after a page reference indicates the number of a note on that page.

Subject Index

Note: 'n.' after a page reference indicates the number of a note on that page.

Printed and bound by CPI Group (UK) Ltd, Croydon, CR0 4YY